T0297215

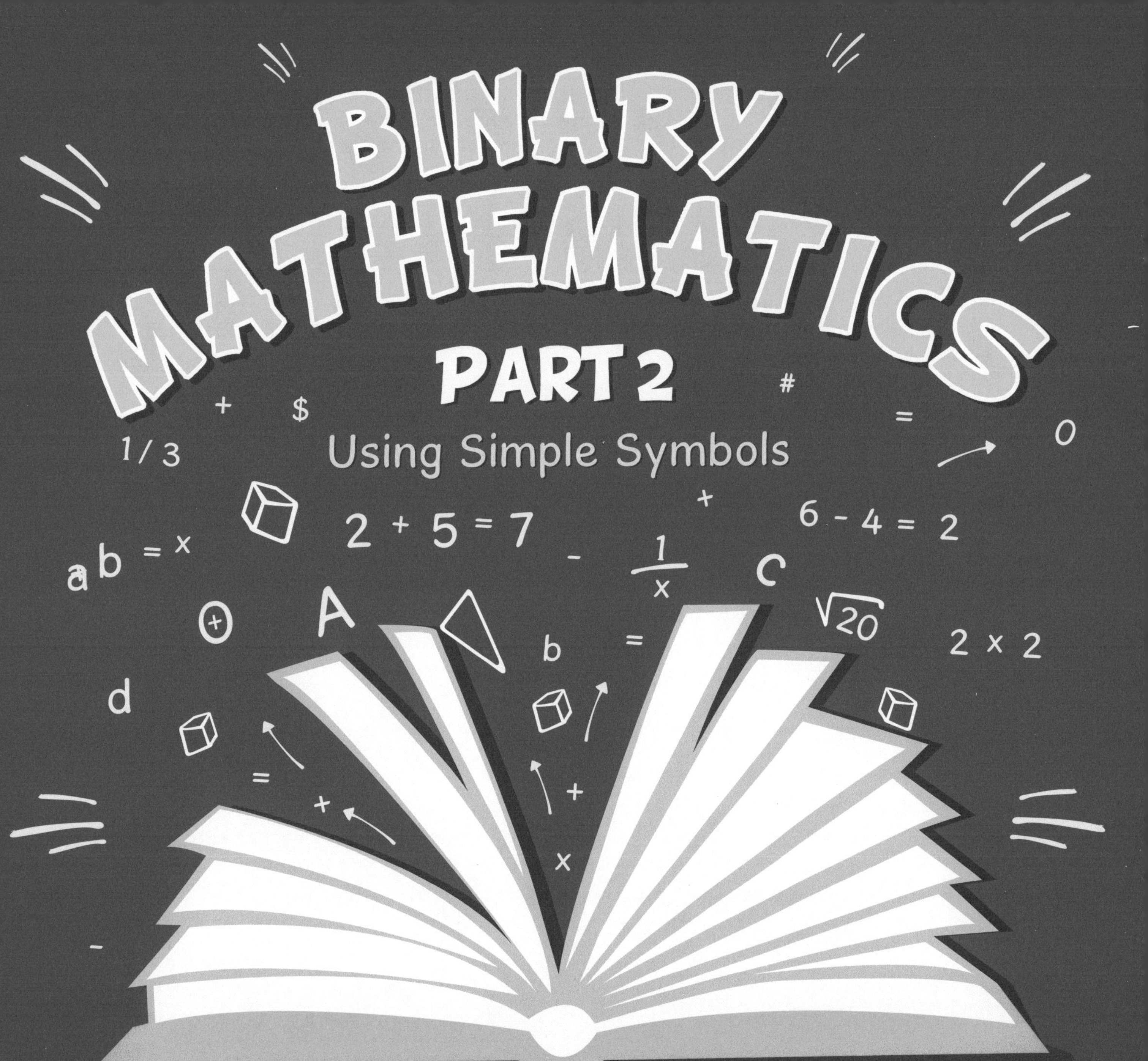

Chester Litvin

Order this book online at www.trafford.com
or email orders@trafford.com

Most Trafford titles are also available at major online book retailers.

Trafford PUBLISHING® www.trafford.com
North America & international
toll-free: 1 888 232 4444 (USA & Canada)
fax: 812 355 4082

Our mission is to efficiently provide the world's finest, most comprehensive book publishing service, enabling every author to experience success. To find out how to publish your book, your way, and have it available worldwide, visit us online at www.trafford.com

ISBN: 978-1-4907-9687-1 (sc)
ISBN: 978-1-4907-9686-4 (e)

Print information available on the last page.

Trafford rev.01/10/2020

From Author

Welcome to mathematics by using simple symbols. Variety of simple symbols can be visual, audio, kinesthetic, olfactory, tactile, musical and etc. The reason that I developed simple symbols approach was fallowing. In the past, I observed students that have difficulty with decimal calculation and regular grammar and wanted to help them. I saw that they done much better when they used binary mathematics and binary grammar. For example, my dyslexic students that, switched to the simple symbols, were having fewer problems with math and grammar. The average grades students that used binary approach also improved in math and grammar. The autistic and hyperactive students, who used binary system were more focused and were able to do better with their school assignments. After working with binary arithmetic they were more effectively used decimal math and with grammar by using regular alphabet. To have process of learning more effective, I was introducing mixture of symbols of diverse modalities.

Psychoconduction is approach of doing mathematics by using simple symbols. I was able to build a system by using simple symbols in many modes of expressions that included visual, audio, and kinesthetic and etc. In my book I explained how to translate simple symbols to variety modes of expression.

I understand that the increased ability of brain by solving the math problems and using grammar has more complex bases. I patented my discovery and called "Psychocondaction", the neurological process of brain restoration. My US patent is in details explaining my hypothesis. In my patent I have listed the researches in the brain stimulation that were using sensory areas.

To correctly identify areas of the brain affected by psychoconduction I done some research. I just mentioned the approximate areas of the brain. The future steps that would prove my hypothesis of brain restoration can be MRI or many other controlling methods to monitor the changes in brain structure. My subjects did great and I am sure that my hypothesis worked. When I was doing the problems with simple symbols, I was enjoying the new approach of doing mathematics and grammar. Try it and see if somehow it would help you.

With respect,

Chester Litvin, PhD

Psychologist

Dedicated to my nephew, David Gimelfarb,
Lost in Costa Rico in 2009.

<u>**Intermediate brain stimulation by psychoconduction**</u>

Chester Litvin, Ph.D., Clinical Psychologist

CONTENTS

ADDITION

Instead of numbers, Litvin's code uses symbols to complete mathematical functions. For example, "x and 0" represent filled and empty positions in Litvin's mathematical configurations. The following examples will give deeper understanding of the addition process using Litvin's Code.

Rules:

- When adding two numbers that have two corresponding positions filled with symbols, we move the symbol forward. When corresponding positions in the first and second numbers are filled with symbols, we move the two symbols to the next position in the answer as one symbol.

- When two symbols from two corresponding positions move to the next position, the next forward position contains a number twice as big as the previous. When symbols get to the following position of the answer, two symbols become one.

Example 2.1

For addition, we use a symbolic representation of digits, where symbols represent binary numbers in assigned positions. To add 3 and 2 we, use symbols in assigned positions, instead of numbers.

First number	3	"x	x	0"
Second number	2	"0	x	0"
The answer is	5	"x	0	x"

	•	•			3
+		•		+	2
	•		•		5

In Example 2.1, the first picture has its first two positions filled with symbols, while the second picture only has its second position filled with a symbol. The first position of the first number is filled with a symbol, while the first position of the second number is empty. In this situation, we simply move the symbol to the first position of the answer.

In the first number, the second position has a symbol and so does the second position of the second number. Since corresponding positions in the two numbers are filled with symbols, we move the two symbols to the next ascending position of the answer, where they are represented by only one symbol.

In Summary:

- When we have two symbols in one position of the answer, we move them from this position, which in our example is the second position, to the third position as one symbol.

- When the two symbols are in parallel positions, we move them to the ascending position of the answer as one symbol.

End.

Example 2.2

In Example 2.2, we have the third corresponding positions in the two numbers that are being added. Therefore, we move them to the next position as one symbol.

First number is 5 "x 0 x"

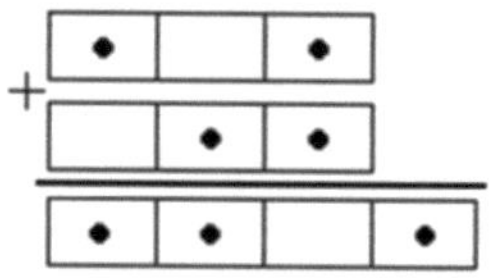

	•		•	
+		•	•	
	•	•		•

+ 5
6
11

Second number is 6 "0 x x"

The answer is 11 "x x 0 x"

If we have two symbols in different positions of the two numbers, we simply move them to the answer.

The first position in the first number is filled with a symbol, while the first position in the second number is empty. We automatically fill with a symbol the first position of the answer. The second position on the first number is empty but the second position on the second number is filled up with a symbol, so we fill with a symbol the second position of the answer. The third parallel position in the first and second numbers is filled with a symbol. This means that the symbol needs to be moved to the fourth position of the answer and the third position in the answer is left empty.

In summary:

- When the first position in the first number is filled with a symbol and not parallel, the second position in the second number is filled with a symbol, then the first two positions of the answer are filled with symbols.

- When the third position in both numbers is filled with a symbol, then we have two symbols present in parallel positions. In this situation, we advance one position forward.

End.

Example 2.3

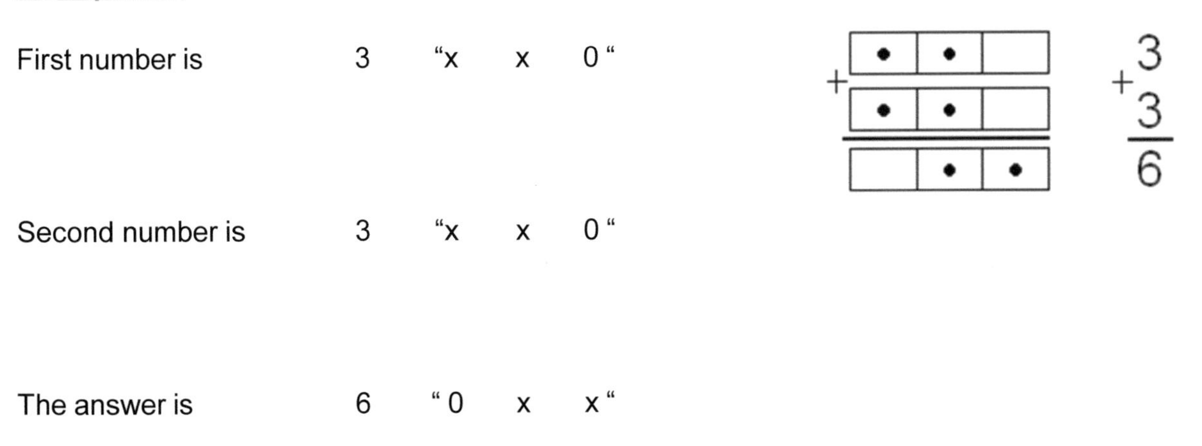

The first positions in the first and second numbers are filled with symbols. We move the symbols to the second position of the answer. The second positions of the first and second numbers are filled with symbols. If the next position in the answer is already occupied with two symbols, then we leave the symbol there and move to the next available ascending position. So we move the symbol forward to the next position of the answer.

In summary.

- When we have two symbols present in parallel positions, we advance one position.

- The two symbols in the first positions of the two numbers become one symbol in the second position of the answer. The second two symbols in the second positions of the numbers become one symbol in the third position of the answer. The same principle is applied for all corresponding positions occupied with symbols.

End.

Example 2.4

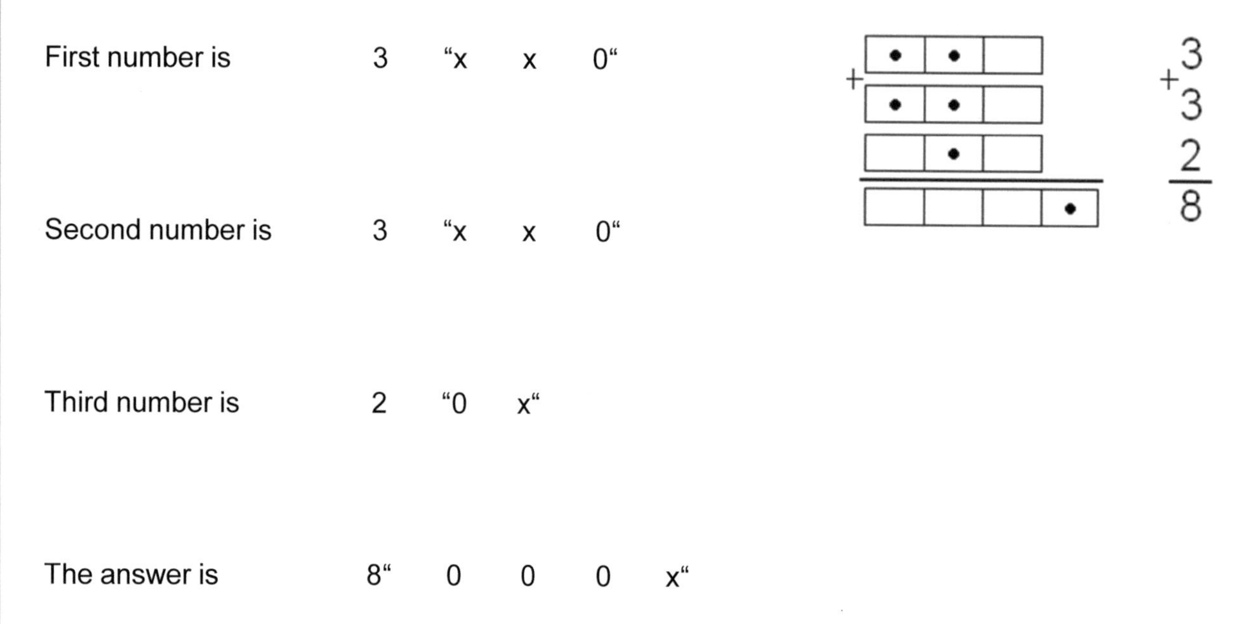

In Figure 2.4, when adding three numbers with matching positions filled with symbols, we have a different set up. The first positions of the first two numbers are filled with symbols. Usually the two symbols in matching positions are supposed to move as a one symbol to the next position of the answer, but in this case, they move two positions forward. We have four symbols in the second position. They come from the second positions of the three numbers and the one symbol transferred from the previous position. When we have four symbols in one position, we move them two positions forward in the answer and fill the fourth position with a symbol. In the second position, we have four symbols. This is an even number of symbols and that's why we move the symbol two times to the fourth position of the answer. In this example, we add up the three numbers. The two numbers are the same and one is different. We count the symbols in the second position of the three numbers. Since the result of this calculation is an even number four, we move the symbols to the next positions, but in half the quantity. In the second position we have four symbols, so we move the symbols two positions forward in the answer. Now we have only one symbol in the fourth position. In parallel position one, we must move until we have one symbol in the answer. We move to the next position of the answer with two symbols and fill it up with one symbol (dot). When we have four symbols, we move two positions forward in the answer. In the first position we are left with two symbols, which inform us to move again to the next position and leave us only with one symbol (dot) in the answer.

In Example 2.4, the first two corresponding positions of the first and second numbers are filled with symbols, but the first position of the third number is empty. When we have two symbols in matching positions, we move these symbols to the next position. Two symbols are represented by one symbol when they get transferred to the next position, leaving the previous position empty. So we move the symbol to the second position. The second position of the first, second and third numbers is filled with symbols. We also have one symbol moved from the first position. Combined together, we have four symbols and we transfer them to the next position. The four symbols in the second position are represented by two symbols in the third position and later, as one symbol in the fourth position. We fill the fourth position of the answer with a symbol. In a different situation, if we had odd numbers greater than one in matching positions, then we would transfer one symbol to the answer and the rest to the next position.

In summary:

- When we have two symbols (dots) present in corresponding positions of the numbers being added, we advance one position.

- When moving the symbol to the next position, which already has three symbols, it creates an even number of symbols, in this case four (dots).

- When we skip the third position, we advance these symbols (dots) forward to the new position, which is represented by one symbol in the fourth position of the answer.

End.

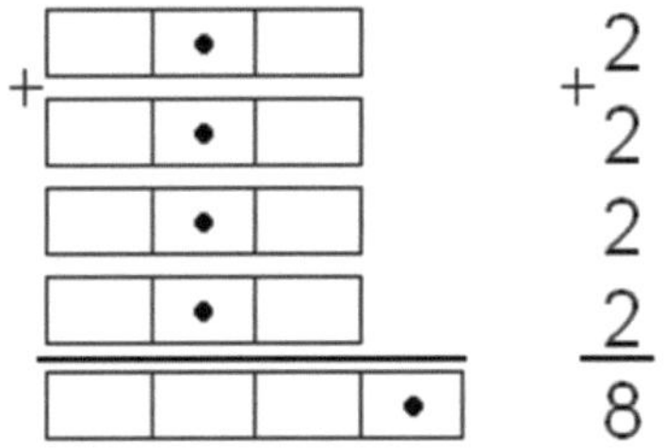

+		•			2
		•			+ 2
		•			2
		•			2
				•	8

Example 2.5

First number is 2 “0 x 0“

Second number is	2	“0	x	0“	
Third number	2	“0	x	0”	
Fourth number	2	“0	x	0”	
The answer is	8	“0	0	0	x“

In Example 2.5, we are adding four numbers. The symbol (dot) in the second position corresponds to all four numbers being added. In other words, the second position in all four numbers is filled with a symbol. We sum up the symbols in the second positions and end up with four symbols. When we move those symbols to the next position, then the next position is represented only by two symbols. Therefore, we move the symbols further. We move these symbols to the next position and continue to move them until the position is represented only by one symbol in the answer.

In summary,

- When we have four symbols (dots) presented in corresponding positions of the numbers being added, we advance two positions in the answer.

- We move the even number of symbols to the advanced position until we can represent those symbols with only one symbol in the answer.

- When we move the symbols to the next position, each advancing position has half the quantity of the symbols before.

- If the position is correct, it is supposed to have only one symbol.

End.

LESSON ONE INTRODUCTION TO SIGNS

There are five different signs. There is an addition sign, division sign, subtraction sign, multiplication sign, and an equal sign. Addition is represented by the sound, **tram**. Division is represented by the sound, **double click**. Subtraction is represented by the sound, **double tram**. Multiplication is represented by the sound, **blick**. The equal sign is represented by the sound, **cling**.

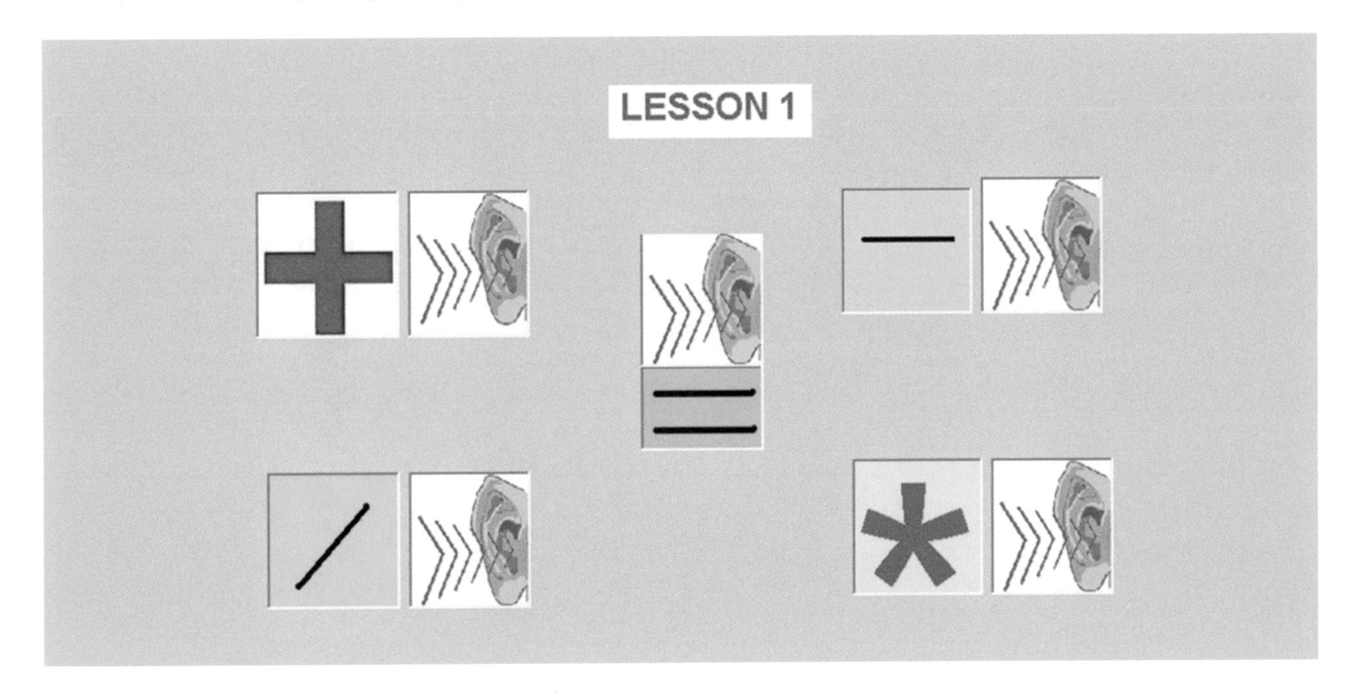

The addition sign is represented by one hand extended in front. The division sign is represented by both hands extended up. The subtraction sign is represented by both hands extended in front. The multiplication sign is represented by one hand raised up. The equal sign is represented by crossing both hands.

MATHEMATICS

ADDITION, EXERCISE 1

In exercise 1, in ***visual*** display of addition we see three pictures of boxes. We want to specify, that between the first and second pictures we see the addition sign. Below the second picture we see the equal line, which separates the problem from the answer. The ***audio*** representation for the three pictures consumes of a **knock, double knock, tram, knock, double knock, cling, left hand, and right hand.**

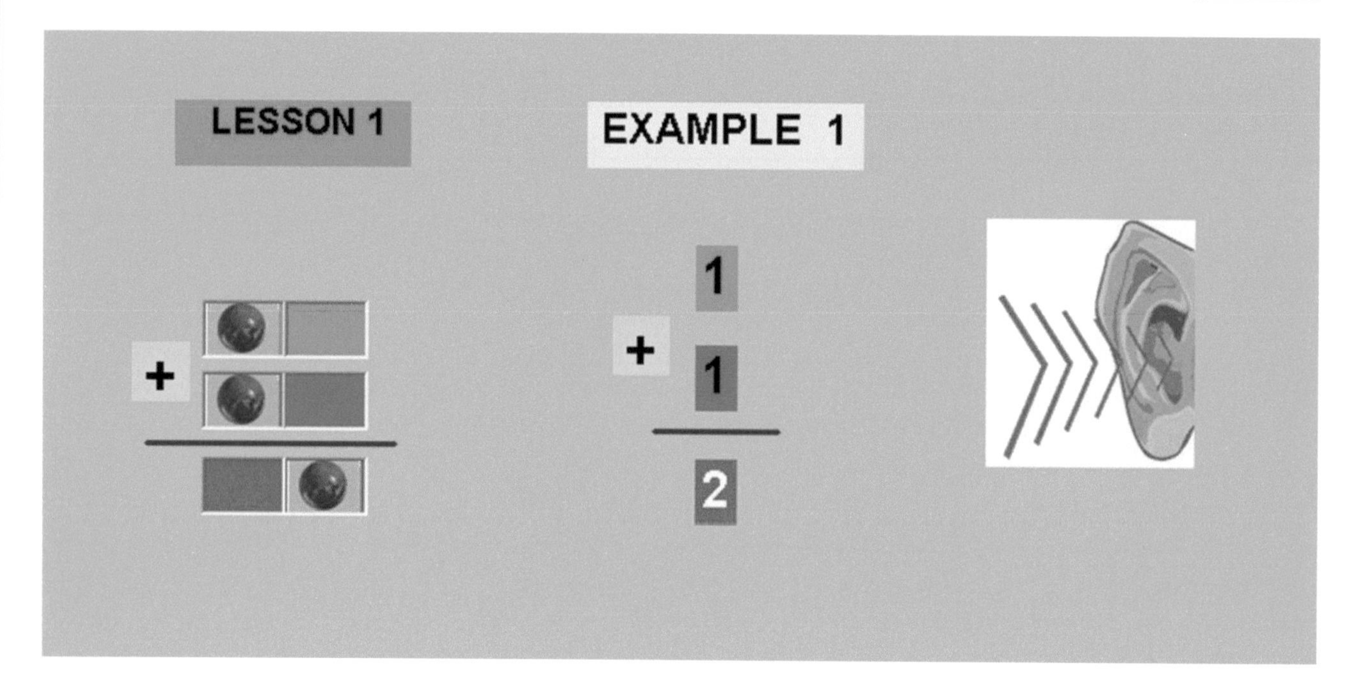

The ***kinesthetic*** representation of the filled up position is announced by clamping of the right hand. An empty position is represented by the clamping of the left hand. The addition sound, tram, is represented by one hand extended in front. The equal sound, cling, is represented by crossing of hands. The kinesthetic representation for both pictures and the answer is the clamping of the right hand, left hand, one hand extended in front, right hand, left hand, crossing of hands, left hand, and right hand.

ADDITION, EXERCISE 1, STEP 1.1

In exercise 1, step 1.1 in ***visual*** display of addition we see three pictures of boxes. We want to specify, that between the first and second pictures we see the addition sign. Below the second picture we see the equal line, which separates the problem from the answer. When we are adding two numbers and the position on the first number is filled with a symbol and the corresponding position on the second number is also with a symbol, we move the two symbols to the following positions in the answer as one symbol. When two symbols form, two corresponding positions move forward to the following position of the answer then the two symbols become one symbol. The ***audio*** representation for the three pictures consumes of a **knock, double knock, tram, knock, double knock, cling, left hand, and right hand.**

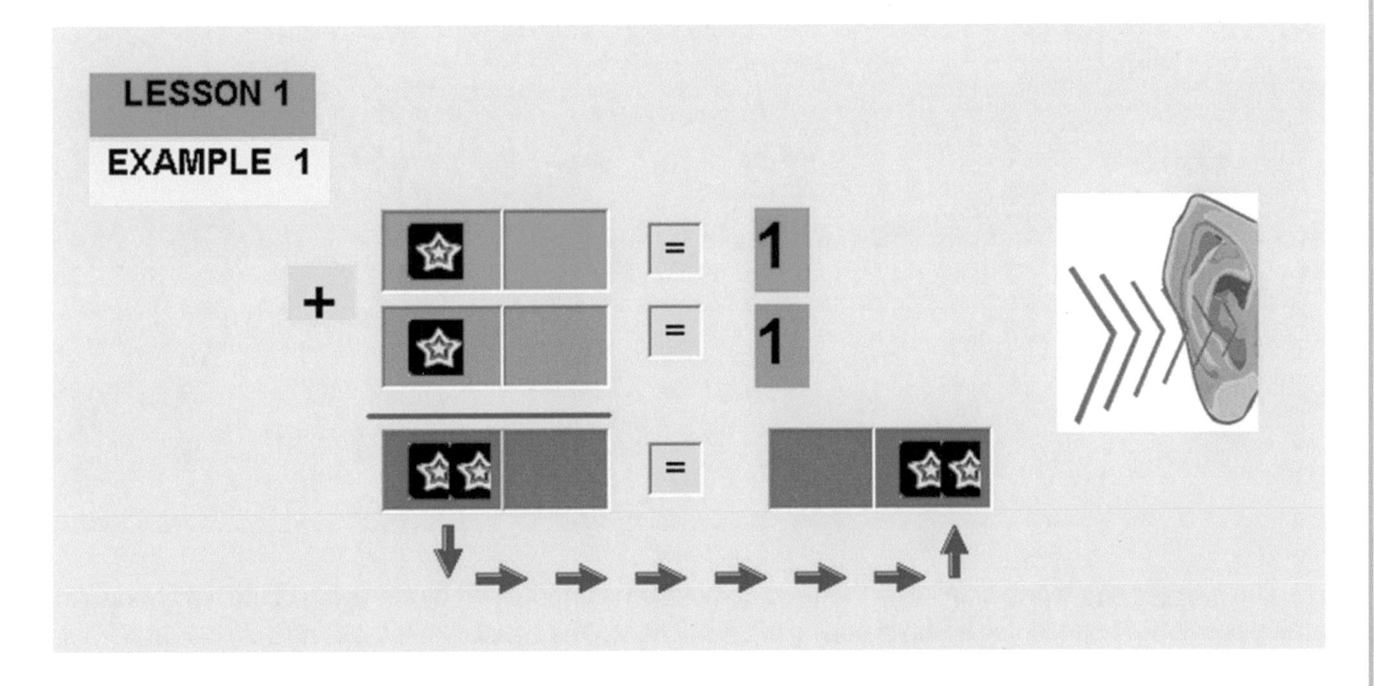

The ***kinesthetic*** representation of the filled up position is announced by clamping of the right hand. An empty position is represented by the clamping of the left hand. The addition sound, tram, is represented by one hand extended in front. The equal sound, cling, is represented by crossing of hands. The kinesthetic representation for both pictures and the answer is the clamping of the right hand, left hand, one hand extended in front, right hand, left hand, crossing of hands, left hand, and right hand.

MATHEMATICS

ADDITION, EXERCISE 1, STEP 1.1

In exercise 1, step 1.1 in ***visual*** display of addition, we see two pictures of boxes with equal sign. On both pictures we see that the first position empty but the second position is filled with a symbol. The first picture has a yellow star in the second position but the second picture has a red ball in the second position. The form and shape of the symbols do not change mathematical value and the numerical value of both pictures is equal to two.

The ***audio*** representation for this picture consumes of a **double knock, knock, and cling.**

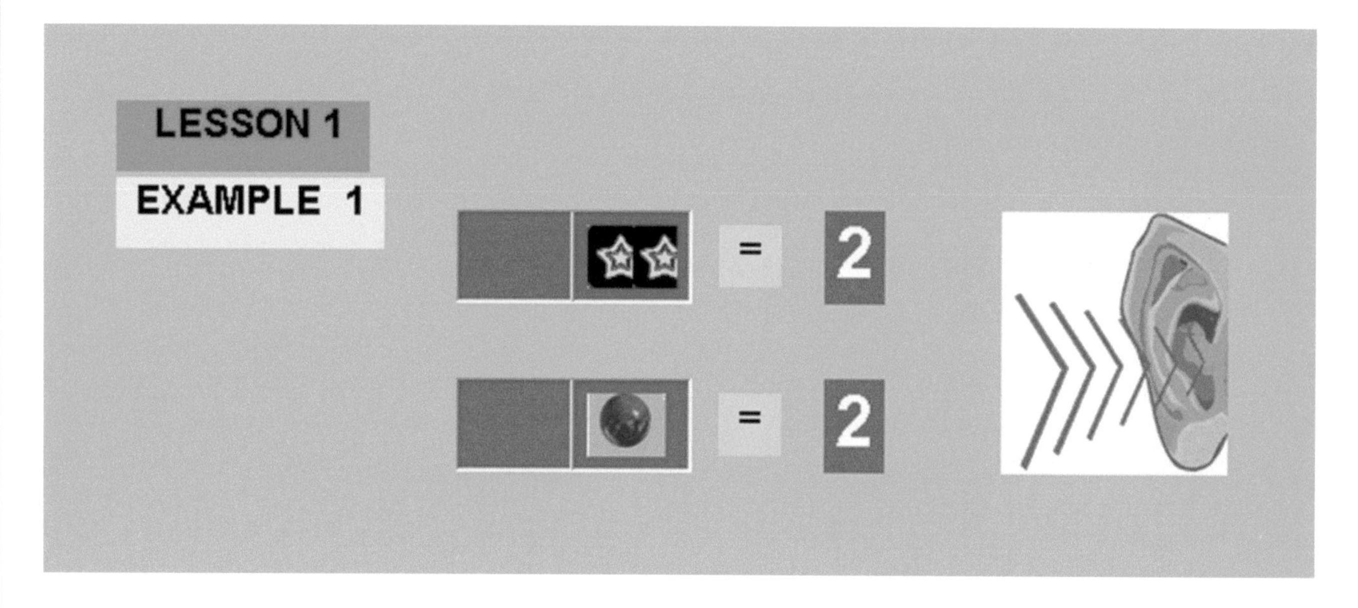

The ***kinesthetic*** representation of the filled up position is announced by clamping of the right hand. An empty position is represented by the clamping of the left hand. The equal sign is crossing both hands in front. The kinesthetic representation for this picture is clamping the left hand, right hand, and crossing of hands.

ADDITION, EXERCISE 1

In exercise 1, in ***visual*** display of addition we see two pictures of boxes. In the bottom we see three pictures with probable answers. Both pictures have a symbol in the first position and the second position is empty. We want to specify, that between the first and second pictures we see the addition sign. Below the second picture we see the equal line, which separates the problem from the answer. The ***audio*** representation for two pictures consumes of a **knock, double knock, tram, knock, double knock, and cling.**

From the probable answers for addition we need to find the right one. The audio signals are: 1) **double knock and double knock.** 2) **knock and knock** 3) **double knock and knock**.

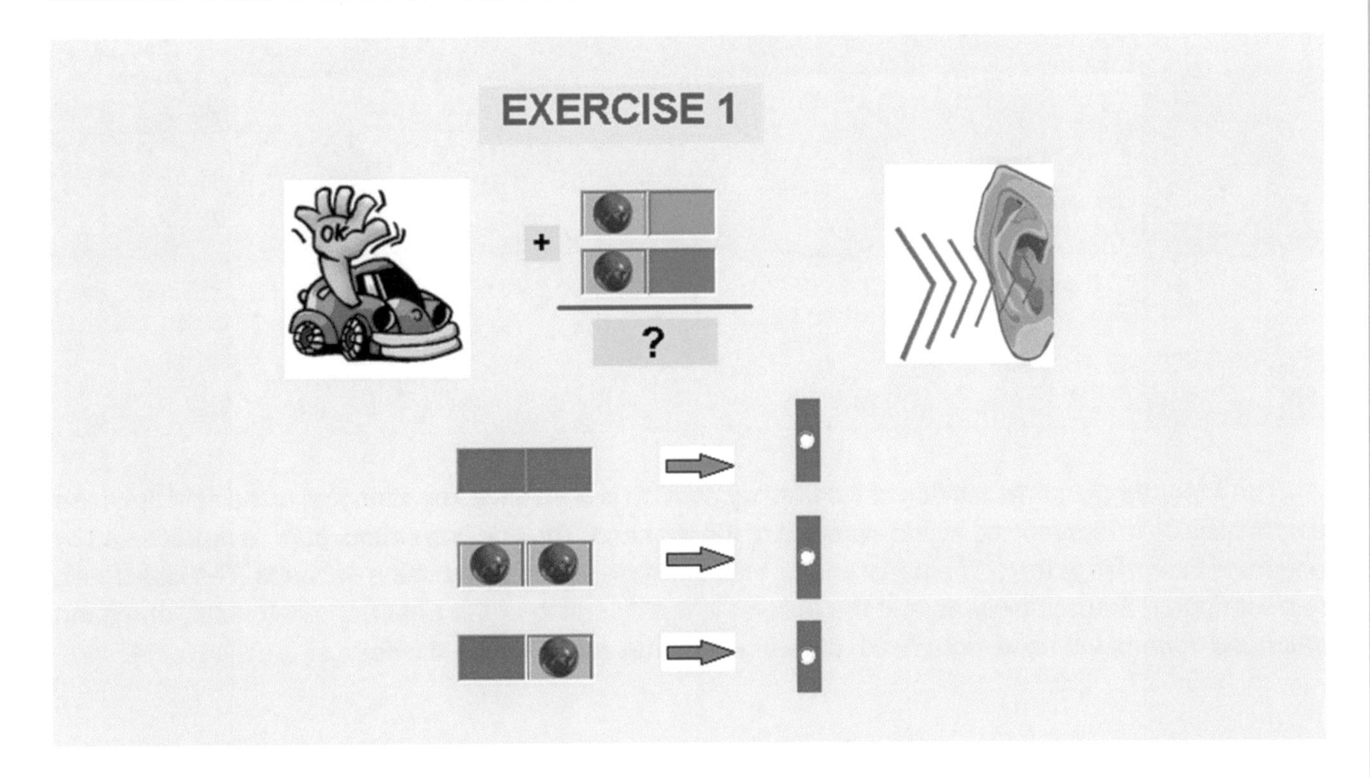

The ***kinesthetic*** representation of the filled up position is announced by clamping of the right hand. An empty position is represented by the clamping of the left hand. The addition sound, tram, is represented by one hand extended in front. The equal sound, cling, is represented by crossing of hands. The kinesthetic representation for both pictures and the answer is the clamping of the right hand, left hand, one hand extended in front, right hand, left hand, crossing of hands, left hand, and right hand.

MATHEMATICS ADDITION, EXERCISE 2

In exercise 2, in ***visual*** display of addition we see three pictures of boxes. We want to specify, that between the first and second pictures we see the addition sign. Below the second picture we see the equal line, which separates the problem from the answer. The ***audio*** representation for the three pictures consumes of a **knock, double knock, tram, double knock, knock, cling, knock, and knock.**

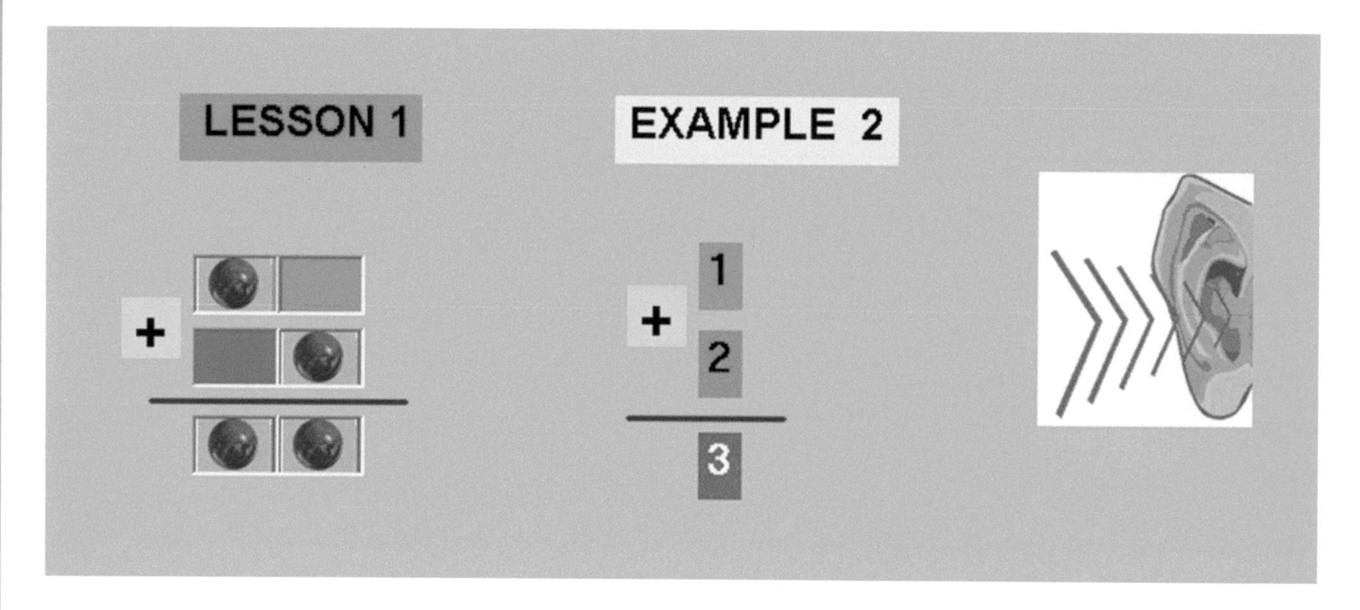

The ***kinesthetic*** representation of the filled up position is announced by clamping of the right hand. An empty position is represented by the clamping of the left hand. The addition sound, tram, is represented by one hand extended in front. The equal sound, cling, is represented by crossing of hands. The kinesthetic representation for both pictures and the answer is the clamping of the right hand, left hand, one hand extended in front, left hand, right hand, crossing of hands, and right hand twice.

ADDITION, EXERCISE 2, STEP 2.1

In exercise 2, in ***visual*** display of addition we see three pictures of boxes. We want to specify, that between the first and second pictures we see the addition sign. Below the second picture we see the equal line, which separates the problem from the answer. The ***audio*** representation for the three pictures consumes of a **knock, double knock, tram, double knock, knock, cling, knock, and knock.**

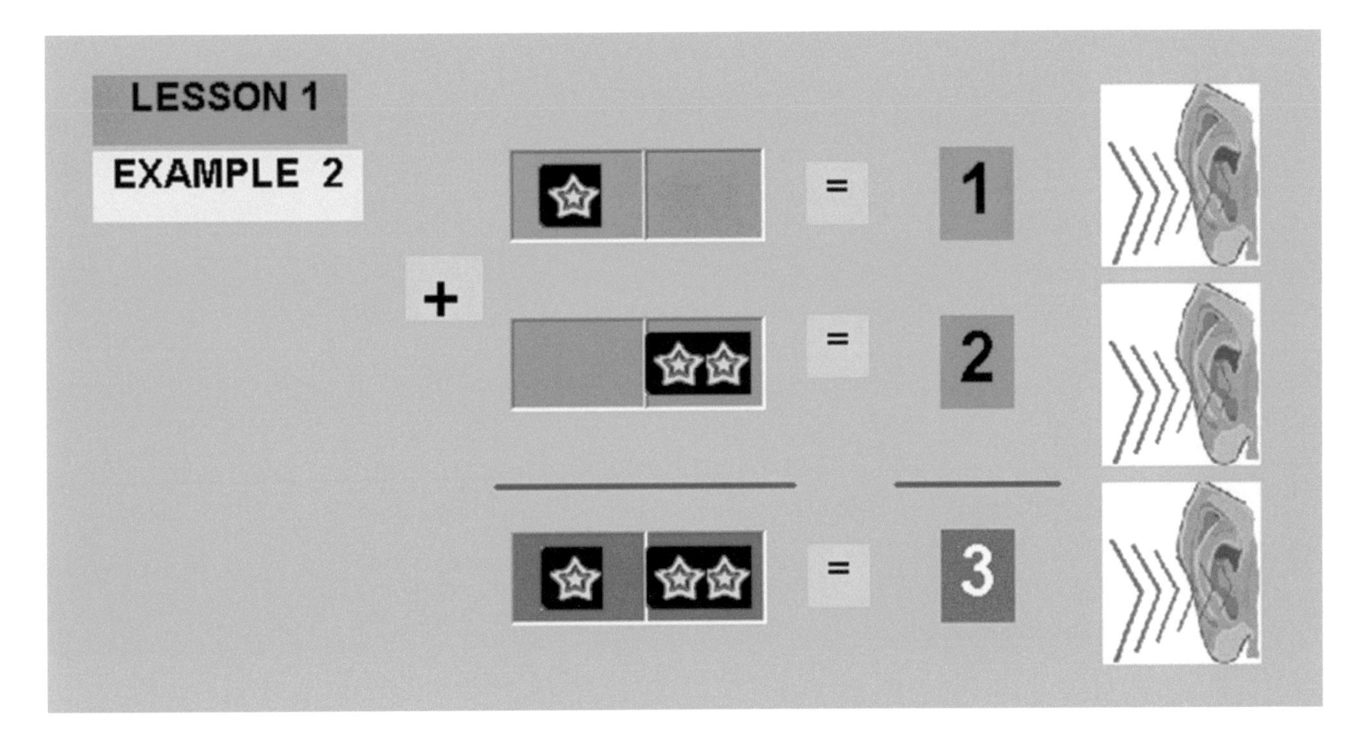

The ***kinesthetic*** representation of the filled up position is announced by clamping of the right hand. An empty position is represented by the clamping of the left hand. The addition sound, tram, is represented by one hand extended in front. The equal sound, cling, is represented by crossing of hands. The kinesthetic representation for both pictures and the answer is the clamping of the right hand, left hand, one hand extended in front, left hand, right hand, crossing of hands, and right hand twice.

MATHEMATICS

ADDITION, EXERCISE 2, STEP 2.2

In exercise 2, step 2.2 in ***visual*** display of addition, we see two pictures of boxes with equal sign. Both pictures have both positions filled with a symbol. The first picture has a yellow star in its positions but the second picture has a red ball in its positions. The form and shape of the symbols do not change mathematical value and the numerical value of both pictures is equal to two.

The ***audio*** representation for this picture consumes of a **knock, knock, and cling.**

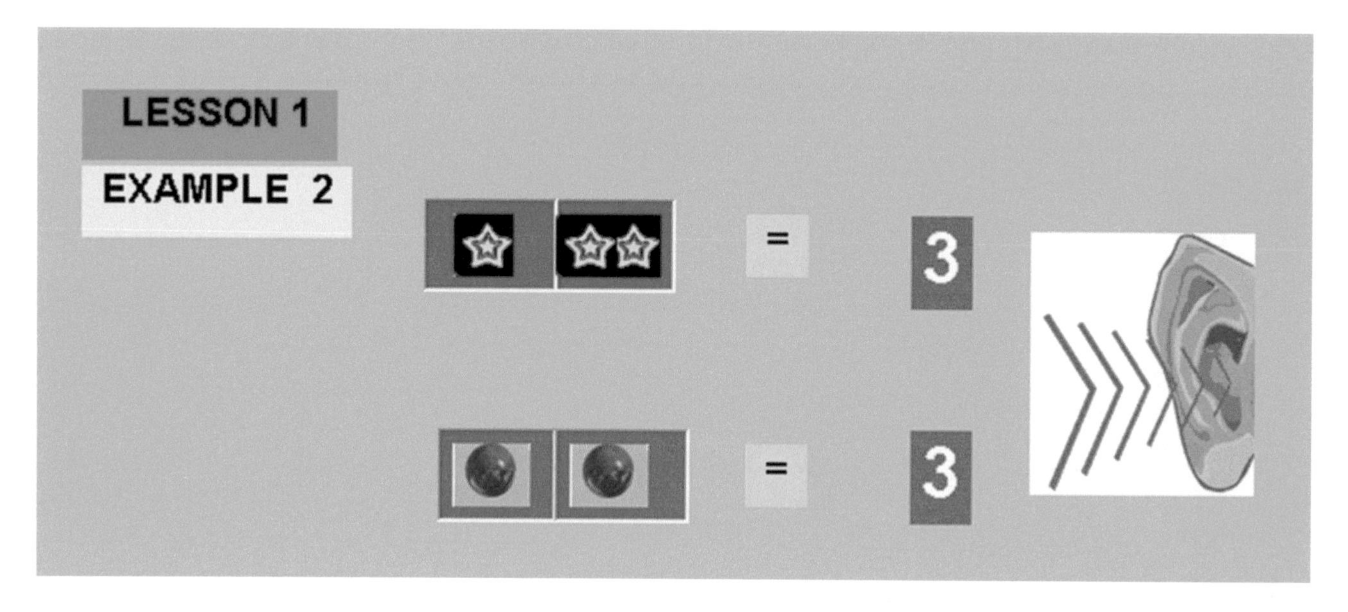

The ***kinesthetic*** representation of the filled up position is announced by clamping of the right hand. An empty position is represented by the clamping of the left hand. The equal sign is crossing both hands in front. The kinesthetic representation for this picture is clamping the right hand twice and crossing of hands.

ADDITION, EXERCISE 2

In exercise 2, in ***visual*** display of addition we see two pictures of boxes. In the bottom we see three pictures with probable answers. We want to specify, that between the first and second pictures we see the addition sign. Below the second picture we see the equal line, which separates the problem from the answer. The ***audio*** representation for two pictures consumes of a **knock, double knock, tram, double knock, knock, and cling.**

From the probable answers for addition we need to find the right one. The audio signals are: 1) **knock and double knock.** 2) **knock and knock** 3) **double knock and knock**.

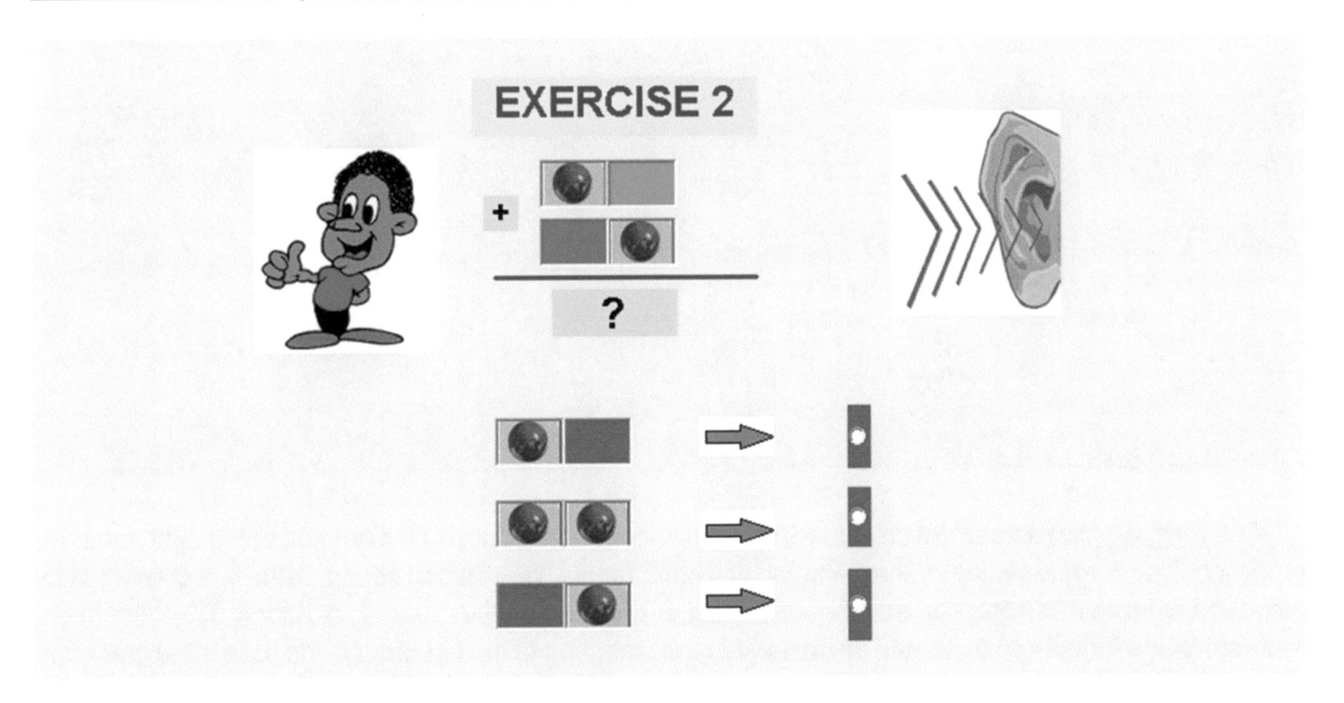

The ***kinesthetic*** representation of the filled up position is announced by clamping of the right hand. An empty position is represented by the clamping of the left hand. The addition sound, tram, is represented by one hand extended in front. The equal sound, cling, is represented by crossing of hands. The kinesthetic representation for both pictures and the answer is the clamping of the right hand, left hand, one hand extended in front, left hand, right hand, crossing of hands, and right hand twice.

MATHEMATICS

ADDITION, EXERCISE 3

In exercise 3, in ***visual*** display of addition we see three pictures of boxes. We want to specify, that between the first and second pictures we see the addition sign. Below the second picture we see the equal line, which separates the problem from the answer. The ***audio*** representation for the three pictures consumes of a **double knock, double knock, tram, double knock, double knock, double knock, and double knock.**

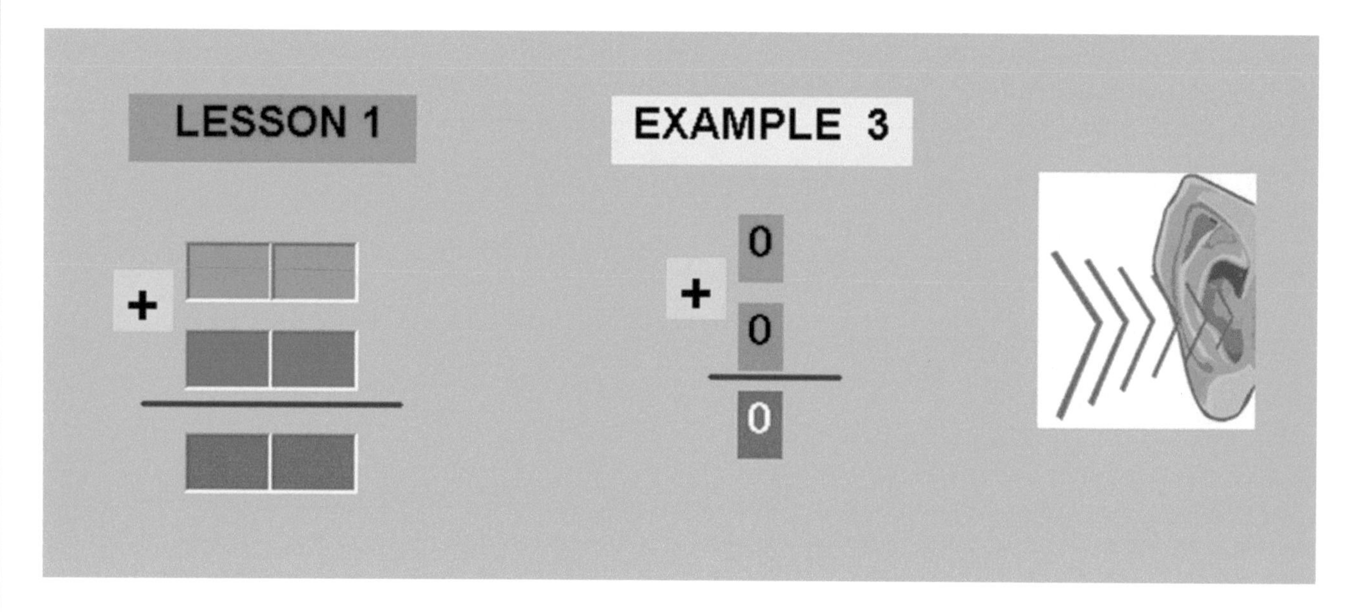

The ***kinesthetic*** representation of the filled up position is announced by clamping of the right hand. An empty position is represented by the clamping of the left hand. The addition sound, tram, is represented by one hand extended in front. The equal sound, cling, is represented by crossing of hands. The kinesthetic representation for both pictures and the answer is the clamping of the left hand twice, one hand extended in front, left hand twice, crossing of hands, and left hand twice.

ADDITION, EXERCISE 3

In exercise 3, in ***visual*** display of addition we see two pictures of boxes. In the bottom we see three pictures with probable answers. We want to specify, that between the first and second pictures we see the addition sign. Below the second picture we see the equal line, which separates the problem from the answer. The ***audio*** representation for two pictures consumes of a **double knock, double knock, tram, double knock, double knock, and cling.**

From the probable answers for addition we need to find the right one. The audio signals are: 1) **knock and double knock.** 2) **double knock and double knock** 3) **double knock and knock**.

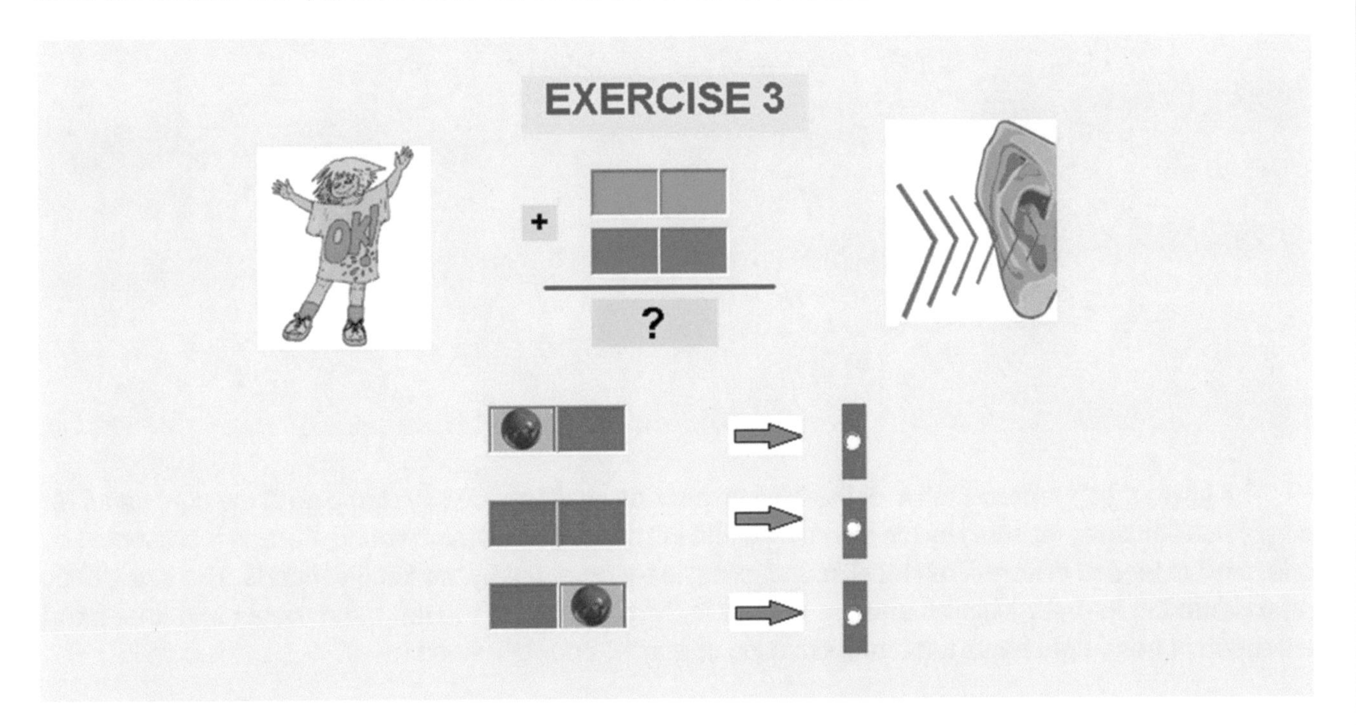

The ***kinesthetic*** representation of the filled up position is announced by clamping of the right hand. An empty position is represented by the clamping of the left hand. The addition sound, tram, is represented by one hand extended in front. The equal sound, cling, is represented by crossing of hands. The kinesthetic representation for both pictures and the answer is the clamping of the left hand twice, one hand extended in front, left hand twice, crossing of hands, and left hand twice.

MATHEMATICS ADDITION, EXERCISE 4

In exercise 4, in ***visual*** display of addition we see three pictures of boxes. We want to specify, that between the first and second pictures we see the addition sign. Below the second picture we see the equal line, which separates the problem from the answer. The ***audio*** representation for the three pictures consumes of a **double knock, knock, tram, knock, double knock, cling, knock, and knock.**

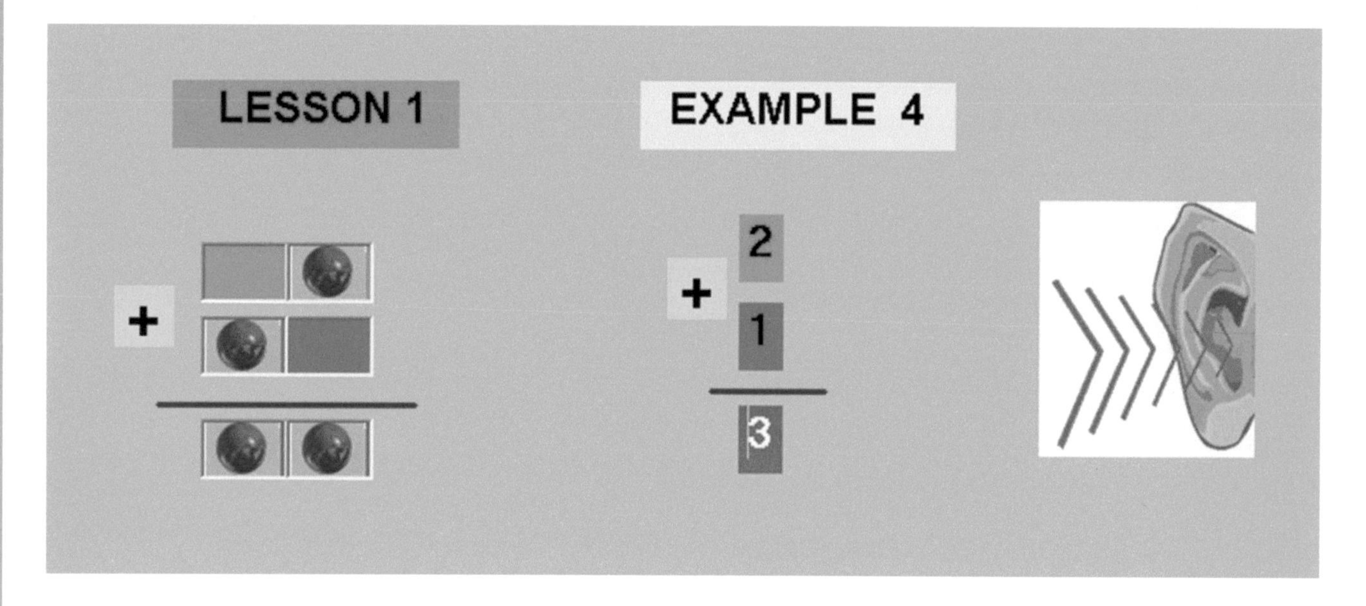

The ***kinesthetic*** representation of the filled up position is announced by clamping of the right hand. An empty position is represented by the clamping of the left hand. The addition sound, tram, is represented by one hand extended in front. The equal sound, cling, is represented by crossing of hands. The kinesthetic representation for both pictures and the answer is the clamping of the left hand, right hand, one hand extended in front, right hand, left hand, crossing of hands, and right hand twice.

ADDITION, EXERCISE 4

In exercise 4, in ***visual*** display of addition we see two pictures of boxes. In the bottom we see three pictures with probable answers. We want to specify, that between the first and second pictures we see the addition sign. Below the second picture we see the equal line, which separates the problem from the answer. The ***audio*** representation for two pictures consumes of a **double knock, knock, tram, knock, double knock, and cling.**

From the probable answers for addition we need to find the right one. The audio signals are: 1) **knock and double knock.** 2) **double knock and knock** 3) **knock and knock**.

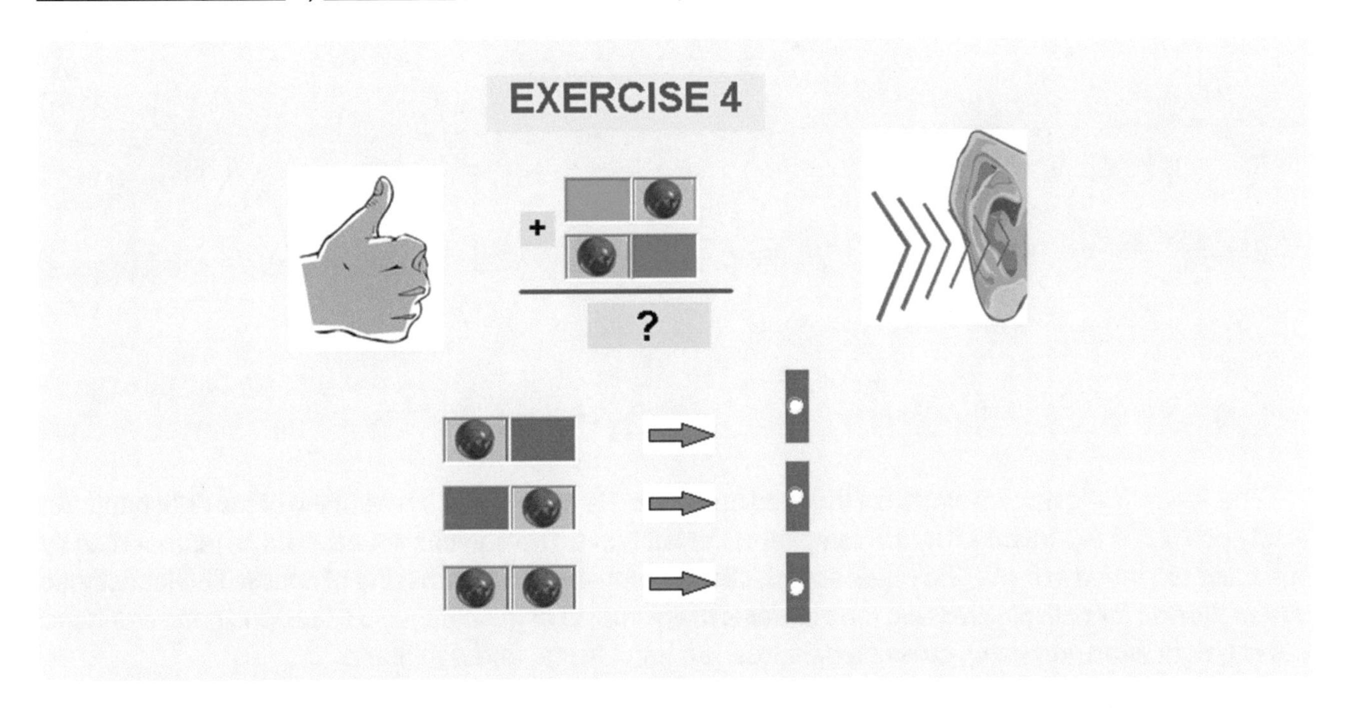

The ***kinesthetic*** representation of the filled up position is announced by clamping of the right hand. An empty position is represented by the clamping of the left hand. The addition sound, tram, is represented by one hand extended in front. The equal sound, cling, is represented by crossing of hands. The kinesthetic representation for both pictures and the answer is the clamping of the left hand, right hand, one hand extended in front, right hand, left hand, crossing of hands, and right hand twice.

MATHEMATICS

ADDITION, EXERCISE 5

In exercise 5, in ***visual*** display of addition we see three pictures of boxes. We want to specify, that between the first and second pictures we see the addition sign. Below the second picture we see the equal line, which separates the problem from the answer. The ***audio*** representation for the three pictures consumes of a **knock, knock, tram, knock, double knock, and cling.**

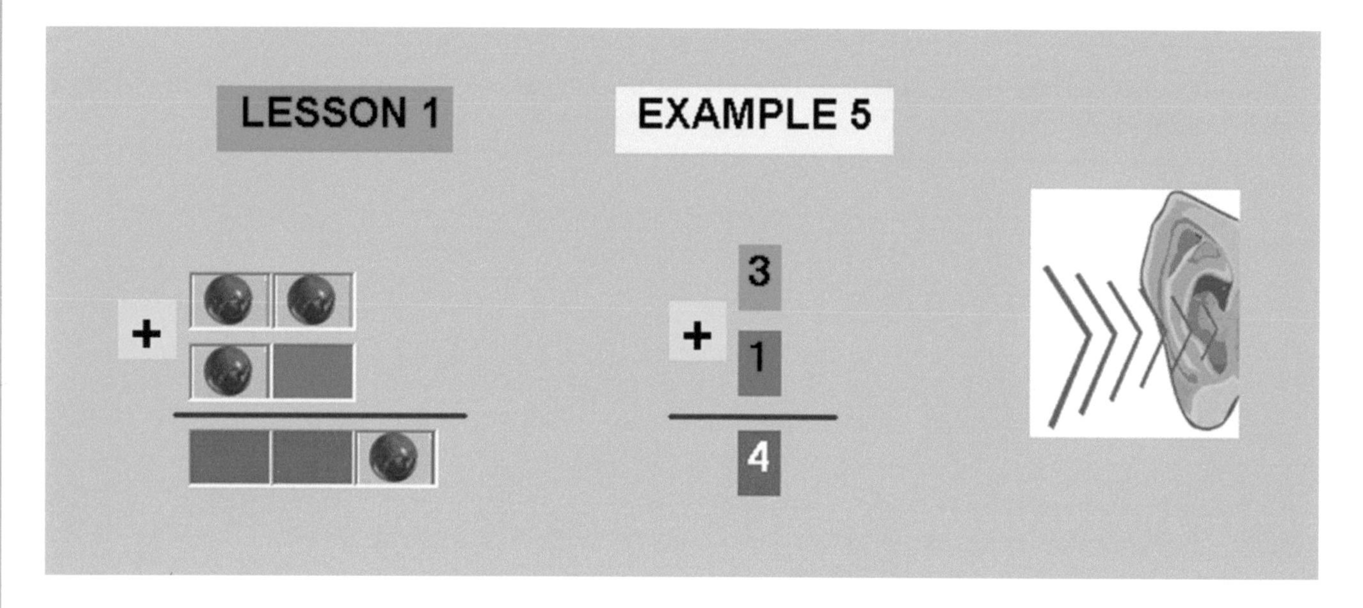

The ***kinesthetic*** representation of the filled up position is announced by clamping of the right hand. An empty position is represented by the clamping of the left hand. The addition sound, tram, is represented by one hand extended in front. The equal sound, cling, is represented by crossing of hands. The kinesthetic representation for both pictures and the answer is the clamping of the right hand twice, one hand extended in front, right hand, left hand, crossing of hands, left hand twice, and right hand.

ADDITION, EXERCISE 5, STEP 5.1

In exercise 5, step 5.1 in ***visual*** display we see two pictures of boxes. In the first picture we see that the both positions are filled with a symbol. This pictures numerical value is 3.

The **audio** representation for the first picture is **knock and knock.**

The second picture has the first position filled with a symbol and the second position is empty.

The **audio** representation for the second picture is **knock and double knock.**

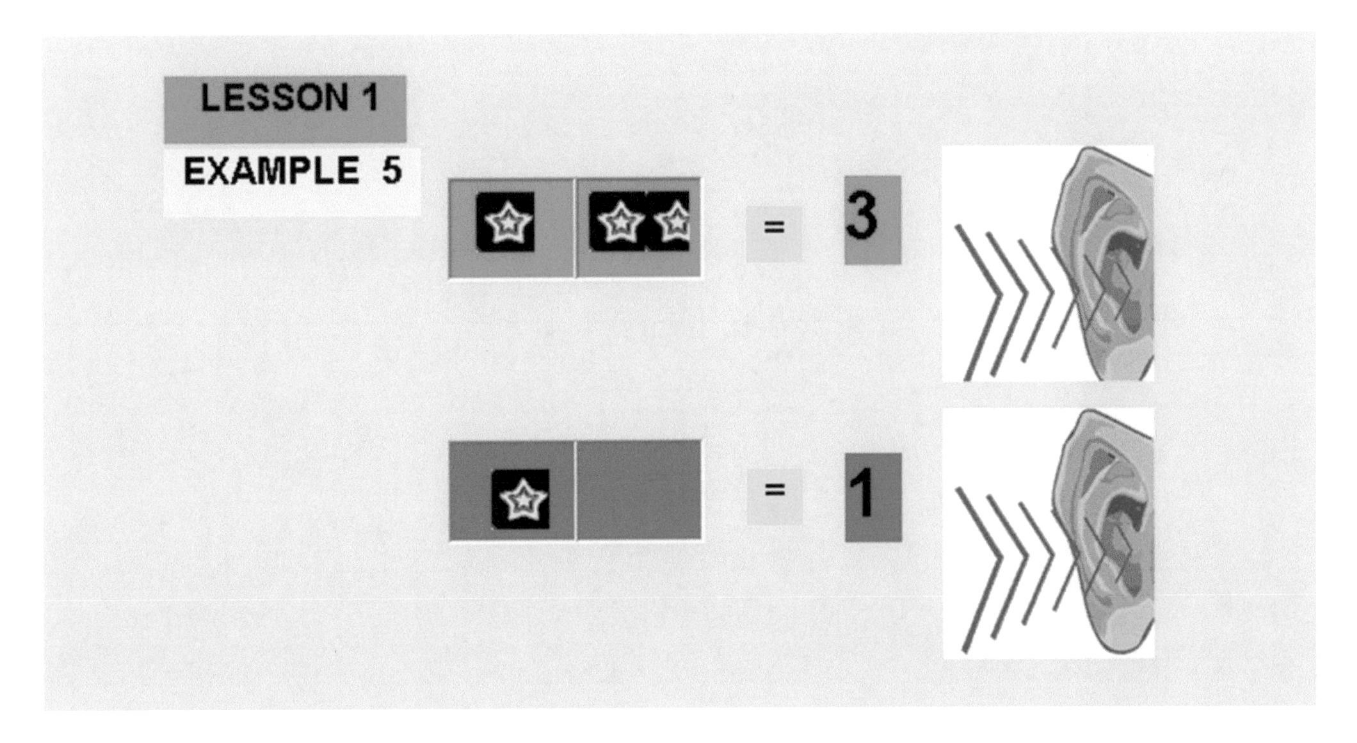

The ***kinesthetic*** representation of the filled up position is announced by clamping of the right hand. An empty position is represented by the clamping of the left hand. The equal sound, cling, is represented by crossing both hands. The kinesthetic representation for the first picture is the clamping of the right hand twice. The kinesthetic representation for the second picture is clamping of right hand and left hand.

MATHEMATICS

ADDITION, EXERCISE 5, STEP 5.2

In exercise 5, step 5.2 in ***visual*** display of addition we see three pictures of boxes. We want to specify, that between the first and second pictures we see the addition sign. Below the second picture we see the equal line, which separates the problem from the answer. When we are adding two numbers and the position on the first number is filled with a symbol and the corresponding position on the second number is also with a symbol, we move the two symbols to the following positions in the answer as one symbol. When two symbols form, two corresponding positions move forward to the following position of the answer then the two symbols become one symbol. The ***audio*** representation for the three pictures consumes of a **knock, knock, tram, knock, double knock, and cling.**

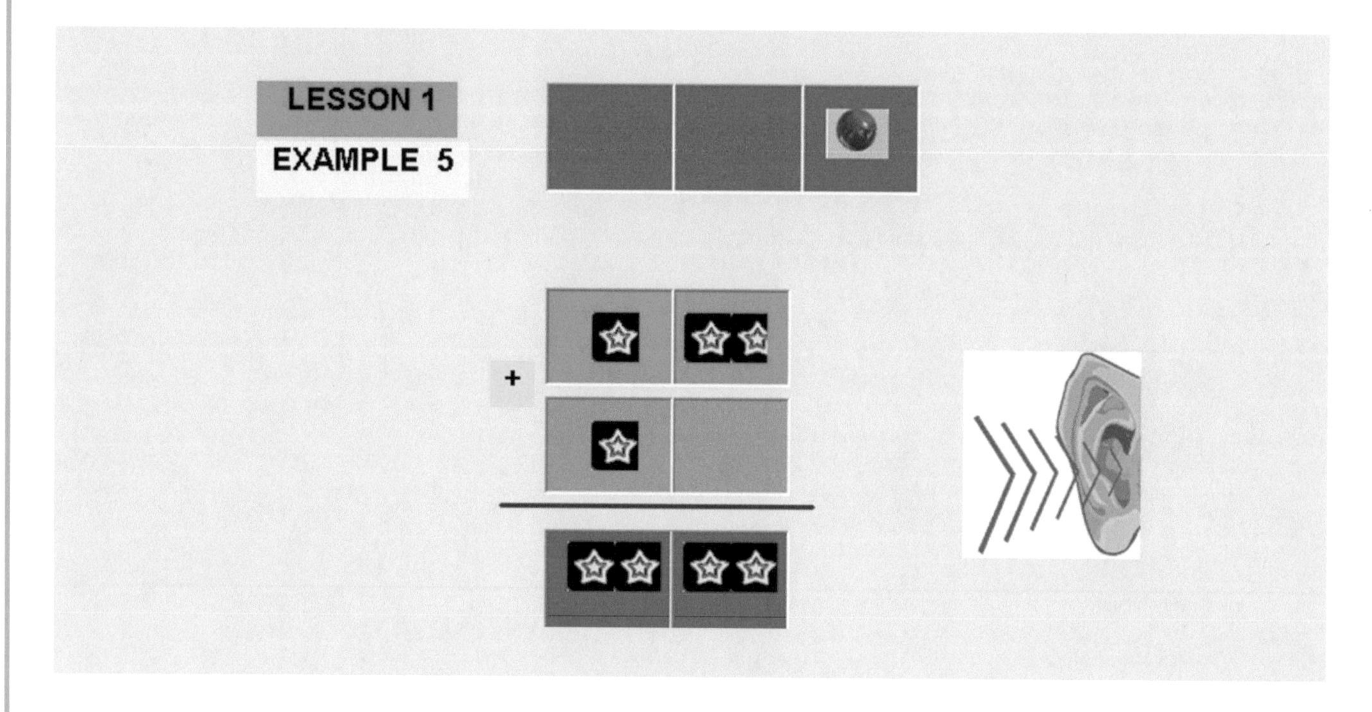

The ***kinesthetic*** representation of the filled up position is announced by clamping of the right hand. An empty position is represented by the clamping of the left hand. The addition sound, tram, is represented by one hand extended in front. The equal sound, cling, is represented by crossing of hands. The kinesthetic representation for both pictures and the answer is the clamping of the right hand twice, one hand extended in front, right hand, left hand, crossing of hands, left hand twice, and right hand.

ADDITION, EXERCISE 5, STEP 5.3

In exercise 5, step 5.3 in ***visual*** display of addition we see two pictures of boxes. In the first picture the first two positions are filled with a symbol and the third position is empty. The second picture has a symbol in the second position and has the first and third positions empty. For purpose of addition we are introducing the concept of moving. When a position passes its maximum capacity the next symbol is simultaneously moved in the ascending position. We are moving from the first to the second position. However the audio representation for the picture does not change because the numerical value does not change.

The ***audio*** representation for the three pictures consumes of a **double knock, knock, and double knock.**

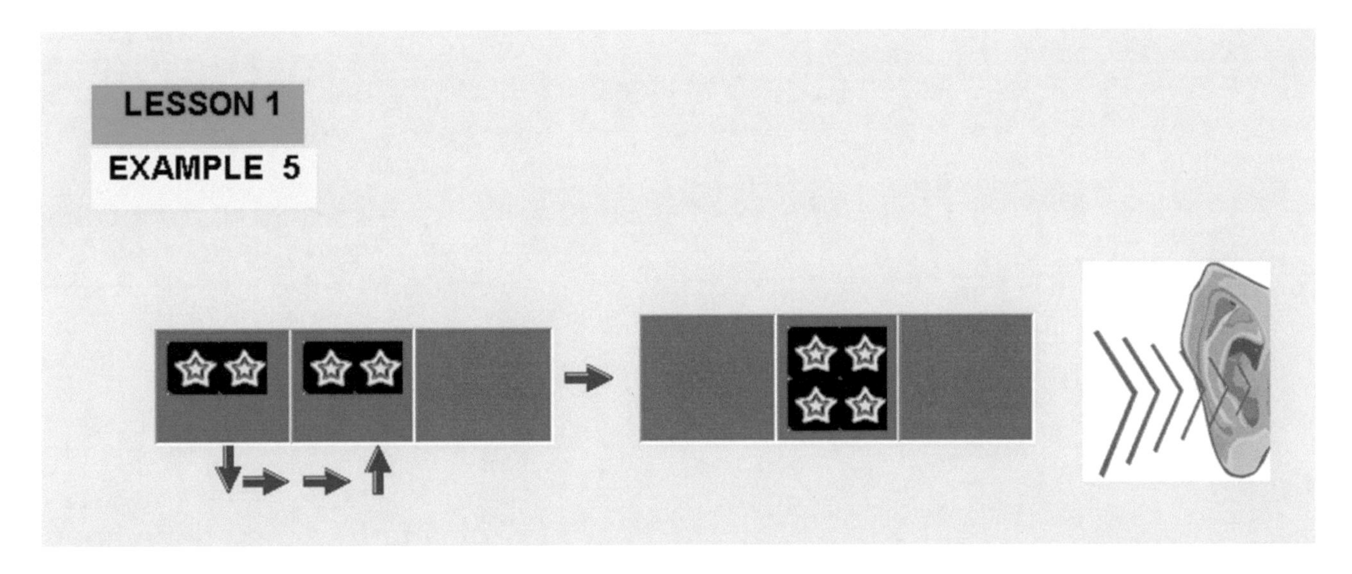

The ***kinesthetic*** representation of the filled up position is announced by clamping of the right hand. An empty position is represented by the clamping of the left hand. The kinesthetic representation for three pictures is the clamping of the left hand, right hand, and left hand.

MATHEMATICS

ADDITION, EXERCISE 5, STEP 5.4

In exercise 5, step 5.4 in ***visual*** display of addition we see two pictures of boxes. The first picture has a symbol in the second position and has the first and third positions empty. The second picture has the third position filled with a symbol. For purpose of addition we are introducing the concept of moving. When a position passes its maximum capacity the next symbol is simultaneously moved in the ascending position. We are moving from the second to the third position. However the audio representation for the picture does not change because the numerical value does not change.

The ***audio*** representation for the three pictures consumes of a **double knock, double knock, and knock.**

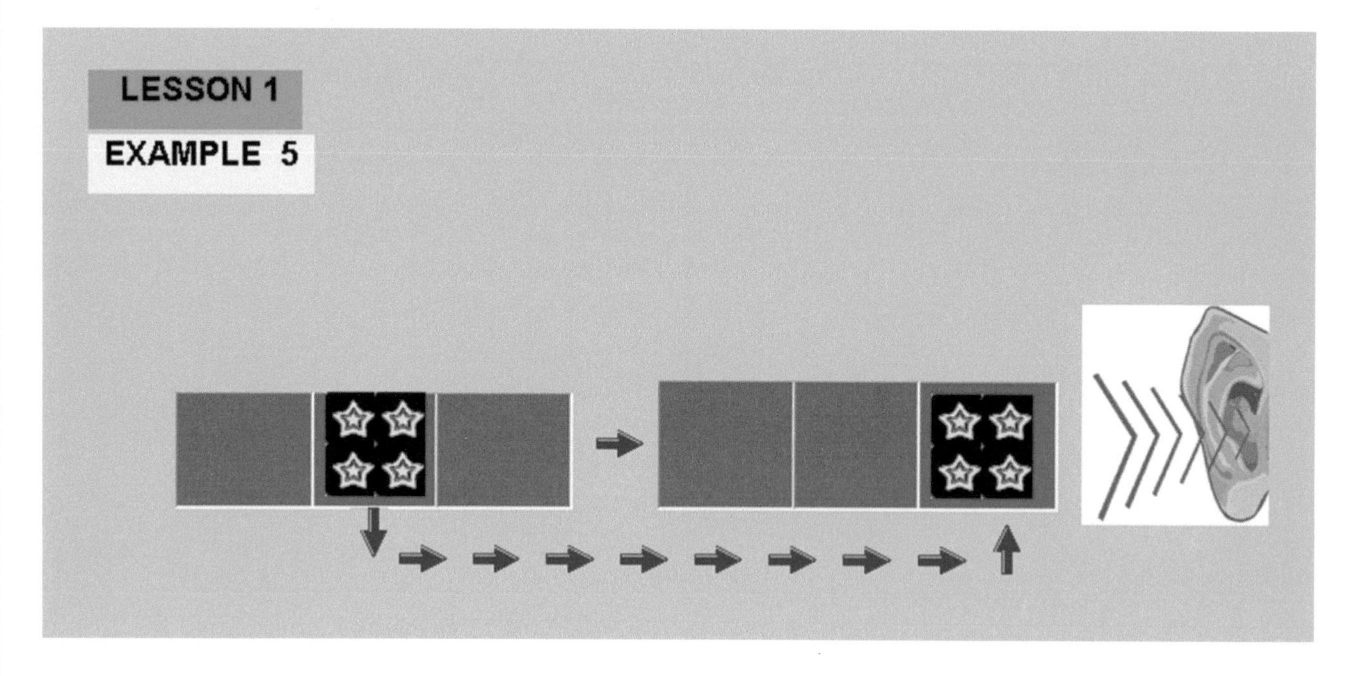

The ***kinesthetic*** representation of the filled up position is announced by clamping of the right hand. An empty position is represented by the clamping of the left hand. The kinesthetic representation for three pictures is the clamping of the left hand twice and right hand.

ADDITION, EXERCISE 5, STEP 5.5

In exercise 5, step 5.5 in ***visual*** display of addition, we see two pictures of boxes with equal sign. Both pictures have the first two positions empty and the last position is filled wit ha symbol. The first picture has a yellow star in its positions but the second picture has a red ball in its positions. The form and shape of the symbols do not change mathematical value and the numerical value of both pictures is equal to four.

The ***audio*** representation for this picture consumes of a **double knock, double knock, knock, and cling.**

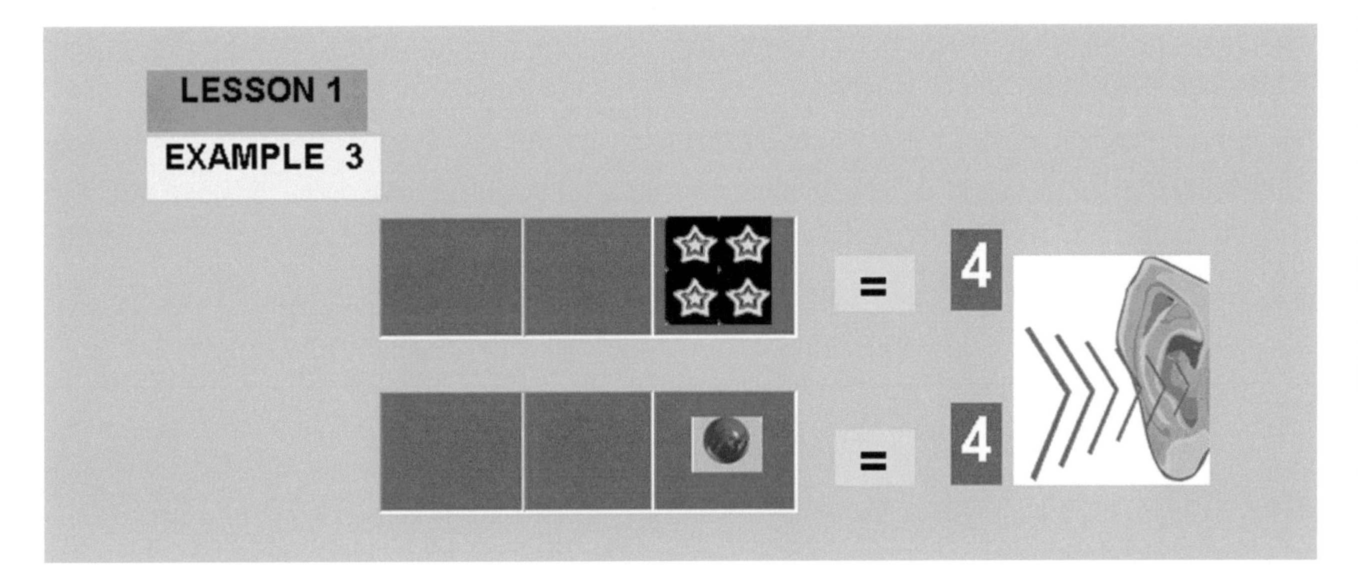

The ***kinesthetic*** representation of the filled up position is announced by clamping of the right hand. An empty position is represented by the clamping of the left hand. The equal sign is crossing both hands in front. The kinesthetic representation for this picture is clamping the left hand twice, right hand, and crossing of hands.

MATHEMATICS ADDITION, EXERCISE 5

In exercise 5, in ***visual*** display of addition we see two pictures of boxes. In the bottom we see three pictures with probable answers. We want to specify, that between the first and second pictures we see the addition sign. Below the second picture we see the equal line, which separates the problem from the answer. The ***audio*** representation for two pictures consumes of a **knock, knock, tram, knock, double knock, and cling.**

From the probable answers for addition we need to find the right one. The audio signals are: 1) **double knock, knock, and double knock.** 2) **double knock, double knock, and knock** 3) **knock, double knock, and double knock**.

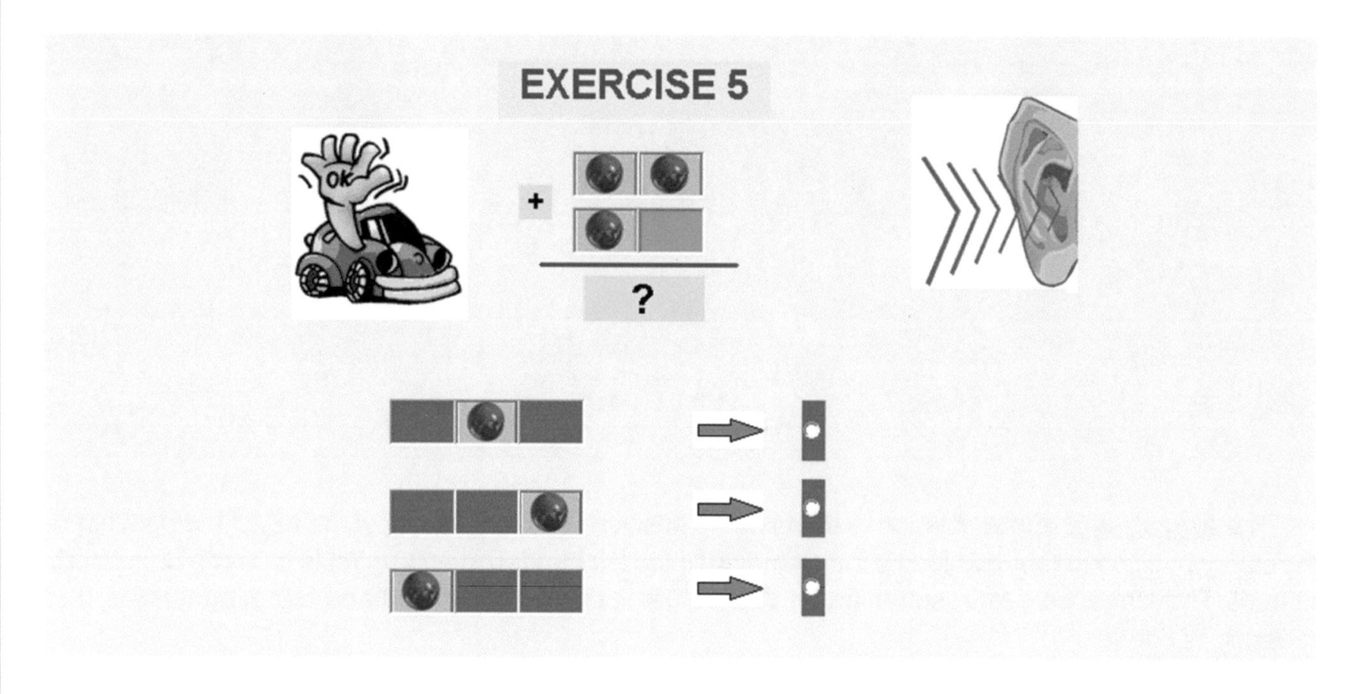

The ***kinesthetic*** representation of the filled up position is announced by clamping of the right hand. An empty position is represented by the clamping of the left hand. The addition sound, tram, is represented by one hand extended in front. The equal sound, cling, is represented by crossing of hands. The kinesthetic representation for both pictures and the answer is the clamping of the right hand twice, one hand extended in front, right hand, left hand, crossing of hands, left hand twice, and right hand.

ADDITION, EXERCISE 6

In exercise 6, in ***visual*** display of addition we see three pictures of boxes. We want to specify, that between the first and second pictures we see the addition sign. Below the second picture we see the equal line, which separates the problem from the answer. When we are adding two numbers and the position on the first number is filled with a symbol and the corresponding position on the second number is also with a symbol, we move the two symbols to the following positions in the answer as one symbol. When two symbols form, two corresponding positions move forward to the following position of the answer then the two symbols become one symbol. The ***audio*** representation for the three pictures consumes of a **double knock, knock, tram, double knock, knock, cling, double knock, double knock, and knock.**

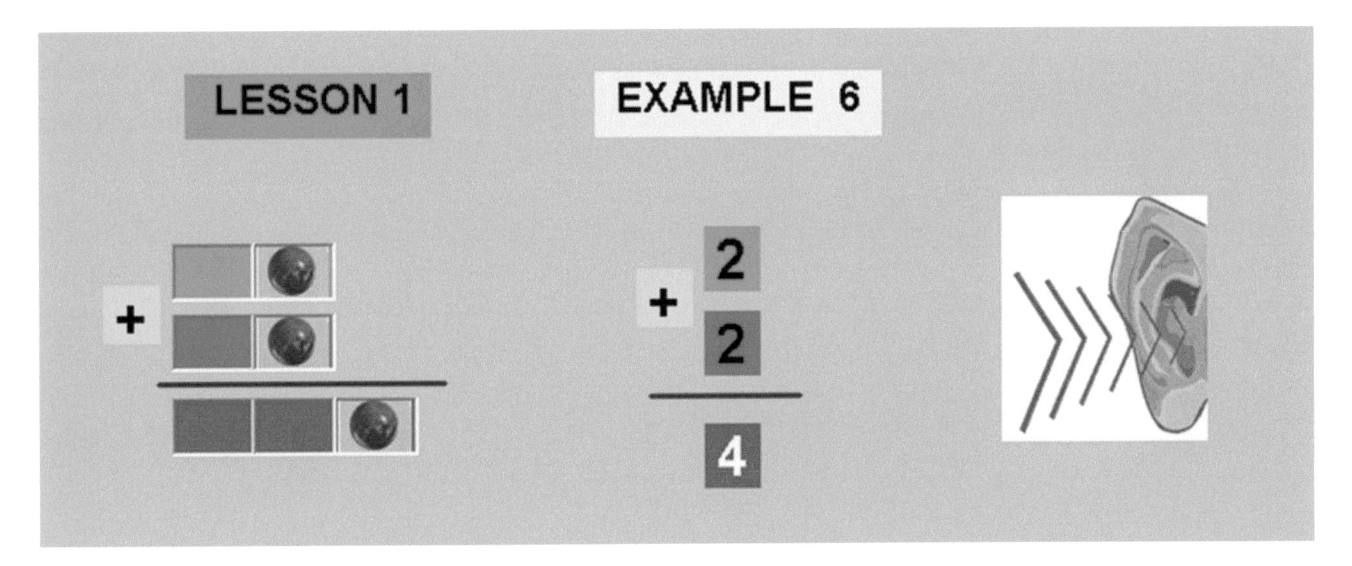

The ***kinesthetic*** representation of the filled up position is announced by clamping of the right hand. An empty position is represented by the clamping of the left hand. The addition sound, tram, is represented by one hand extended in front. The equal sound, cling, is represented by crossing of hands. The kinesthetic representation for both pictures and the answer is the clamping of the left hand, right hand, one hand extended in front, left hand, right hand, crossing of hands, left hand twice, and right hand.

MATHEMATICS

ADDITION, EXERCISE 6, STEP 6.1

In exercise 6, step 6.1 in ***visual*** display of addition we see three pictures of boxes. We want to specify, that between the first and second pictures we see the addition sign. Below the second picture we see the equal line, which separates the problem from the answer. The ***audio*** representation for the three pictures consumes of a **double knock, knock, tram, double knock, knock, cling, double knock, double knock, and knock.**

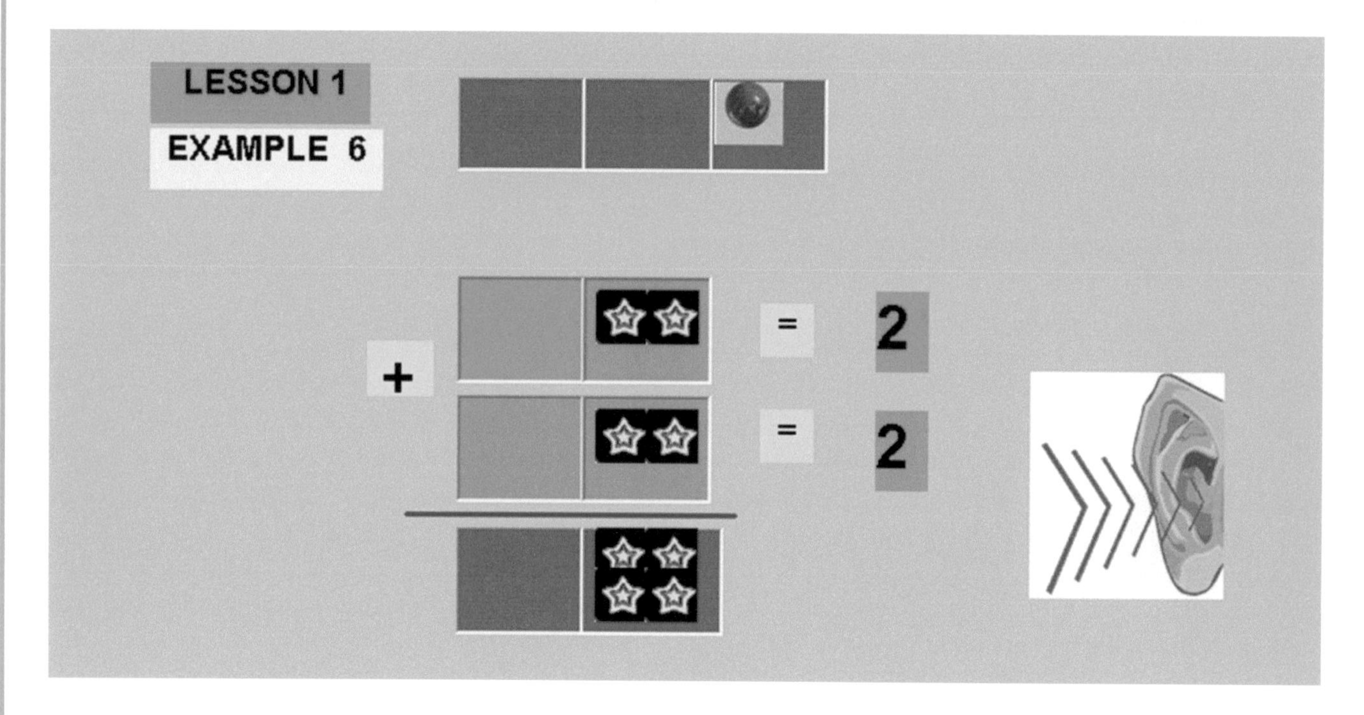

The ***kinesthetic*** representation of the filled up position is announced by clamping of the right hand. An empty position is represented by the clamping of the left hand. The addition sound, tram, is represented by one hand extended in front. The equal sound, cling, is represented by crossing of hands. The kinesthetic representation for both pictures and the answer is the clamping of the left hand, right hand, one hand extended in front, left hand, right hand, crossing of hands, left hand twice, and right hand.

ADDITION, EXERCISE 6, STEP 6.2

In exercise 6, step 6.2 in ***visual*** display of addition we see two pictures of boxes. The first picture has a symbol in the second position and has the first and third positions empty. The second picture has the third position filled with a symbol. For purpose of addition we are introducing the concept of moving. When a position passes its maximum capacity the next symbol is simultaneously moved in the ascending position. We are moving from the second to the third position. However the audio representation for the picture does not change because the numerical value does not change.

The ***audio*** representation for the three pictures consumes of a **double knock, double knock, and knock.**

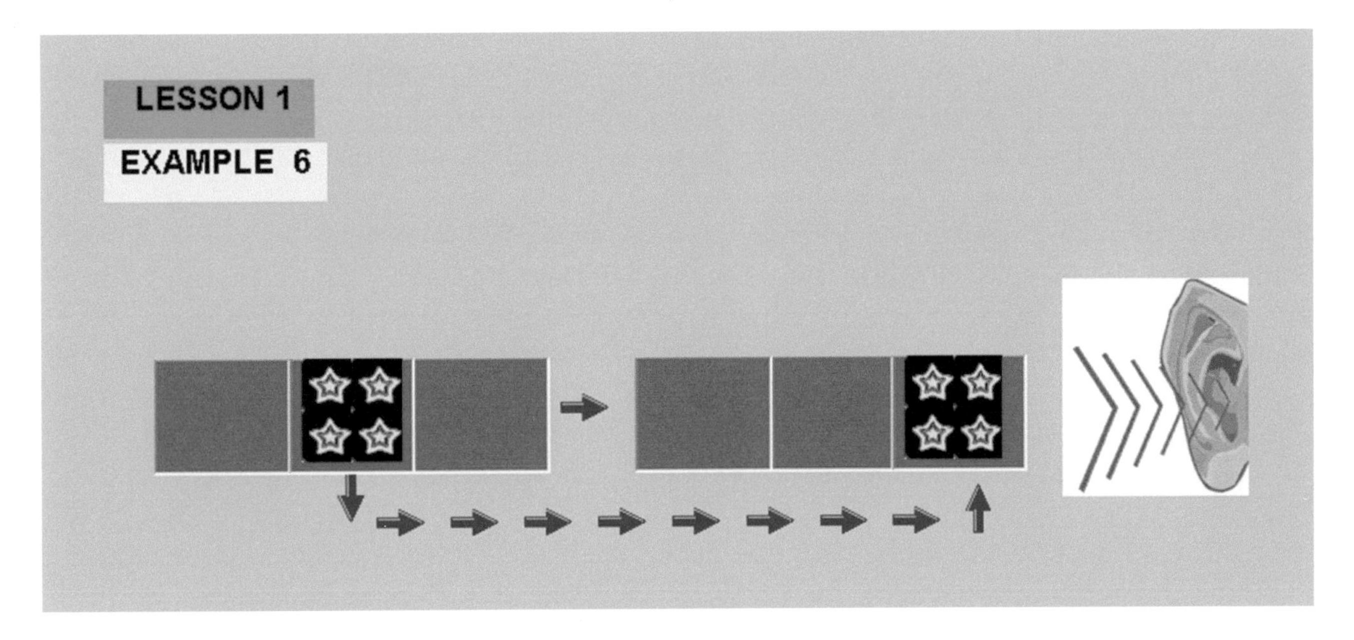

The ***kinesthetic*** representation of the filled up position is announced by clamping of the right hand. An empty position is represented by the clamping of the left hand. The kinesthetic representation for three pictures is the clamping of the left hand twice and right hand.

MATHEMATICS

ADDITION, EXERCISE 6, STEP 6.3

In exercise 6, step 6.3 in ***visual*** display of addition, we see two pictures of boxes with equal sign. Both pictures have the first two positions empty and the last position is filled with a symbol. The first picture has a yellow star in its positions but the second picture has a red ball in its positions. The form and shape of the symbols do not change mathematical value and the numerical value of both pictures is equal to four.

The ***audio*** representation for this picture consumes of a **double knock, double knock, knock, and cling.**

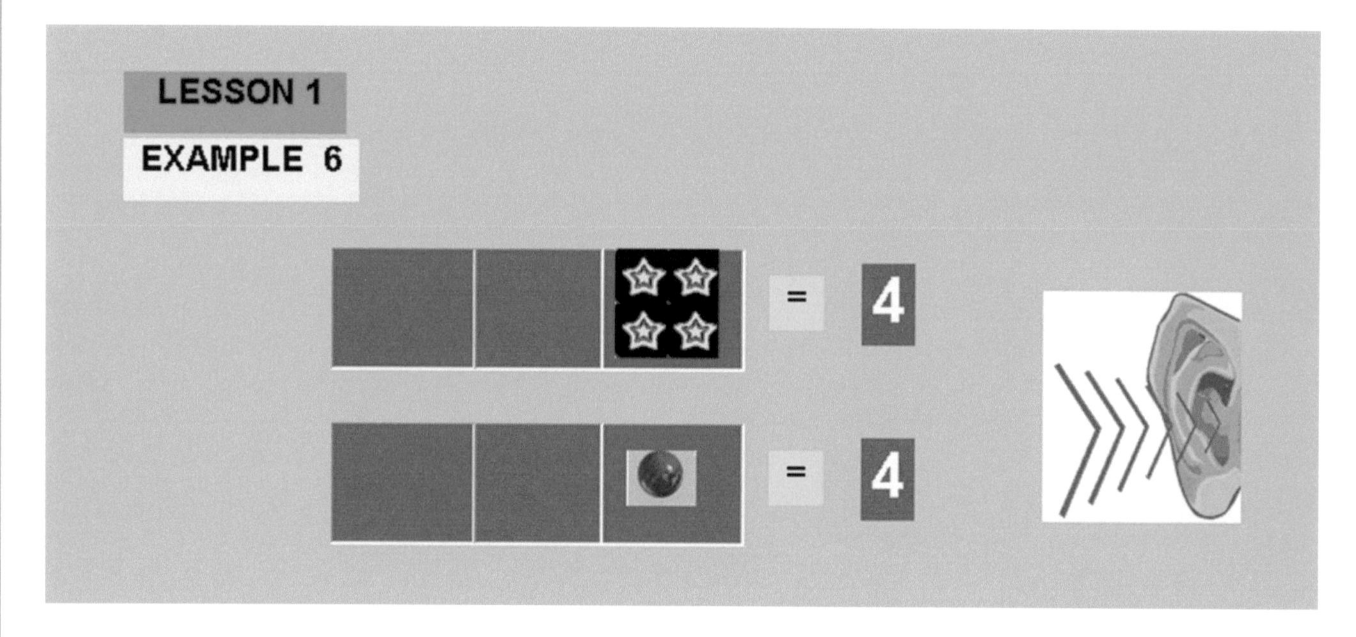

The ***kinesthetic*** representation of the filled up position is announced by clamping of the right hand. An empty position is represented by the clamping of the left hand. The equal sign is crossing both hands in front. The kinesthetic representation for this picture is clamping the left hand twice, right hand, and crossing of hands.

ADDITION, EXERCISE 6

In exercise 6, in ***visual*** display of addition we see two pictures of boxes. In the bottom we see three pictures with probable answers. We want to specify, that between the first and second pictures we see the addition sign. Below the second picture we see the equal line, which separates the problem from the answer. The ***audio*** representation for two pictures consumes of a **double knock, knock, tram, double knock, knock, and cling.**

From the probable answers for addition we need to find the right one. The audio signals are: 1) **double knock, knock, and double knock.** 2) **knock, double knock, and double knock** 3) **double knock, double knock, and knock**.

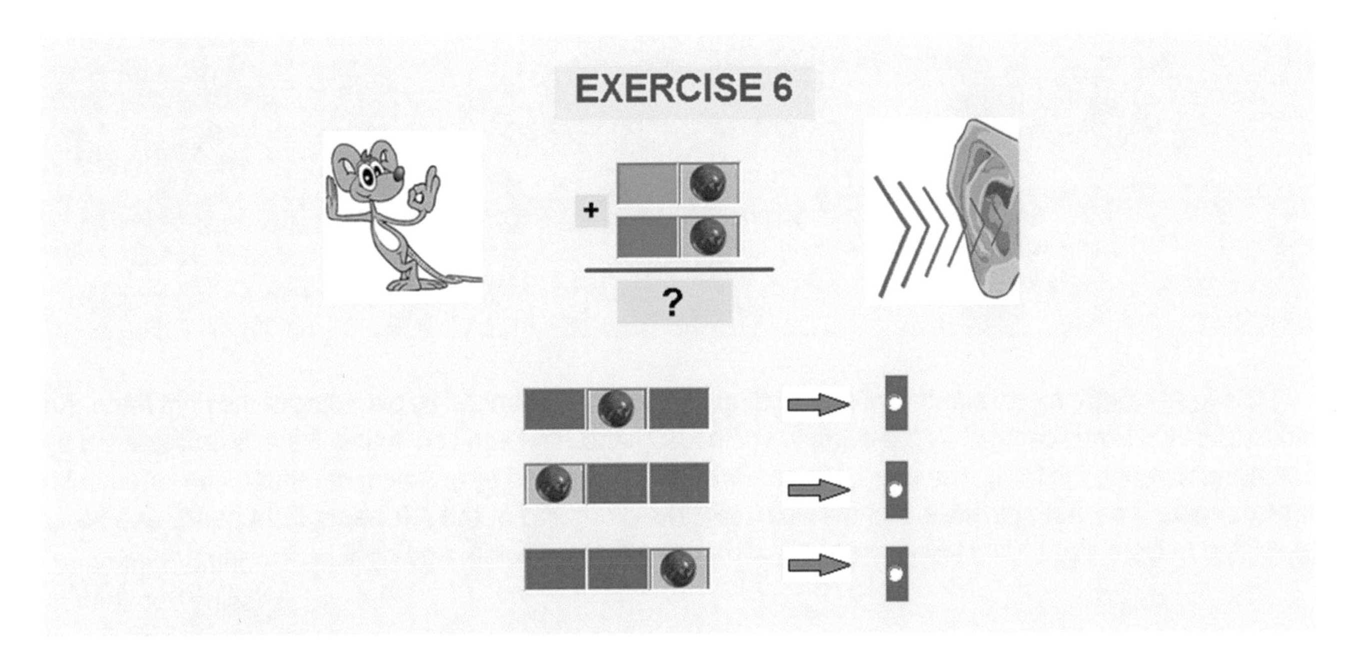

The ***kinesthetic*** representation of the filled up position is announced by clamping of the right hand. An empty position is represented by the clamping of the left hand. The addition sound, tram, is represented by one hand extended in front. The equal sound, cling, is represented by crossing of hands. The kinesthetic representation for both pictures and the answer is the clamping of the left hand, right hand, one hand extended in front, left hand, right hand, crossing of hands, left hand twice, and right hand.

MATHEMATICS ADDITION, EXERCISE 7

In exercise 7, in ***visual*** display of addition we see three pictures of boxes. We want to specify, that between the first and second pictures we see the addition sign. Below the second picture we see the equal line, which separates the problem from the answer. The ***audio*** representation for the three pictures consumes of a **double knock, knock, tram, knock, knock, cling, knock, double knock, and knock.**

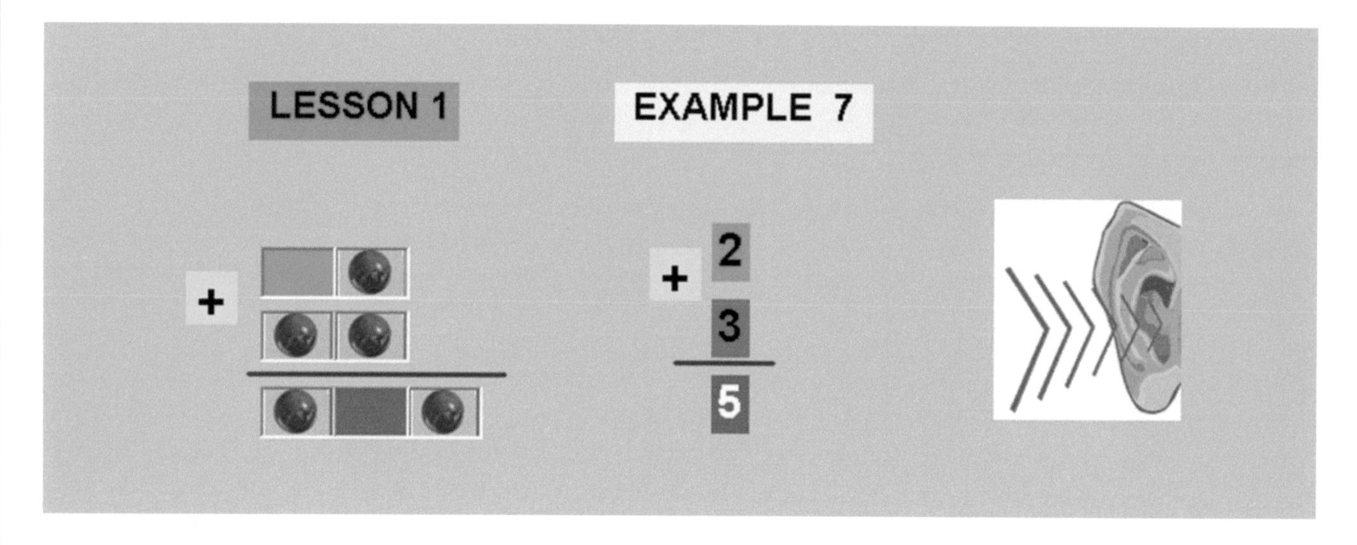

The ***kinesthetic*** representation of the filled up position is announced by clamping of the right hand. An empty position is represented by the clamping of the left hand. The addition sound, tram, is represented by one hand extended in front. The equal sound, cling, is represented by crossing of hands. The kinesthetic representation for both pictures and the answer is the clamping of the left hand, right hand, one hand extended in front, right hand twice, crossing of hands, left hand twice, and right hand.

ADDITION, EXERCISE 7, STEP 7.1

In exercise 7, step 7.1 in ***visual*** display we see two pictures of boxes. In the first picture we see that the first position is empty and the second position is filled with a symbol. This pictures numerical value is 2.

The **audio** representation for the first picture is **double knock and knock.**

The second picture has both its positions filled with a symbol. The **audio** representation for the second picture is **knock and knock.**

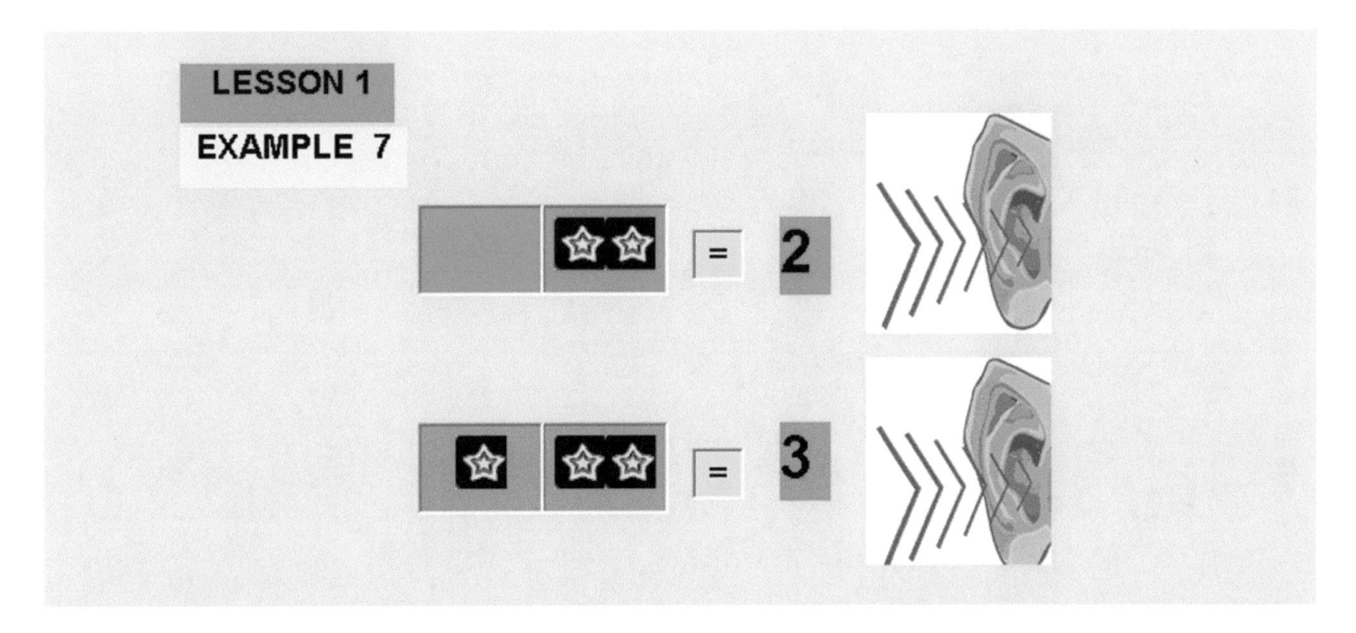

The ***kinesthetic*** representation of the filled up position is announced by clamping of the right hand. An empty position is represented by the clamping of the left hand. The equal sound, cling, is represented by crossing both hands. The kinesthetic representation for the first picture is the clamping of the left hand and right hand. The kinesthetic representation for the second picture is clamping of right hand twice.

MATHEMATICS ADDITION, EXERCISE 7, STEP 7.2

In exercise 7, step 7.2 in ***visual*** display of addition we see three pictures of boxes. We want to specify, that between the first and second pictures we see the addition sign. Below the second picture we see the equal line, which separates the problem from the answer. When we are adding two numbers and the position on the first number is filled with a symbol and the corresponding position on the second number is also with a symbol, we move the two symbols to the following positions in the answer as one symbol. When two symbols form, two corresponding positions move forward to the following position of the answer then the two symbols become one symbol. The ***audio*** representation for the three pictures consumes of a **double knock, knock, tram, knock, knock, cling, knock, double knock, and knock.**

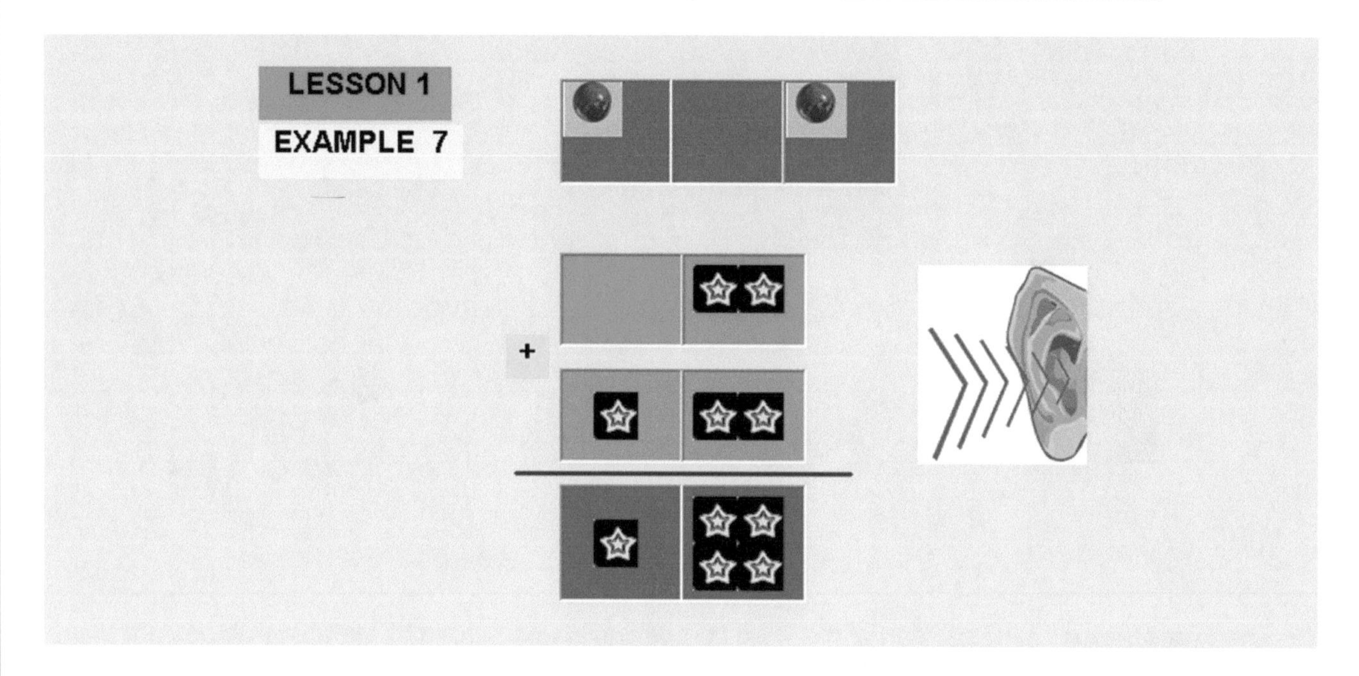

The ***kinesthetic*** representation of the filled up position is announced by clamping of the right hand. An empty position is represented by the clamping of the left hand. The addition sound, tram, is represented by one hand extended in front. The equal sound, cling, is represented by crossing of hands. The kinesthetic representation for both pictures and the answer is the clamping of the left hand, right hand, one hand extended in front, right hand twice, crossing of hands, left hand twice, and right hand.

ADDITION, EXERCISE 7, STEP 7.3

In exercise 7, step 7.3 in ***visual*** display of addition we see two pictures of boxes. The first picture has a symbol in the first and second positions and has the third position empty. The second picture has the first and third positions filled with a symbol. For purpose of addition we are introducing the concept of moving. When a position passes its maximum capacity the next symbol is simultaneously moved in the ascending position. We are moving from the second to the third position. However the audio representation for the picture does not change because the numerical value does not change.

The ***audio*** representation for the three pictures consumes of a **knock, double knock, and knock.**

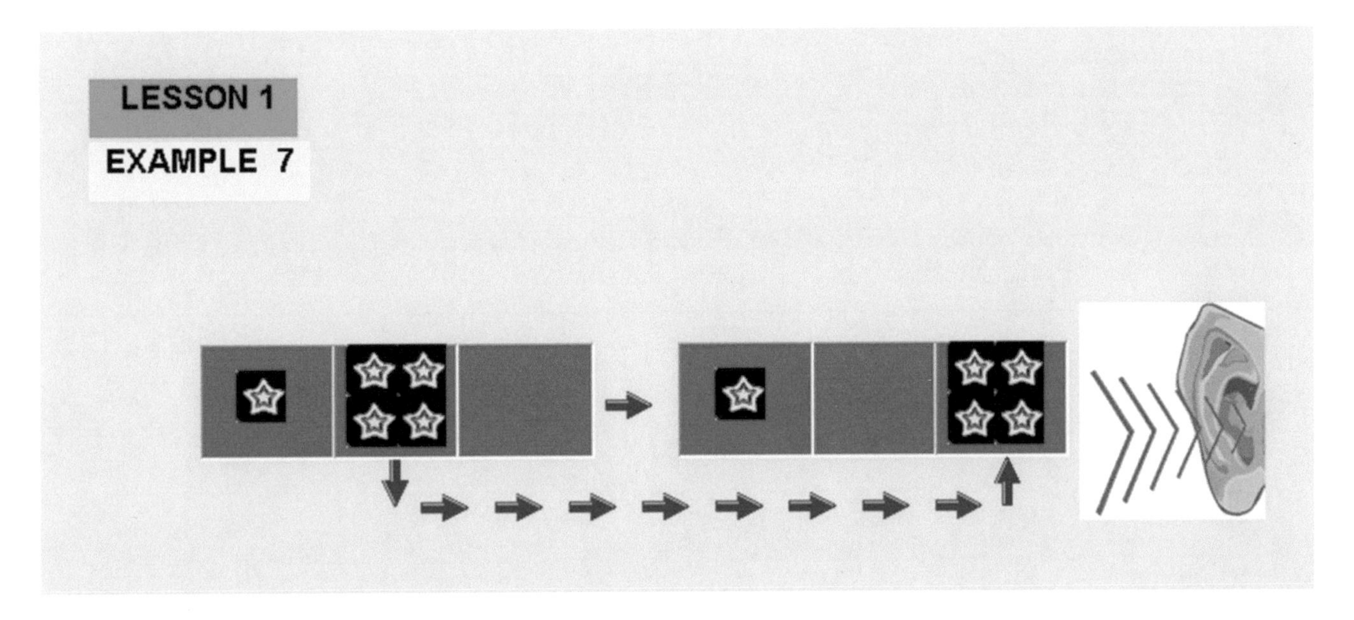

The ***kinesthetic*** representation of the filled up position is announced by clamping of the right hand. An empty position is represented by the clamping of the left hand. The kinesthetic representation for three pictures is the clamping of the right hand, left hand, and right hand.

MATHEMATICS

ADDITION, EXERCISE 7, STEP 7.4

In exercise 7, step 7.4 in ***visual*** display of addition, we see two pictures of boxes with equal sign. Both pictures have the first and third positions filled with a symbol and the second position is empty. The first picture has a yellow star in its positions but the second picture has a red ball in its positions. The form and shape of the symbols do not change mathematical value and the numerical value of both pictures is equal to five.

The ***audio*** representation for this picture consumes of a **knock, double knock, knock, and cling.**

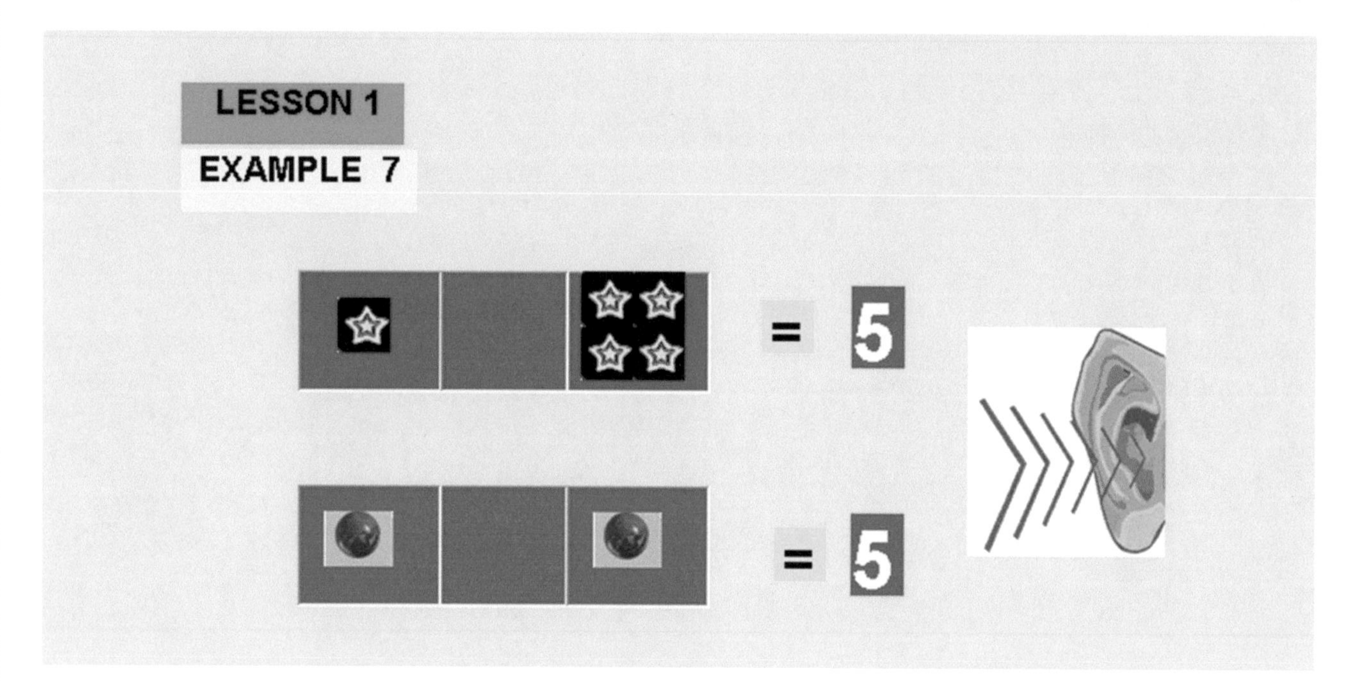

The ***kinesthetic*** representation of the filled up position is announced by clamping of the right hand. An empty position is represented by the clamping of the left hand. The equal sign is crossing both hands in front. The kinesthetic representation for this picture is clamping the right hand, left hand, right hand, and crossing of hands.

ADDITION, EXERCISE 7

In exercise 7, in ***visual*** display of addition we see two pictures of boxes. In the bottom we see three pictures with probable answers. We want to specify, that between the first and second pictures we see the addition sign. Below the second picture we see the equal line, which separates the problem from the answer. The ***audio*** representation for two pictures consumes of a **double knock, knock, tram, knock, knock, and cling.**

From the probable answers for addition we need to find the right one. The audio signals are: 1) **double knock, knock, and double knock.** 2) **knock, double knock, and knock** 3) **double knock, double knock, and knock**.

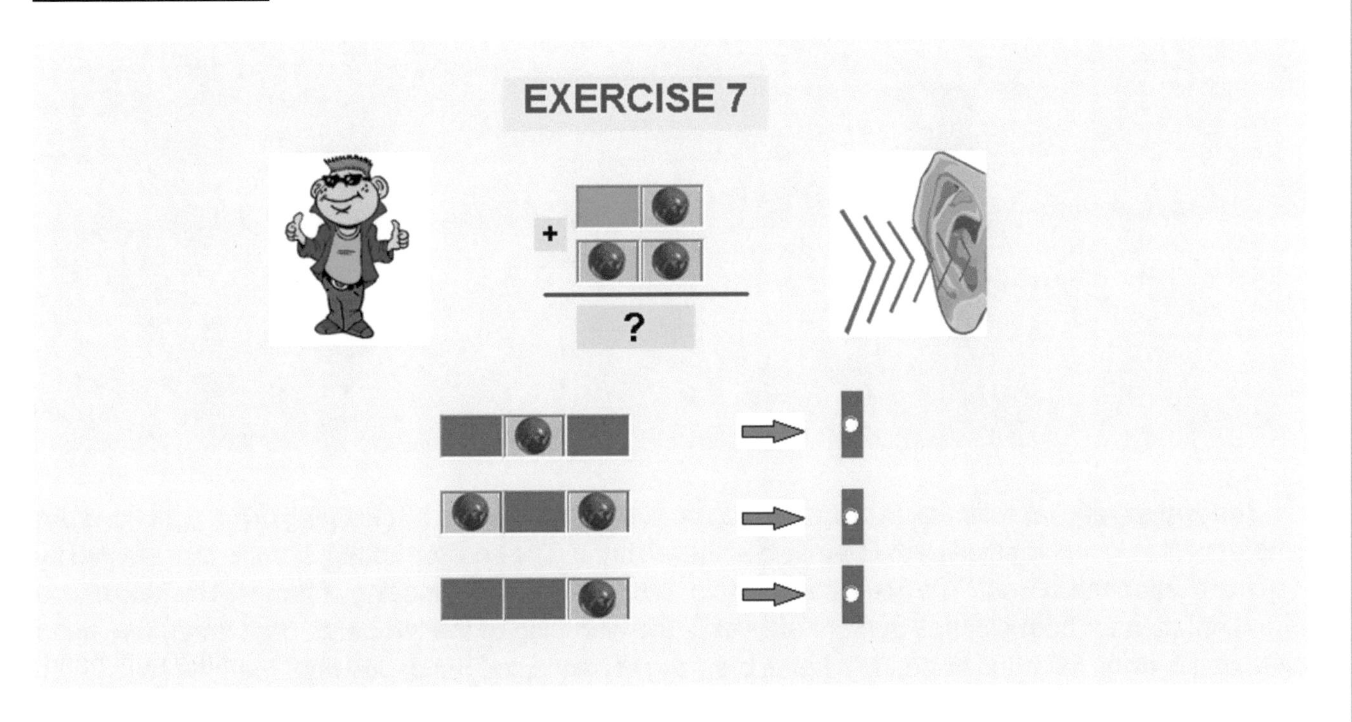

The ***kinesthetic*** representation of the filled up position is announced by clamping of the right hand. An empty position is represented by the clamping of the left hand. The addition sound, tram, is represented by one hand extended in front. The equal sound, cling, is represented by crossing of hands. The kinesthetic representation for both pictures and the answer is the clamping of the left hand, right hand, one hand extended in front, right hand twice, crossing of hands, left hand twice, and right hand.

MATHEMATICS **ADDITION, EXERCISE 8**

In exercise 8, in ***visual*** display of addition we see two pictures of boxes. We want to specify, that between the first and second pictures we see the addition sign. Below the second picture we see the equal line, which separates the problem from the answer. The ***audio*** representation for two pictures consumes of a **double knock, knock, tram, double knock, double knock, knock, cling, double knock, knock, and knock.**

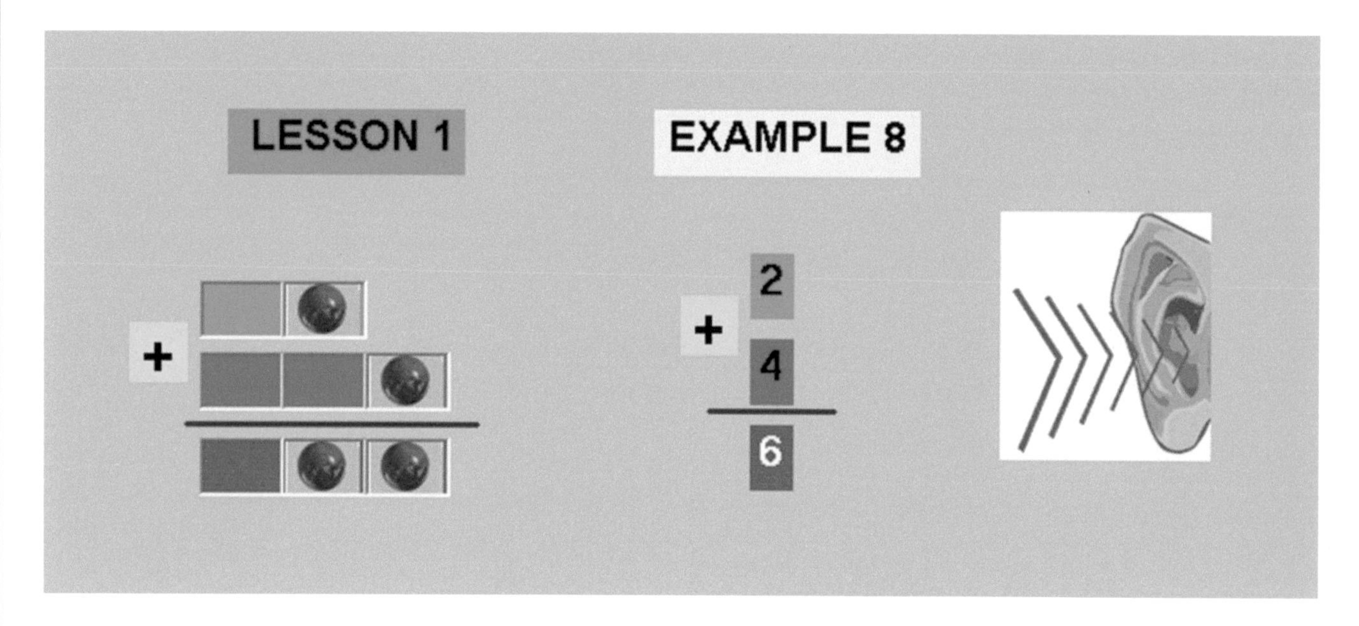

The ***kinesthetic*** representation of the filled up position is announced by clamping of the right hand. An empty position is represented by the clamping of the left hand. The addition sound, tram, is represented by one hand extended in front. The equal sound, cling, is represented by crossing of hands. The kinesthetic representation for both pictures and the answer is the clamping of the left hand, right hand, one hand extended in front, left hand twice, right hand, crossing of hands, left hand, and right hand twice.

ADDITION, EXERCISE 8, STEP 8.1

In exercise 8, step 8.1 in ***visual*** display of addition we see two pictures of boxes. We want to specify, that between the first and second pictures we see the addition sign. Below the second picture we see the equal line, which separates the problem from the answer. When we are adding two numbers and the position on the first number is filled with a symbol and the corresponding position on the second number is also with a symbol, we move the two symbols to the following positions in the answer as one symbol. When two symbols form, two corresponding positions move forward to the following position of the answer then the two symbols become one symbol. The ***audio*** representation for two pictures consumes of a **double knock, knock, tram, double knock, double knock, knock, cling, double knock, knock, and knock.**

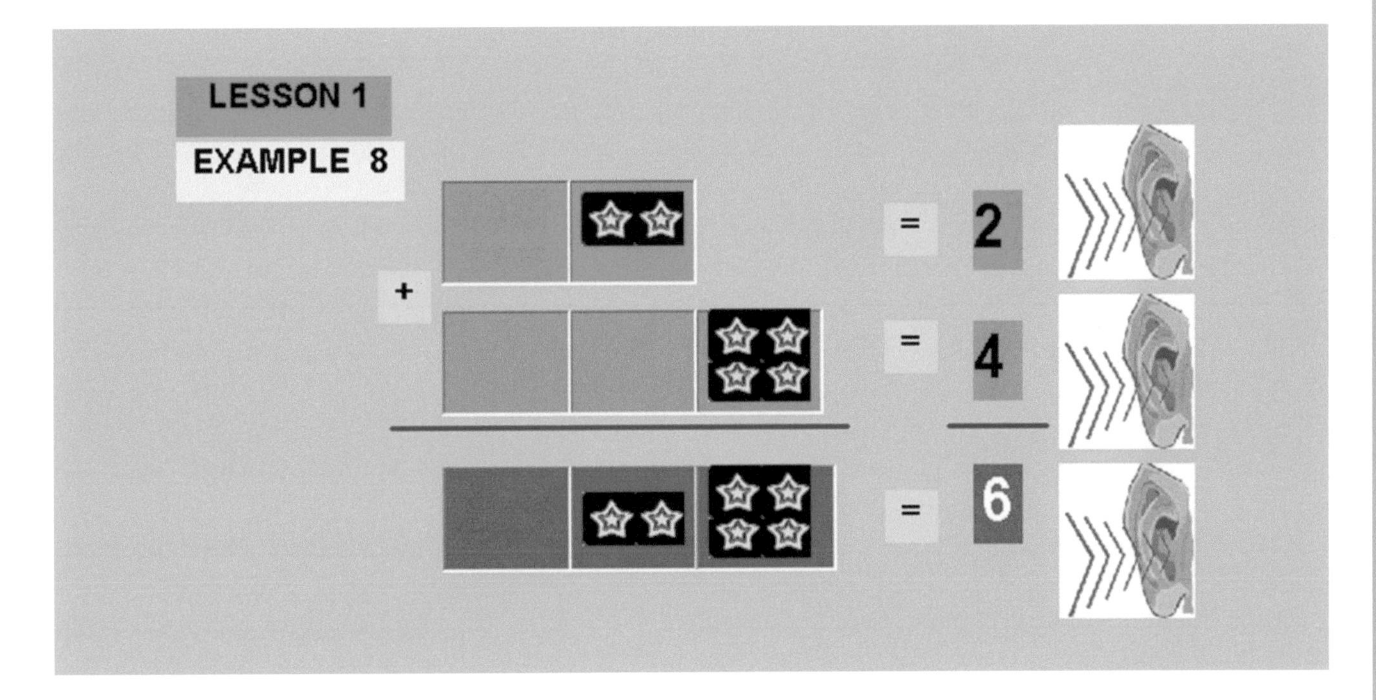

The ***kinesthetic*** representation of the filled up position is announced by clamping of the right hand. An empty position is represented by the clamping of the left hand. The addition sound, tram, is represented by one hand extended in front. The equal sound, cling, is represented by crossing of hands. The kinesthetic representation for both pictures and the answer is the clamping of the left hand, right hand, one hand extended in front, left hand twice, right hand, crossing of hands, left hand, and right hand twice.

MATHEMATICS ADDITION, EXERCISE 8, STEP 8.2

In exercise 8, step 8.2 in ***visual*** display of addition, we see two pictures of boxes with equal sign. Both pictures have the second and third positions filled with a symbol and the first position is empty. The first picture has a yellow star in its positions but the second picture has a red ball in its positions. The form and shape of the symbols do not change mathematical value and the numerical value of both pictures is equal to six.

The ***audio*** representation for this picture consumes of a **double knock, knock, knock, and cling.**

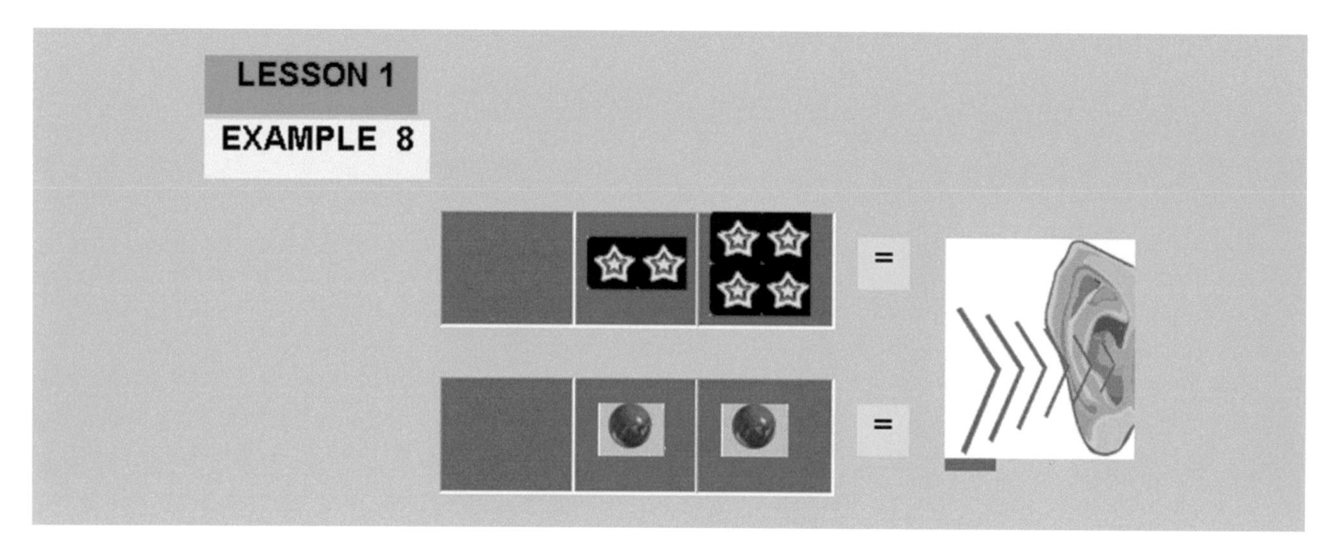

The ***kinesthetic*** representation of the filled up position is announced by clamping of the right hand. An empty position is represented by the clamping of the left hand. The equal sign is crossing both hands in front. The kinesthetic representation for this picture is clamping the left hand, right hand twice, and crossing of hands.

ADDITION, EXERCISE 8

In exercise 8, in ***visual*** display of addition we see two pictures of boxes. In the bottom we see three pictures with probable answers. We want to specify, that between the first and second pictures we see the addition sign. Below the second picture we see the equal line, which separates the problem from the answer. The ***audio*** representation for two pictures consumes of a **double knock, knock, tram, double knock, double knock, knock, and cling.**

From the probable answers for addition we need to find the right one. The audio signals are: 1) **double knock, knock, and knock.** 2) **knock, double knock, and double knock** 3) **double knock, knock, and double knock**.

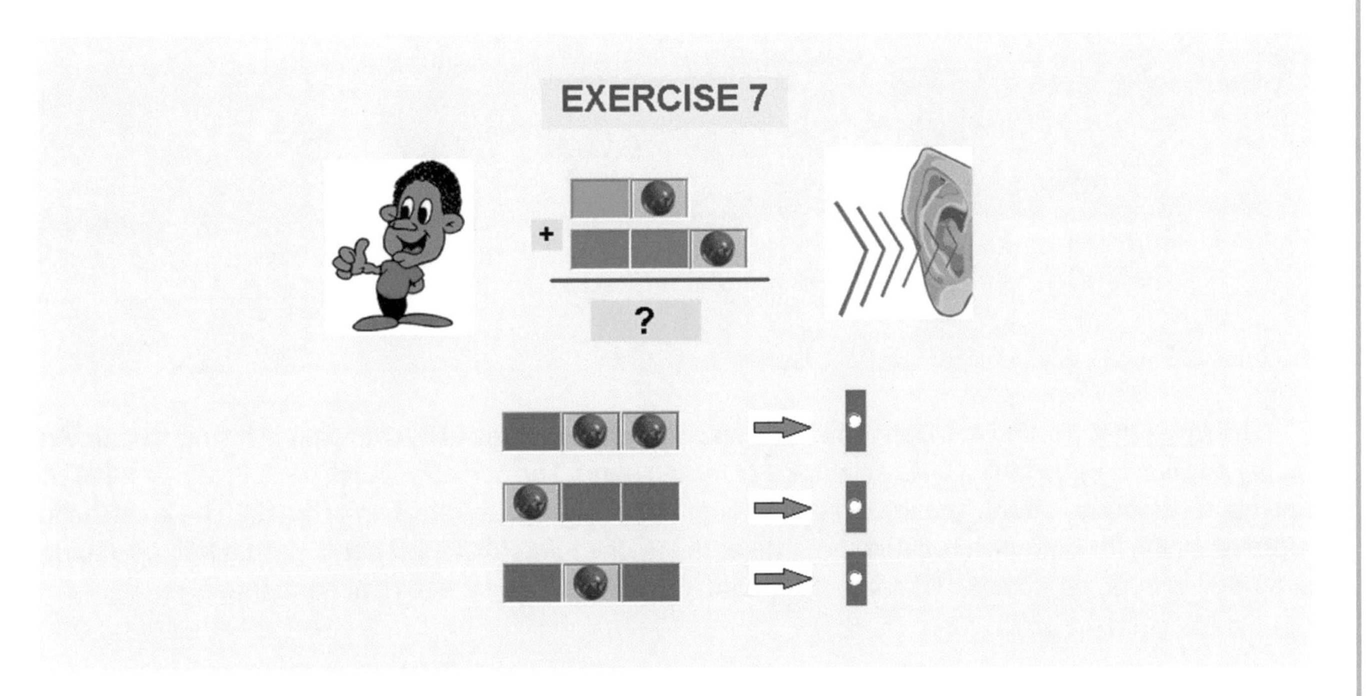

The ***kinesthetic*** representation of the filled up position is announced by clamping of the right hand. An empty position is represented by the clamping of the left hand. The addition sound, tram, is represented by one hand extended in front. The equal sound, cling, is represented by crossing of hands. The kinesthetic representation for both pictures and the answer is the clamping of the left hand, right hand, one hand extended in front, left hand twice, right hand, crossing of hands, left hand, and right hand twice.

MATHEMATICS ADDITION, EXERCISE 9

In exercise 9, in ***visual*** display of addition we see three pictures of boxes. We want to specify, that between the first and second pictures we see the addition sign. Below the second picture we see the equal line, which separates the problem from the answer. The ***audio*** representation for the three pictures consumes of a **double knock, knock, tram, knock, double knock, knock, cling, knock, knock, and knock.**

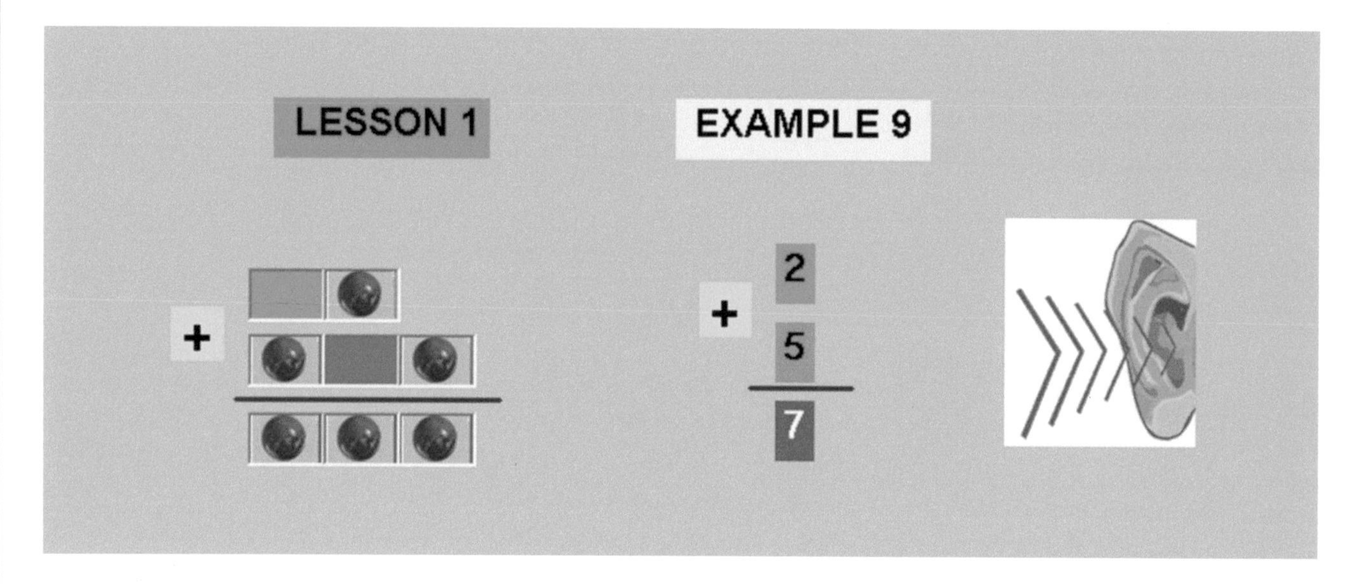

The ***kinesthetic*** representation of the filled up position is announced by clamping of the right hand. An empty position is represented by the clamping of the left hand. The addition sound, tram, is represented by one hand extended in front. The equal sound, cling, is represented by crossing of hands. The kinesthetic representation for both pictures and the answer is the clamping of the left hand, right hand, one hand extended in front, right hand, left hand, right hand, crossing of hands, and right hand thrice.

ADDITION, EXERCISE 9, STEP 9.1

In exercise 9, step 9.1 in ***visual*** display of addition we see three pictures of boxes. We want to specify, that between the first and second pictures we see the addition sign. Below the second picture we see the equal line, which separates the problem from the answer. When we are adding two numbers and the position on the first number is filled with a symbol and the corresponding position on the second number is also with a symbol, we move the two symbols to the following positions in the answer as one symbol. When two symbols form, two corresponding positions move forward to the following position of the answer then the two symbols become one symbol. The ***audio*** representation for the three pictures consumes of a **double knock, knock, tram, knock, double knock, knock, cling, knock, knock, and knock.**

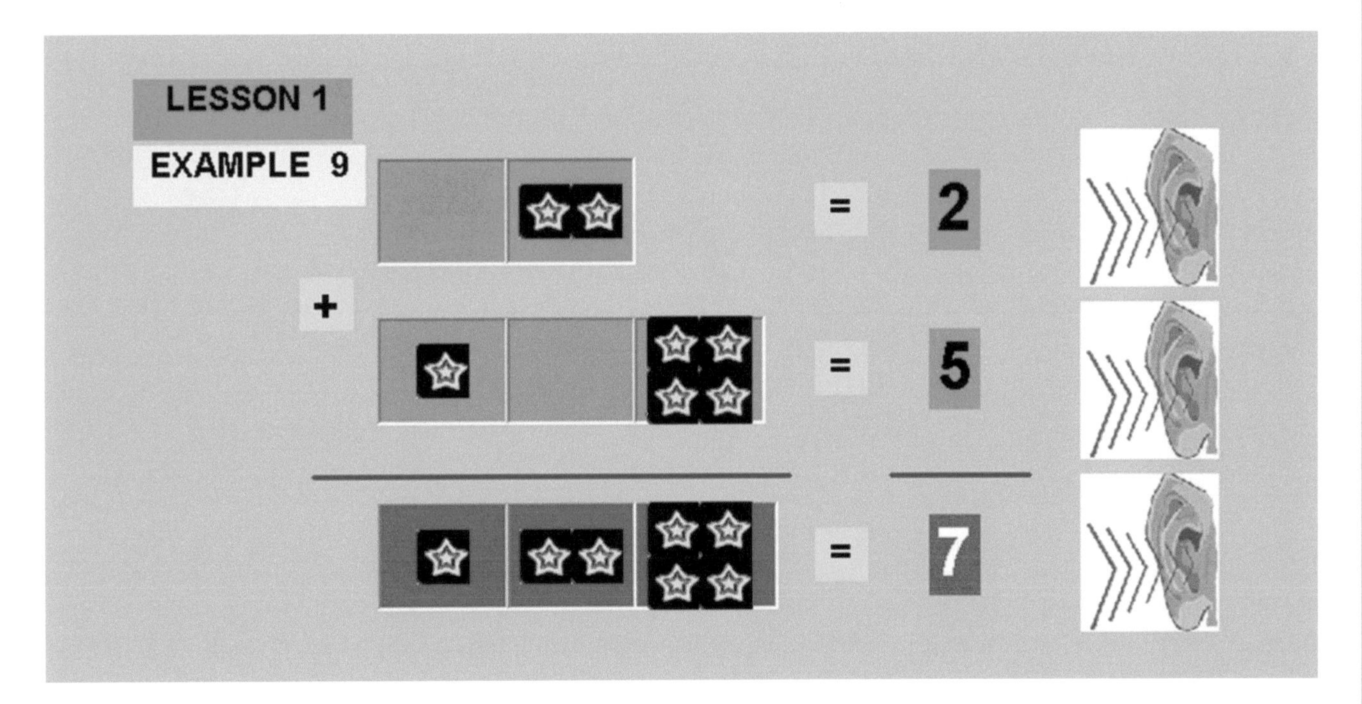

The ***kinesthetic*** representation of the filled up position is announced by clamping of the right hand. An empty position is represented by the clamping of the left hand. The addition sound, tram, is represented by one hand extended in front. The equal sound, cling, is represented by crossing of hands. The kinesthetic representation for both pictures and the answer is the clamping of the left hand, right hand, one hand extended in front, right hand, left hand, right hand, crossing of hands, and right hand thrice.

MATHEMATICS

ADDITION, EXERCISE 9, STEP 9.2

In exercise 9, step 9.2 in ***visual*** display of addition, we see two pictures of boxes with equal sign. Both pictures have all three of their positions filled with a symbol. The first picture has a yellow star in its positions but the second picture has a red ball in its positions. The form and shape of the symbols do not change mathematical value and the numerical value of both pictures is equal to seven.

The ***audio*** representation for this picture consumes of a **knock, knock, knock, and cling.**

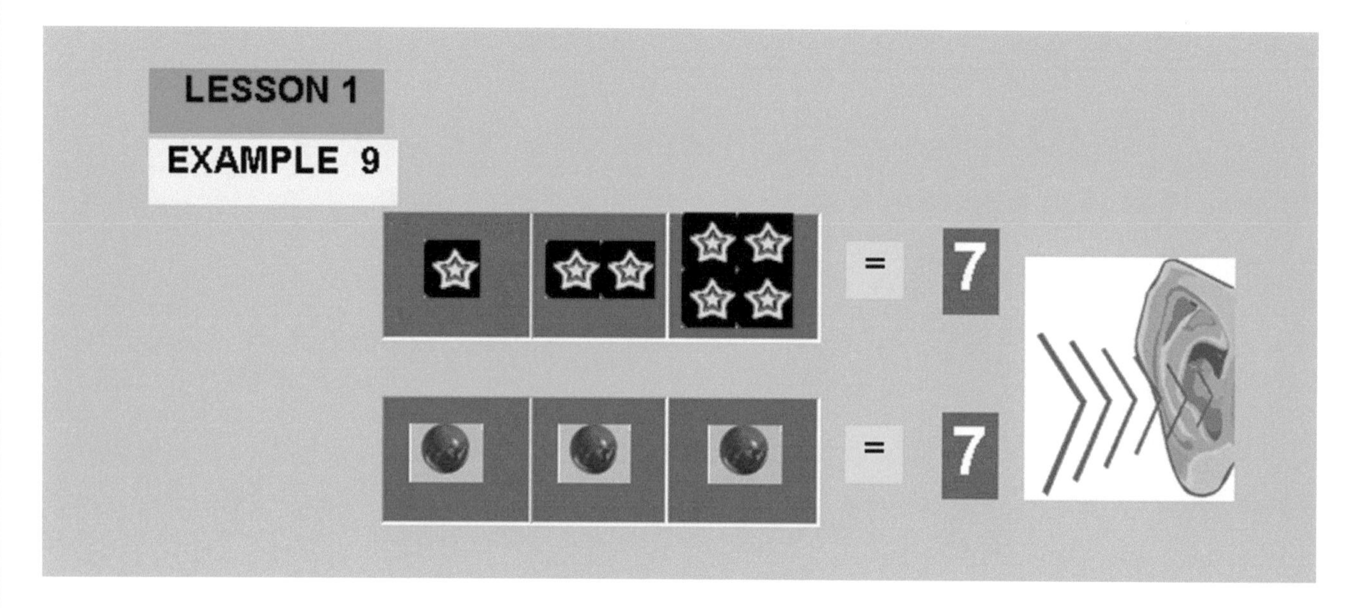

The ***kinesthetic*** representation of the filled up position is announced by clamping of the right hand. An empty position is represented by the clamping of the left hand. The equal sign is crossing both hands in front. The kinesthetic representation for this picture is clamping the right hand thrice and crossing of hands.

ADDITION, EXERCISE 9

In exercise 9, in ***visual*** display of addition we see two pictures of boxes. In the bottom we see three pictures with probable answers. We want to specify, that between the first and second pictures we see the addition sign. Below the second picture we see the equal line, which separates the problem from the answer. The ***audio*** representation for two pictures consumes of a **double knock, knock, tram, knock, double knock, knock, and cling.**

From the probable answers for addition we need to find the right one. The audio signals are: 1) **double knock, knock, and knock.** 2) **knock, knock, and knock** 3) **knock, double knock, and knock**.

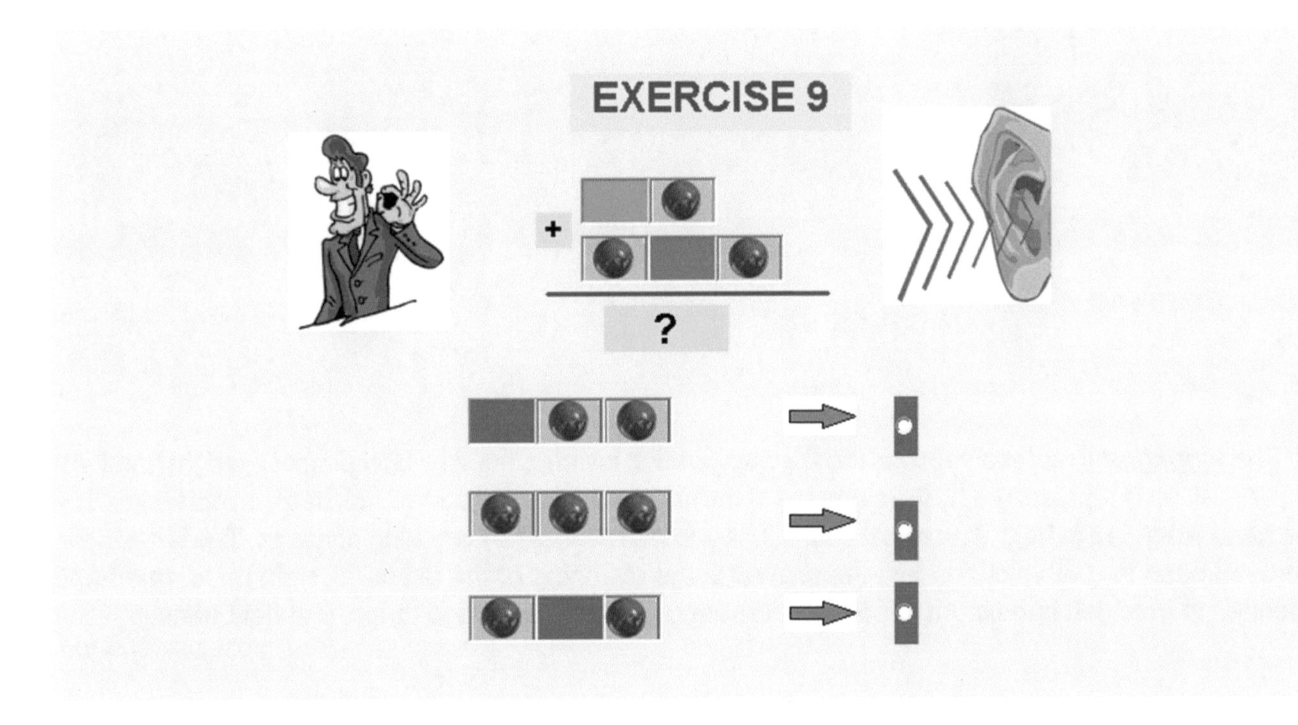

The ***kinesthetic*** representation of the filled up position is announced by clamping of the right hand. An empty position is represented by the clamping of the left hand. The addition sound, tram, is represented by one hand extended in front. The equal sound, cling, is represented by crossing of hands. The kinesthetic representation for both pictures and the answer is the clamping of the left hand, right hand, one hand extended in front, right hand, left hand, right hand, crossing of hands, and right hand thrice.

MATHEMATICS

ADDITION, EXERCISE 10

In exercise 10, in ***visual*** display of addition we see three pictures of boxes. We want to specify, that between the first and second pictures we see the addition sign. Below the second picture we see the equal line, which separates the problem from the answer. The ***audio*** representation for the three pictures consumes of a **double knock, knock, tram, double knock, knock, knock, cling, double knock, double knock, double knock, and knock.**

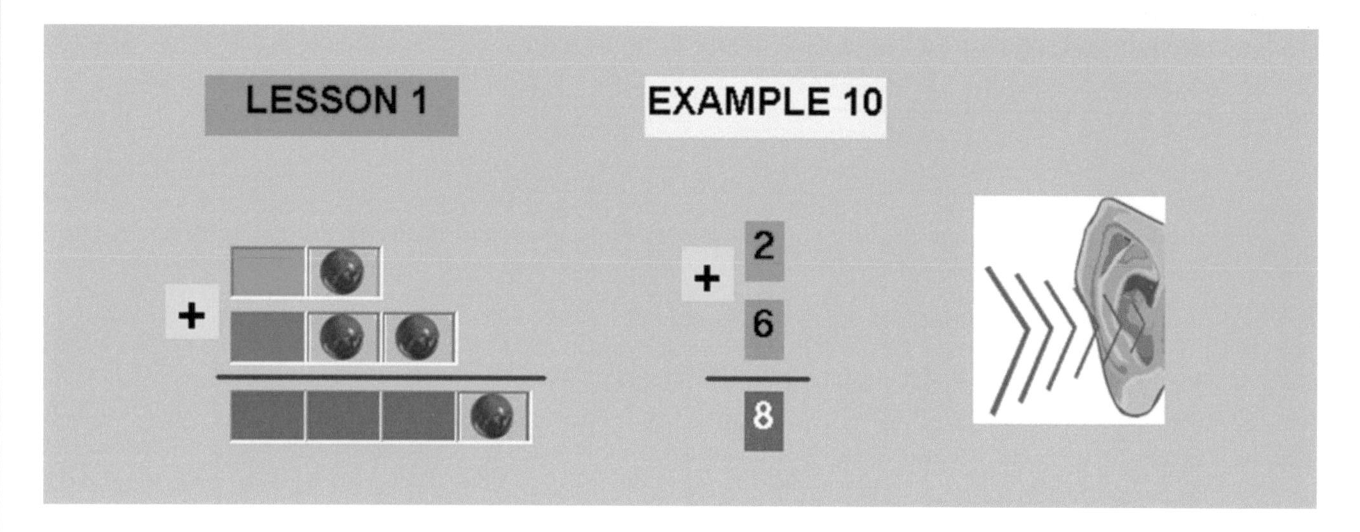

The ***kinesthetic*** representation of the filled up position is announced by clamping of the right hand. An empty position is represented by the clamping of the left hand. The addition sound, tram, is represented by one hand extended in front. The equal sound, cling, is represented by crossing of hands. The kinesthetic representation for both pictures and the answer is the clamping of the left hand, right hand, one hand extended in front, left hand, right hand twice, crossing of hands, left hand thrice, and right hand.

ADDITION, EXERCISE 10, STEP 10.1

In exercise 10, step 10.1 in ***visual*** display of addition we see three pictures of boxes. We want to specify, that between the first and second pictures we see the addition sign. Below the second picture we see the equal line, which separates the problem from the answer. When we are adding two numbers and the position on the first number is filled with a symbol and the corresponding position on the second number is also with a symbol, we move the two symbols to the following positions in the answer as one symbol. When two symbols form, two corresponding positions move forward to the following position of the answer then the two symbols become one symbol. The ***audio*** representation for the three pictures consumes of a **double knock, knock, tram, double knock, knock, knock, cling, double knock, double knock, double knock, and knock.**

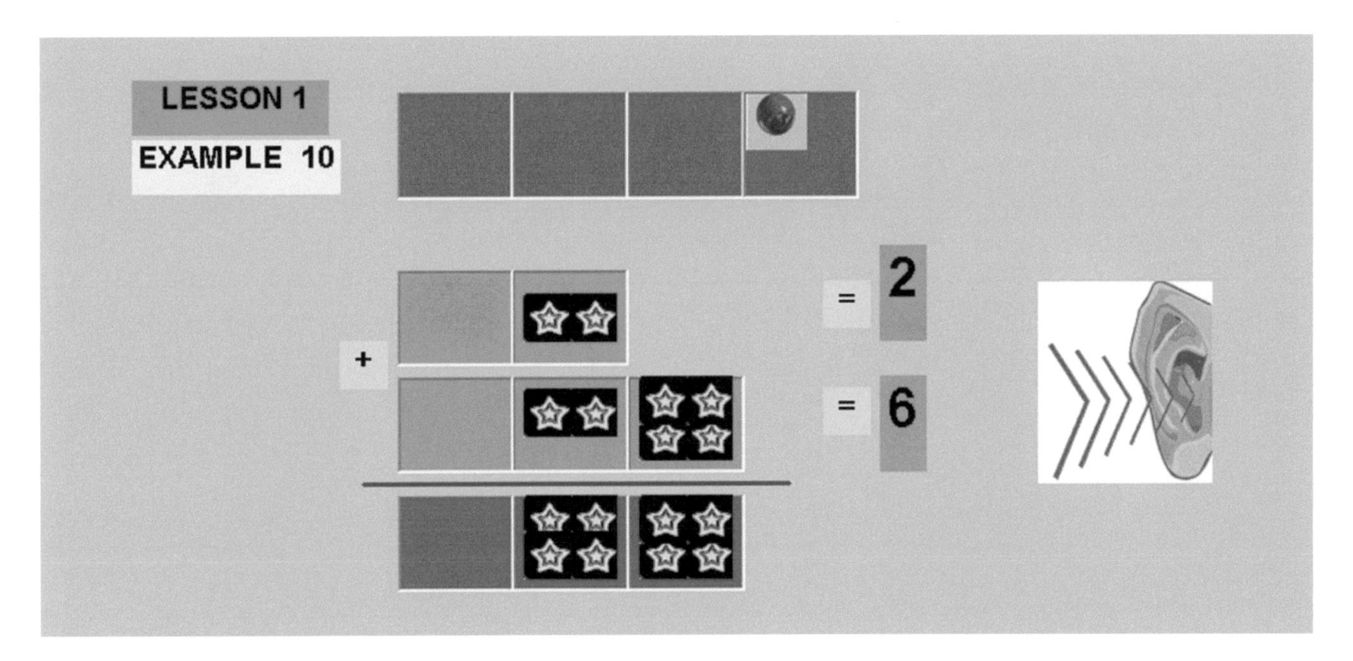

The ***kinesthetic*** representation of the filled up position is announced by clamping of the right hand. An empty position is represented by the clamping of the left hand. The addition sound, tram, is represented by one hand extended in front. The equal sound, cling, is represented by crossing of hands. The kinesthetic representation for both pictures and the answer is the clamping of the left hand, right hand, one hand extended in front, left hand, right hand twice, crossing of hands, left hand thrice, and right hand.

MATHEMATICS ADDITION, EXERCISE 10, STEP 10.2

In exercise 10, step 10.2 in ***visual*** display of addition we see two pictures of boxes. The first picture has a symbol in the second and third positions and has the first position empty. The second picture has the third position filled with a symbol. For purpose of addition we are introducing the concept of moving. When a position passes its maximum capacity the next symbol is simultaneously moved in the ascending position. We are moving from the second to the third position. However the audio representation for the picture does not change because the numerical value does not change.

The ***audio*** representation for the three pictures consumes of a **double knock, double knock, double knock, and knock.**

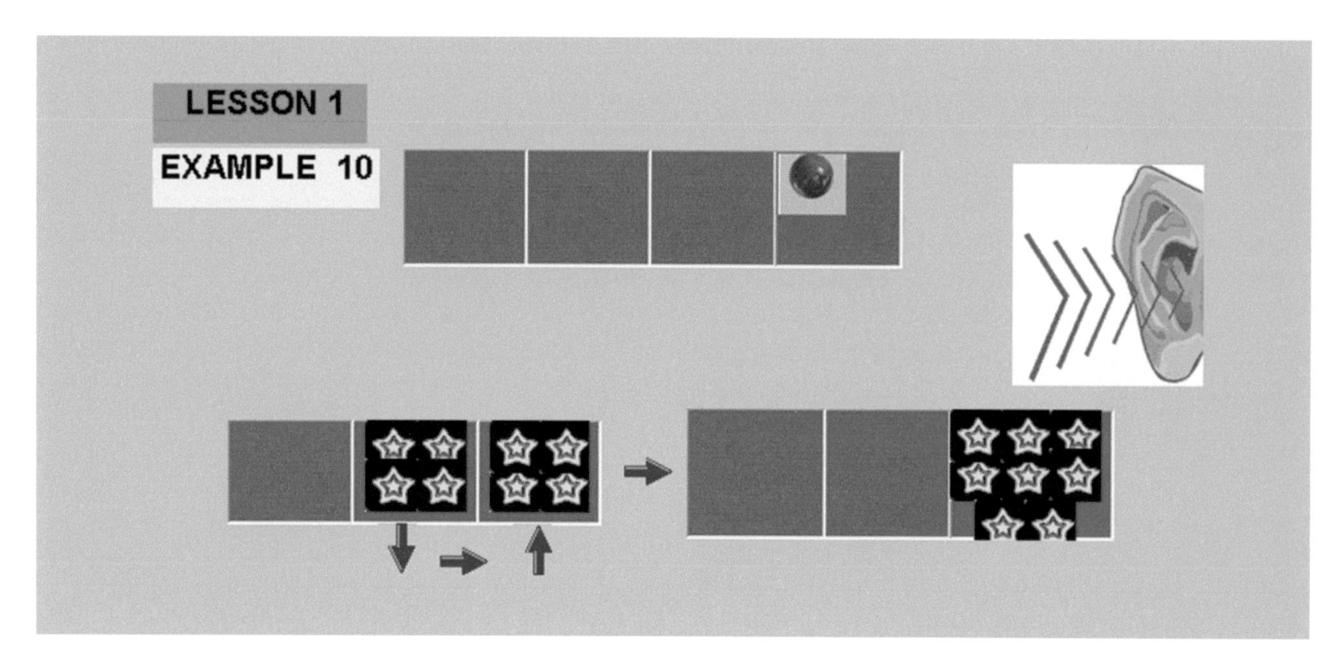

The ***kinesthetic*** representation of the filled up position is announced by clamping of the right hand. An empty position is represented by the clamping of the left hand. The kinesthetic representation for three pictures is the clamping of the left hand thrice and right hand.

ADDITION, EXERCISE 10, STEP 10.3

In exercise 10, step 10.3 in ***visual*** display of addition we see two pictures of boxes. The first picture has a symbol in the third position and has the first and second positions empty. The second picture has the fourth position filled with a symbol. For purpose of addition we are introducing the concept of moving. When a position passes its maximum capacity the next symbol is simultaneously moved in the ascending position. We are moving from the second to the third position. However the audio representation for the picture does not change because the numerical value does not change.

The ***audio*** representation for the three pictures consumes of a **double knock, double knock, double knock, and knock.**

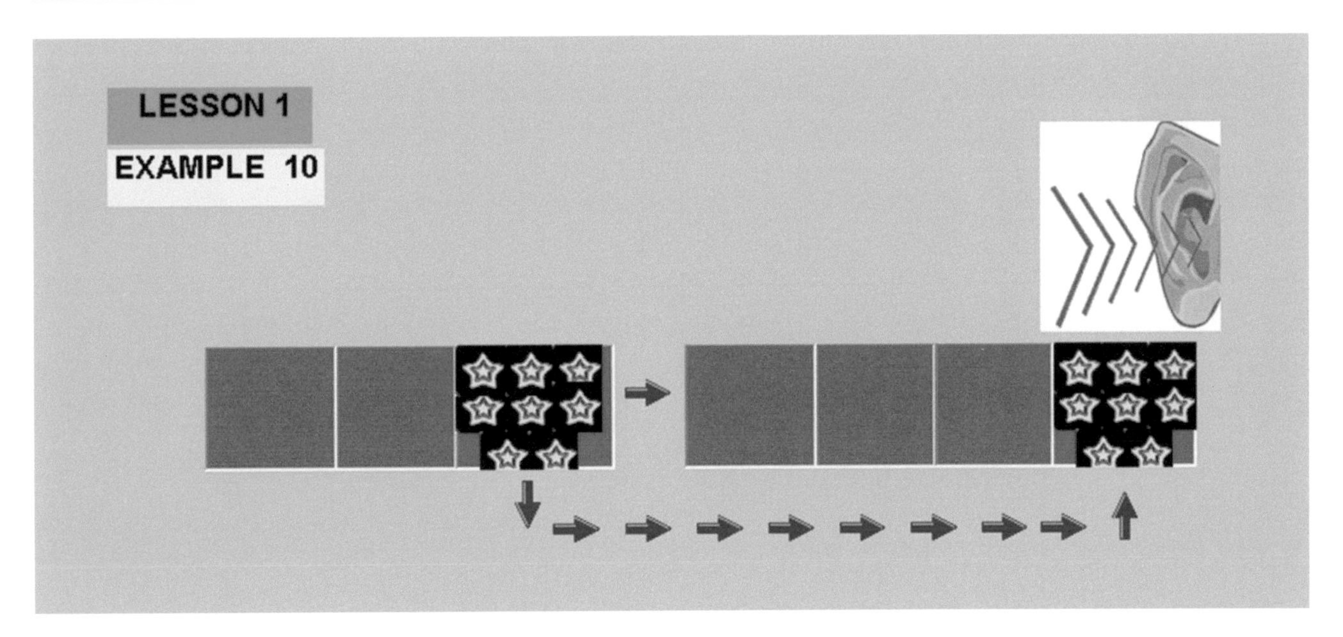

The ***kinesthetic*** representation of the filled up position is announced by clamping of the right hand. An empty position is represented by the clamping of the left hand. The kinesthetic representation for three pictures is the clamping of the left hand thrice and right hand.

MATHEMATICS

ADDITION, EXERCISE 10, STEP 10.4

In exercise 10, step 10.4 in ***visual*** display of addition, we see two pictures of boxes with equal sign. Both pictures have the first three of their positions empty and the last position is filled with a symbol. The first picture has a yellow star in its positions but the second picture has a red ball in its positions. The form and shape of the symbols do not change mathematical value and the numerical value of both pictures is equal to eight.

The ***audio*** representation for this picture consumes of a **double knock, double knock, double knock, knock, and cling.**

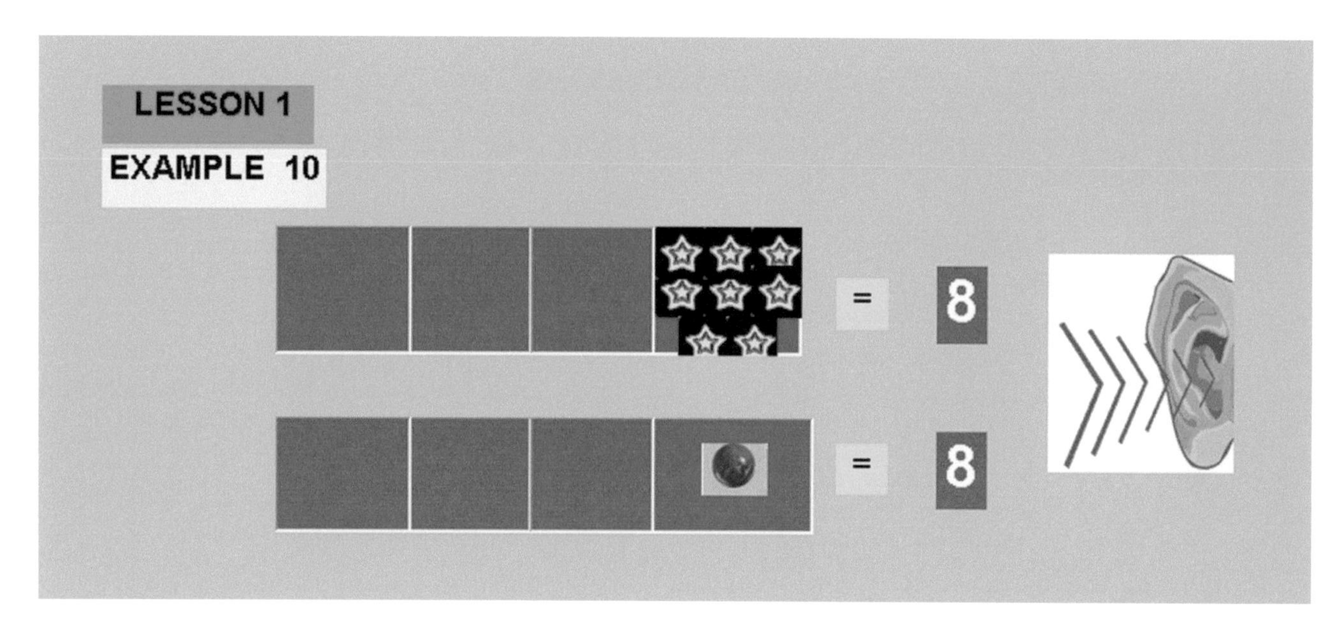

The ***kinesthetic*** representation of the filled up position is announced by clamping of the right hand. An empty position is represented by the clamping of the left hand. The equal sign is crossing both hands in front. The kinesthetic representation for this picture is clamping the left hand thrice, right hand, and crossing of hands.

ADDITION, EXERCISE 10

In exercise 10, in ***visual*** display of addition we see two pictures of boxes. In the bottom we see three pictures with probable answers. We want to specify, that between the first and second pictures we see the addition sign. Below the second picture we see the equal line, which separates the problem from the answer. The ***audio*** representation for two pictures consumes of a **double knock, knock, tram, double knock, knock, knock, and cling.**

From the probable answers for addition we need to find the right one. The audio signals are: 1) **knock, double knock, double knock, and double knock.** 2) **double knock, double knock, double knock, and knock** 3) **double knock, knock, double knock, and double knock**.

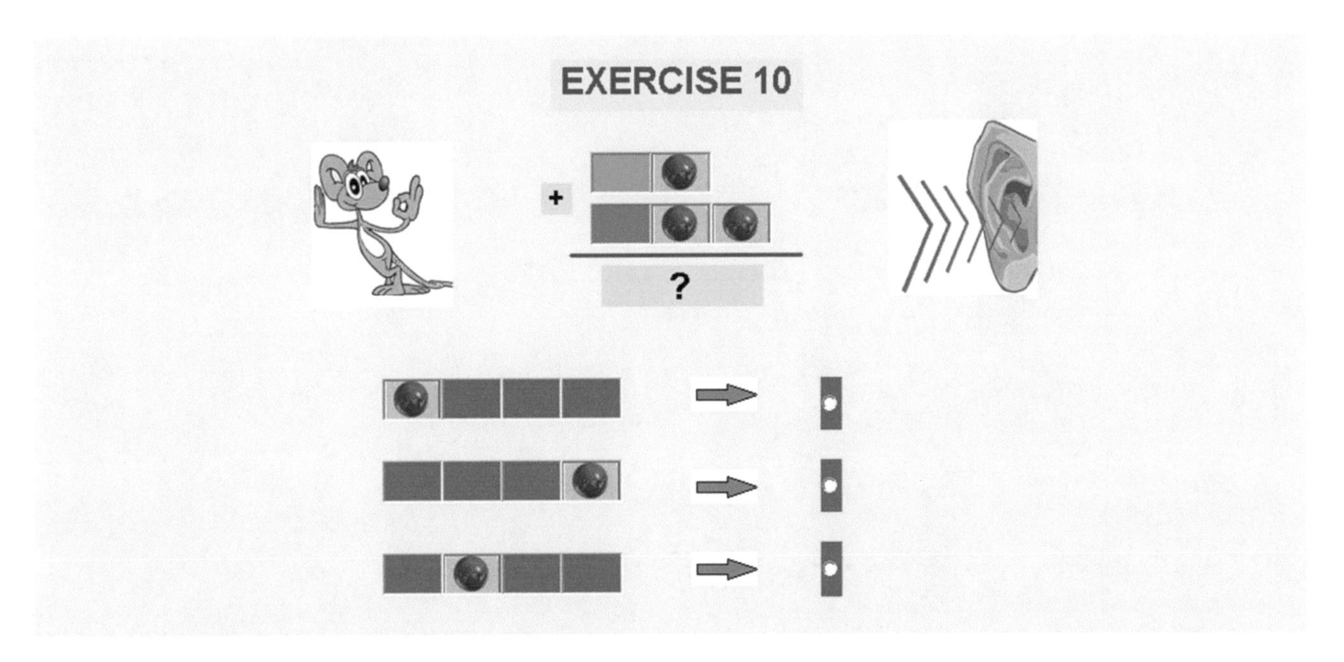

The ***kinesthetic*** representation of the filled up position is announced by clamping of the right hand. An empty position is represented by the clamping of the left hand. The addition sound, tram, is represented by one hand extended in front. The equal sound, cling, is represented by crossing of hands. The kinesthetic representation for both pictures and the answer is the clamping of the left hand, right hand, one hand extended in front, left hand, right hand twice, crossing of hands, left hand thrice, and right hand.

MATHEMATICS ADDITION, EXERCISE 11, STEP 11.1

In exercise 11, step 11.1 in ***visual*** display of addition we see three pictures of boxes. We want to specify, that between the first and second pictures we see the addition sign. Below the second picture we see the equal line, which separates the problem from the answer. When we are adding two numbers and the position on the first number is filled with a symbol and the corresponding position on the second number is also with a symbol, we move the two symbols to the following positions in the answer as one symbol. When two symbols form, two corresponding positions move forward to the following position of the answer then the two symbols become one symbol. The ***audio*** representation for the three pictures consumes of a **double knock, knock, tram, knock, knock, knock, cling, knock, double knock, double knock, and knock.**

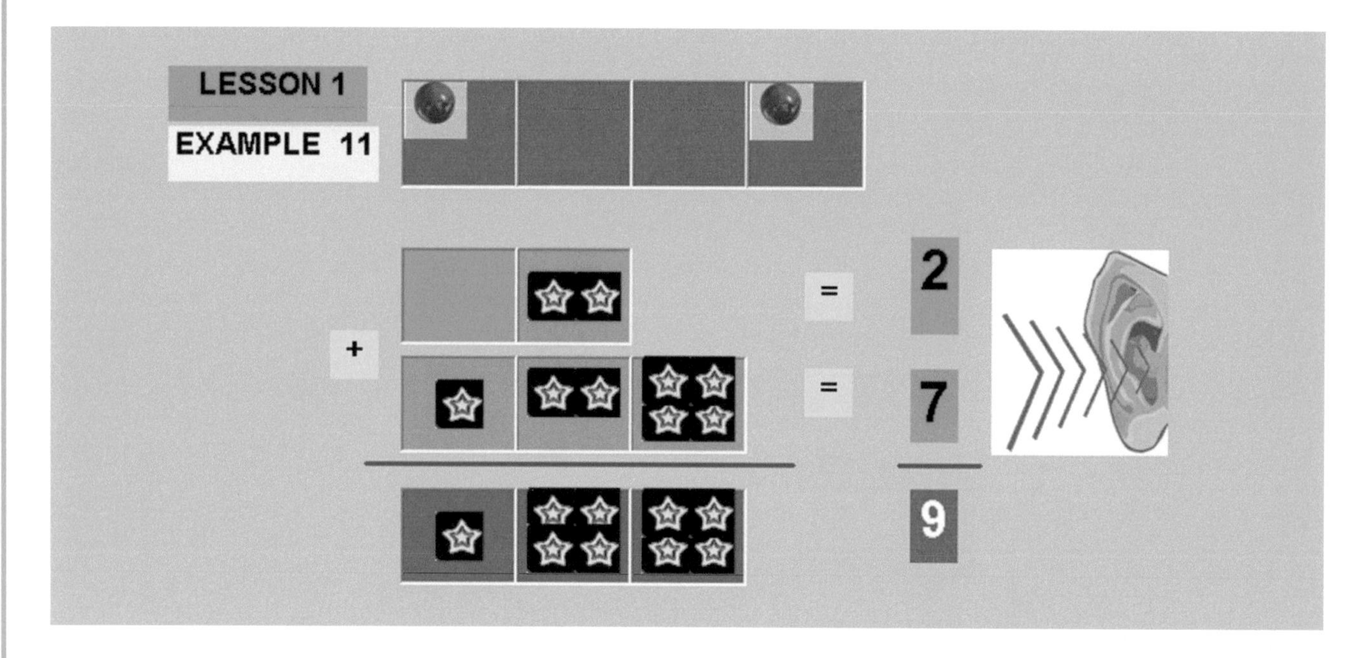

The ***kinesthetic*** representation of the filled up position is announced by clamping of the right hand. An empty position is represented by the clamping of the left hand. The addition sound, tram, is represented by one hand extended in front. The equal sound, cling, is represented by crossing of hands. The kinesthetic representation for both pictures and the answer is the clamping of the left hand, right hand, one hand extended in front, right hand thrice, crossing of hands, right hand, left hand twice, and right hand.

ADDITION, EXERCISE 11, STEP 11.2

In exercise 11, step 11.2 in ***visual*** display of addition we see two pictures of boxes. The first picture has a symbol in all three positions. The second picture has the first and third positions filled with symbols. For purpose of addition we are introducing the concept of moving. When a position passes its maximum capacity the next symbol is simultaneously moved in the ascending position. We are moving from the second to the third position. However the audio representation for the picture does not change because the numerical value does not change.

The ***audio*** representation for the three pictures consumes of a **knock, double knock, double knock, and knock.**

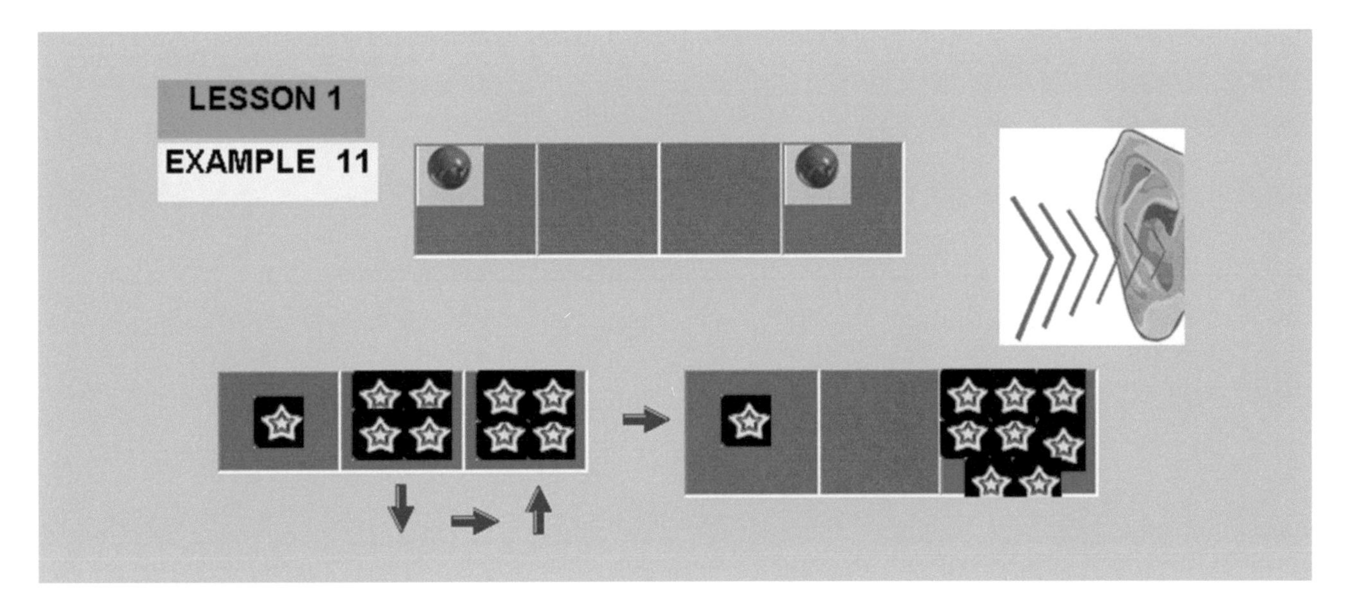

The ***kinesthetic*** representation of the filled up position is announced by clamping of the right hand. An empty position is represented by the clamping of the left hand. The kinesthetic representation for three pictures is the clamping of the right hand, left hand twice, and right hand.

MATHEMATICS

ADDITION, EXERCISE 11, STEP 11.3

In exercise 11, step 11.3 in ***visual*** display of addition we see two pictures of boxes. The first picture has a symbol in the first and third positions. The second picture has the first and fourth positions filled with symbols. For purpose of addition we are introducing the concept of moving. When a position passes its maximum capacity the next symbol is simultaneously moved in the ascending position. We are moving from the second to the third position. However the audio representation for the picture does not change because the numerical value does not change.

The ***audio*** representation for the three pictures consumes of a **knock, double knock, double knock, and knock.**

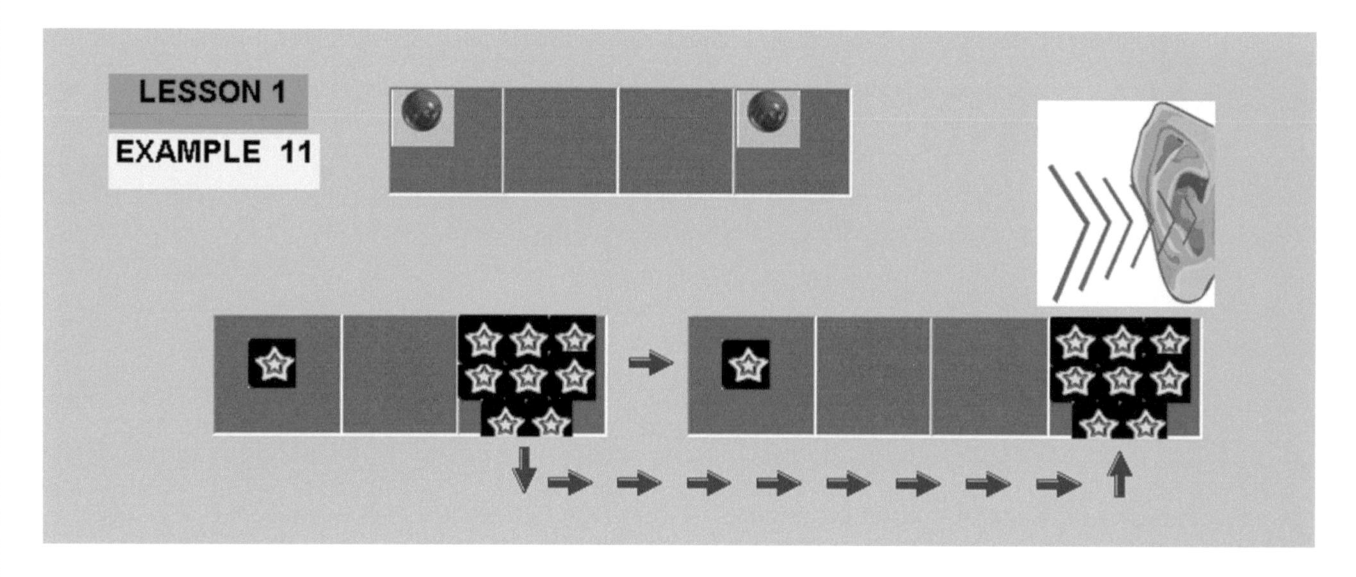

The ***kinesthetic*** representation of the filled up position is announced by clamping of the right hand. An empty position is represented by the clamping of the left hand. The kinesthetic representation for three pictures is the clamping of the right hand, left hand twice, and right hand.

ADDITION, EXERCISE 11, STEP 11.3

In exercise 11, step 11.3 in ***visual*** display of addition we see two pictures of boxes. The first picture has a symbol in the first and third positions. The second picture has the first and fourth positions filled with symbols. For purpose of addition we are introducing the concept of moving. When a position passes its maximum capacity the next symbol is simultaneously moved in the ascending position. We are moving from the second to the third position. However the audio representation for the picture does not change because the numerical value does not change.

The ***audio*** representation for the three pictures consumes of a **knock, double knock, double knock, and knock.**

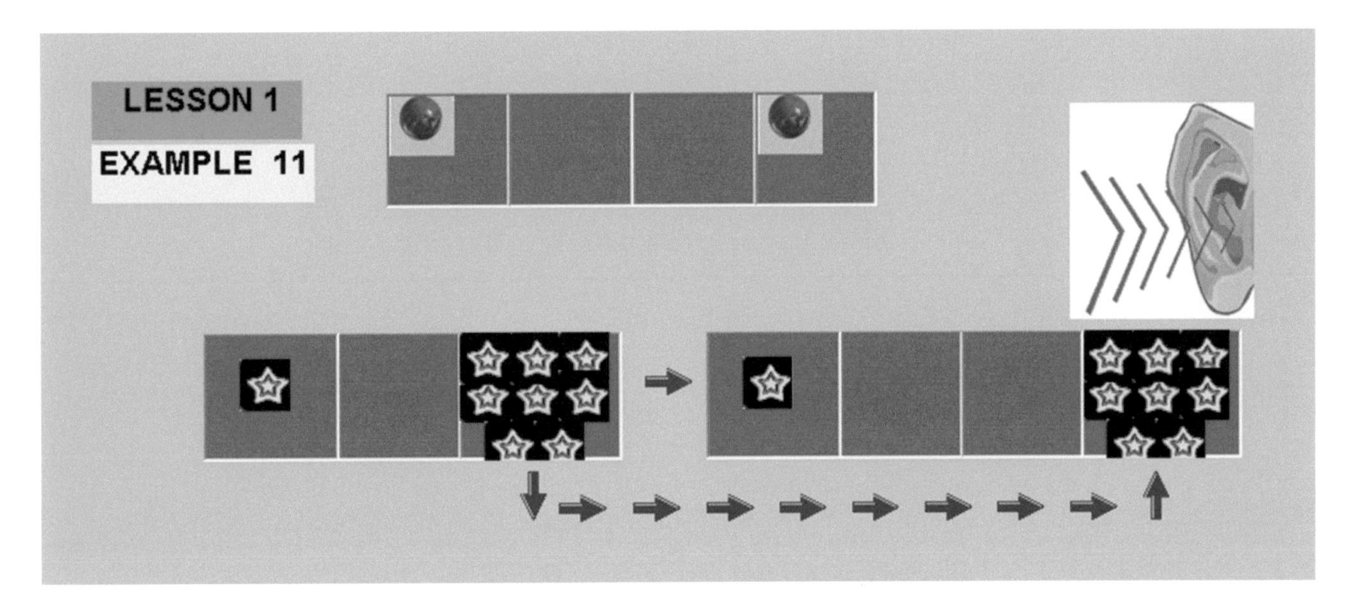

The ***kinesthetic*** representation of the filled up position is announced by clamping of the right hand. An empty position is represented by the clamping of the left hand. The kinesthetic representation for three pictures is the clamping of the right hand, left hand twice, and right hand.

MATHEMATICS

ADDITION, EXERCISE 11

In exercise 11, in ***visual*** display of addition we see two pictures of boxes. In the bottom we see three pictures with probable answers. We want to specify, that between the first and second pictures we see the addition sign. Below the second picture we see the equal line, which separates the problem from the answer. The ***audio*** representation for two pictures consumes of a **double knock, knock, tram, knock, knock, knock, and cling.**

From the probable answers for addition we need to find the right one. The audio signals are: 1) **knock, double knock, knock, and double knock.** 2) **double knock, knock, knock, and double knock** 3) **knock, double knock, double knock, and knock**.

The ***kinesthetic*** representation of the filled up position is announced by clamping of the right hand. An empty position is represented by the clamping of the left hand. The addition sound, tram, is represented by one hand extended in front. The equal sound, cling, is represented by crossing of hands. The kinesthetic representation for both pictures and the answer is the clamping of the left hand, right hand, one hand extended in front, right hand thrice, crossing of hands, right hand, left hand twice, and right hand.

ADDITION, EXERCISE 12

In exercise 12, in ***visual*** display of addition we see three pictures of boxes. In We want to specify, that between the first and second pictures we see the addition sign. Bellow the second picture we see the equal line, which separates the problem from the answer. The ***audio*** representation for the three pictures consumes of a **knock, knock, tram, double knock, double knock, cling, knock, and knock.**

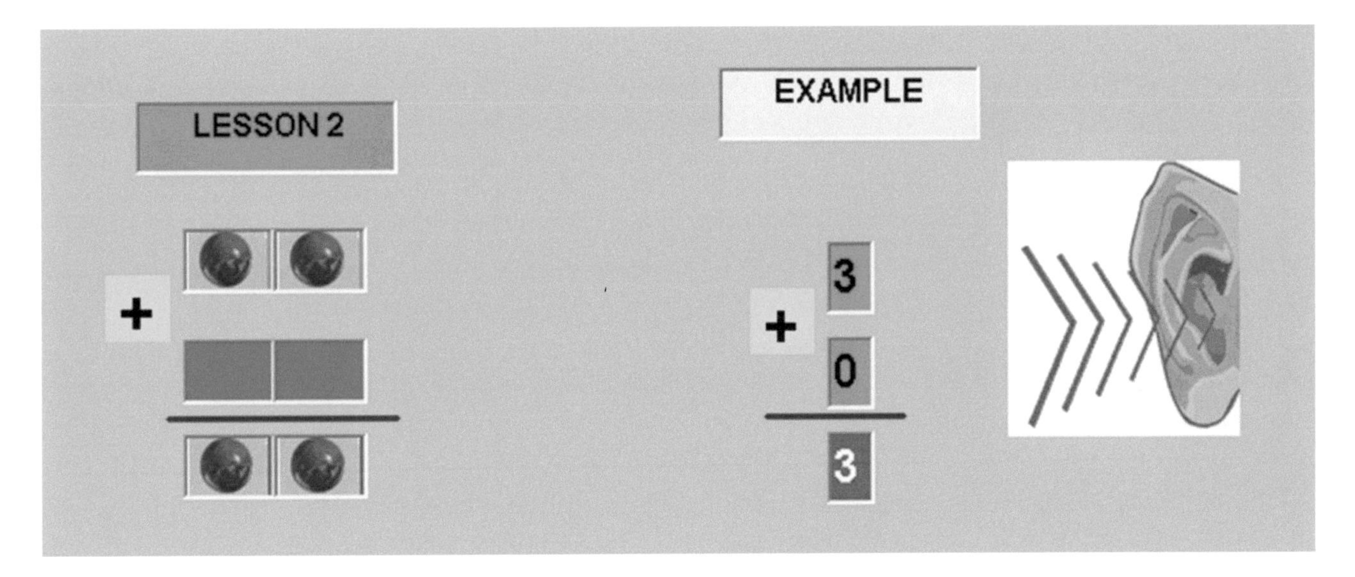

The ***kinesthetic*** representation of the filled up position is announced by clamping of the right hand. An empty position is represented by the clamping of the left hand. The addition sound, tram, is represented by one hand extended in front. The equal sound, cling, is represented by crossing of hands. The kinesthetic representation for both pictures and the answer is the clamping of the right hand twice, one hand extended in front, left hand twice, crossing of hands, right hand twice and left hand.

MATHEMATICS

ADDITION, EXERCISE 12

In exercise 12, in ***visual*** display of addition we see two pictures of boxes. In the bottom we see three pictures with probable answers. We want to specify, that between the first and second pictures we see the addition sign. Bellow the second picture we see the equal line, which separates the problem from the answer. The ***audio*** representation for two pictures consumes of a **knock, knock, tram, double knock, double knock, and cling.**

From the probable answers for addition we need to find the right one. The audio signals are: 1) **knock, knock, and double knock.** 2) **knock, double knock, and knock** 3) **double knock, knock, and knock**.

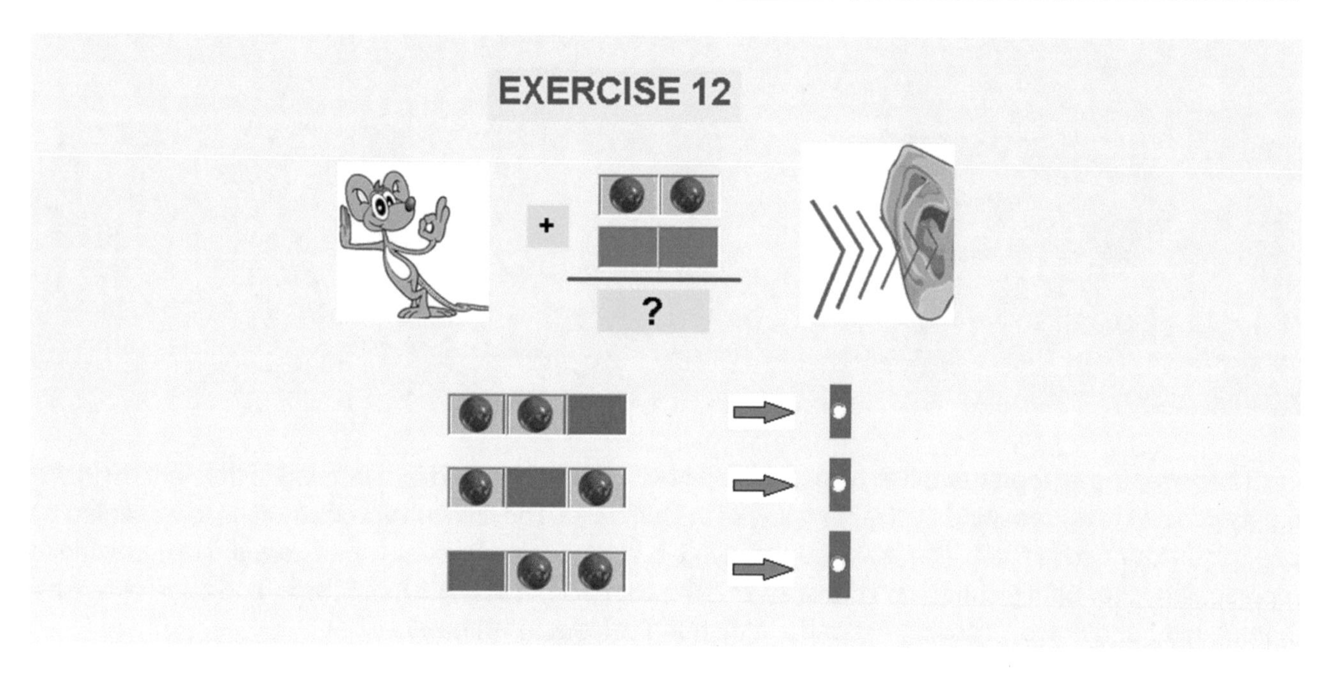

The ***kinesthetic*** representation of the filled up position is announced by clamping of the right hand. An empty position is represented by the clamping of the left hand. The addition sound, tram, is represented by one hand extended in front. The equal sound, cling, is represented by crossing of hands. The kinesthetic representation for both pictures and the answer is the clamping of the right hand twice, one hand extended in front, left hand twice, crossing of hands, right hand twice and left hand.

ADDITION, EXERCISE 13

In exercise 13, in ***visual*** display of addition we see three pictures of boxes. We want to specify, that between the first and second pictures we see the addition sign. Below the second picture we see the equal line, which separates the problem from the answer. The ***audio*** representation for the three pictures consumes of a **knock, knock, tram, knock, double knock, cling, double knock, double knock, and knock.**

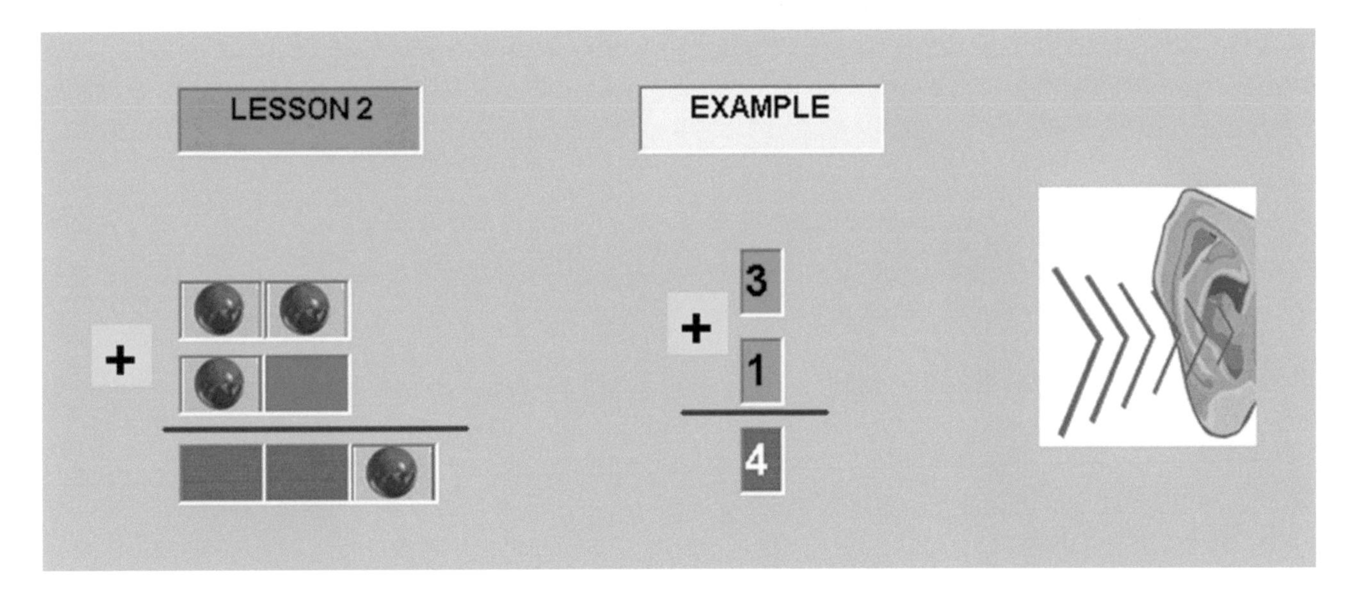

The ***kinesthetic*** representation of the filled up position is announced by clamping of the right hand. An empty position is represented by the clamping of the left hand. The addition sound, tram, is represented by one hand extended in front. The equal sound, cling, is represented by crossing of hands. The kinesthetic representation for both pictures and the answer is the clamping of the right hand twice, one hand extended in front, right hand, left hand, crossing of hands, left hand twice and right hand.

MATHEMATICS ADDITION, EXERCISE 13, STEP 13.1

In exercise 13, step 13.1 in ***visual*** display of addition we see three pictures of boxes. We want to specify, that between the first and second pictures we see the addition sign. Below the second picture we see the equal line, which separates the problem from the answer. When we are adding two numbers and the position on the first number is filled with a symbol and the corresponding position on the second number is also with a symbol, we move the two symbols to the following positions in the answer as one symbol. When two symbols form, two corresponding positions move forward to the following position of the answer then the two symbols become one symbol. The ***audio*** representation for the three pictures consumes of a **knock, knock, tram, knock, double knock, cling, double knock, double knock, and knock.**

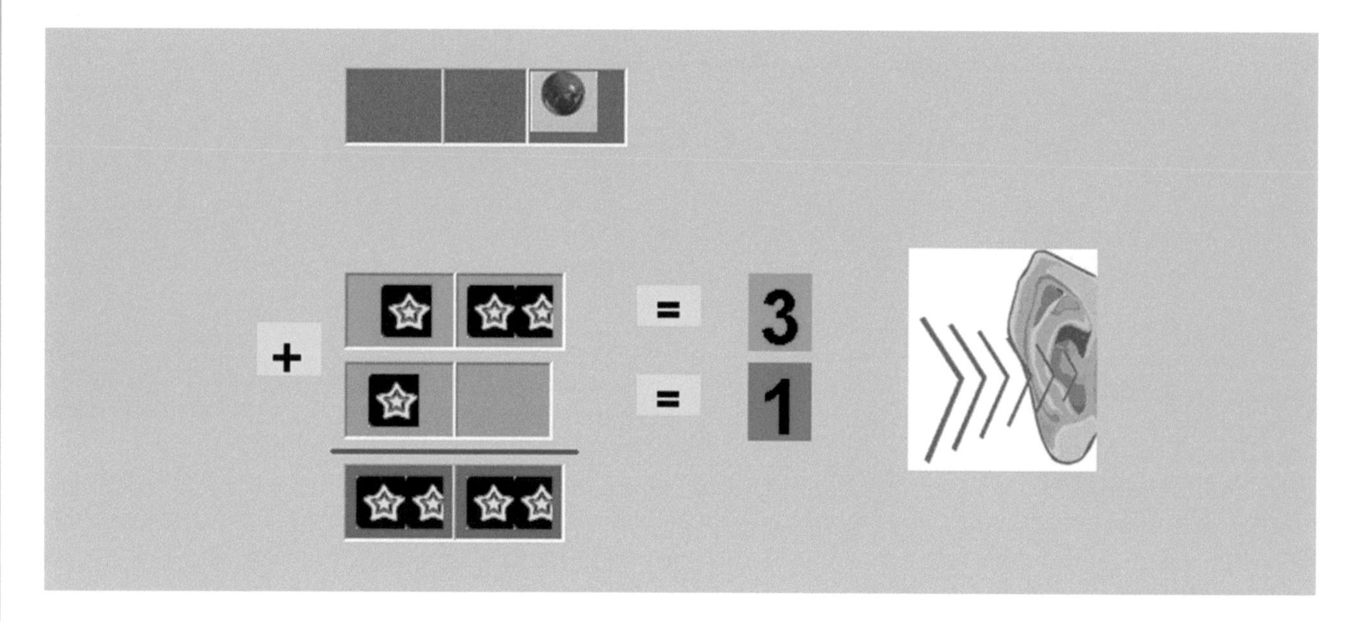

The ***kinesthetic*** representation of the filled up position is announced by clamping of the right hand. An empty position is represented by the clamping of the left hand. The addition sound, tram, is represented by one hand extended in front. The equal sound, cling, is represented by crossing of hands. The kinesthetic representation for both pictures and the answer is the clamping of the right hand twice, one hand extended in front, right hand, left hand, crossing of hands, left hand twice and right hand.

ADDITION, EXERCISE 13, STEP 13.2

In exercise 13, step 13.2 in ***visual*** display of addition we see two pictures of boxes. In the first picture the two positions are filled with a symbol. The second picture has a symbol in the second position. For purpose of addition we are introducing the concept of moving. When a position passes its maximum capacity the next symbol is simultaneously moved in the ascending position. We are moving from the first to the second position. However the audio representation for the picture does not change because the numerical value does not change.

The ***audio*** representation for the three pictures consumes of a **double knock, knock, and knock.**

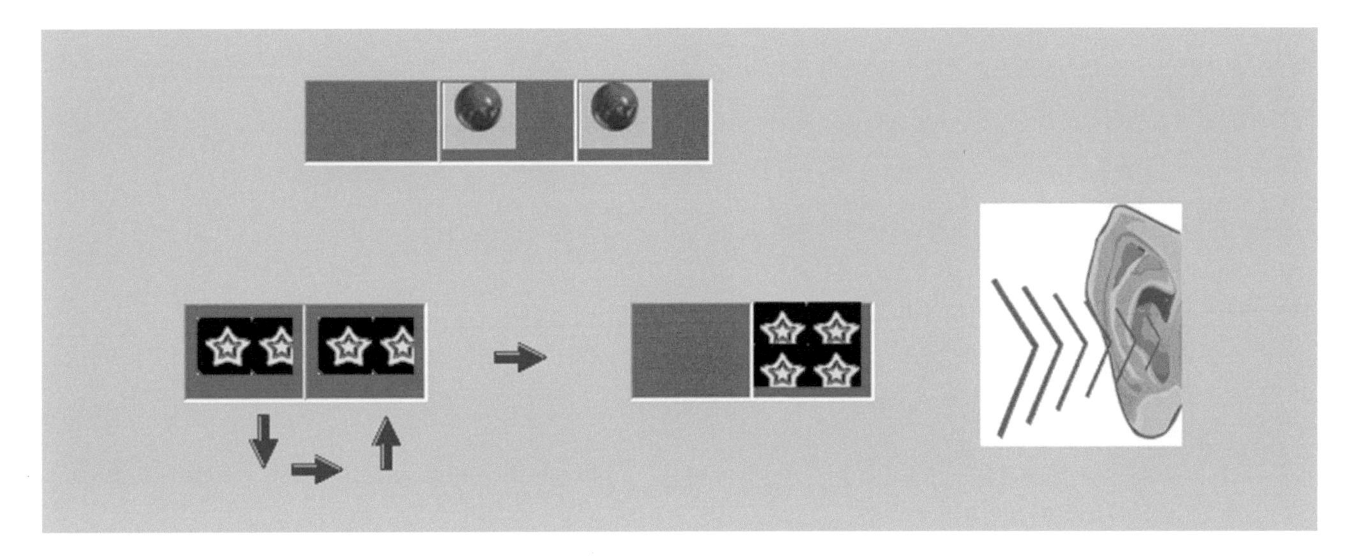

The ***kinesthetic*** representation of the filled up position is announced by clamping of the right hand. An empty position is represented by the clamping of the left hand. The kinesthetic representation for three pictures is the clamping of the left hand and right hand twice.

MATHEMATICS

ADDITION, EXERCISE 13, STEP 13.3

In exercise 13, step 13.3 in ***visual*** display of addition we see two pictures of boxes. The first picture has a symbol in the second position. The second picture has the third position filled with a symbol. For purpose of addition we are introducing the concept of moving. When a position passes its maximum capacity the next symbol is simultaneously moved in the ascending position. We are moving from the first to the second position. However the audio representation for the picture does not change because the numerical value does not change.

The ***audio*** representation for the three pictures consumes of a **double knock, double knock, and knock.**

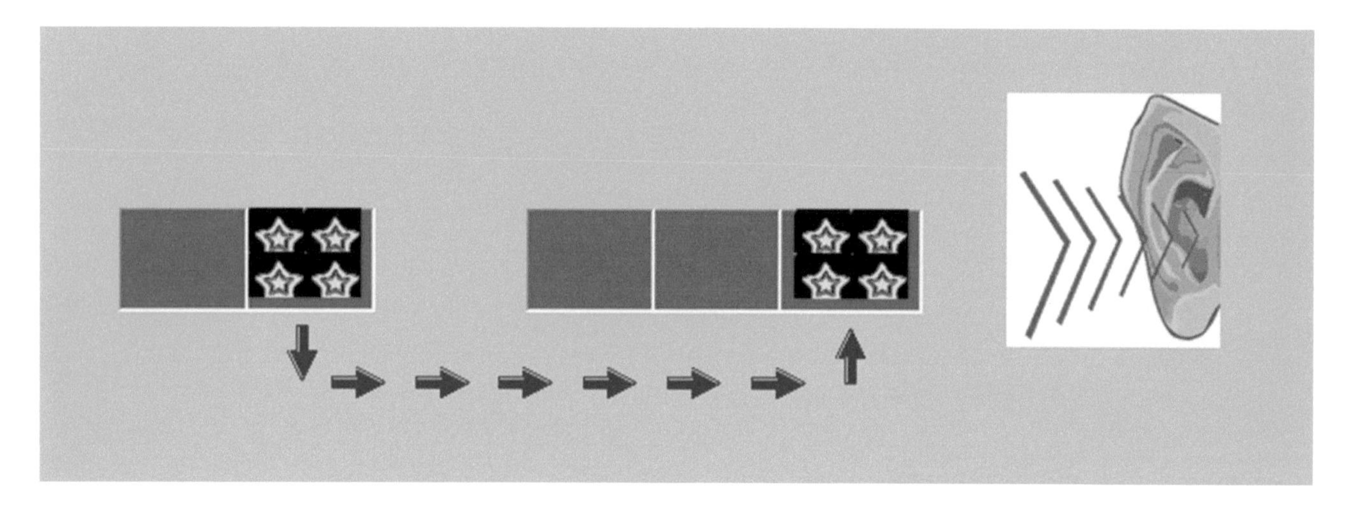

The ***kinesthetic*** representation of the filled up position is announced by clamping of the right hand. An empty position is represented by the clamping of the left hand. The kinesthetic representation for three pictures is the clamping of the left hand twice and right hand.

ADDITION, EXERCISE 13, STEP 13.4

In exercise 13, step 13.4 in ***visual*** display of addition, we see two pictures of boxes with equal sign. Both pictures have the first two positions empty and the last position filled with a symbol. The first picture has a yellow star in its positions but the second picture has a red ball in its positions. The form and shape of the symbols do not change mathematical value and the numerical value of both pictures is equal to four.

The ***audio*** representation for this picture consumes of a **double knock, double knock, knock, and cling.**

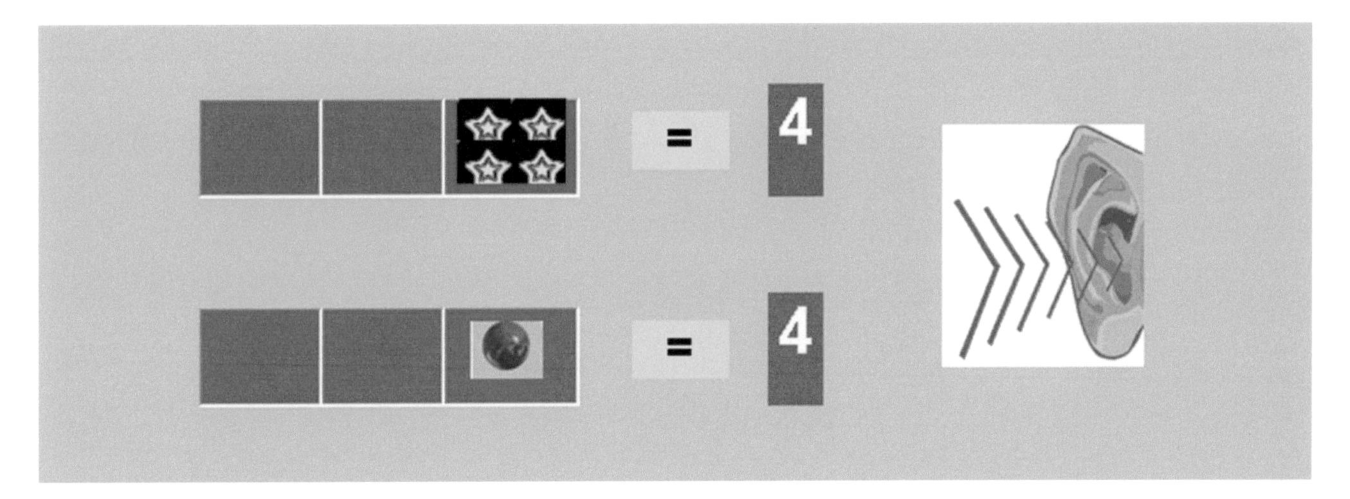

The ***kinesthetic*** representation of the filled up position is announced by clamping of the right hand. An empty position is represented by the clamping of the left hand. The equal sign is crossing both hands in front. The kinesthetic representation for this picture is clamping the left hand twice, right hand, and crossing of hands.

MATHEMATICS ADDITION, EXERCISE 13

In exercise 13, in ***visual*** display of addition we see two pictures of boxes. In the bottom we see three pictures with probable answers. We want to specify, that between the first and second pictures we see the addition sign. Bellow the second picture we see the equal line, which separates the problem from the answer. The ***audio*** representation for two pictures consumes of a **knock, knock, tram, knock, double knock, and cling.**

From the probable answers for addition we need to find the right one. The audio signals are: 1) **knock, double knock, and double knock.** 2) **double knock, knock, and double knock** 3) **double knock, double knock, and knock**.

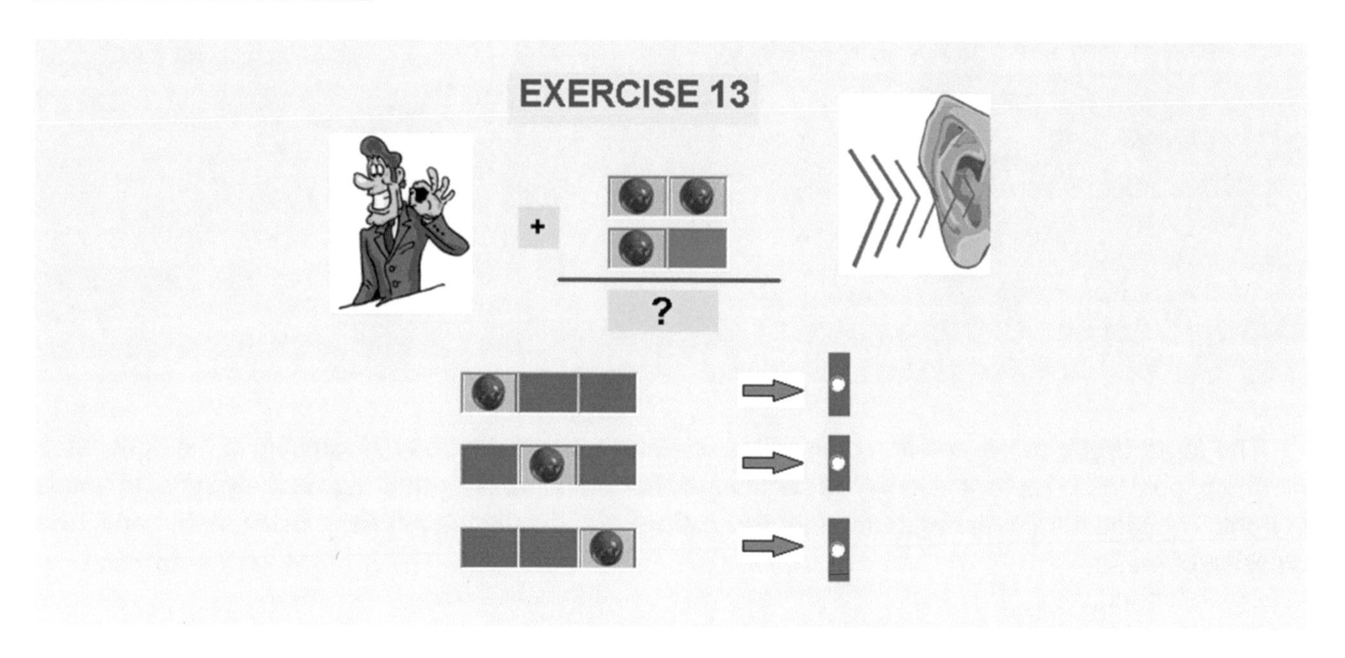

The ***kinesthetic*** representation of the filled up position is announced by clamping of the right hand. An empty position is represented by the clamping of the left hand. The addition sound, tram, is represented by one hand extended in front. The equal sound, cling, is represented by crossing of hands. The kinesthetic representation for both pictures and the answer is the clamping of the right hand twice, one hand extended in front, right hand, left hand, crossing of hands, left hand twice and right hand.

ADDITION, EXERCISE 14

In exercise 14, in ***visual*** display of addition we see three pictures of boxes. We want to specify, that between the first and second pictures we see the addition sign. Below the second picture we see the equal line, which separates the problem from the answer. The ***audio*** representation for the three pictures consumes of a **knock, knock, tram, double knock, knock, cling, knock, double knock, and knock.**

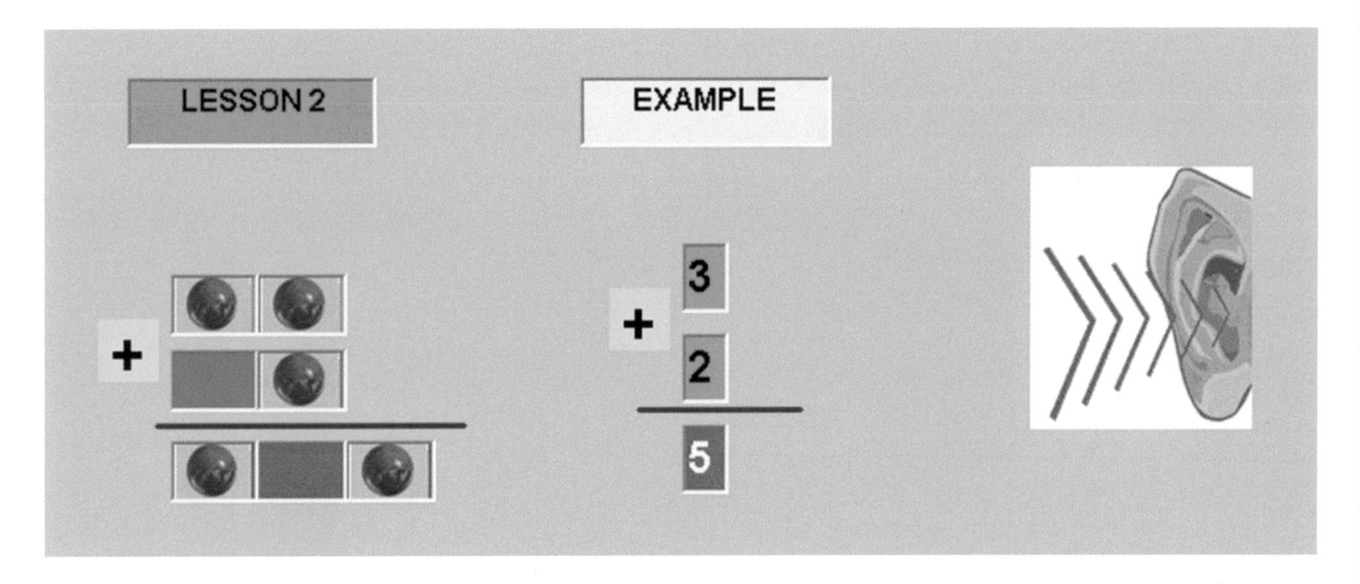

The ***kinesthetic*** representation of the filled up position is announced by clamping of the right hand. An empty position is represented by the clamping of the left hand. The addition sound, tram, is represented by one hand extended in front. The equal sound, cling, is represented by crossing of hands. The kinesthetic representation for both pictures and the answer is the clamping of the right hand twice, one hand extended in front, left hand, right hand, crossing of hands, right hand, left hand, and right hand.

MATHEMATICS

ADDITION, EXERCISE 14, STEP 14.1

In exercise 14, in ***visual*** display of addition we see three pictures of boxes. We want to specify, that between the first and second pictures we see the addition sign. Below the second picture we see the equal line, which separates the problem from the answer. When we are adding two numbers and the position on the first number is filled with a symbol and the corresponding position on the second number is also with a symbol, we move the two symbols to the following positions in the answer as one symbol. When two symbols form, two corresponding positions move forward to the following position of the answer then the two symbols become one symbol. The ***audio*** representation for the three pictures consumes of a **knock, knock, tram, double knock, knock, cling, knock, double knock, and knock.**

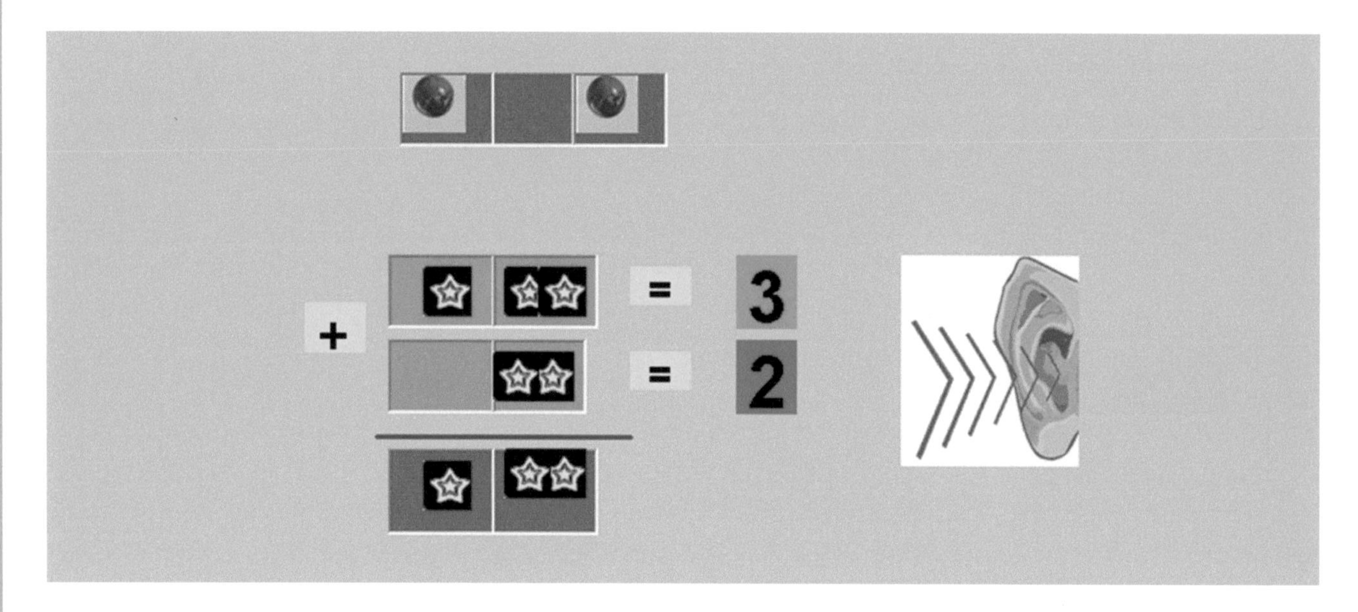

The ***kinesthetic*** representation of the filled up position is announced by clamping of the right hand. An empty position is represented by the clamping of the left hand. The addition sound, tram, is represented by one hand extended in front. The equal sound, cling, is represented by crossing of hands. The kinesthetic representation for both pictures and the answer is the clamping of the right hand twice, one hand extended in front, left hand, right hand, crossing of hands, right hand, left hand, and right hand.

ADDITION, EXERCISE 14, STEP 14.2

In exercise 14, step 14.2 in ***visual*** display of addition we see two pictures of boxes. The first picture has a symbol in the both positions. The second picture has the first and third positions filled with a symbol. For purpose of addition we are introducing the concept of moving. When a position passes its maximum capacity the next symbol is simultaneously moved in the ascending position. We are moving from the first to the second position. However the audio representation for the picture does not change because the numerical value does not change.

The ***audio*** representation for the three pictures consumes of a **double knock, knock, and knock.**

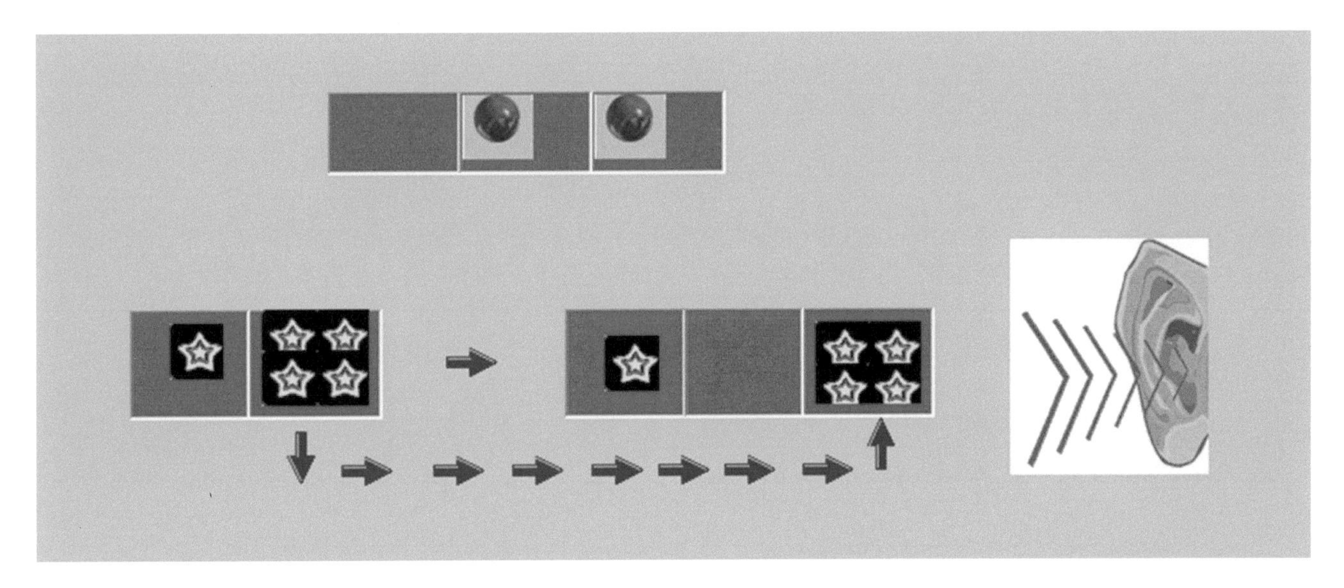

The ***kinesthetic*** representation of the filled up position is announced by clamping of the right hand. An empty position is represented by the clamping of the left hand. The kinesthetic representation for three pictures is the clamping of the left hand and right hand twice.

MATHEMATICS

ADDITION, EXERCISE 14, STEP 14.3

In exercise 14, step 14.3 in ***visual*** display of addition, we see two pictures of boxes with equal sign. Both pictures have the first position filled with a symbol, the second position empty, and the last position filled with a symbol. The first picture has a yellow star in its positions but the second picture has a red ball in its positions. The form and shape of the symbols do not change mathematical value and the numerical value of both pictures is equal to five.

The ***audio*** representation for this picture consumes of a **knock, double knock, knock, and cling.**

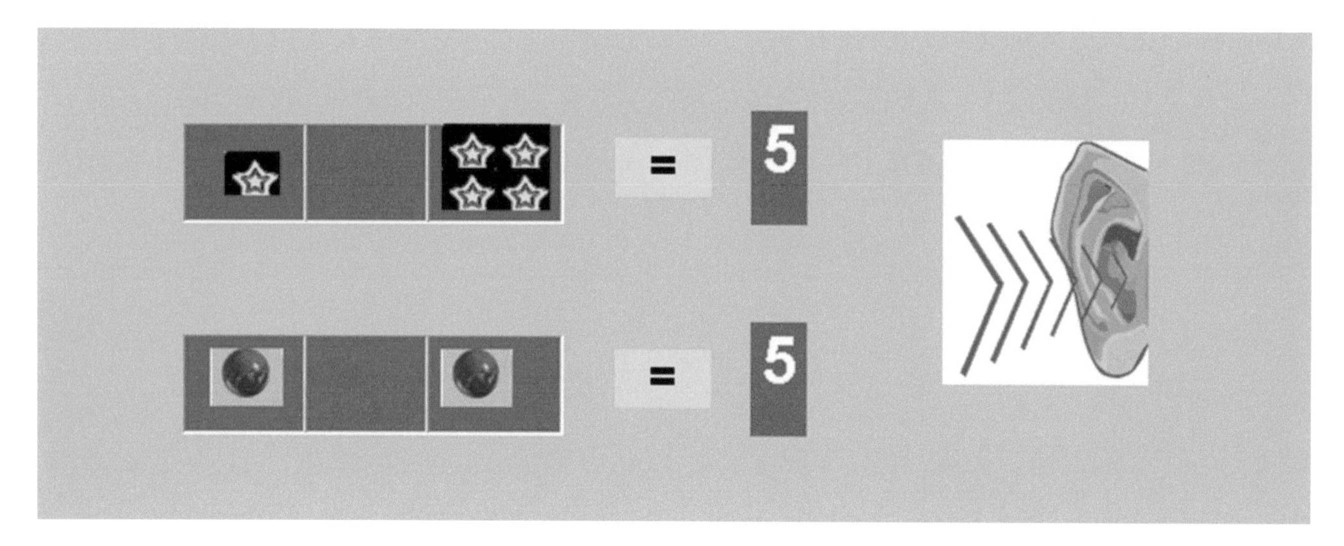

The ***kinesthetic*** representation of the filled up position is announced by clamping of the right hand. An empty position is represented by the clamping of the left hand. The equal sign is crossing both hands in front. The kinesthetic representation for this picture is clamping the right hand, left hand, right hand, and crossing of hands.

ADDITION, EXERCISE 14

In exercise 14, in ***visual*** display of addition we see two pictures of boxes. In the bottom we see three pictures with probable answers. We want to specify, that between the first and second pictures we see the addition sign. Bellow the second picture we see the equal line, which separates the problem from the answer. The ***audio*** representation for two pictures consumes of a **knock, knock, tram, double knock, knock, and cling.**

From the probable answers for addition we need to find the right one. The audio signals are: 1) **double knock, knock, and knock.** 2) **knock, double knock, and knock** 3) **knock, knock, and double knock**.

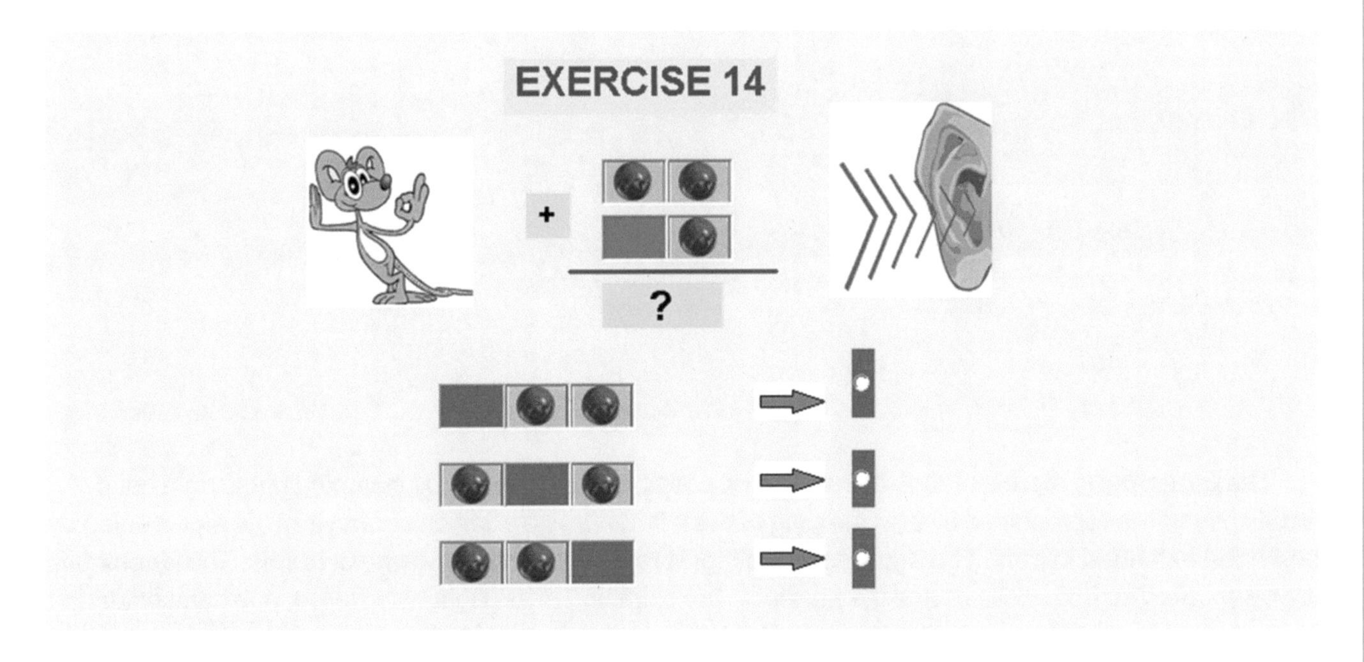

The ***kinesthetic*** representation of the filled up position is announced by clamping of the right hand. An empty position is represented by the clamping of the left hand. The addition sound, tram, is represented by one hand extended in front. The equal sound, cling, is represented by crossing of hands. The kinesthetic representation for both pictures and the answer is the clamping of the right hand twice, one hand extended in front, left hand, right hand, crossing of hands, right hand, left hand, and right hand.

MATHEMATICS ADDITION, EXERCISE 15

In exercise 15, in ***visual*** display of addition we see three pictures of boxes. We want to specify, that between the first and second pictures we see the addition sign. Below the second picture we see the equal line, which separates the problem from the answer. The ***audio*** representation for the three pictures consumes of a **knock, knock, tram, knock, knock, cling, double knock, knock, and knock.**

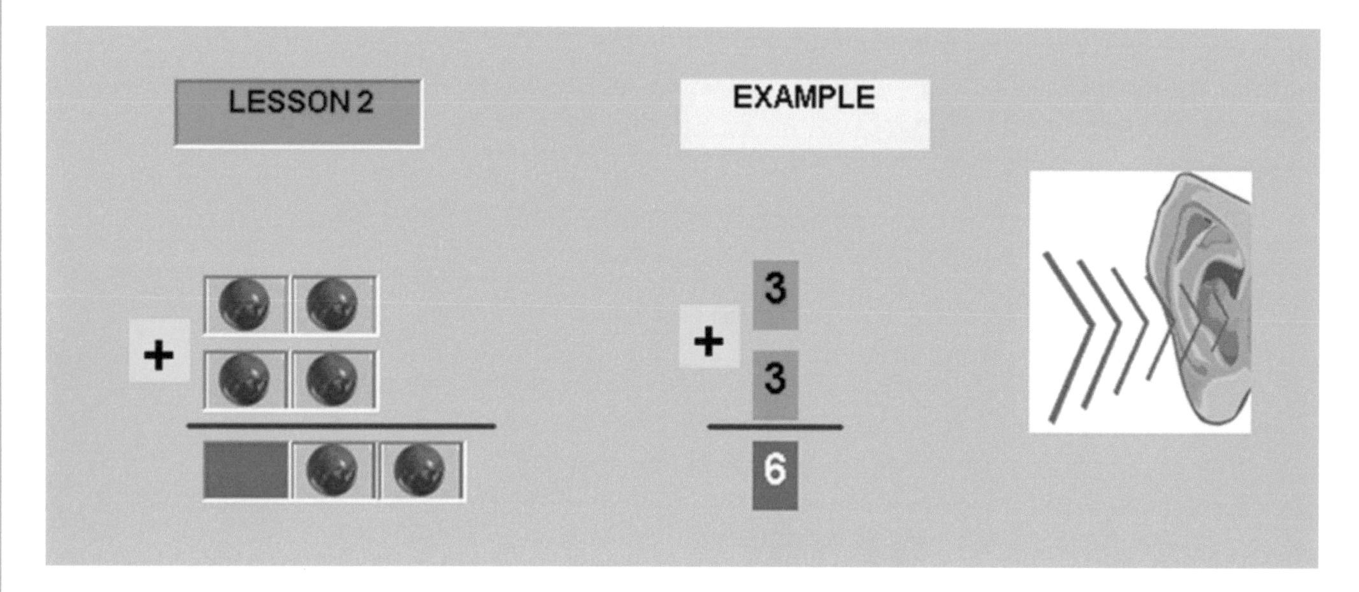

The ***kinesthetic*** representation of the filled up position is announced by clamping of the right hand. An empty position is represented by the clamping of the left hand. The addition sound, tram, is represented by one hand extended in front. The equal sound, cling, is represented by crossing of hands. The kinesthetic representation for both pictures and the answer is the clamping of the right hand twice, one hand extended in front, right hand twice, crossing of hands, left hand, and right hand twice.

ADDITION, EXERCISE 15, STEP 15.1

In exercise 15, step 15.1 in ***visual*** display of addition we see three pictures of boxes. We want to specify, that between the first and second pictures we see the addition sign. Below the second picture we see the equal line, which separates the problem from the answer. When we are adding two numbers and the position on the first number is filled with a symbol and the corresponding position on the second number is also with a symbol, we move the two symbols to the following positions in the answer as one symbol. When two symbols form, two corresponding positions move forward to the following position of the answer then the two symbols become one symbol. The ***audio*** representation for the three pictures consumes of a **knock, knock, tram, knock, knock, cling, double knock, knock, and knock.**

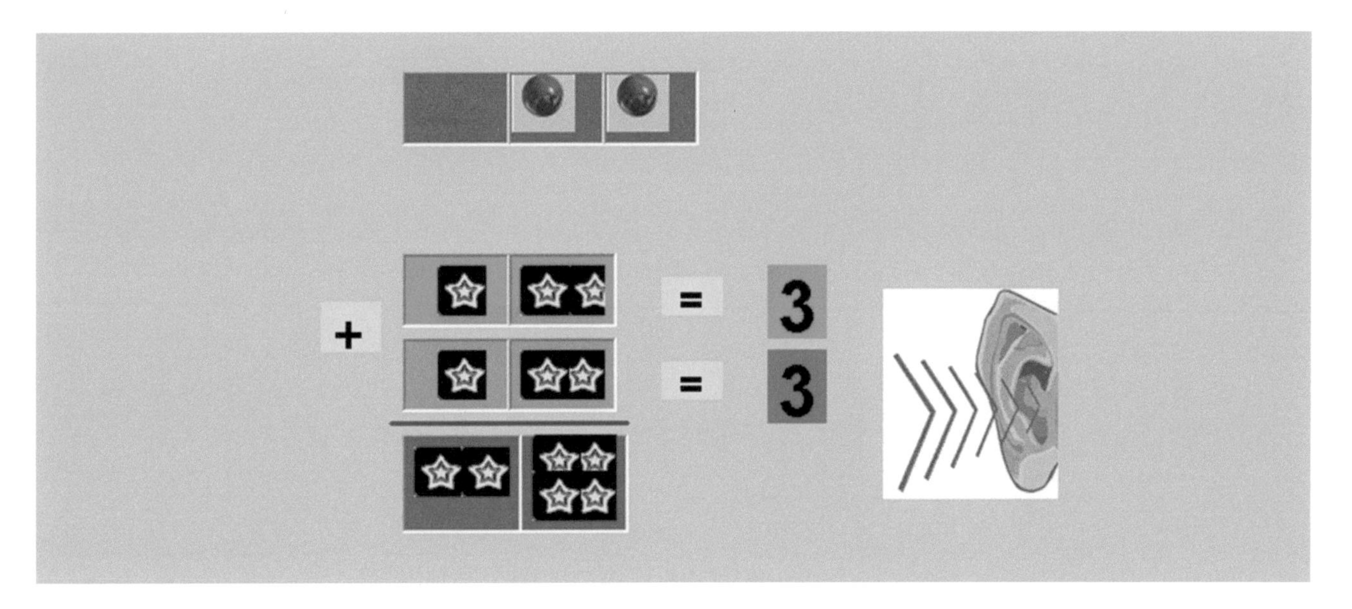

The ***kinesthetic*** representation of the filled up position is announced by clamping of the right hand. An empty position is represented by the clamping of the left hand. The addition sound, tram, is represented by one hand extended in front. The equal sound, cling, is represented by crossing of hands. The kinesthetic representation for both pictures and the answer is the clamping of the right hand twice, one hand extended in front, right hand twice, crossing of hands, left hand, and right hand twice.

MATHEMATICS

ADDITION, EXERCISE 15, STEP 15.2

In exercise 15, step 15.2 in ***visual*** display of addition we see two pictures of boxes. The first picture has a symbol in the both positions. The second picture has the third position filled with a symbol. For purpose of addition we are introducing the concept of moving. When a position passes its maximum capacity the next symbol is simultaneously moved in the ascending position. We are moving from the first to the second position. However the audio representation for the picture does not change because the numerical value does not change.

The ***audio*** representation for the three pictures consumes of a **double knock, knock, and knock.**

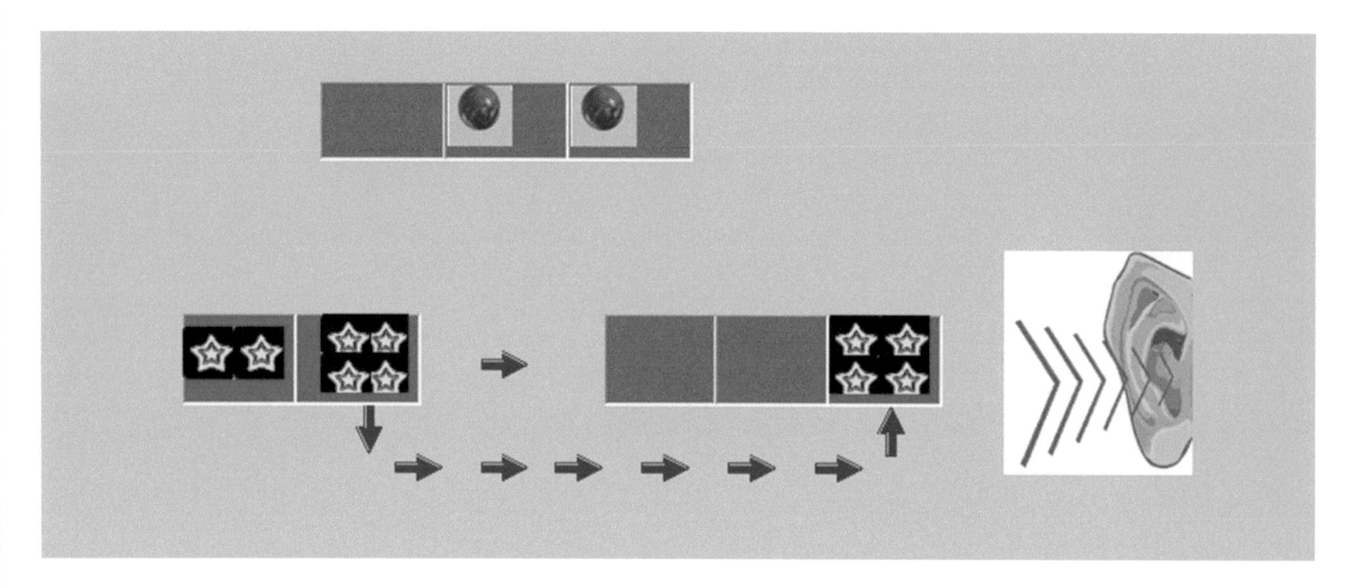

The ***kinesthetic*** representation of the filled up position is announced by clamping of the right hand. An empty position is represented by the clamping of the left hand. The kinesthetic representation for three pictures is the clamping of the left hand and right hand twice.

ADDITION, EXERCISE 15, STEP 15.3

In exercise 15, step 15.3 in ***visual*** display of addition we see two pictures of boxes. The first picture has a symbol in the first position. The second picture has the second and third positions filled with a symbol. For purpose of addition we are introducing the concept of moving. When a position passes its maximum capacity the next symbol is simultaneously moved in the ascending position. We are moving from the first to the second position. However the audio representation for the picture does not change because the numerical value does not change.

The ***audio*** representation for the three pictures consumes of a **double knock, knock, and knock.**

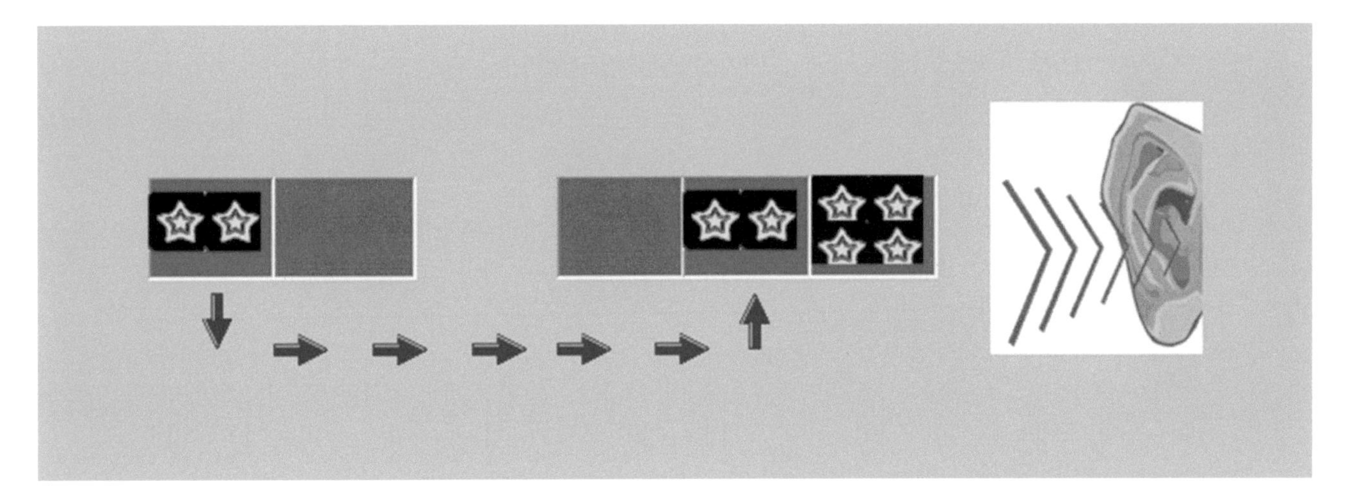

The ***kinesthetic*** representation of the filled up position is announced by clamping of the right hand. An empty position is represented by the clamping of the left hand. The kinesthetic representation for three pictures is the clamping of the left hand and right hand twice.

MATHEMATICS

ADDITION, EXERCISE 15, STEP 15.4

In exercise 15, step 15.4 in ***visual*** display of addition, we see two pictures of boxes with equal sign. Both pictures have the second and third positions filled with a symbol and the first position is empty. The first picture has a yellow star in its positions but the second picture has a red ball in its positions. The form and shape of the symbols do not change mathematical value and the numerical value of both pictures is equal to six.

The ***audio*** representation for this picture consumes of a **double knock, knock, knock, and cling.**

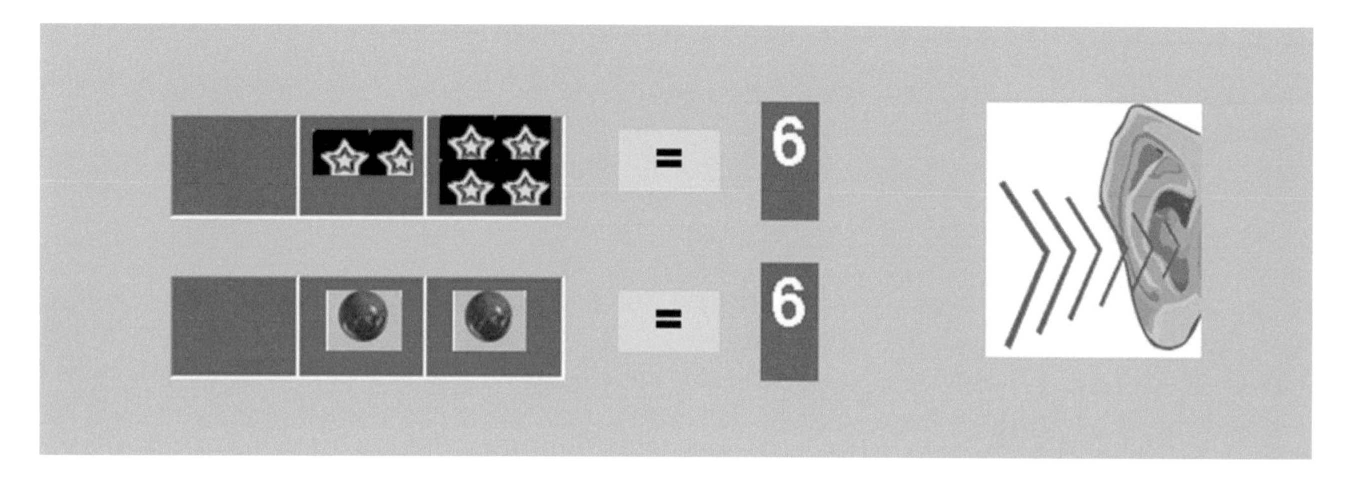

The ***kinesthetic*** representation of the filled up position is announced by clamping of the right hand. An empty position is represented by the clamping of the left hand. The equal sign is crossing both hands in front. The kinesthetic representation for this picture is clamping the left hand, right hand twice, and crossing of hands.

ADDITION, EXERCISE 15

In exercise 15, in ***visual*** display of addition we see two pictures of boxes. In the bottom we see three pictures with probable answers. We want to specify, that between the first and second pictures we see the addition sign. Bellow the second picture we see the equal line, which separates the problem from the answer. The ***audio*** representation for two pictures consumes of a **knock, knock, tram, knock, knock, and cling.**

From the probable answers for addition we need to find the right one. The audio signals are: 1) **knock, knock, and double knock.** 2) **double knock, knock, and knock** 3) **knock, double knock, and knock**.

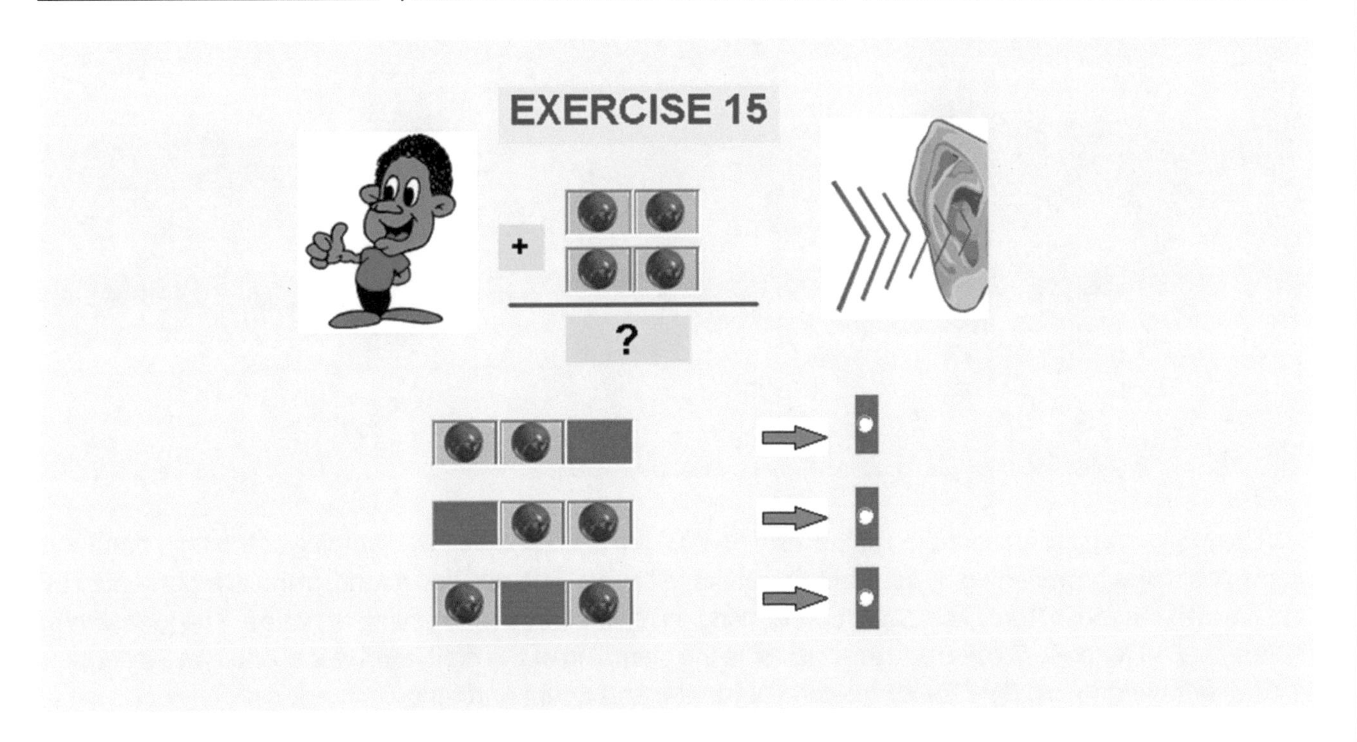

The ***kinesthetic*** representation of the filled up position is announced by clamping of the right hand. An empty position is represented by the clamping of the left hand. The addition sound, tram, is represented by one hand extended in front. The equal sound, cling, is represented by crossing of hands. The kinesthetic representation for both pictures and the answer is the clamping of the right hand twice, one hand extended in front, right hand twice, crossing of hands, left hand, and right hand twice.

MATHEMATICS ADDITION, EXERCISE 16

In exercise 16, in ***visual*** display of addition we see three pictures of boxes. We want to specify, that between the first and second pictures we see the addition sign. Below the second picture we see the equal line, which separates the problem from the answer. The ***audio*** representation for the three pictures consumes of a **knock, knock, tram, double knock, double knock, knock, cling, knock, knock, and knock.**

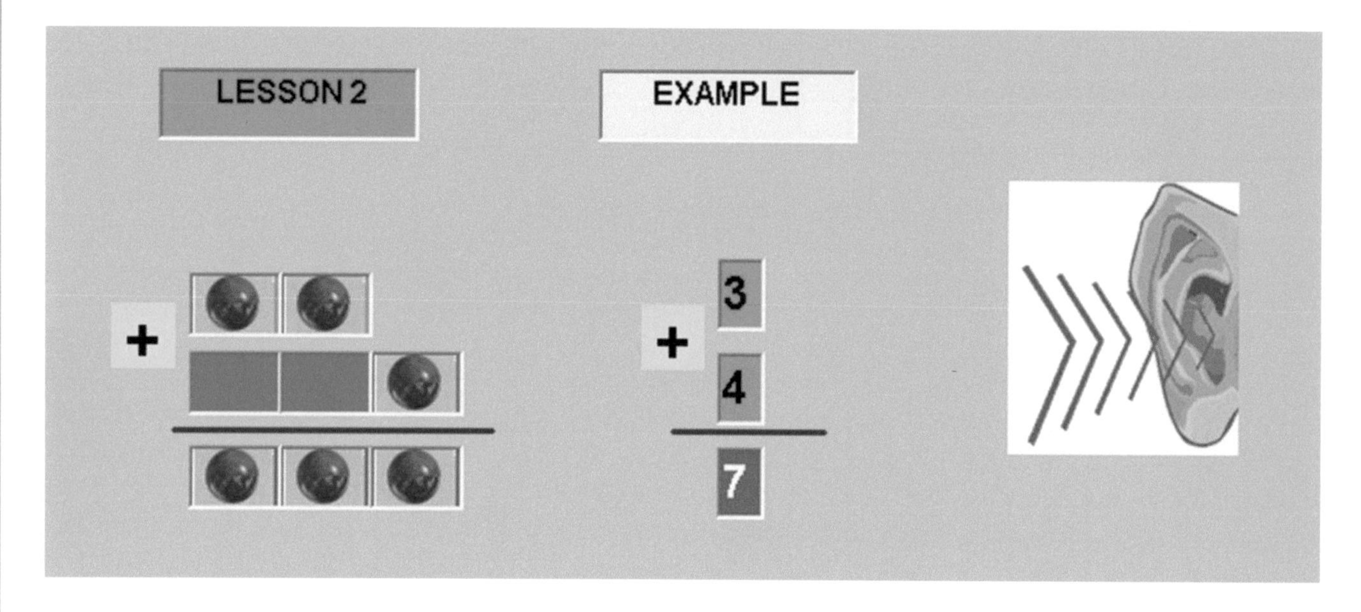

The ***kinesthetic*** representation of the filled up position is announced by clamping of the right hand. An empty position is represented by the clamping of the left hand. The addition sound, tram, is represented by one hand extended in front. The equal sound, cling, is represented by crossing of hands. The kinesthetic representation for both pictures and the answer is the clamping of the right hand twice, one hand extended in front, left hand twice, right hand, crossing of hands, and right hand thrice.

ADDITION, EXERCISE 16, STEP 16.1

In exercise 16, step 16.1 in ***visual*** display of addition we see three pictures of boxes. We want to specify, that between the first and second pictures we see the addition sign. Below the second picture we see the equal line, which separates the problem from the answer. When we are adding two numbers and the position on the first number is filled with a symbol and the corresponding position on the second number is also with a symbol, we move the two symbols to the following positions in the answer as one symbol. When two symbols form, two corresponding positions move forward to the following position of the answer then the two symbols become one symbol. The ***audio*** representation for the three pictures consumes of a **knock, knock, tram, double knock, double knock, knock, cling, knock, knock, and knock.**

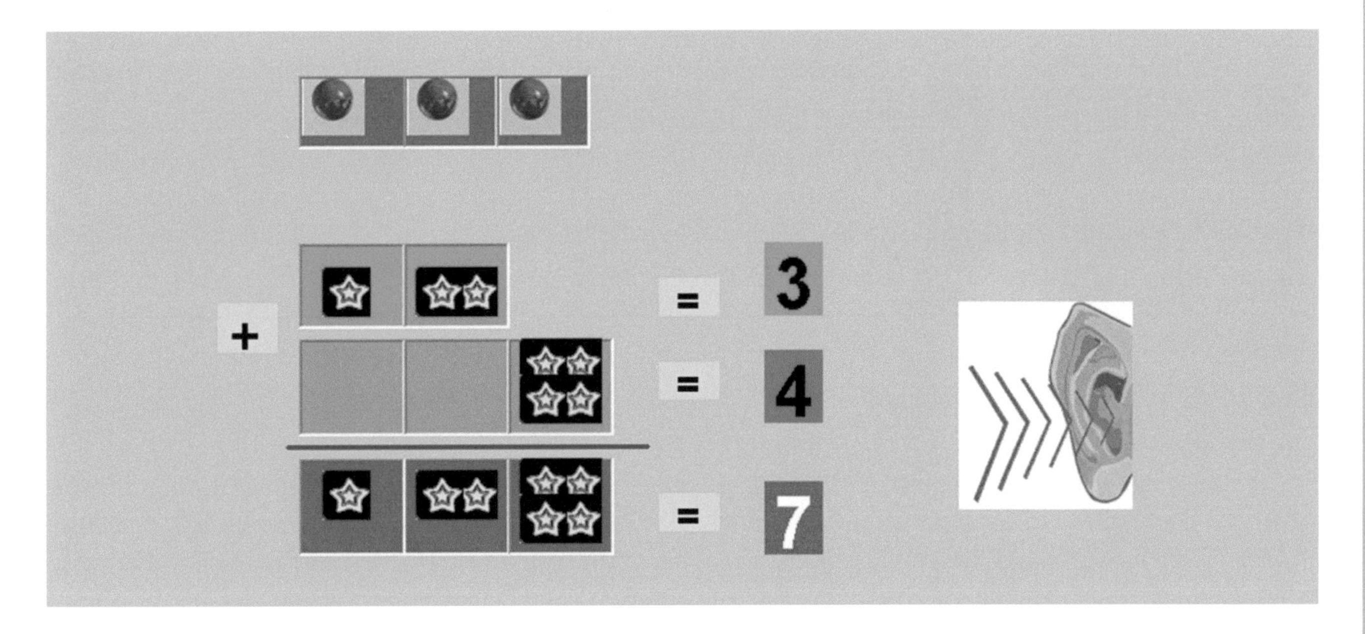

The ***kinesthetic*** representation of the filled up position is announced by clamping of the right hand. An empty position is represented by the clamping of the left hand. The addition sound, tram, is represented by one hand extended in front. The equal sound, cling, is represented by crossing of hands. The kinesthetic representation for both pictures and the answer is the clamping of the right hand twice, one hand extended in front, left hand twice, right hand, crossing of hands, and right hand thrice.

MATHEMATICS

ADDITION, EXERCISE 16, STEP 16.2

In exercise 16, step 16.2 in ***visual*** display of addition, we see two pictures of boxes with equal sign. Both pictures have all three of their positions filled with a symbol. The first picture has a yellow star in its positions but the second picture has a red ball in its positions. The form and shape of the symbols do not change mathematical value and the numerical value of both pictures is equal to seven.

The ***audio*** representation for this picture consumes of a **knock, knock, knock, and cling.**

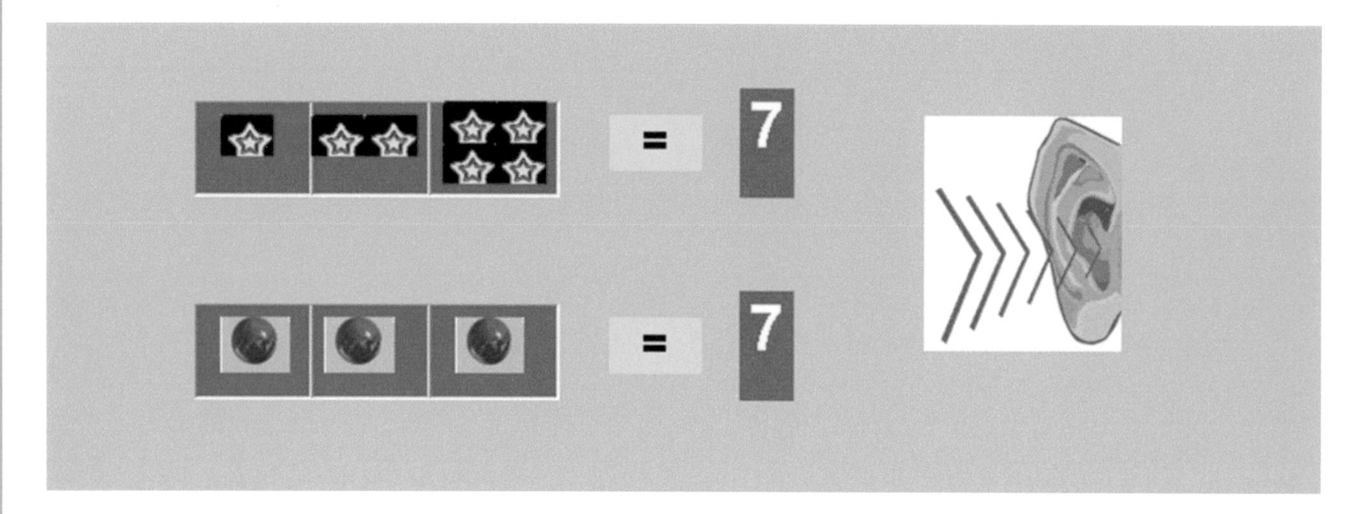

The ***kinesthetic*** representation of the filled up position is announced by clamping of the right hand. An empty position is represented by the clamping of the left hand. The equal sign is crossing both hands in front. The kinesthetic representation for this picture is clamping the right hand thrice and crossing of hands.

ADDITION, EXERCISE 16

In exercise 16, in ***visual*** display of addition we see two pictures of boxes. In the bottom we see three pictures with probable answers. We want to specify, that between the first and second pictures we see the addition sign. Bellow the second picture we see the equal line, which separates the problem from the answer. The ***audio*** representation for two pictures consumes of a **knock, knock, tram, double knock, double knock, knock, and cling.**

From the probable answers for addition we need to find the right one. The audio signals are: 1) **knock, knock, and knock.** 2) **double knock, knock, and knock** 3) **double knock, double knock, and knock**.

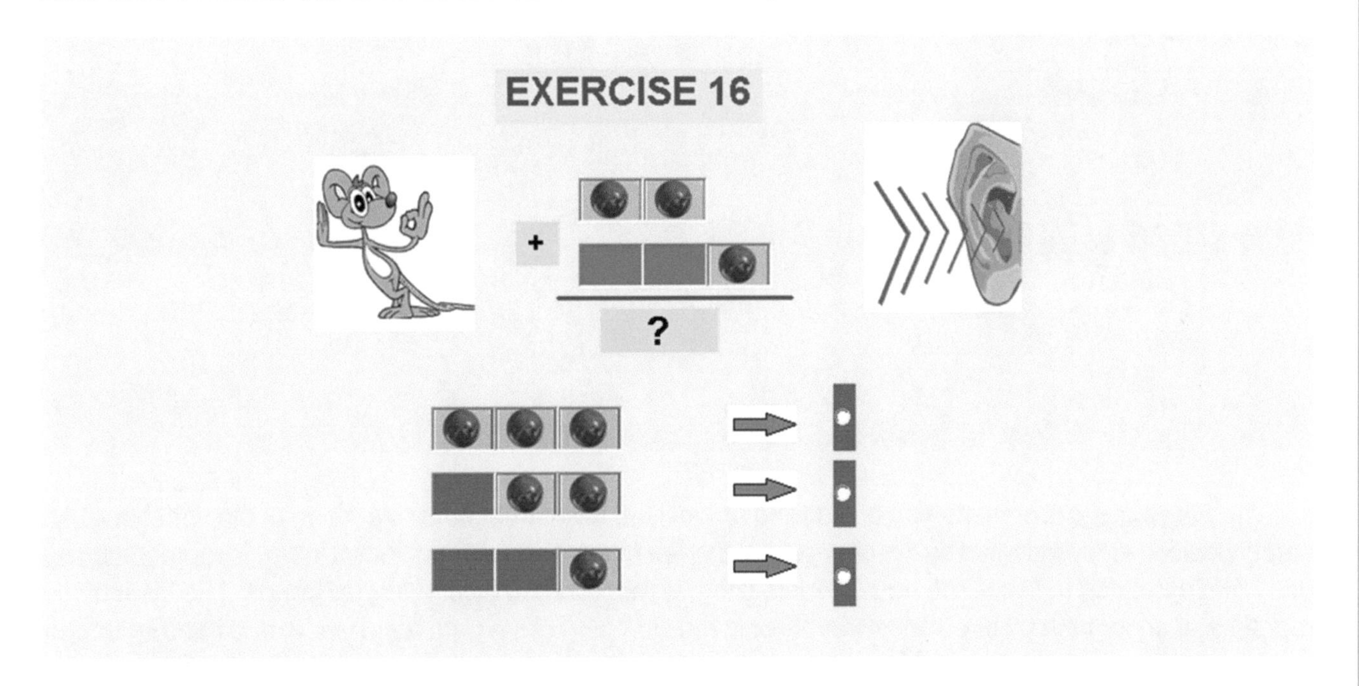

The ***kinesthetic*** representation of the filled up position is announced by clamping of the right hand. An empty position is represented by the clamping of the left hand. The addition sound, tram, is represented by one hand extended in front. The equal sound, cling, is represented by crossing of hands. The kinesthetic representation for both pictures and the answer is the clamping of the right hand twice, one hand extended in front, left hand twice, right hand, crossing of hands, and right hand thrice.

MATHEMATICS

ADDITION, EXERCISE 17

In exercise 17, in ***visual*** display of addition we see three pictures of boxes. We want to specify, that between the first and second pictures we see the addition sign. Below the second picture we see the equal line, which separates the problem from the answer. The ***audio*** representation for the three pictures consumes of a **knock, knock, tram, knock, double knock, knock, cling, double knock, double knock, double knock, and knock.**

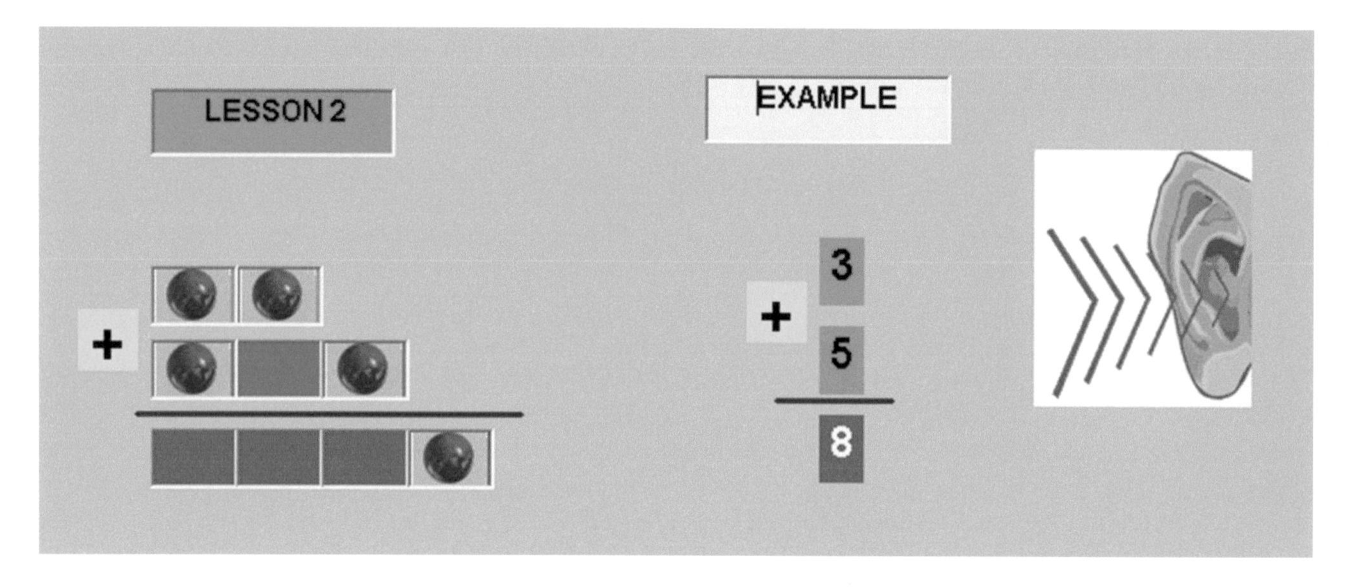

The ***kinesthetic*** representation of the filled up position is announced by clamping of the right hand. An empty position is represented by the clamping of the left hand. The addition sound, tram, is represented by one hand extended in front. The equal sound, cling, is represented by crossing of hands. The kinesthetic representation for both pictures and the answer is the clamping of the right hand twice, one hand extended in front, right hand, left hand, right hand, crossing of hands, left hand thrice, and right hand.

ADDITION, EXERCISE 17, STEP 17.1

In exercise 17, in ***visual*** display of addition we see three pictures of boxes. We want to specify, that between the first and second pictures we see the addition sign. Below the second picture we see the equal line, which separates the problem from the answer. When we are adding two numbers and the position on the first number is filled with a symbol and the corresponding position on the second number is also with a symbol, we move the two symbols to the following positions in the answer as one symbol. When two symbols form, two corresponding positions move forward to the following position of the answer then the two symbols become one symbol. The ***audio*** representation for the three pictures consumes of a **knock, knock, tram, knock, double knock, knock, cling, double knock, double knock, double knock, and knock.**

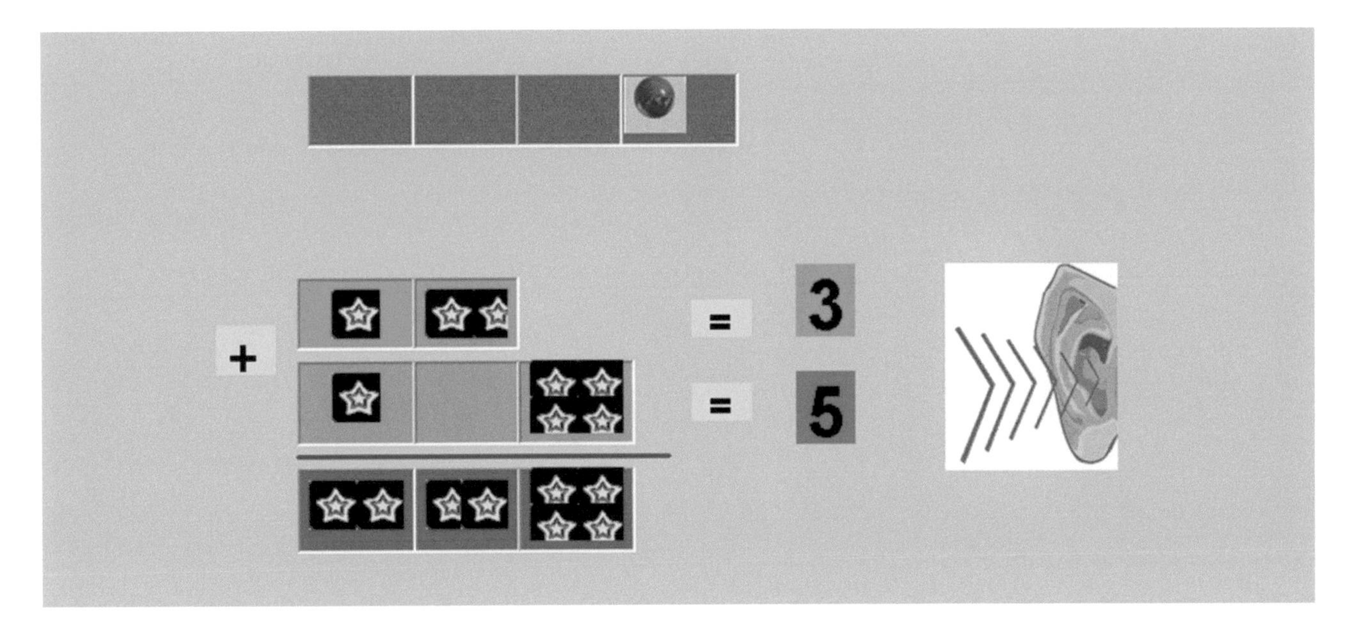

The ***kinesthetic*** representation of the filled up position is announced by clamping of the right hand. An empty position is represented by the clamping of the left hand. The addition sound, tram, is represented by one hand extended in front. The equal sound, cling, is represented by crossing of hands. The kinesthetic representation for both pictures and the answer is the clamping of the right hand twice, one hand extended in front, right hand, left hand, right hand, crossing of hands, left hand thrice, and right hand.

MATHEMATICS ADDITION, EXERCISE 17, STEP 17.2

In exercise 17, step 17.2 in ***visual*** display of addition we see two pictures of boxes. The first picture has a symbol in all three positions. The second picture has the second and third positions filled with a symbol. For purpose of addition we are introducing the concept of moving. When a position passes its maximum capacity the next symbol is simultaneously moved in the ascending position. We are moving from the first to the second position. However the audio representation for the picture does not change because the numerical value does not change.

The ***audio*** representation for the three pictures consumes of a **double knock, double knock, double knock, and knock.**

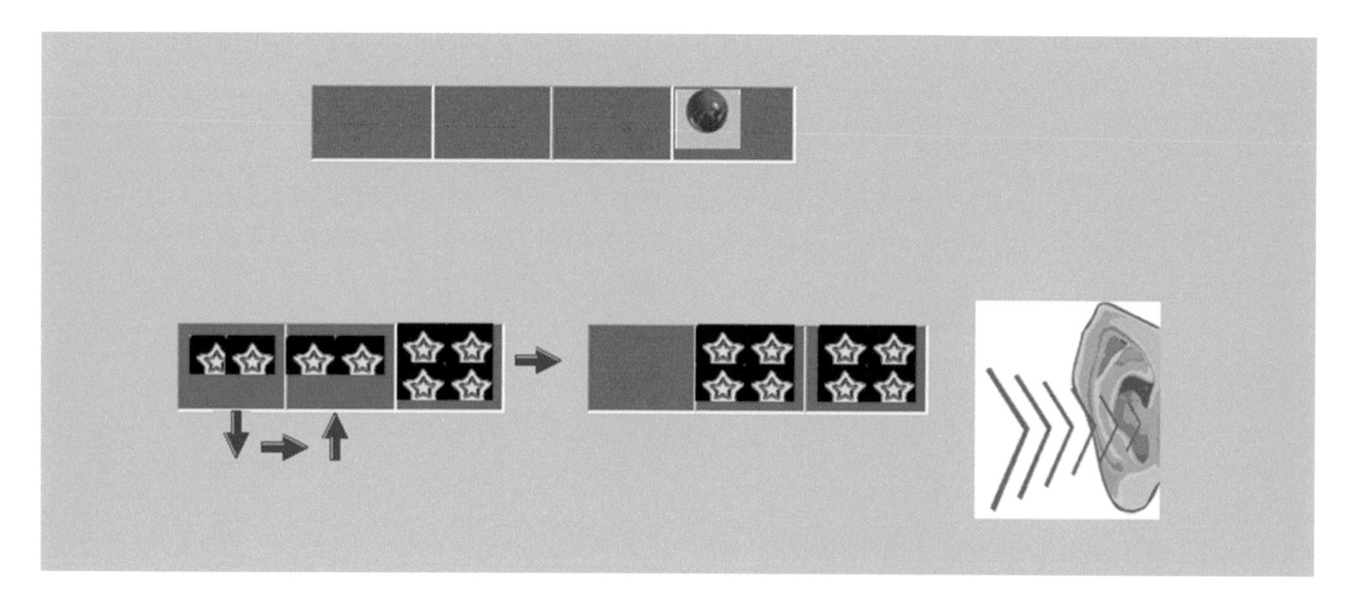

The ***kinesthetic*** representation of the filled up position is announced by clamping of the right hand. An empty position is represented by the clamping of the left hand. The kinesthetic representation for three pictures is the clamping of the left hand thrice and right hand.

ADDITION, EXERCISE 17, STEP 17.3

In exercise 17, step 17.3 in ***visual*** display of addition we see two pictures of boxes. The first picture has a symbol in the second and third positions. The second picture has the third position filled with a symbol. For purpose of addition we are introducing the concept of moving. When a position passes its maximum capacity the next symbol is simultaneously moved in the ascending position. We are moving from the first to the second position. However the audio representation for the picture does not change because the numerical value does not change.

The ***audio*** representation for the three pictures consumes of a **double knock, double knock, double knock, and knock.**

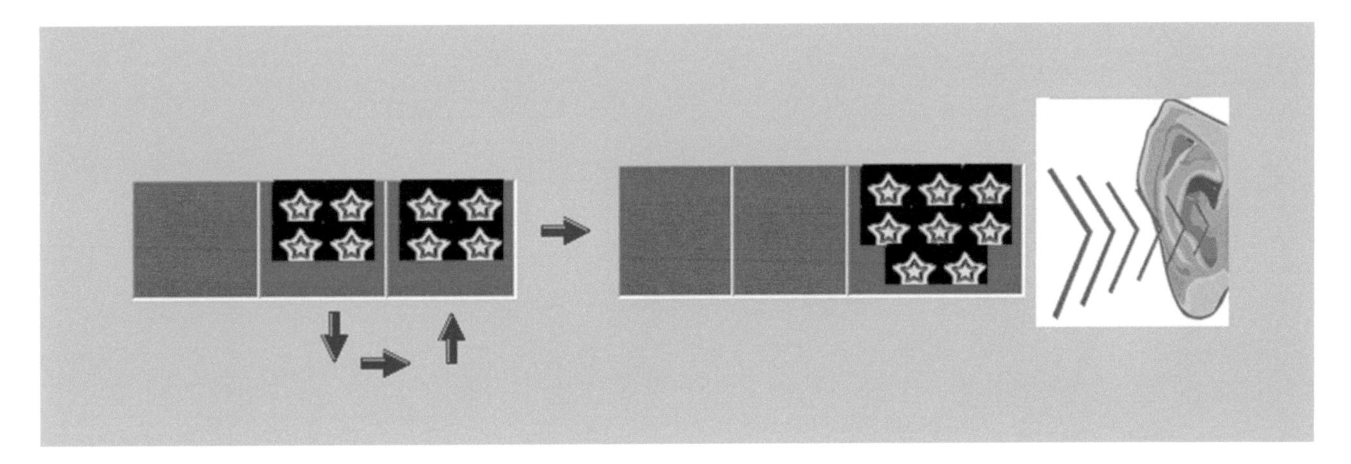

The ***kinesthetic*** representation of the filled up position is announced by clamping of the right hand. An empty position is represented by the clamping of the left hand. The kinesthetic representation for three pictures is the clamping of the left hand thrice and right hand.

MATHEMATICS ADDITION, EXERCISE 17, STEP 17.4

In exercise 17, step 17.4 in ***visual*** display of addition we see two pictures of boxes. The first picture has a symbol in the third position. The second picture has the fourth position filled with a symbol. For purpose of addition we are introducing the concept of moving. When a position passes its maximum capacity the next symbol is simultaneously moved in the ascending position. We are moving from the first to the second position. However the audio representation for the picture does not change because the numerical value does not change.

The ***audio*** representation for the three pictures consumes of a **double knock, double knock, double knock, and knock.**

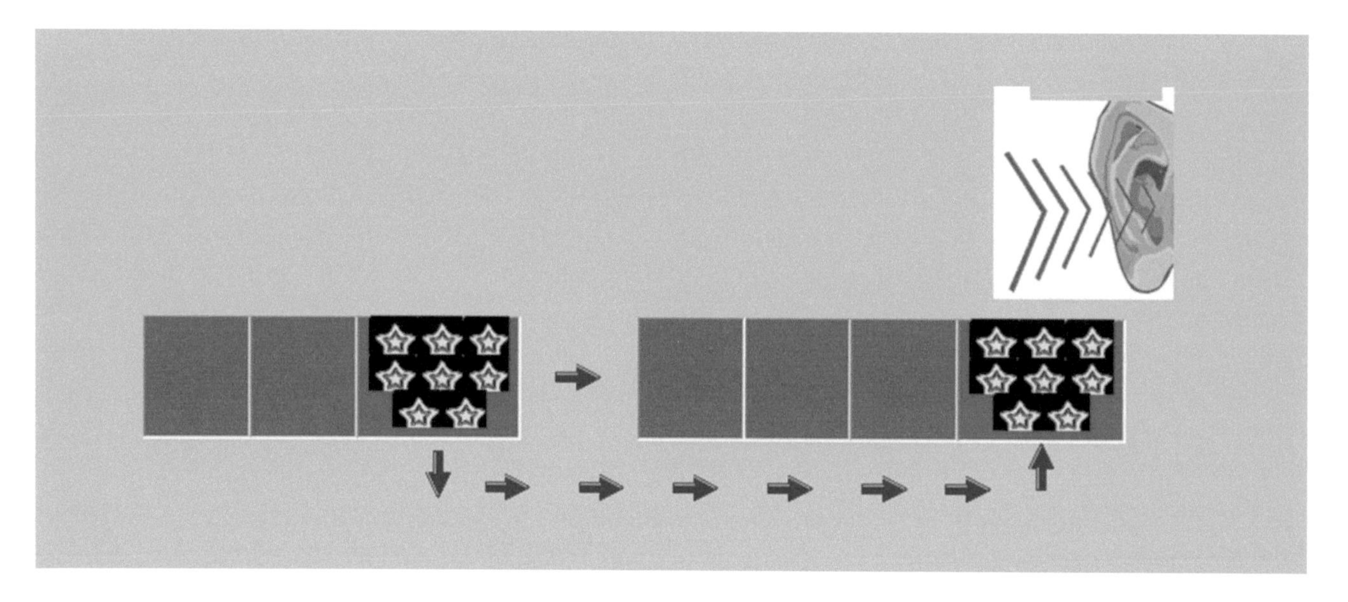

The ***kinesthetic*** representation of the filled up position is announced by clamping of the right hand. An empty position is represented by the clamping of the left hand. The kinesthetic representation for three pictures is the clamping of the left hand thrice and right hand.

ADDITION, EXERCISE 17, STEP 17.5

In exercise 17, step 17.5 in ***visual*** display of addition, we see two pictures of boxes with equal sign. Both pictures have the first three of their positions empty and the last position is filled with a symbol. The first picture has a yellow star in its positions but the second picture has a red ball in its positions. The form and shape of the symbols do not change mathematical value and the numerical value of both pictures is equal to eight.

The ***audio*** representation for this picture consumes of a **double knock, double knock, double knock, knock, and cling.**

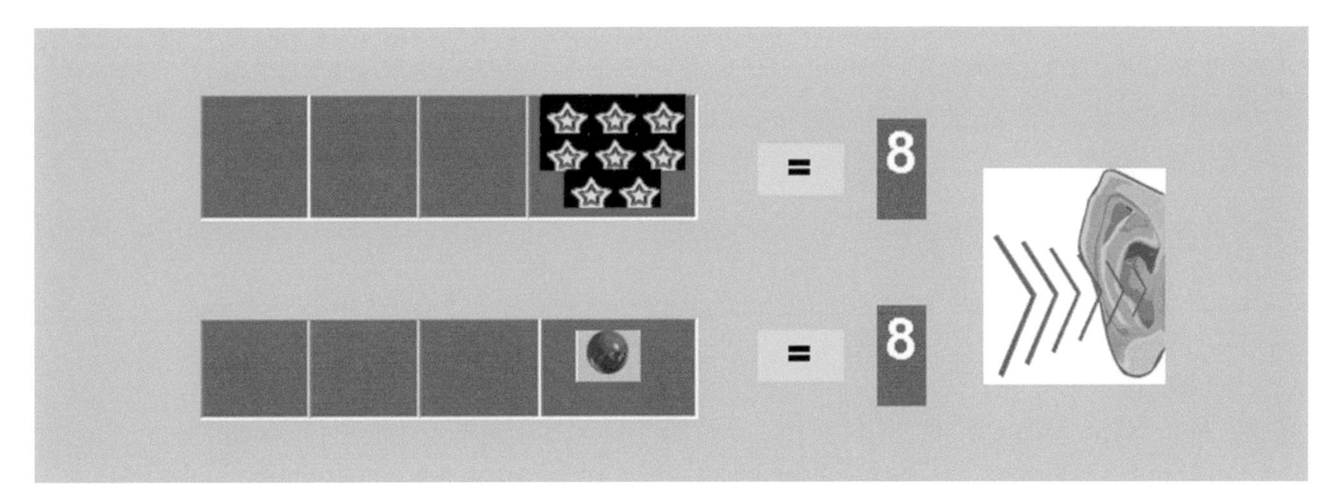

The ***kinesthetic*** representation of the filled up position is announced by clamping of the right hand. An empty position is represented by the clamping of the left hand. The equal sign is crossing both hands in front. The kinesthetic representation for this picture is clamping the left hand thrice, right hand, and crossing of hands.

MATHEMATICS

ADDITION, EXERCISE 17

In exercise 17, in ***visual*** display of addition we see two pictures of boxes. In the bottom we see three pictures with probable answers. We want to specify, that between the first and second pictures we see the addition sign. Bellow the second picture we see the equal line, which separates the problem from the answer. The ***audio*** representation for two pictures consumes of a **knock, knock, tram, knock, double knock, knock, and cling.**

From the probable answers for addition we need to find the right one. The audio signals are: 1) **double knock, knock, double knock, and double knock.** 2) **double knock, double knock, double knock, and knock** 3) **knock, double knock, double knock, and double knock**.

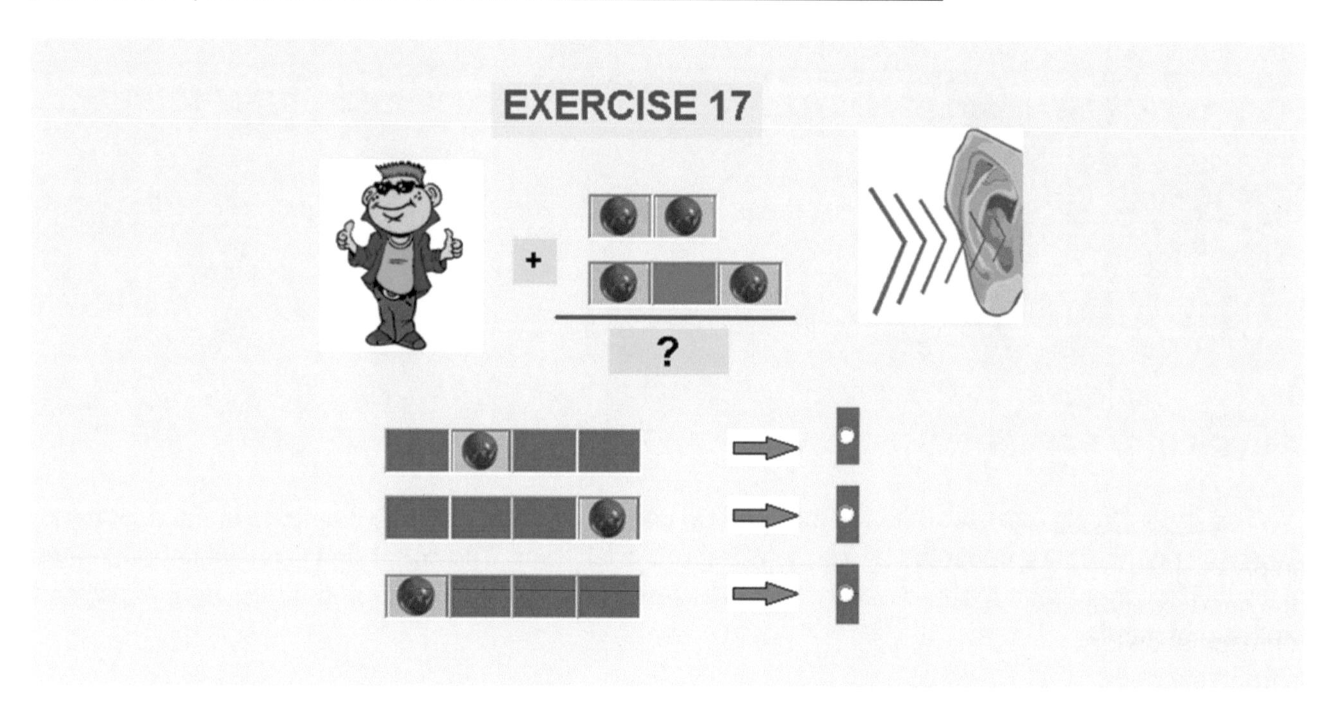

The ***kinesthetic*** representation of the filled up position is announced by clamping of the right hand. An empty position is represented by the clamping of the left hand. The addition sound, tram, is represented by one hand extended in front. The equal sound, cling, is represented by crossing of hands. The kinesthetic representation for both pictures and the answer is the clamping of the right hand twice, one hand extended in front, right hand, left hand, right hand, crossing of hands, left hand thrice, and right hand.

ADDITION, EXERCISE 18

In exercise 18, in ***visual*** display of addition we see three pictures of boxes. We want to specify, that between the first and second pictures we see the addition sign. Below the second picture we see the equal line, which separates the problem from the answer. The ***audio*** representation for the three pictures consumes of a **knock, knock, tram, double knock, knock, knock, cling, knock, double knock, double knock, and knock.**

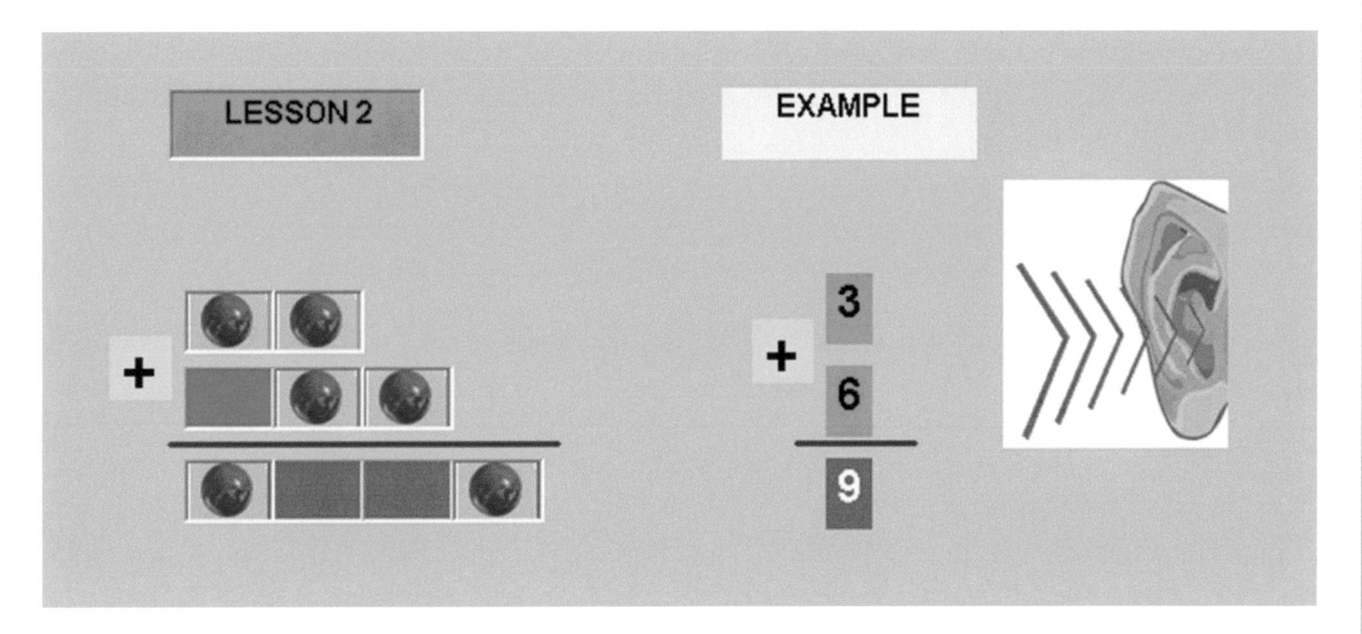

The ***kinesthetic*** representation of the filled up position is announced by clamping of the right hand. An empty position is represented by the clamping of the left hand. The addition sound, tram, is represented by one hand extended in front. The equal sound, cling, is represented by crossing of hands. The kinesthetic representation for both pictures and the answer is the clamping of the right hand twice, one hand extended in front, left hand, right hand twice, crossing of hands, right hand, left hand twice, and right hand.

MATHEMATICS

ADDITION, EXERCISE 18, STEP 18.1

In exercise 18, step 18.1 in ***visual*** display of addition we see three pictures of boxes. We want to specify, that between the first and second pictures we see the addition sign. Below the second picture we see the equal line, which separates the problem from the answer. When we are adding two numbers and the position on the first number is filled with a symbol and the corresponding position on the second number is also with a symbol, we move the two symbols to the following positions in the answer as one symbol. When two symbols form, two corresponding positions move forward to the following position of the answer then the two symbols become one symbol. The ***audio*** representation for the three pictures consumes of a **knock, knock, tram, double knock, knock, knock, cling, knock, double knock, double knock, and knock.**

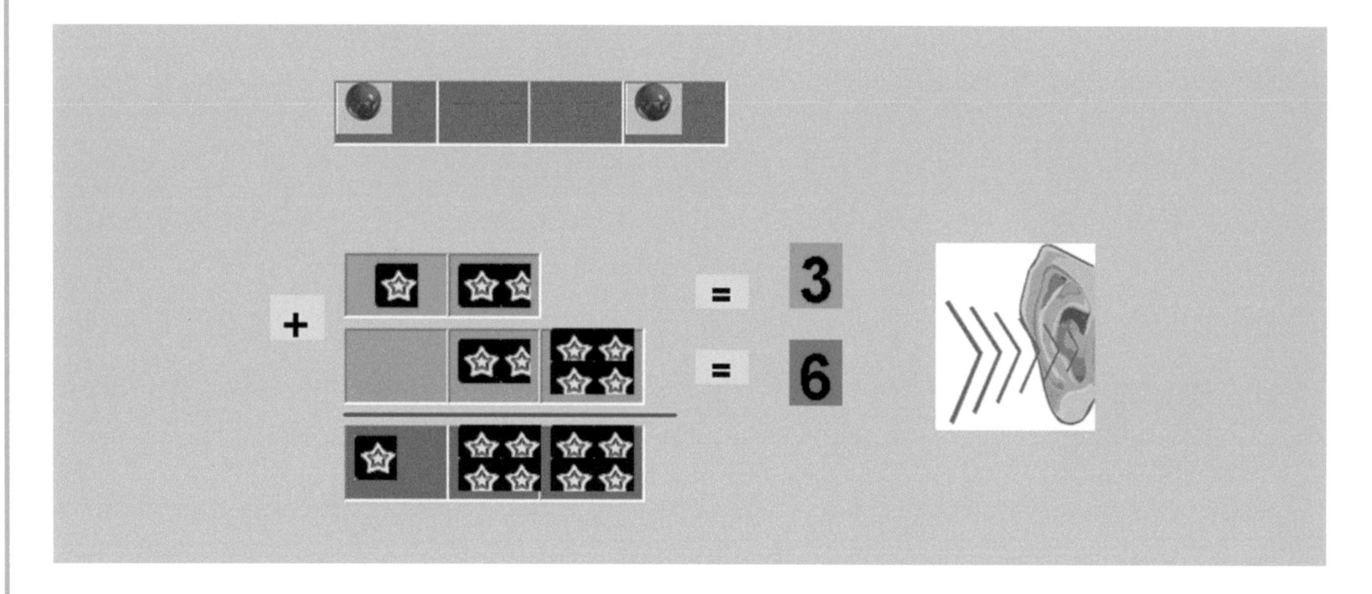

The ***kinesthetic*** representation of the filled up position is announced by clamping of the right hand. An empty position is represented by the clamping of the left hand. The addition sound, tram, is represented by one hand extended in front. The equal sound, cling, is represented by crossing of hands. The kinesthetic representation for both pictures and the answer is the clamping of the right hand twice, one hand extended in front, left hand, right hand twice, crossing of hands, right hand, left hand twice, and right hand.

ADDITION, EXERCISE 18, STEP 18.2

In exercise 18, step 18.2 in ***visual*** display of addition we see two pictures of boxes. The first picture has a symbol in all three positions. The second picture has the first and fourth positions filled with a symbol. For purpose of addition we are introducing the concept of moving. When a position passes its maximum capacity the next symbol is simultaneously moved in the ascending position. We are moving from the first to the second position. However the audio representation for the picture does not change because the numerical value does not change.

The ***audio*** representation for the three pictures consumes of a **knock, double knock, double knock, and knock.**

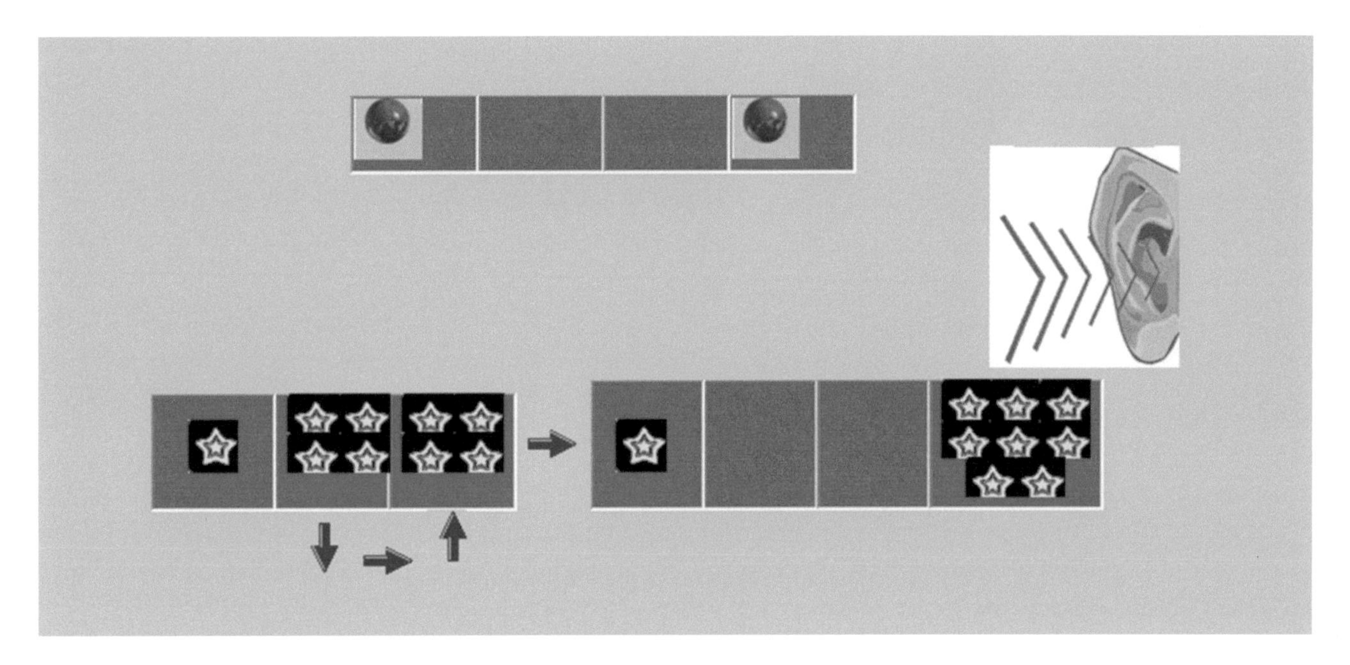

The ***kinesthetic*** representation of the filled up position is announced by clamping of the right hand. An empty position is represented by the clamping of the left hand. The kinesthetic representation for three pictures is the clamping of the right hand, left hand twice, and right hand.

MATHEMATICS ADDITION, EXERCISE 18, STEP 18.3

In exercise 18, step 18.3 in ***visual*** display of addition, we see two pictures of boxes with equal sign. Both pictures have the first and fourth positions filled with a symbol. The first picture has a yellow star in its positions but the second picture has a red ball in its positions. The form and shape of the symbols do not change mathematical value and the numerical value of both pictures is equal to nine.

The ***audio*** representation for this picture consumes of a **knock, double knock, double knock, knock, and cling.**

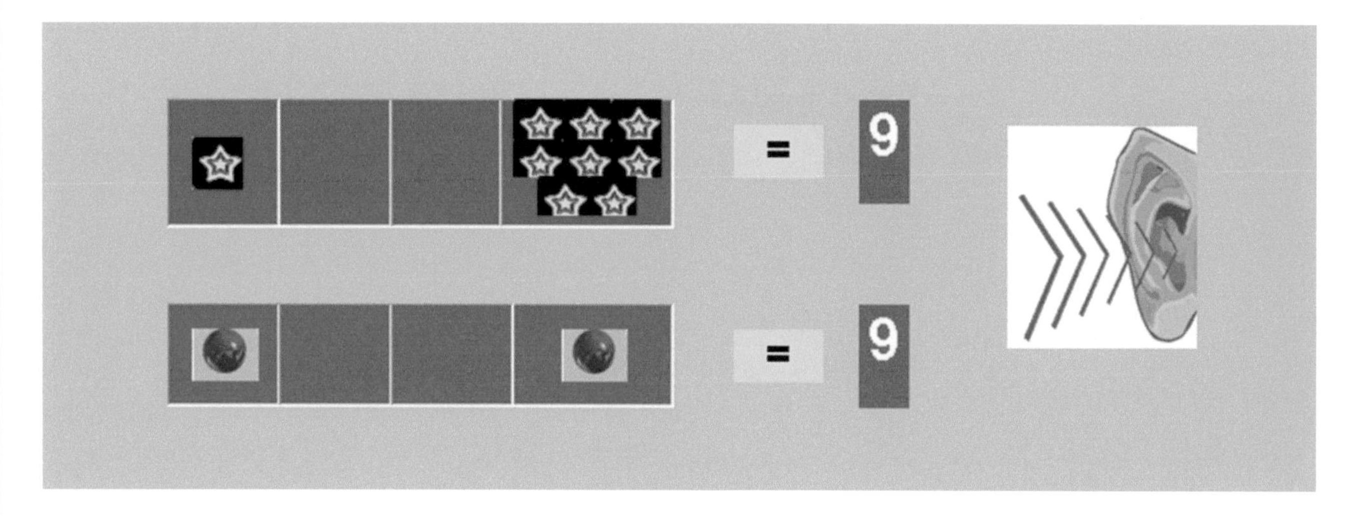

The ***kinesthetic*** representation of the filled up position is announced by clamping of the right hand. An empty position is represented by the clamping of the left hand. The equal sign is crossing both hands in front. The kinesthetic representation for this picture is clamping the right hand, left hand twice, right hand, and crossing of hands.

ADDITION, EXERCISE 18

In exercise 18, in ***visual*** display of addition we see two pictures of boxes. In the bottom we see three pictures with probable answers. We want to specify, that between the first and second pictures we see the addition sign. Bellow the second picture we see the equal line, which separates the problem from the answer. The ***audio*** representation for two pictures consumes of a **knock, knock, tram, double knock, knock, knock, and cling.**

From the probable answers for addition we need to find the right one. The audio signals are: 1) **double knock, knock, double knock, and knock.** 2) **knock, double knock, double knock, and knock** 3) **double knock, knock, double knock, and knock**.

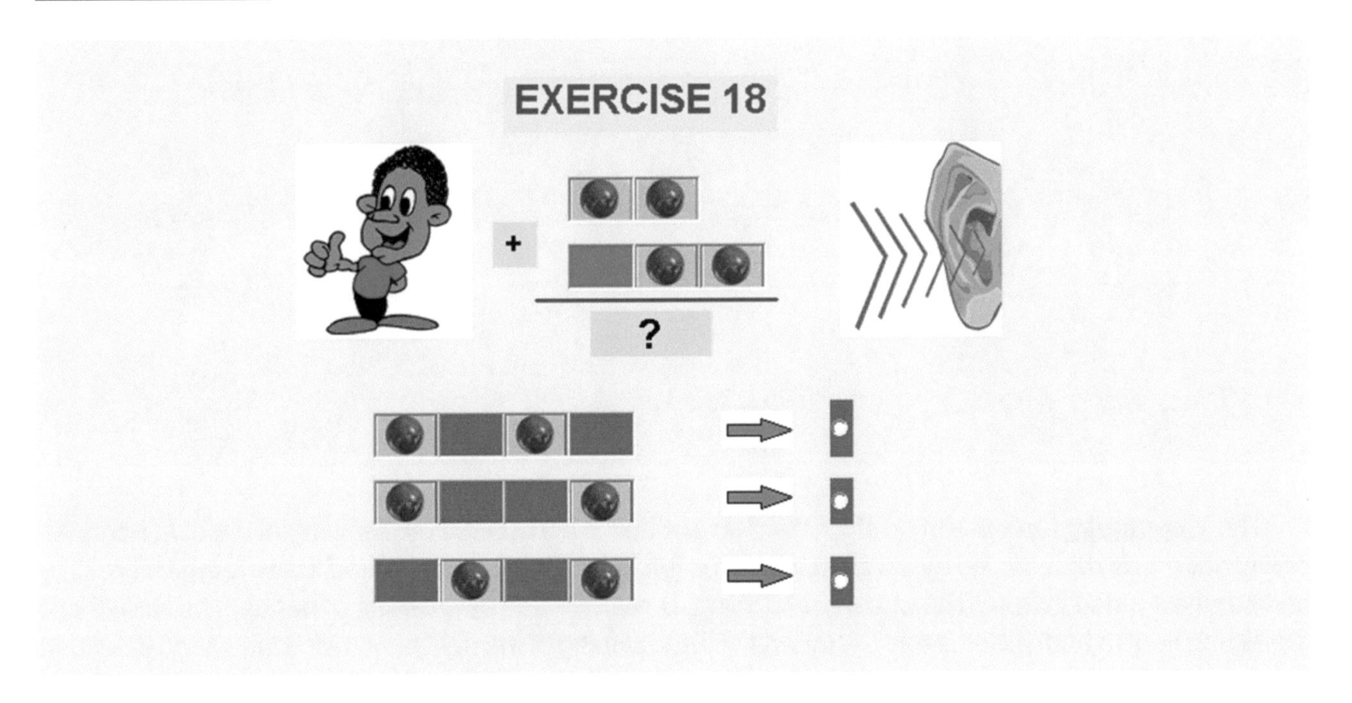

The ***kinesthetic*** representation of the filled up position is announced by clamping of the right hand. An empty position is represented by the clamping of the left hand. The addition sound, tram, is represented by one hand extended in front. The equal sound, cling, is represented by crossing of hands. The kinesthetic representation for both pictures and the answer is the clamping of the right hand twice, one hand extended in front, left hand, right hand twice, crossing of hands, right hand, left hand twice, and right hand.

MATHEMATICS ADDITION, EXERCISE 19

In exercise 19, in ***visual*** display of addition we see three pictures of boxes. We want to specify, that between the first and second pictures we see the addition sign. Below the second picture we see the equal line, which separates the problem from the answer. The ***audio*** representation for the three pictures consumes of a **knock, knock, tram, knock, knock, knock, cling, double knock, knock, double knock, and knock.**

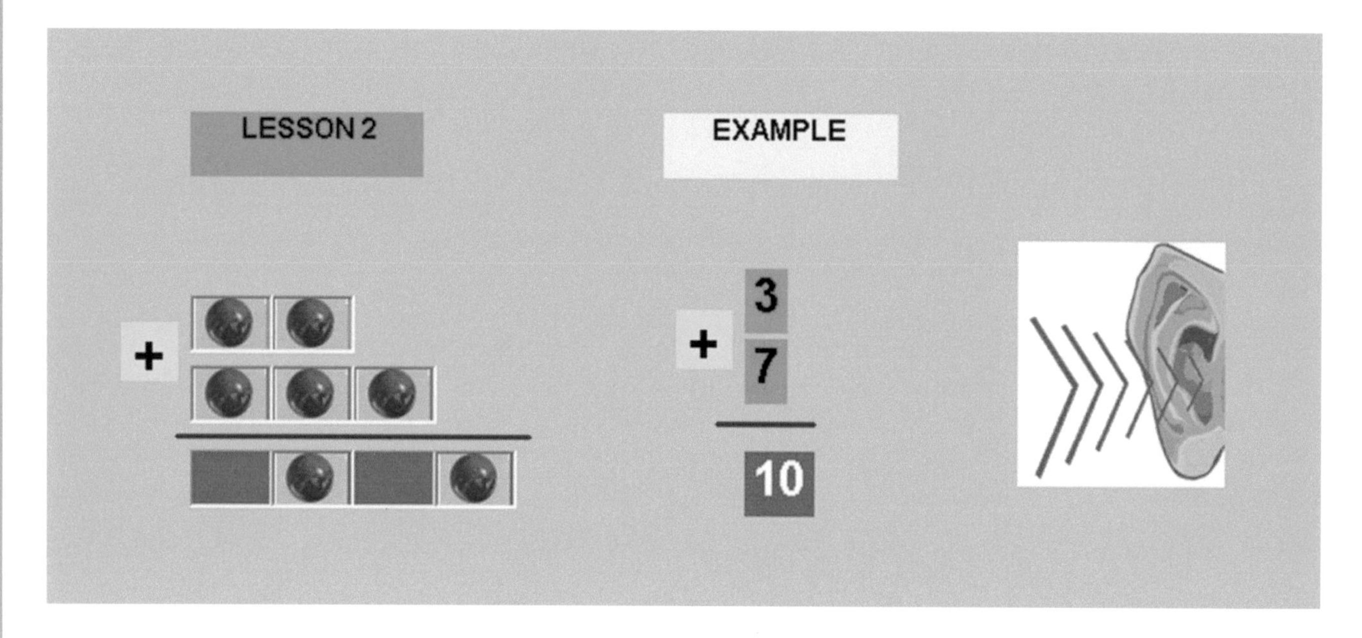

The ***kinesthetic*** representation of the filled up position is announced by clamping of the right hand. An empty position is represented by the clamping of the left hand. The addition sound, tram, is represented by one hand extended in front. The equal sound, cling, is represented by crossing of hands. The kinesthetic representation for both pictures and the answer is the clamping of the right hand twice, one hand extended in front, right hand thrice, crossing of hands, left hand, right hand, left hand, and right hand.

ADDITION, EXERCISE 19, STEP 19.1

In exercise 19, in ***visual*** display of addition we see three pictures of boxes. We want to specify, that between the first and second pictures we see the addition sign. Below the second picture we see the equal line, which separates the problem from the answer. When we are adding two numbers and the position on the first number is filled with a symbol and the corresponding position on the second number is also with a symbol, we move the two symbols to the following positions in the answer as one symbol. When two symbols form, two corresponding positions move forward to the following position of the answer then the two symbols become one symbol. The ***audio*** representation for the three pictures consumes of a **knock, knock, tram, knock, knock, knock, cling, double knock, knock, double knock, and knock.**

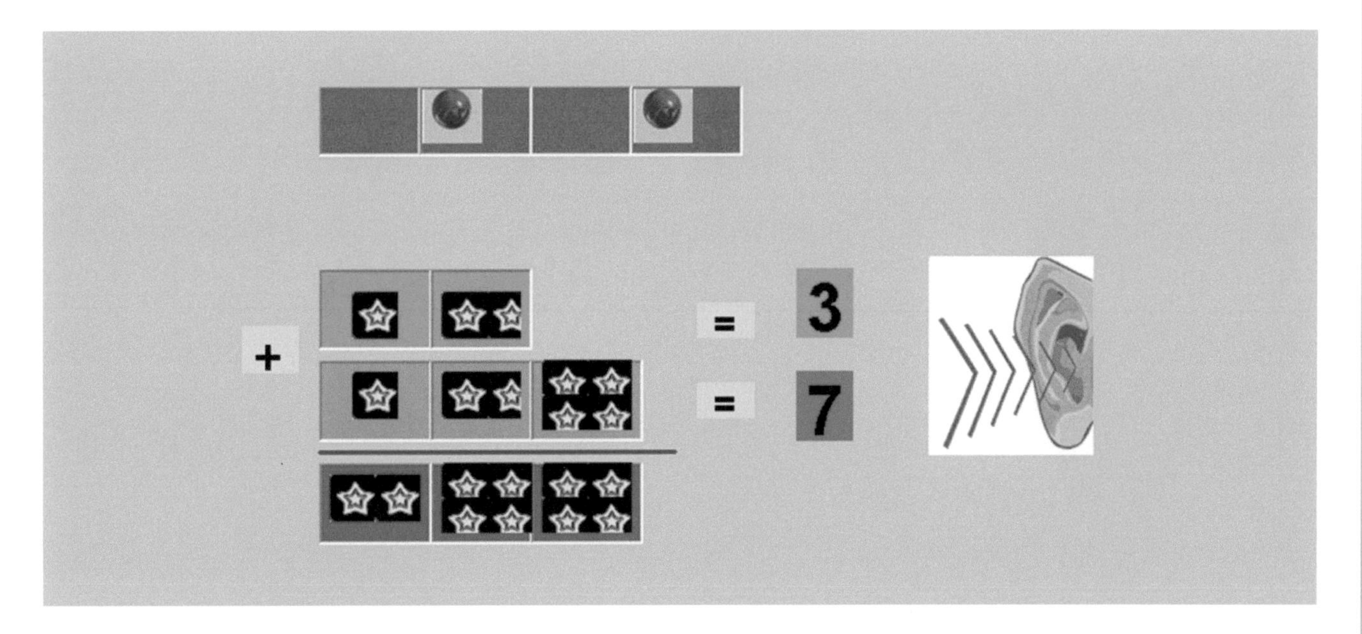

The ***kinesthetic*** representation of the filled up position is announced by clamping of the right hand. An empty position is represented by the clamping of the left hand. The addition sound, tram, is represented by one hand extended in front. The equal sound, cling, is represented by crossing of hands. The kinesthetic representation for both pictures and the answer is the clamping of the right hand twice, one hand extended in front, right hand thrice, crossing of hands, left hand, right hand, left hand, and right hand.

MATHEMATICS ADDITION, EXERCISE 19, STEP 19.2

In exercise 19, step 19.2 in ***visual*** display of addition we see two pictures of boxes. The first picture has a symbol in all three positions. The second picture has the first and fourth positions filled with a symbol. For purpose of addition we are introducing the concept of moving. When a position passes its maximum capacity the next symbol is simultaneously moved in the ascending position. We are moving from the first to the second position. However the audio representation for the picture does not change because the numerical value does not change.

The ***audio*** representation for the three pictures consumes of a **double knock, knock, double knock, and knock.**

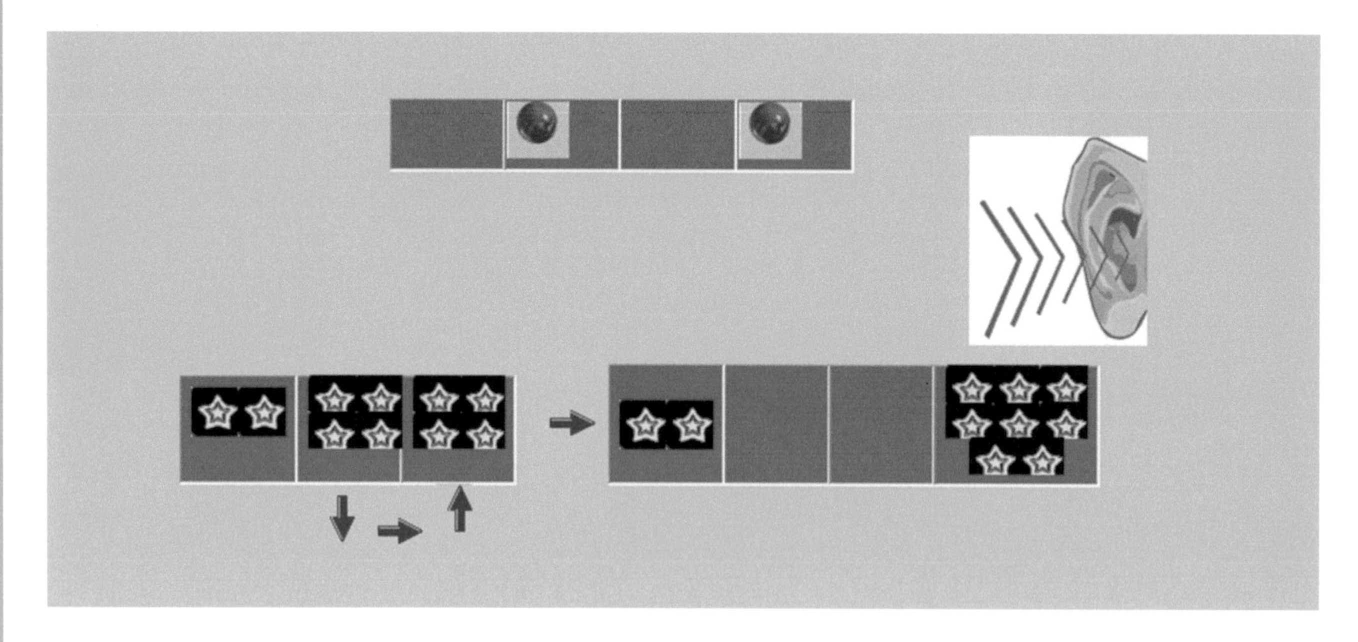

The ***kinesthetic*** representation of the filled up position is announced by clamping of the right hand. An empty position is represented by the clamping of the left hand. The kinesthetic representation for three pictures is the clamping of the left hand, right hand, left hand, and right hand.

ADDITION, EXERCISE 19, STEP 19.3

In exercise 19, step 19.3 in ***visual*** display of addition we see two pictures of boxes. The first picture has a symbol in the first and fourth positions. The second picture has the second and fourth positions filled with a symbol. For purpose of addition we are introducing the concept of moving. When a position passes its maximum capacity the next symbol is simultaneously moved in the ascending position. We are moving from the first to the second position. However the audio representation for the picture does not change because the numerical value does not change.

The ***audio*** representation for the three pictures consumes of a **double knock, knock, double knock, and knock.**

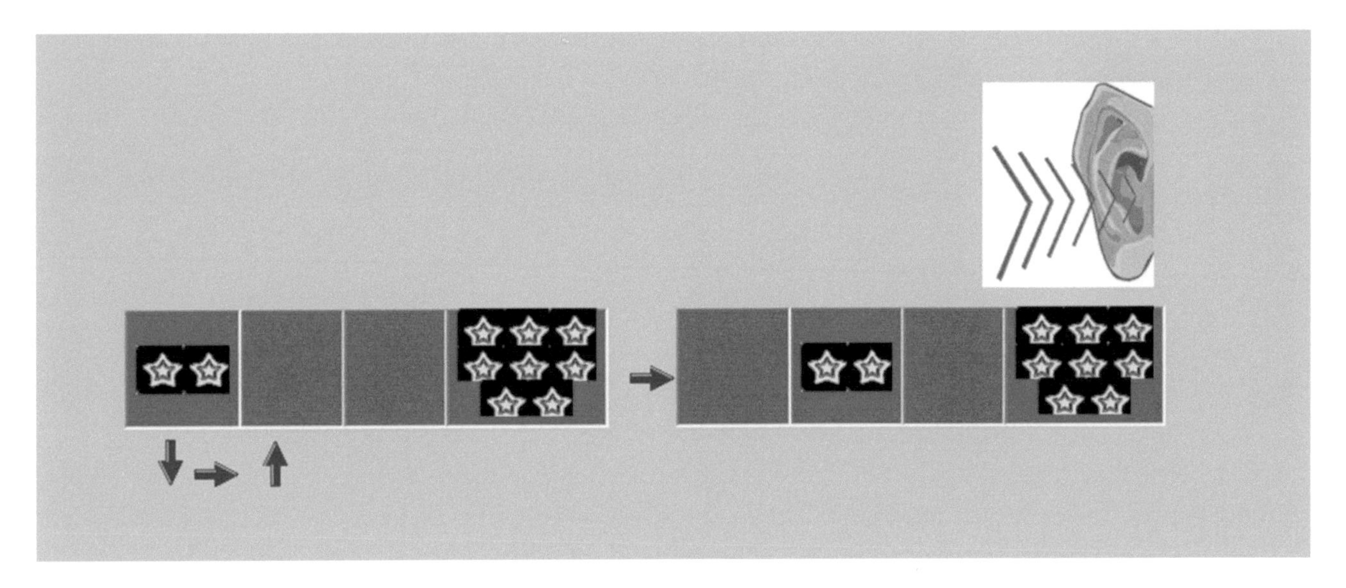

The ***kinesthetic*** representation of the filled up position is announced by clamping of the right hand. An empty position is represented by the clamping of the left hand. The kinesthetic representation for three pictures is the clamping of the left hand, right hand, left hand, and right hand.

MATHEMATICS ADDITION, EXERCISE 19, STEP 19.4

In exercise 19, step 19.4 in ***visual*** display of addition, we see two pictures of boxes with equal sign. Both pictures have the second and fourth positions filled with a symbol. The first picture has a yellow star in its positions but the second picture has a red ball in its positions. The form and shape of the symbols do not change mathematical value and the numerical value of both pictures is equal to ten.

The ***audio*** representation for this picture consumes of a **double knock, knock, double knock, knock, and cling.**

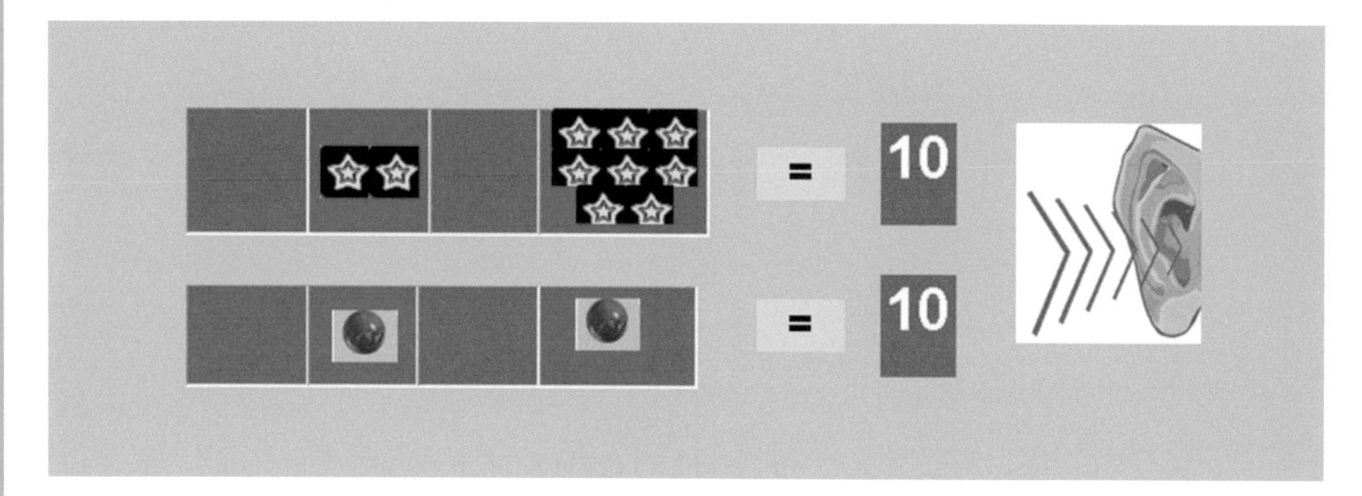

The ***kinesthetic*** representation of the filled up position is announced by clamping of the right hand. An empty position is represented by the clamping of the left hand. The equal sign is crossing both hands in front. The kinesthetic representation for this picture is clamping the left hand, right hand, left hand, right hand, and crossing of hands.

ADDITION, EXERCISE 19

In exercise 19, in ***visual*** display of addition we see two pictures of boxes. In the bottom we see three pictures with probable answers. We want to specify, that between the first and second pictures we see the addition sign. Bellow the second picture we see the equal line, which separates the problem from the answer. The ***audio*** representation for two pictures consumes of a **knock, knock, tram, knock, knock, knock, and cling.**

From the probable answers for addition we need to find the right one. The audio signals are: 1) **knock, double knock, knock, and double knock.** 2) **knock, double knock, double knock, and knock** 3) **double knock, knock, double knock, and knock**.

The ***kinesthetic*** representation of the filled up position is announced by clamping of the right hand. An empty position is represented by the clamping of the left hand. The addition sound, tram, is represented by one hand extended in front. The equal sound, cling, is represented by crossing of hands. The kinesthetic representation for both pictures and the answer is the clamping of the right hand twice, one hand extended in front, right hand thrice, crossing of hands, left hand, right hand, left hand, and right hand.

MATHEMATICS ADDITION, EXERCISE 20

In exercise 20, in ***visual*** display of addition we see three pictures of boxes. We want to specify, that between the first and second pictures we see the addition sign. Below the second picture we see the equal line, which separates the problem from the answer. The ***audio*** representation for the three pictures consumes of a **double knock, double knock, knock, tram, double knock, double knock, double knock, cling, double knock, double knock, and knock.**

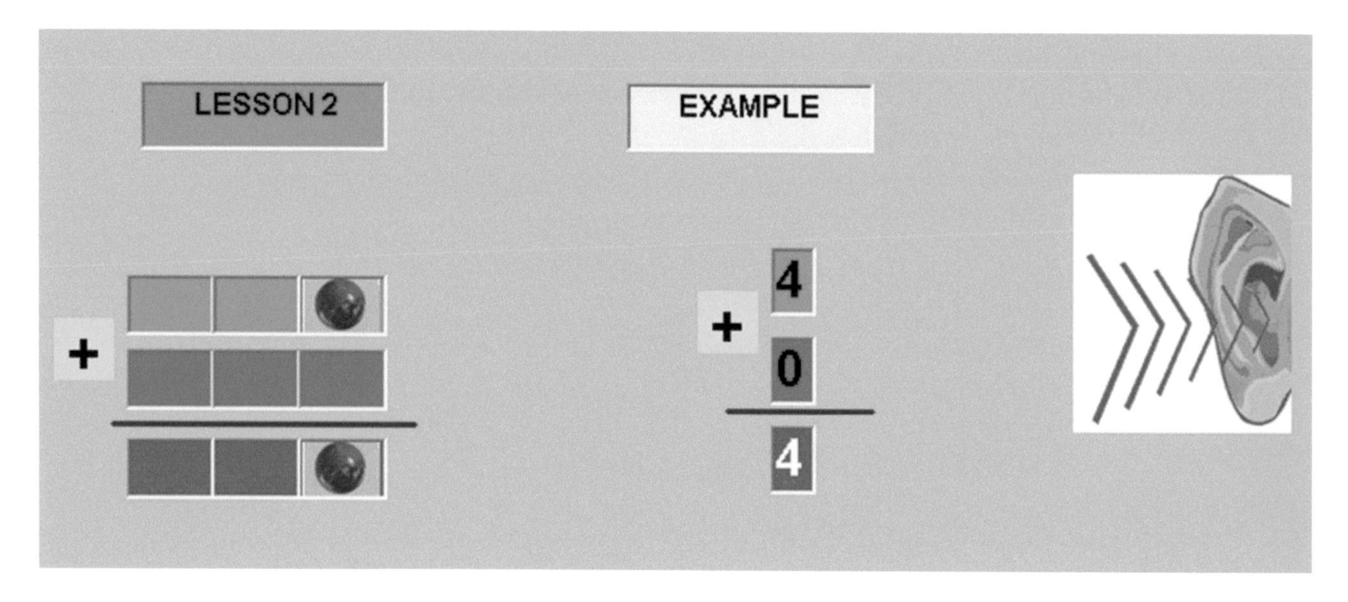

The ***kinesthetic*** representation of the filled up position is announced by clamping of the right hand. An empty position is represented by the clamping of the left hand. The addition sound, tram, is represented by one hand extended in front. The equal sound, cling, is represented by crossing of hands. The kinesthetic representation for both pictures and the answer is the clamping of the left hand twice, right hand, one hand extended in front, left hand thrice, crossing of hands, left hand twice, and right hand.

ADDITION, EXERCISE 21

In exercise 21, in ***visual*** display of addition we see three pictures of boxes. We want to specify, that between the first and second pictures we see the addition sign. Below the second picture we see the equal line, which separates the problem from the answer. The ***audio*** representation for the three pictures consumes of a **double knock, double knock, knock, tram, knock, double knock, double knock, cling, knock, double knock, and knock.**

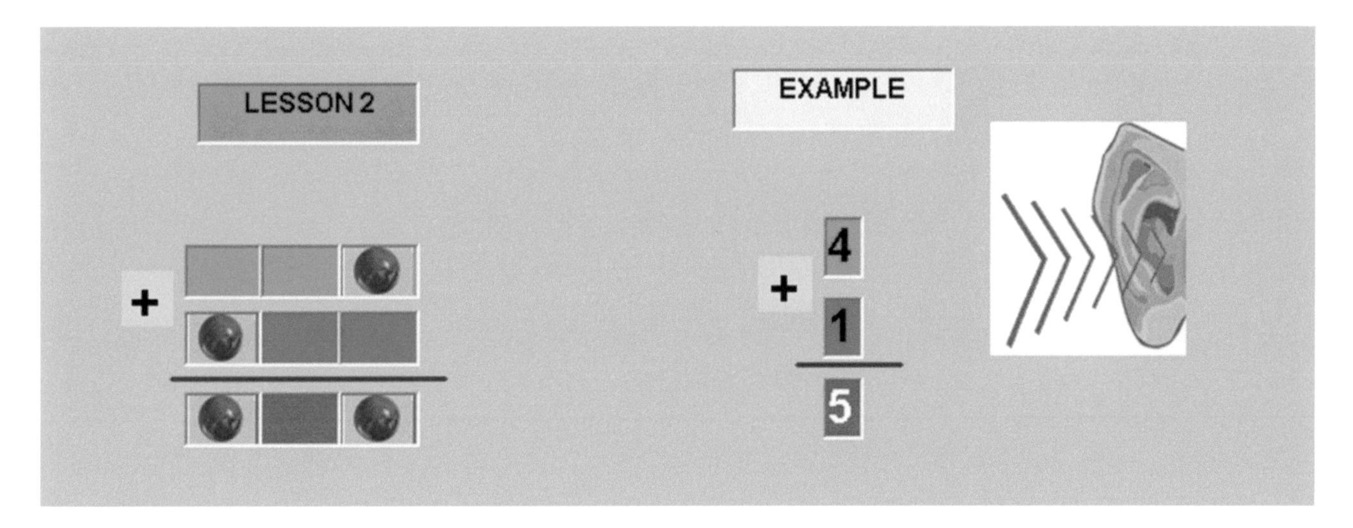

The ***kinesthetic*** representation of the filled up position is announced by clamping of the right hand. An empty position is represented by the clamping of the left hand. The addition sound, tram, is represented by one hand extended in front. The equal sound, cling, is represented by crossing of hands. The kinesthetic representation for both pictures and the answer is the clamping of the left hand twice, right hand, one hand extended in front, right hand, left hand twice, crossing of hands, right hand, left hand, and right hand.

MATHEMATICS ADDITION, EXERCISE 21

In exercise 21, in ***visual*** display of addition we see two pictures of boxes. In the bottom we see three pictures with probable answers. We want to specify, that between the first and second pictures we see the addition sign. Bellow the second picture we see the equal line, which separates the problem from the answer. The ***audio*** representation for two pictures consumes of a **double knock, double knock, knock, tram, knock, double knock, double knock, and cling.**

From the probable answers for addition we need to find the right one. The audio signals are: 1) **knock, double knock, and knock.** 2) **knock, knock, and double knock** 3) **double knock, knock, and knock**.

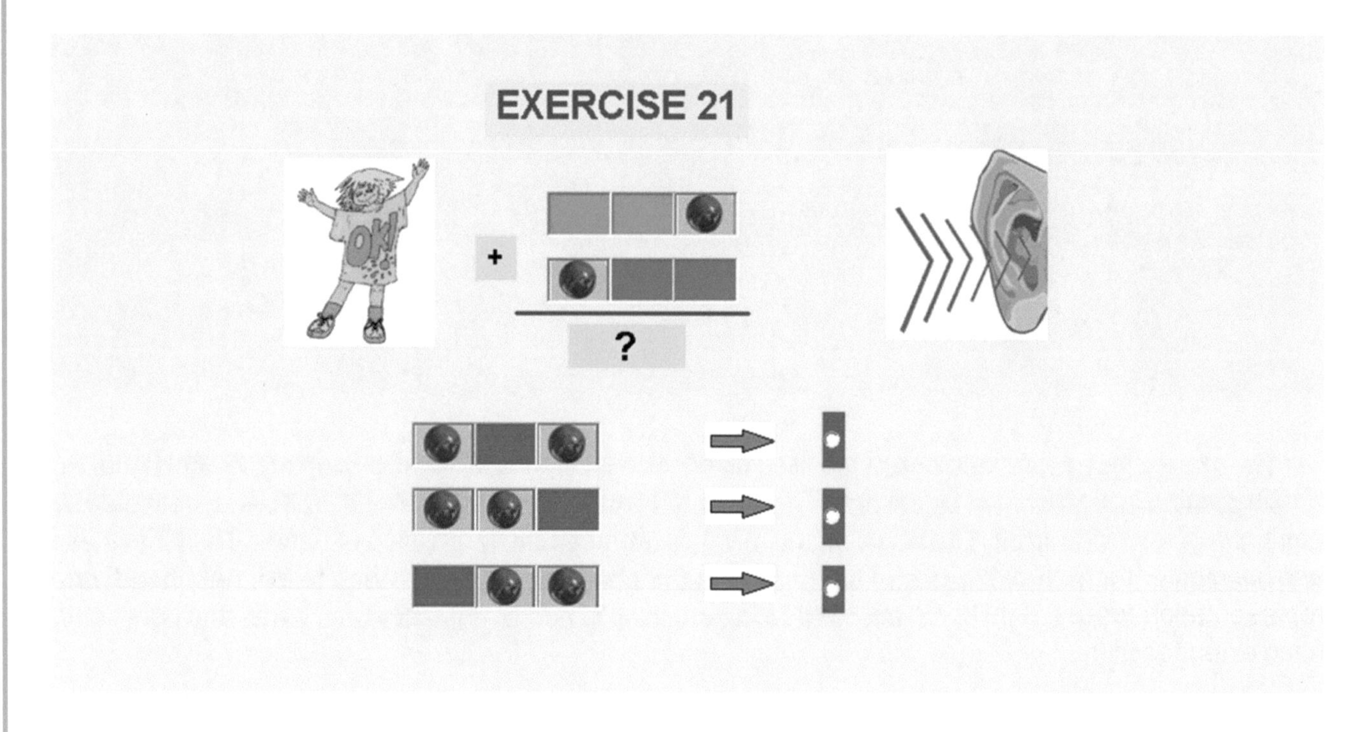

The ***kinesthetic*** representation of the filled up position is announced by clamping of the right hand. An empty position is represented by the clamping of the left hand. The addition sound, tram, is represented by one hand extended in front. The equal sound, cling, is represented by crossing of hands. The kinesthetic representation for both pictures and the answer is the clamping of the left hand twice, right hand, one hand extended in front, right hand, left hand twice, crossing of hands, right hand, left hand, and right hand.

ADDITION, EXERCISE 22

In exercise 22, in ***visual*** display of addition we see three pictures of boxes. We want to specify, that between the first and second pictures we see the addition sign. Below the second picture we see the equal line, which separates the problem from the answer. The ***audio*** representation for the three pictures consumes of a **double knock, double knock, knock, tram, double knock, knock, double knock, and cling.**

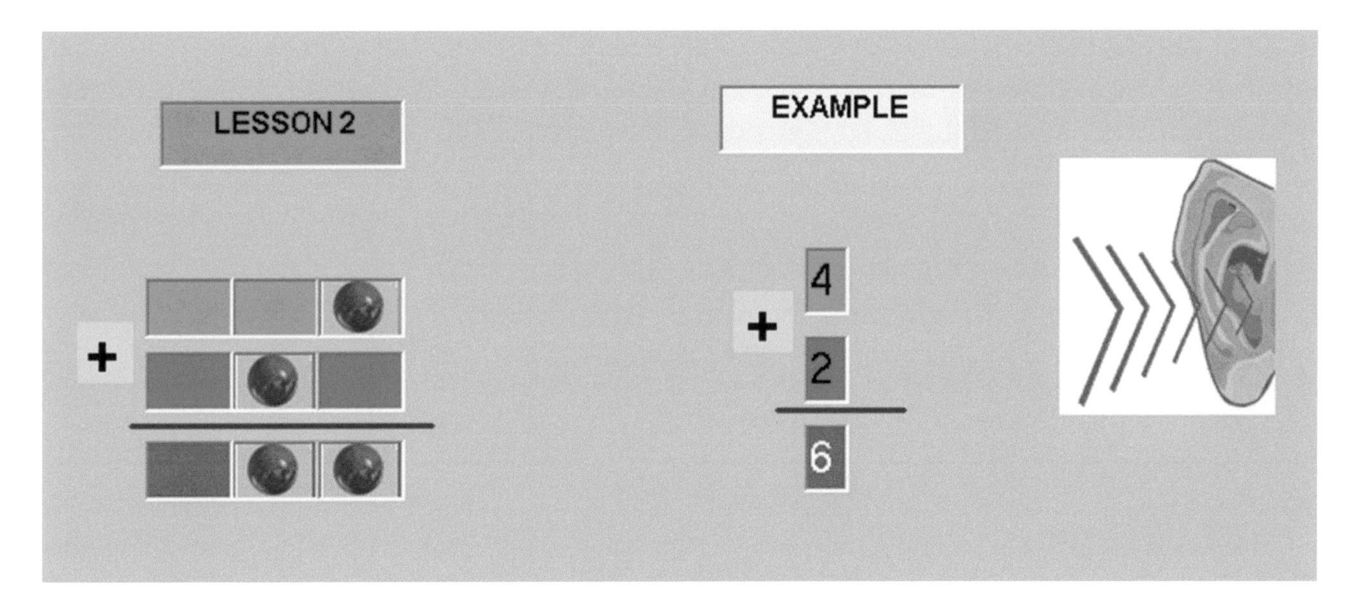

The ***kinesthetic*** representation of the filled up position is announced by clamping of the right hand. An empty position is represented by the clamping of the left hand. The addition sound, tram, is represented by one hand extended in front. The equal sound, cling, is represented by crossing of hands. The kinesthetic representation for both pictures and the answer is the clamping of the left hand twice, right hand, one hand extended in front, left hand, right hand, left hand, crossing of hands, left hand, and right hand twice.

MATHEMATICS

ADDITION, EXERCISE 22, STEP 22.1

In exercise 22, step 22.1 in ***visual*** display of addition we see three pictures of boxes. We want to specify, that between the first and second pictures we see the addition sign. Below the second picture we see the equal line, which separates the problem from the answer. When we are adding two numbers and the position on the first number is filled with a symbol and the corresponding position on the second number is also with a symbol, we move the two symbols to the following positions in the answer as one symbol. When two symbols form, two corresponding positions move forward to the following position of the answer then the two symbols become one symbol. The ***audio*** representation for the three pictures consumes of a **double knock, double knock, knock, tram, double knock, knock, double knock, and cling.**

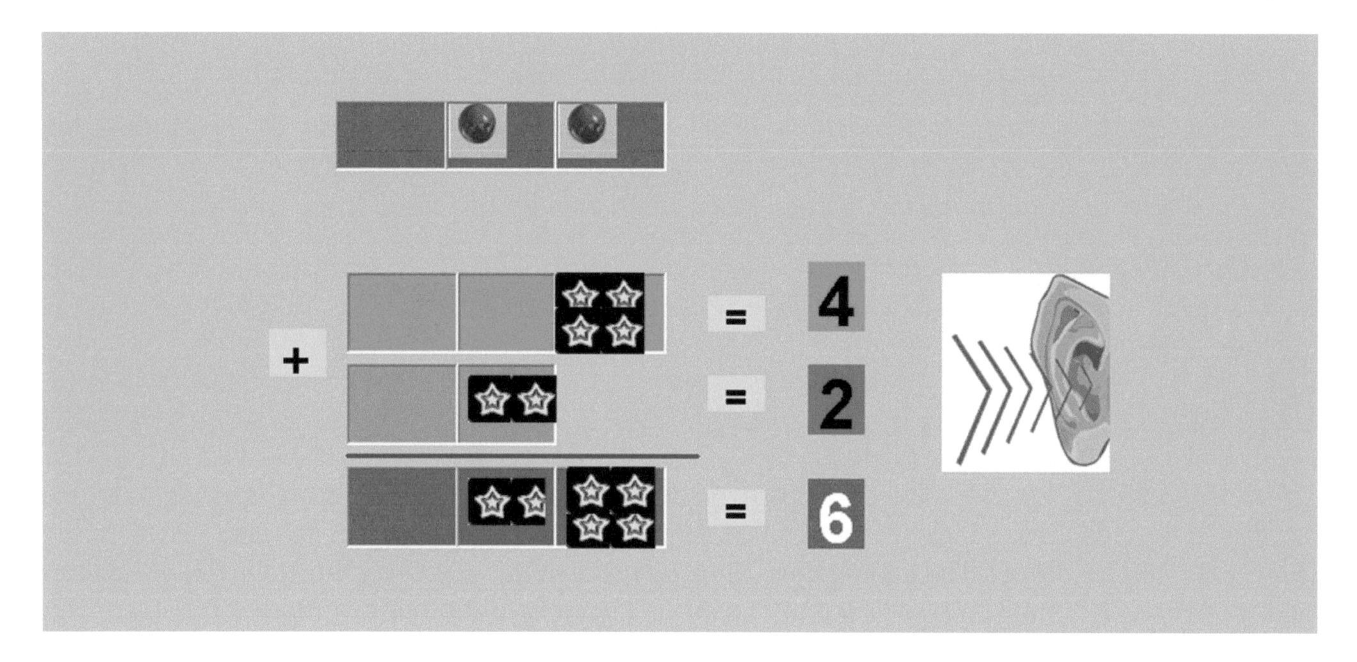

The ***kinesthetic*** representation of the filled up position is announced by clamping of the right hand. An empty position is represented by the clamping of the left hand. The addition sound, tram, is represented by one hand extended in front. The equal sound, cling, is represented by crossing of hands. The kinesthetic representation for both pictures and the answer is the clamping of the left hand twice, right hand, one hand extended in front, left hand, right hand, left hand, crossing of hands, left hand, and right hand twice.

ADDITION, EXERCISE 22, STEP 22.2

In exercise 22, step 22.2 in ***visual*** display of addition, we see two pictures of boxes with equal sign. Both pictures have the second and third positions filled with a symbol and the first position is empty. The first picture has a yellow star in its positions but the second picture has a red ball in its positions. The form and shape of the symbols do not change mathematical value and the numerical value of both pictures is equal to six.

The ***audio*** representation for this picture consumes of a **double knock, knock, knock, and cling.**

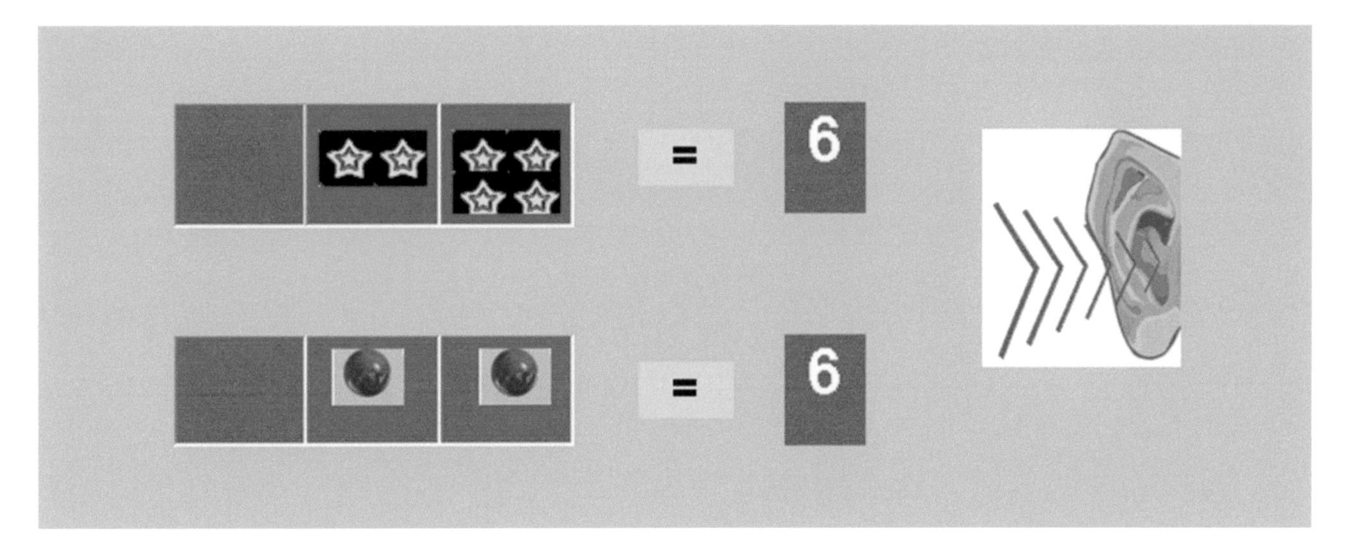

The ***kinesthetic*** representation of the filled up position is announced by clamping of the right hand. An empty position is represented by the clamping of the left hand. The equal sign is crossing both hands in front. The kinesthetic representation for this picture is clamping the left hand, right hand twice, and crossing of hands.

MATHEMATICS

ADDITION, EXERCISE 22

In exercise 22, in ***visual*** display of addition we see two pictures of boxes. In the bottom we see three pictures with probable answers. We want to specify, that between the first and second pictures we see the addition sign. Bellow the second picture we see the equal line, which separates the problem from the answer. The ***audio*** representation for two pictures consumes of a **double knock, double knock, knock, tram, double knock, knock, double knock, and cling.**

From the probable answers for addition we need to find the right one. The audio signals are: 1) **knock, knock, and double knock.** 2) **double knock, knock, and knock** 3) **knock, double knock, and knock**.

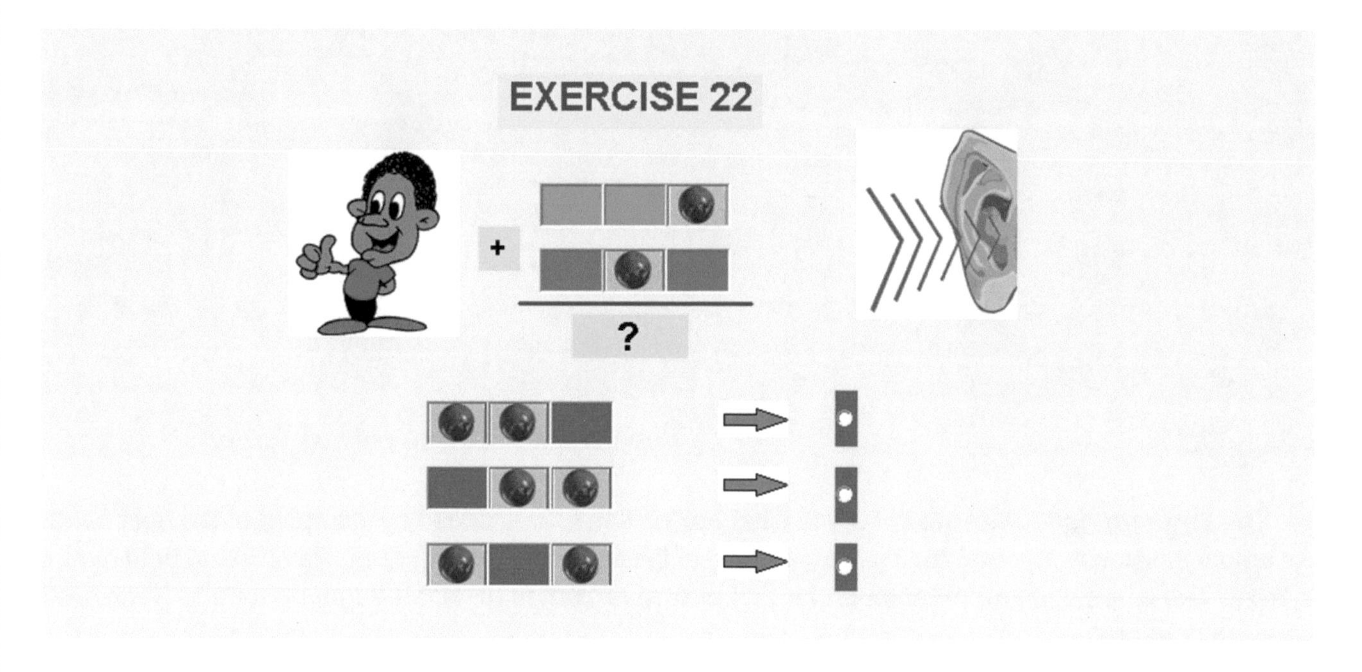

The ***kinesthetic*** representation of the filled up position is announced by clamping of the right hand. An empty position is represented by the clamping of the left hand. The addition sound, tram, is represented by one hand extended in front. The equal sound, cling, is represented by crossing of hands. The kinesthetic representation for both pictures and the answer is the clamping of the left hand twice, right hand, one hand extended in front, left hand, right hand, left hand, crossing of hands, left hand, and right hand twice.

ADDITION, EXERCISE 23

In exercise 23, in ***visual*** display of addition we see three pictures of boxes. We want to specify, that between the first and second pictures we see the addition sign. Below the second picture we see the equal line, which separates the problem from the answer. The ***audio*** representation for the three pictures consumes of a **double knock, double knock, knock, tram, knock, knock, cling, knock, knock, and knock.**

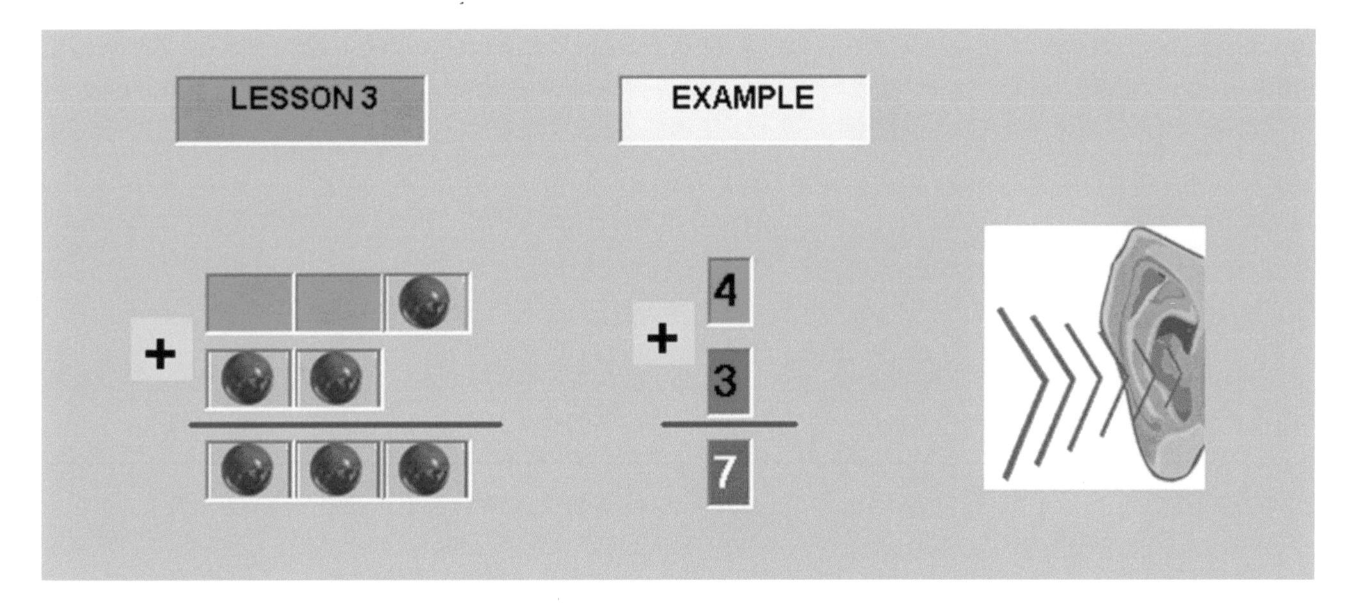

The ***kinesthetic*** representation of the filled up position is announced by clamping of the right hand. An empty position is represented by the clamping of the left hand. The addition sound, tram, is represented by one hand extended in front. The equal sound, cling, is represented by crossing of hands. The kinesthetic representation for both pictures and the answer is the clamping of the left hand twice, right hand, one hand extended in front, right hand twice, crossing of hands, and right hand thrice.

MATHEMATICS

ADDITION, EXERCISE 23, STEP 23.1

In exercise 23, step 23.1 in ***visual*** display of addition we see three pictures of boxes. We want to specify, that between the first and second pictures we see the addition sign. Below the second picture we see the equal line, which separates the problem from the answer. When we are adding two numbers and the position on the first number is filled with a symbol and the corresponding position on the second number is also with a symbol, we move the two symbols to the following positions in the answer as one symbol. When two symbols form, two corresponding positions move forward to the following position of the answer then the two symbols become one symbol. The ***audio*** representation for the three pictures consumes of a **double knock, double knock, knock, tram, knock, knock, cling, knock, knock, and knock.**

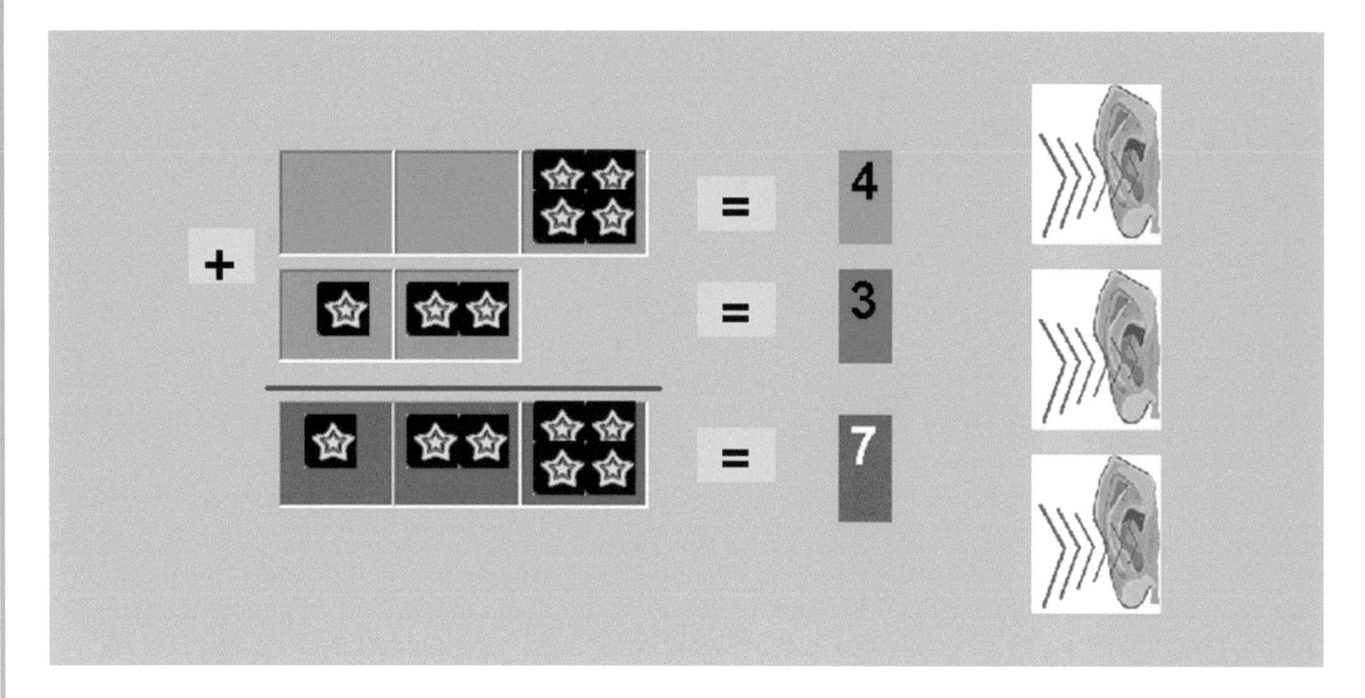

The ***kinesthetic*** representation of the filled up position is announced by clamping of the right hand. An empty position is represented by the clamping of the left hand. The addition sound, tram, is represented by one hand extended in front. The equal sound, cling, is represented by crossing of hands. The kinesthetic representation for both pictures and the answer is the clamping of the left hand twice, right hand, one hand extended in front, right hand twice, crossing of hands, and right hand thrice.

ADDITION, EXERCISE 17, STEP 17.2

In exercise 17, step 17.2 in ***visual*** display of addition, we see two pictures of boxes with equal sign. Both pictures have all three of their positions filled with a symbol. The first picture has a yellow star in its positions but the second picture has a red ball in its positions. The form and shape of the symbols do not change mathematical value and the numerical value of both pictures is equal to seven.

The ***audio*** representation for this picture consumes of a **knock, knock, knock, and cling.**

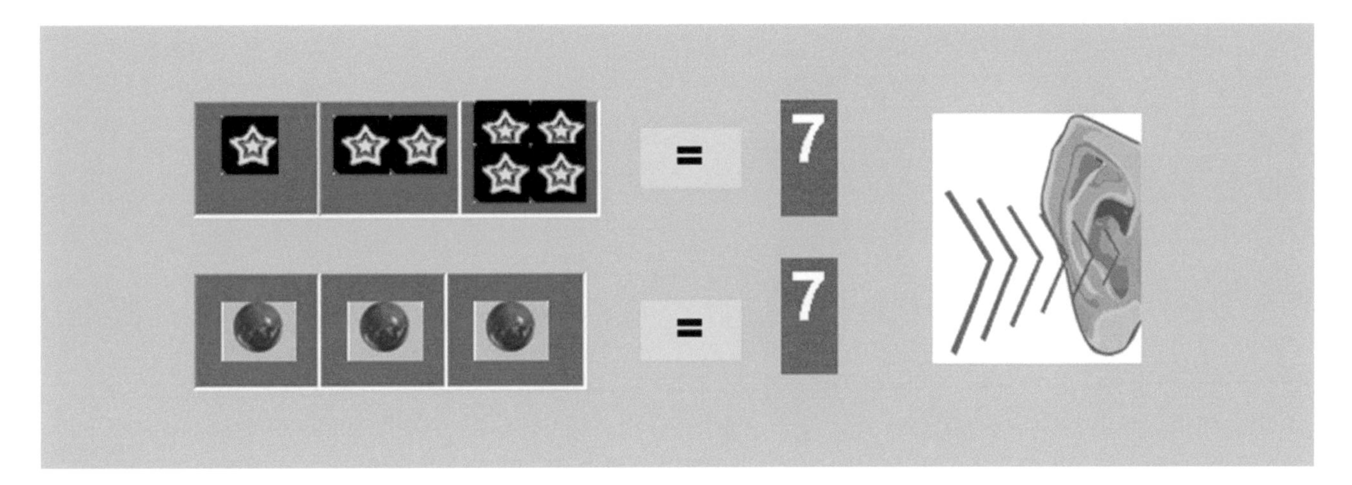

The ***kinesthetic*** representation of the filled up position is announced by clamping of the right hand. An empty position is represented by the clamping of the left hand. The equal sign is crossing both hands in front. The kinesthetic representation for this picture is clamping the right hand thrice and crossing of hands.

MATHEMATICS

ADDITION, EXERCISE 23

In exercise 23, in ***visual*** display of addition we see two pictures of boxes. In the bottom we see three pictures with probable answers. We want to specify, that between the first and second pictures we see the addition sign. Bellow the second picture we see the equal line, which separates the problem from the answer. The ***audio*** representation for two pictures consumes of a **double knock, double knock, knock, tram, knock, knock, and cling.**

From the probable answers for addition we need to find the right one. The audio signals are: 1) **knock, knock, and knock.** 2) **double knock, knock, and knock** 3) **knock, double knock, and knock**.

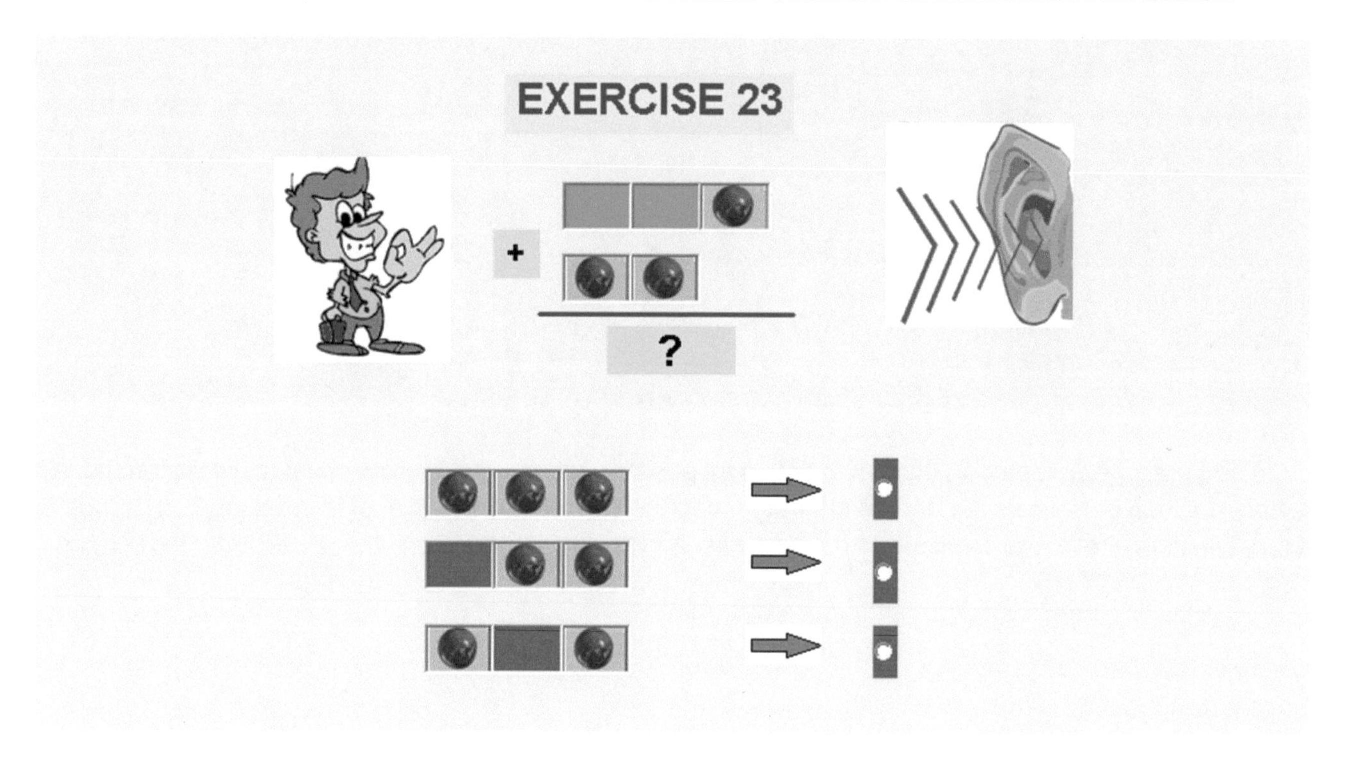

The ***kinesthetic*** representation of the filled up position is announced by clamping of the right hand. An empty position is represented by the clamping of the left hand. The addition sound, tram, is represented by one hand extended in front. The equal sound, cling, is represented by crossing of hands. The kinesthetic representation for both pictures and the answer is the clamping of the left hand twice, right hand, one hand extended in front, right hand twice, crossing of hands, and right hand thrice.

ADDITION, EXERCISE 24

In exercise 24, in ***visual*** display of addition we see three pictures of boxes. We want to specify, that between the first and second pictures we see the addition sign. Below the second picture we see the equal line, which separates the problem from the answer. The ***audio*** representation for the three pictures consumes of a **double knock, double knock, knock, tram, double knock, double knock, knock, cling, double knock, double knock, double knock, and knock.**

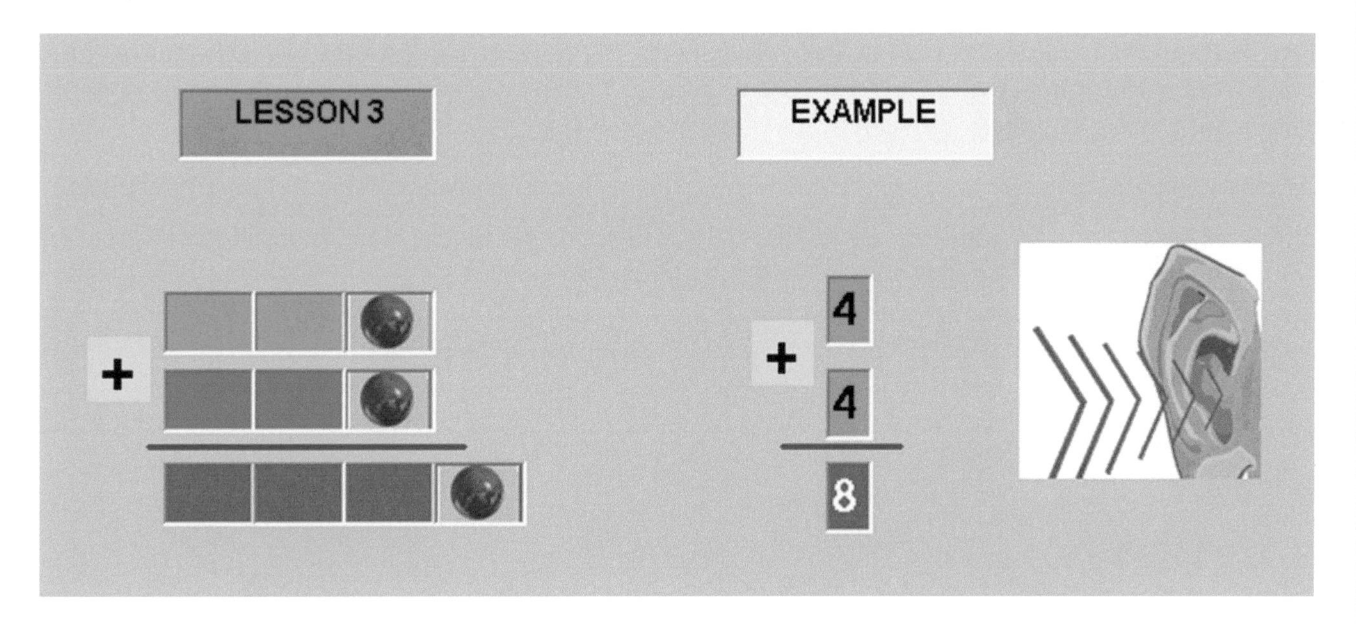

The ***kinesthetic*** representation of the filled up position is announced by clamping of the right hand. An empty position is represented by the clamping of the left hand. The addition sound, tram, is represented by one hand extended in front. The equal sound, cling, is represented by crossing of hands. The kinesthetic representation for both pictures and the answer is the clamping of the left hand twice, right hand, one hand extended in front, left hand twice, right hand, crossing of hands, left hand thrice, and right hand.

MATHEMATICS ADDITION, EXERCISE 24, STEP 24.1

In exercise 24, step 24.1 in ***visual*** display of addition we see three pictures of boxes. We want to specify, that between the first and second pictures we see the addition sign. Below the second picture we see the equal line, which separates the problem from the answer. When we are adding two numbers and the position on the first number is filled with a symbol and the corresponding position on the second number is also with a symbol, we move the two symbols to the following positions in the answer as one symbol. When two symbols form, two corresponding positions move forward to the following position of the answer then the two symbols become one symbol. The ***audio*** representation for the three pictures consumes of a **double knock, double knock, knock, tram, double knock, double knock, knock, cling, double knock, double knock, double knock, and knock.**

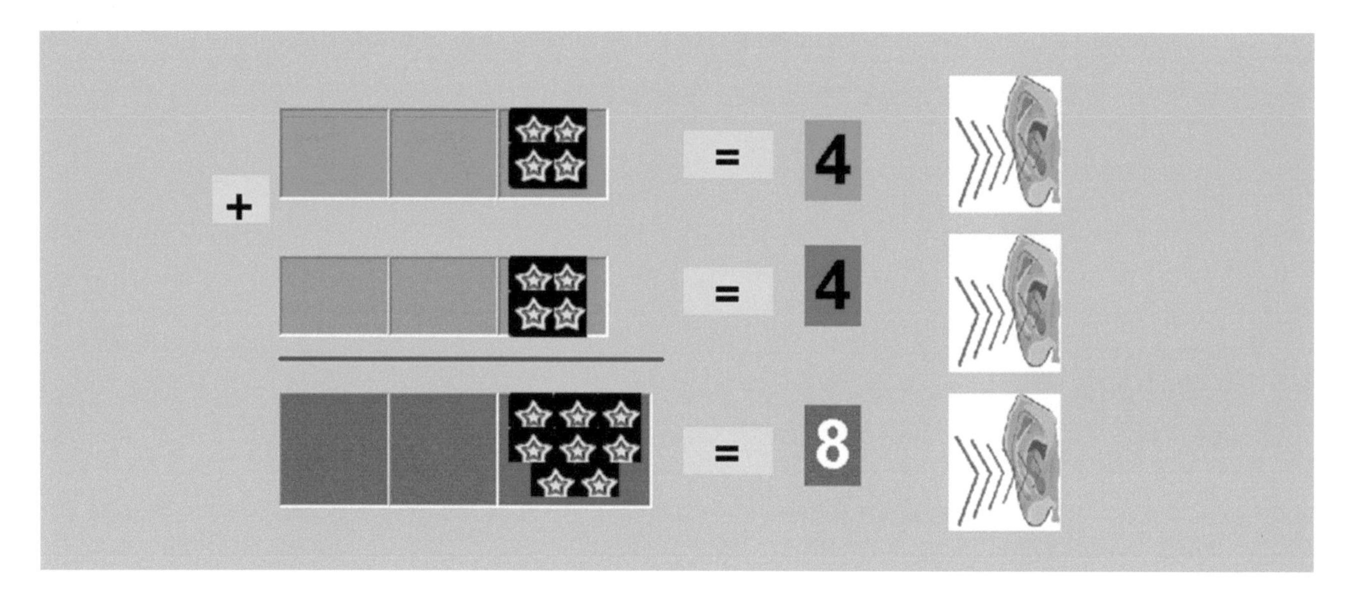

The ***kinesthetic*** representation of the filled up position is announced by clamping of the right hand. An empty position is represented by the clamping of the left hand. The addition sound, tram, is represented by one hand extended in front. The equal sound, cling, is represented by crossing of hands. The kinesthetic representation for both pictures and the answer is the clamping of the left hand twice, right hand, one hand extended in front, left hand twice, right hand, crossing of hands, left hand thrice, and right hand.

ADDITION, EXERCISE 24, STEP 24.2

In exercise 24, step 24.2 in ***visual*** display of addition we see two pictures of boxes. The first picture has a symbol in the third position. The second picture has the fourth position filled with a symbol. For purpose of addition we are introducing the concept of moving. When a position passes its maximum capacity the next symbol is simultaneously moved in the ascending position. We are moving from the first to the second position. However the audio representation for the picture does not change because the numerical value does not change.

The ***audio*** representation for the three pictures consumes of a **double knock, double knock, double knock, and knock.**

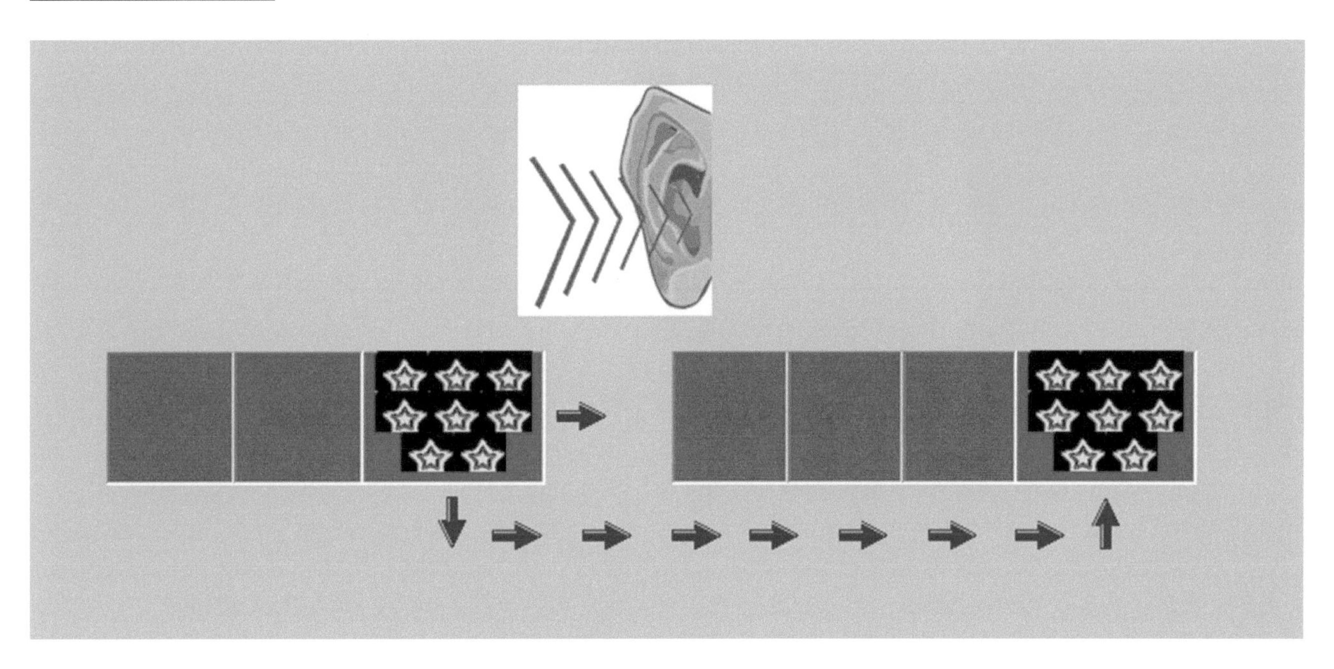

The ***kinesthetic*** representation of the filled up position is announced by clamping of the right hand. An empty position is represented by the clamping of the left hand. The kinesthetic representation for three pictures is the clamping of the left hand thrice and right hand.

MATHEMATICS ADDITION, EXERCISE 24, STEP 24.3

In exercise 24, step 24.3 in ***visual*** display of addition, we see two pictures of boxes with equal sign. Both pictures have the first three of their positions empty and the last position is filled with a symbol. The first picture has a yellow star in its positions but the second picture has a red ball in its positions. The form and shape of the symbols do not change mathematical value and the numerical value of both pictures is equal to eight.

The ***audio*** representation for this picture consumes of a **double knock, double knock, double knock, knock, and cling.**

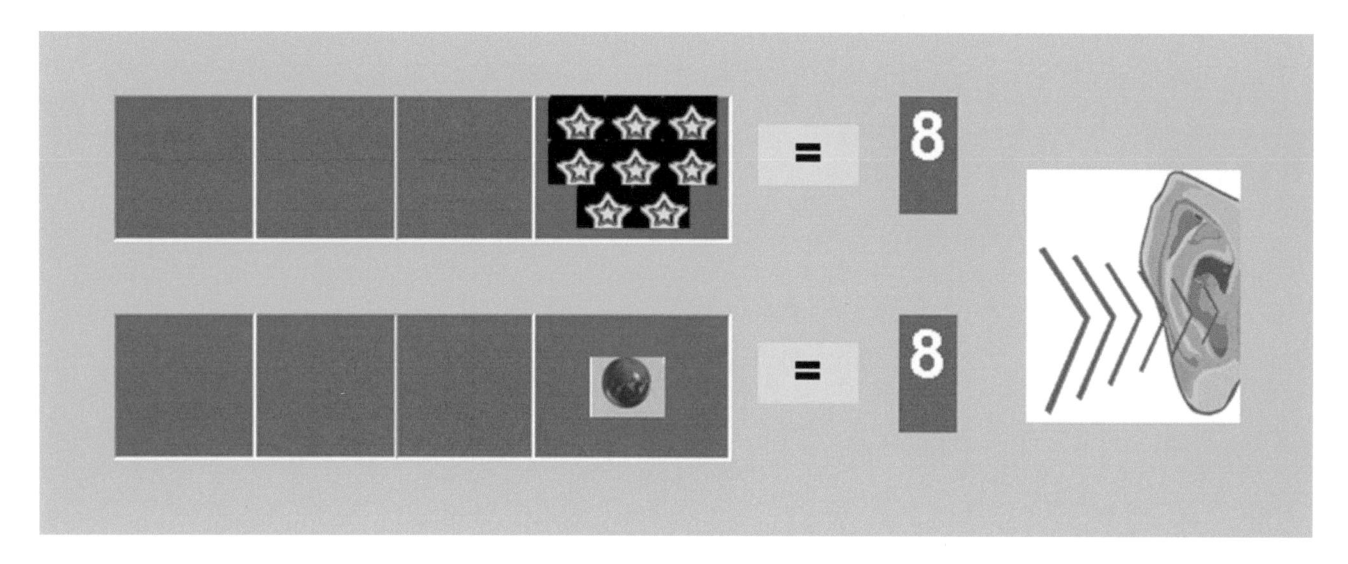

The ***kinesthetic*** representation of the filled up position is announced by clamping of the right hand. An empty position is represented by the clamping of the left hand. The equal sign is crossing both hands in front. The kinesthetic representation for this picture is clamping the left hand thrice, right hand, and crossing of hands.

ADDITION, EXERCISE 24

In exercise 24, in ***visual*** display of addition we see two pictures of boxes. In the bottom we see three pictures with probable answers. We want to specify, that between the first and second pictures we see the addition sign. Bellow the second picture we see the equal line, which separates the problem from the answer. The ***audio*** representation for two pictures consumes of a **double knock, double knock, knock, tram, double knock, double knock, knock, and cling.**

From the probable answers for addition we need to find the right one. The audio signals are: 1) **double knock, knock, double knock, and double knock.** 2) **double knock, double knock, knock, and double knock** 3) **double knock, double knock, double knock, and knock**.

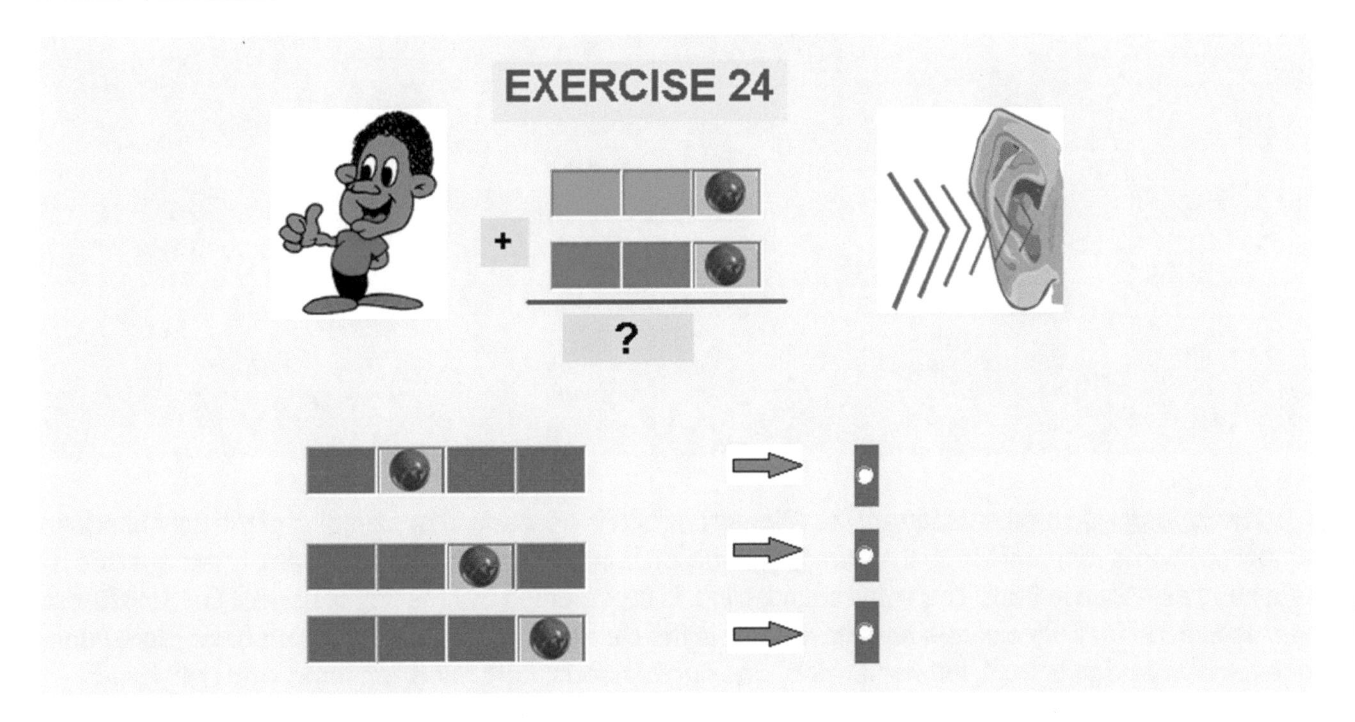

The ***kinesthetic*** representation of the filled up position is announced by clamping of the right hand. An empty position is represented by the clamping of the left hand. The addition sound, tram, is represented by one hand extended in front. The equal sound, cling, is represented by crossing of hands. The kinesthetic representation for both pictures and the answer is the clamping of the left hand twice, right hand, one hand extended in front, left hand twice, right hand, crossing of hands, left hand thrice, and right hand.

MATHEMATICS ADDITION, EXERCISE 25

In exercise 25, in ***visual*** display of addition we see three pictures of boxes. We want to specify, that between the first and second pictures we see the addition sign. Below the second picture we see the equal line, which separates the problem from the answer. The ***audio*** representation for the three pictures consumes of a **knock, double knock, knock, tram, double knock, double knock, cling, knock, double knock, and knock.**

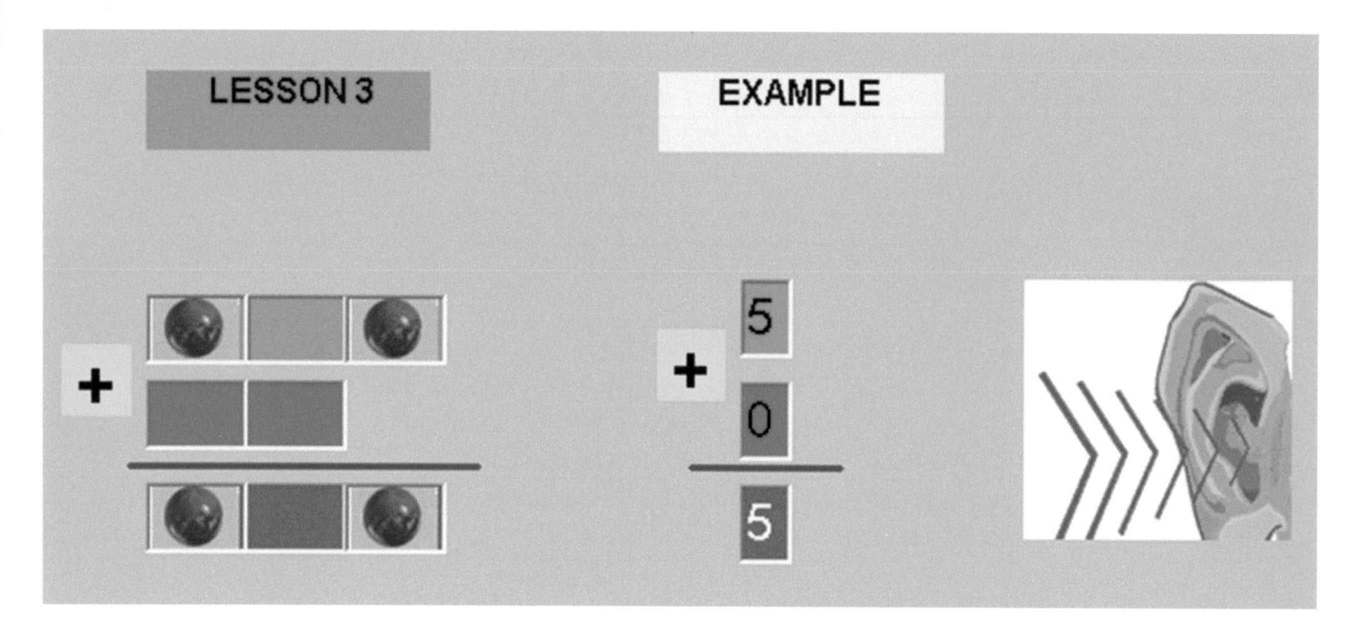

The ***kinesthetic*** representation of the filled up position is announced by clamping of the right hand. An empty position is represented by the clamping of the left hand. The addition sound, tram, is represented by one hand extended in front. The equal sound, cling, is represented by crossing of hands. The kinesthetic representation for both pictures and the answer is the clamping of the right hand, left hand, right hand, one hand extended in front, left hand twice, crossing of hands, right hand, left hand, and right hand.

ADDITION, EXERCISE 25

In exercise 25, in ***visual*** display of addition we see two pictures of boxes. In the bottom we see three pictures with probable answers. We want to specify, that between the first and second pictures we see the addition sign. Bellow the second picture we see the equal line, which separates the problem from the answer. The ***audio*** representation for two pictures consumes of a **knock, double knock, knock, tram, double knock, double knock, and cling.**

From the probable answers for addition we need to find the right one. The audio signals are: 1) **double knock, knock, knock.** 2) **knock, knock, and double knock** 3) **knock, double knock, and knock**.

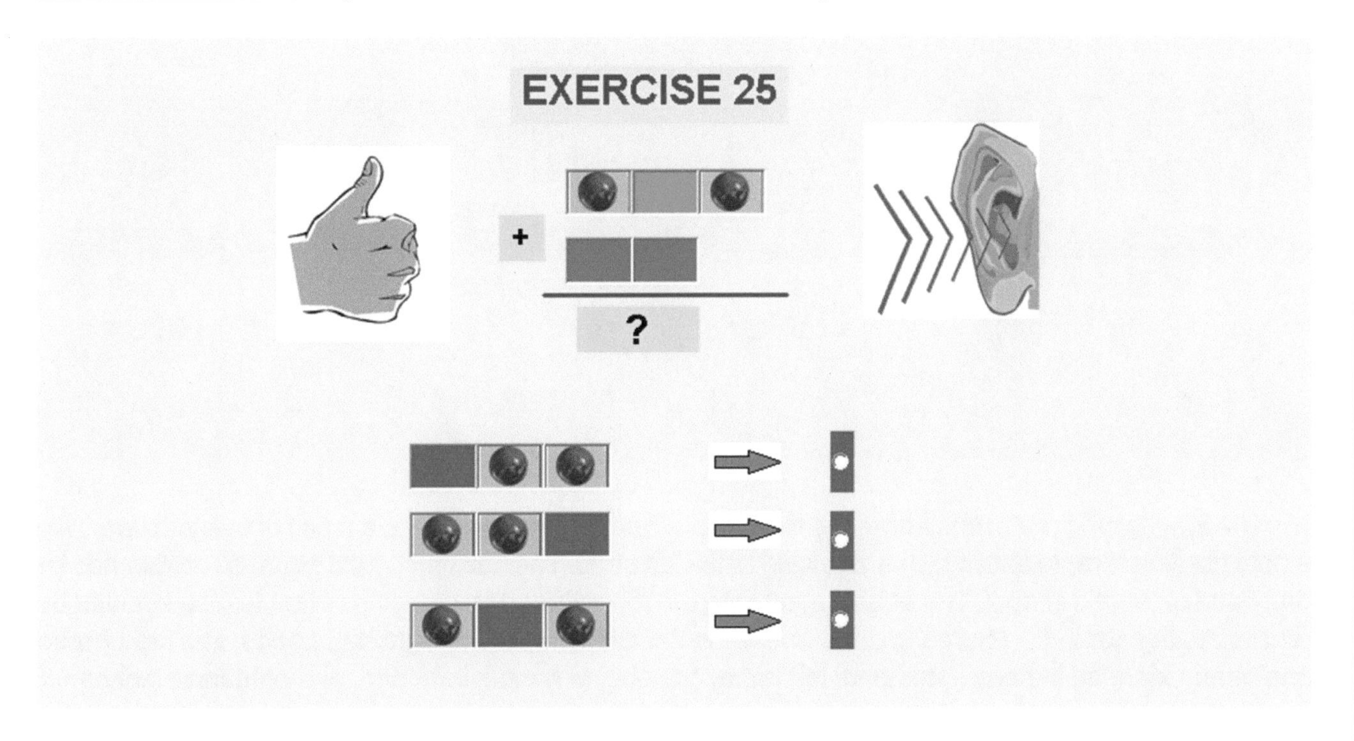

The ***kinesthetic*** representation of the filled up position is announced by clamping of the right hand. An empty position is represented by the clamping of the left hand. The addition sound, tram, is represented by one hand extended in front. The equal sound, cling, is represented by crossing of hands. The kinesthetic representation for both pictures and the answer is the clamping of the right hand, left hand, right hand, one hand extended in front, left hand twice, crossing of hands, right hand, left hand, and right hand.

MATHEMATICS

ADDITION, EXERCISE 26

In exercise 26, in ***visual*** display of addition we see three pictures of boxes. We want to specify, that between the first and second pictures we see the addition sign. Below the second picture we see the equal line, which separates the problem from the answer. The ***audio*** representation for two pictures consumes of a **knock, double knock, knock, tram, knock, double knock, cling, double knock, knock, and knock.**

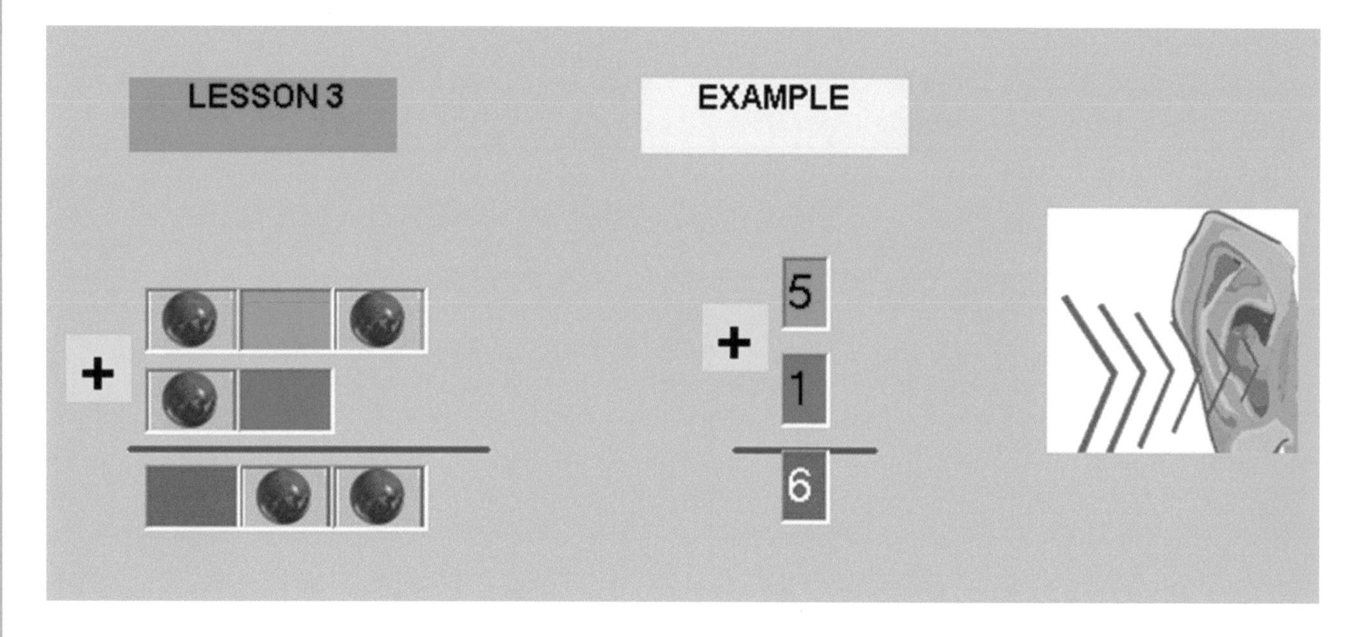

The ***kinesthetic*** representation of the filled up position is announced by clamping of the right hand. An empty position is represented by the clamping of the left hand. The addition sound, tram, is represented by one hand extended in front. The equal sound, cling, is represented by crossing of hands. The kinesthetic representation for both pictures and the answer is the clamping of the right hand, left hand, right hand, one hand extended in front, right hand, left hand, crossing of hands, left hand, and right hand twice.

ADDITION, EXERCISE 26, STEP 26.1

In exercise 26, in ***visual*** display of addition we see three pictures of boxes. We want to specify, that between the first and second pictures we see the addition sign. Below the second picture we see the equal line, which separates the problem from the answer. When we are adding two numbers and the position on the first number is filled with a symbol and the corresponding position on the second number is also with a symbol, we move the two symbols to the following positions in the answer as one symbol. When two symbols form, two corresponding positions move forward to the following position of the answer then the two symbols become one symbol. The ***audio*** representation for two pictures consumes of a **knock, double knock, knock, tram, knock, double knock, cling, double knock, knock, and knock.**

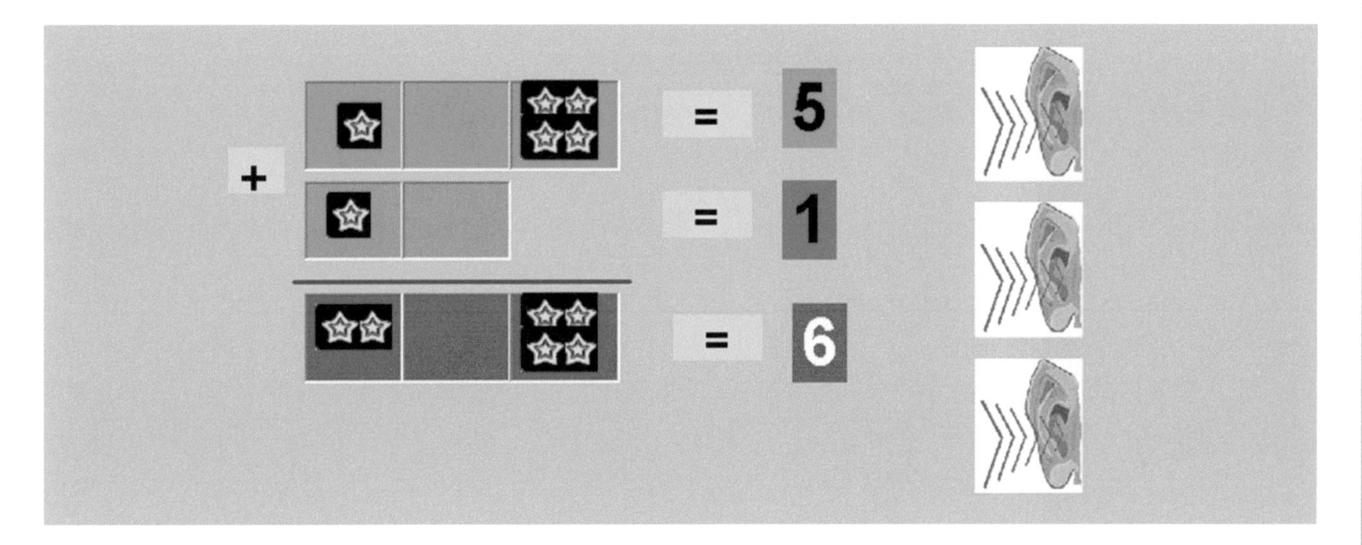

The ***kinesthetic*** representation of the filled up position is announced by clamping of the right hand. An empty position is represented by the clamping of the left hand. The addition sound, tram, is represented by one hand extended in front. The equal sound, cling, is represented by crossing of hands. The kinesthetic representation for both pictures and the answer is the clamping of the right hand, left hand, right hand, one hand extended in front, right hand, left hand, crossing of hands, left hand, and right hand twice.

MATHEMATICS

ADDITION, EXERCISE 26, STEP 26.2

In exercise 26, step 26.2 in ***visual*** display of addition we see two pictures of boxes. The first picture has a symbol in the first and third positions. The second picture has the second and third positions filled with a symbol. For purpose of addition we are introducing the concept of moving. When a position passes its maximum capacity the next symbol is simultaneously moved in the ascending position. We are moving from the first to the second position. However the audio representation for the picture does not change because the numerical value does not change.

The ***audio*** representation for the three pictures consumes of a **double knock, knock, and knock.**

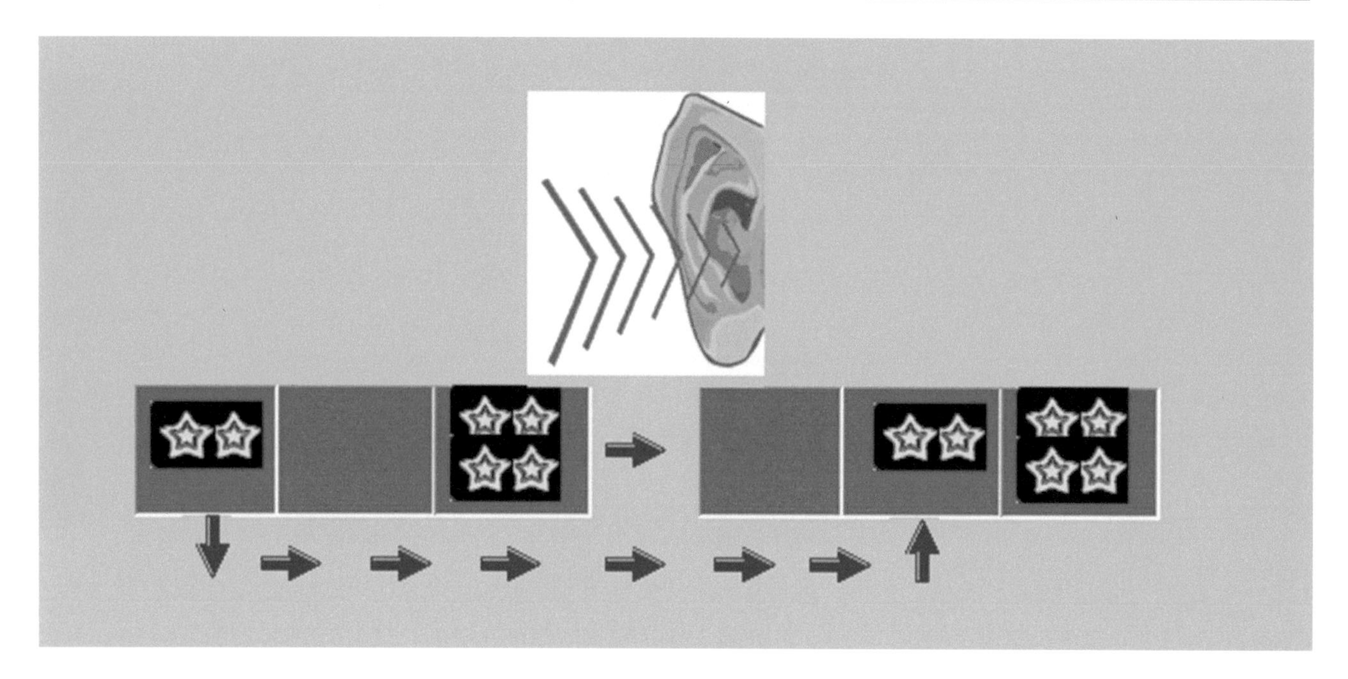

The ***kinesthetic*** representation of the filled up position is announced by clamping of the right hand. An empty position is represented by the clamping of the left hand. The kinesthetic representation for three pictures is the clamping of the left hand and right hand twice.

ADDITION, EXERCISE 26, STEP 26.3

In exercise 26, step 26.3 in ***visual*** display of addition, we see two pictures of boxes with equal sign. Both pictures have the second and third positions filled with a symbol and the first position is empty. The first picture has a yellow star in its positions but the second picture has a red ball in its positions. The form and shape of the symbols do not change mathematical value and the numerical value of both pictures is equal to six.

The ***audio*** representation for this picture consumes of a **double knock, knock, knock, and cling.**

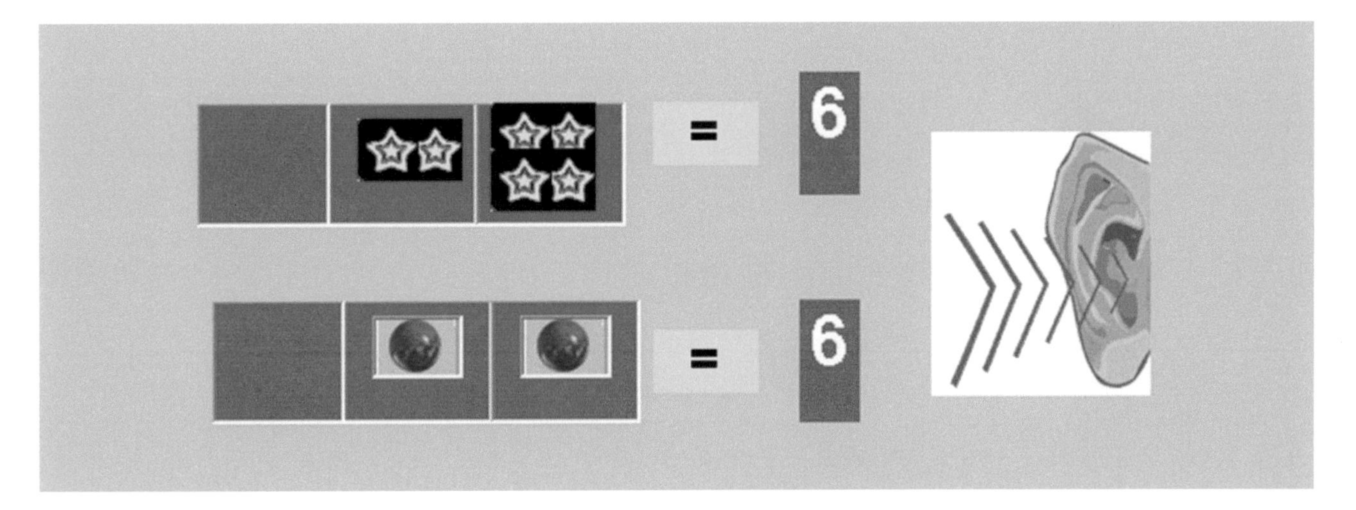

The ***kinesthetic*** representation of the filled up position is announced by clamping of the right hand. An empty position is represented by the clamping of the left hand. The equal sign is crossing both hands in front. The kinesthetic representation for this picture is clamping the left hand, right hand twice, and crossing of hands.

MATHEMATICS

ADDITION, EXERCISE 26

In exercise 26, in ***visual*** display of addition we see two pictures of boxes. In the bottom we see three pictures with probable answers. We want to specify, that between the first and second pictures we see the addition sign. Bellow the second picture we see the equal line, which separates the problem from the answer. The ***audio*** representation for two pictures consumes of a **knock, double knock, knock, tram, knock, double knock, and cling.**

From the probable answers for addition we need to find the right one. The audio signals are: 1) **knock, knock, double knock.** 2) **knock, double knock, and knock** 3) **double knock, knock, and knock**.

The ***kinesthetic*** representation of the filled up position is announced by clamping of the right hand. An empty position is represented by the clamping of the left hand. The addition sound, tram, is represented by one hand extended in front. The equal sound, cling, is represented by crossing of hands. The kinesthetic representation for both pictures and the answer is the clamping of the right hand, left hand, right hand, one hand extended in front, right hand, left hand, crossing of hands, left hand, and right hand twice.

ADDITION, EXERCISE 27

In exercise 27, in ***visual*** display of addition we see three pictures of boxes. We want to specify, that between the first and second pictures we see the addition sign. Below the second picture we see the equal line, which separates the problem from the answer. The ***audio*** representation for the three pictures consumes of a **knock, double knock, knock, tram, double knock, knock, cling, knock, knock, and knock.**

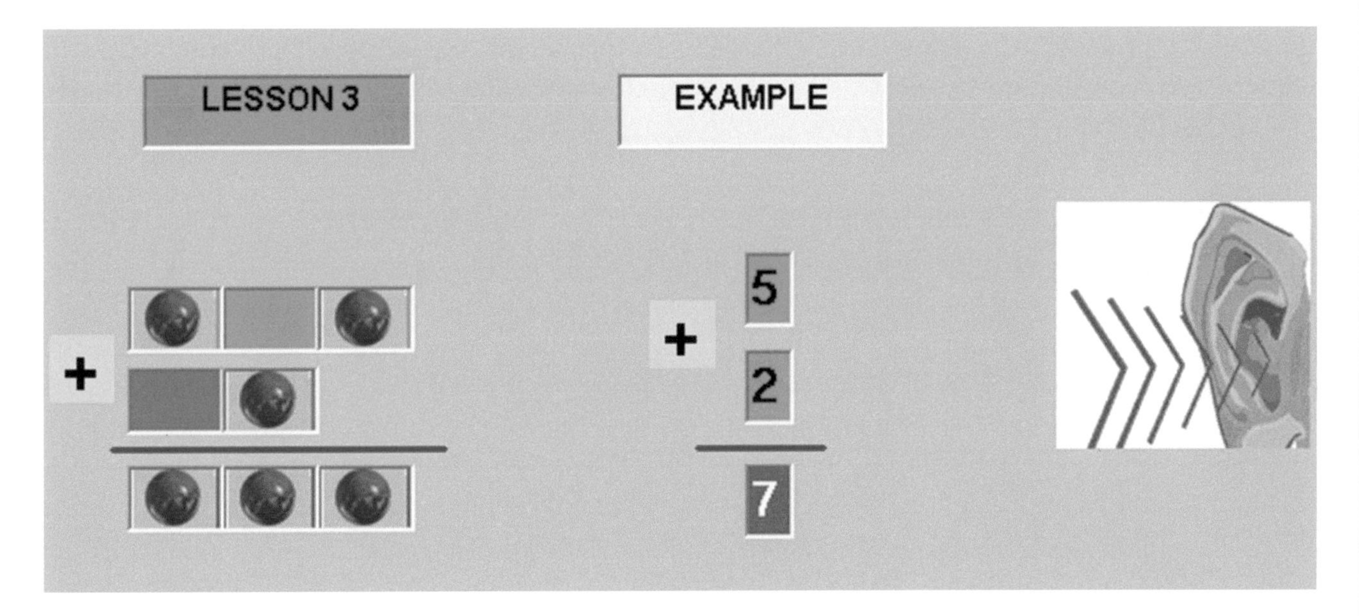

The ***kinesthetic*** representation of the filled up position is announced by clamping of the right hand. An empty position is represented by the clamping of the left hand. The addition sound, tram, is represented by one hand extended in front. The equal sound, cling, is represented by crossing of hands. The kinesthetic representation for both pictures and the answer is the clamping of the right hand, left hand, right hand, one hand extended in front, left hand, right hand, crossing of hands, right hand thrice.

MATHEMATICS

ADDITION, EXERCISE 27, STEP 27.1

In exercise 27, step 27.1 in ***visual*** display of addition we see three pictures of boxes. We want to specify, that between the first and second pictures we see the addition sign. Below the second picture we see the equal line, which separates the problem from the answer. When we are adding two numbers and the position on the first number is filled with a symbol and the corresponding position on the second number is also with a symbol, we move the two symbols to the following positions in the answer as one symbol. When two symbols form, two corresponding positions move forward to the following position of the answer then the two symbols become one symbol. The ***audio*** representation for the three pictures consumes of a **knock, double knock, knock, tram, double knock, knock, cling, knock, knock, and knock.**

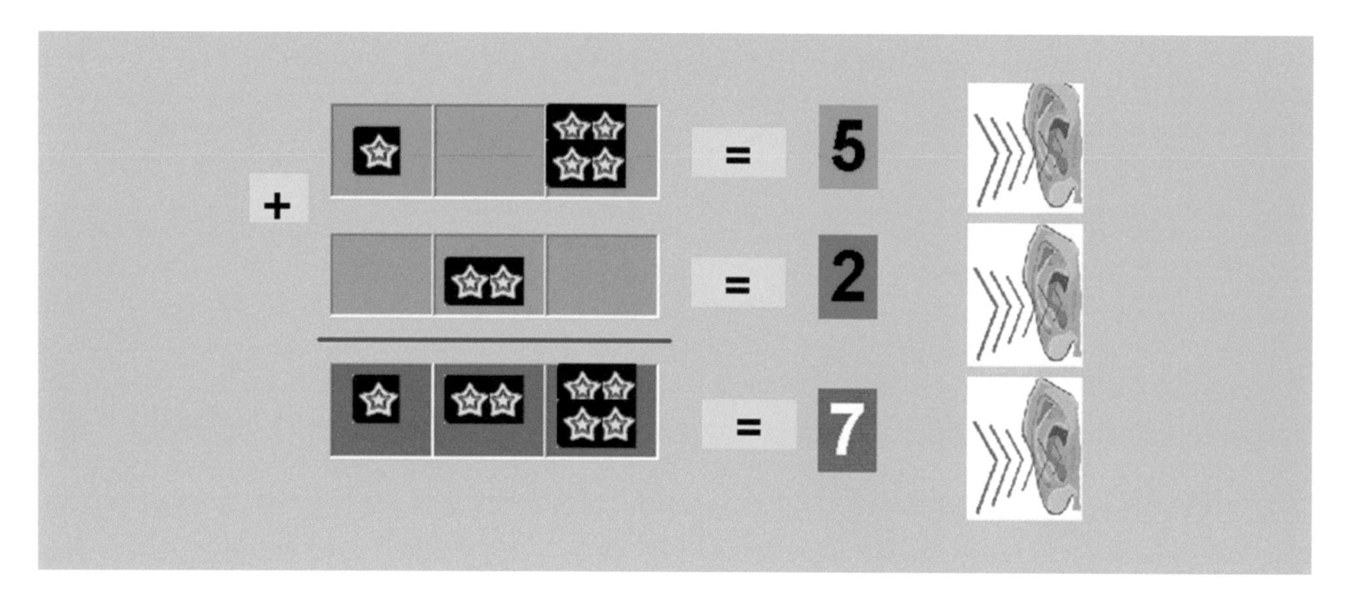

The ***kinesthetic*** representation of the filled up position is announced by clamping of the right hand. An empty position is represented by the clamping of the left hand. The addition sound, tram, is represented by one hand extended in front. The equal sound, cling, is represented by crossing of hands. The kinesthetic representation for both pictures and the answer is the clamping of the right hand, left hand, right hand, one hand extended in front, left hand, right hand, crossing of hands, right hand thrice.

ADDITION, EXERCISE 27, STEP 27.2

In exercise 27, step 27.2 in ***visual*** display of addition, we see two pictures of boxes with equal sign. Both pictures have all three of their positions filled with a symbol. The first picture has a yellow star in its positions but the second picture has a red ball in its positions. The form and shape of the symbols do not change mathematical value and the numerical value of both pictures is equal to seven.

The ***audio*** representation for this picture consumes of a **knock, knock, knock, and cling.**

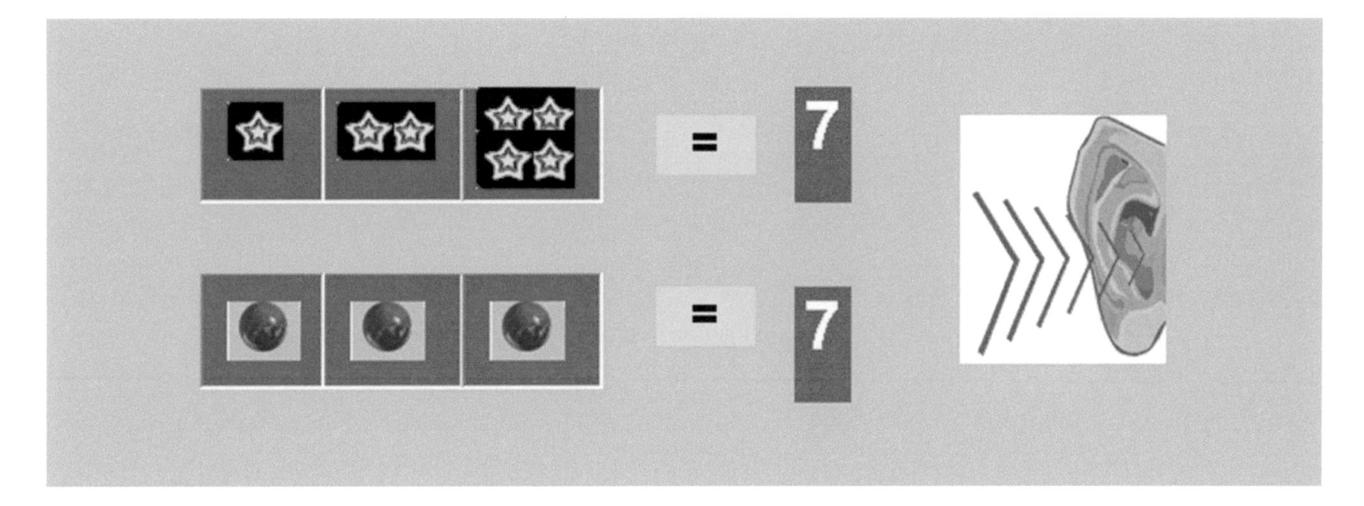

The ***kinesthetic*** representation of the filled up position is announced by clamping of the right hand. An empty position is represented by the clamping of the left hand. The equal sign is crossing both hands in front. The kinesthetic representation for this picture is clamping the right hand thrice and crossing of hands.

MATHEMATICS

ADDITION, EXERCISE 27

In exercise 27, in ***visual*** display of addition we see two pictures of boxes. In the bottom we see three pictures with probable answers. We want to specify, that between the first and second pictures we see the addition sign. Bellow the second picture we see the equal line, which separates the problem from the answer. The ***audio*** representation for two pictures consumes of a **knock, double knock, knock, tram, double knock, knock, and cling.**

From the probable answers for addition we need to find the right one. The audio signals are: 1) **knock, double knock, knock.** 2) **knock, knock, and knock** 3) **double knock, knock, and knock**.

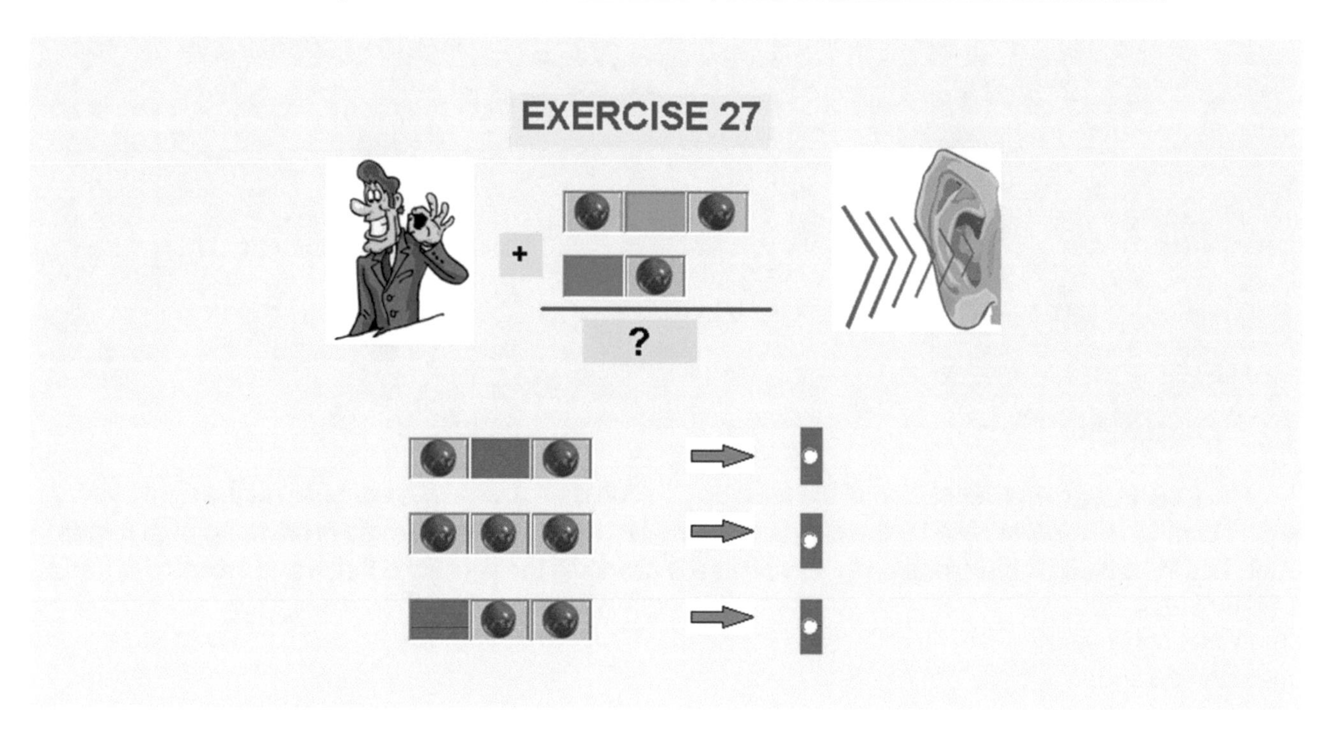

The ***kinesthetic*** representation of the filled up position is announced by clamping of the right hand. An empty position is represented by the clamping of the left hand. The addition sound, tram, is represented by one hand extended in front. The equal sound, cling, is represented by crossing of hands. The kinesthetic representation for both pictures and the answer is the clamping of the right hand, left hand, right hand, one hand extended in front, left hand, right hand, crossing of hands, right hand thrice.

ADDITION, EXERCISE 28

In exercise 28, in ***visual*** display of addition we see three pictures of boxes. We want to specify, that between the first and second pictures we see the addition sign. Below the second picture we see the equal line, which separates the problem from the answer. The ***audio*** representation for the three pictures consumes of a **knock, double knock, knock, tram, knock, knock, cling, double knock, double knock, double knock, and knock.**

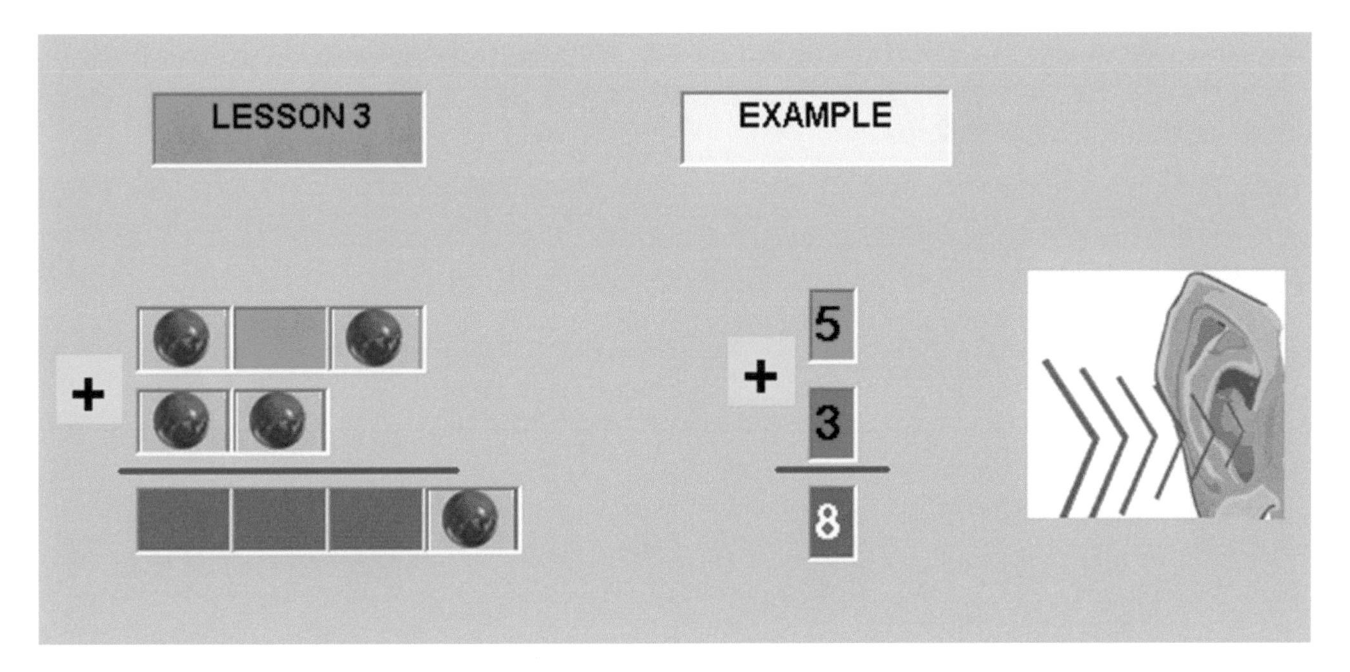

The ***kinesthetic*** representation of the filled up position is announced by clamping of the right hand. An empty position is represented by the clamping of the left hand. The addition sound, tram, is represented by one hand extended in front. The equal sound, cling, is represented by crossing of hands. The kinesthetic representation for both pictures and the answer is the clamping of the right hand, left hand, right hand, one hand extended in front, right hand twice, crossing of hands, left hand thrice, and right hand.

MATHEMATICS

ADDITION, EXERCISE 28, STEP 28.1

In exercise 28, step 28.1 in ***visual*** display of addition we see three pictures of boxes. We want to specify, that between the first and second pictures we see the addition sign. Below the second picture we see the equal line, which separates the problem from the answer. When we are adding two numbers and the position on the first number is filled with a symbol and the corresponding position on the second number is also with a symbol, we move the two symbols to the following positions in the answer as one symbol. When two symbols form, two corresponding positions move forward to the following position of the answer then the two symbols become one symbol. The ***audio*** representation for the three pictures consumes of a **knock, double knock, knock, tram, knock, knock, cling, double knock, double knock, double knock, and knock.**

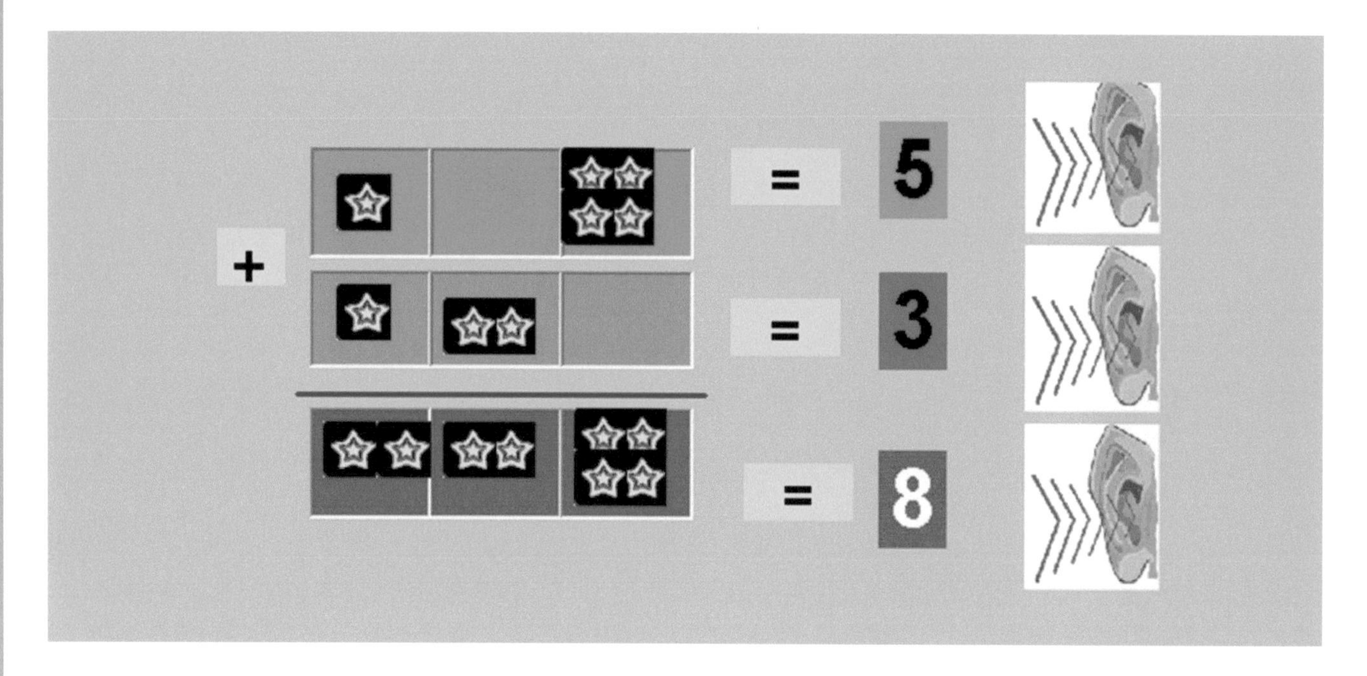

The ***kinesthetic*** representation of the filled up position is announced by clamping of the right hand. An empty position is represented by the clamping of the left hand. The addition sound, tram, is represented by one hand extended in front. The equal sound, cling, is represented by crossing of hands. The kinesthetic representation for both pictures and the answer is the clamping of the right hand, left hand, right hand, one hand extended in front, right hand twice, crossing of hands, left hand thrice, and right hand.

ADDITION, EXERCISE 28, STEP 28.2

In exercise 28, step 28.2 in ***visual*** display of addition we see two pictures of boxes. The first picture has a symbol in all three positions. The second picture has the second and third positions filled with a symbol. For purpose of addition we are introducing the concept of moving. When a position passes its maximum capacity the next symbol is simultaneously moved in the ascending position. We are moving from the first to the second position. However the audio representation for the picture does not change because the numerical value does not change.

The ***audio*** representation for the three pictures consumes of a **double knock, double knock, double knock, and knock.**

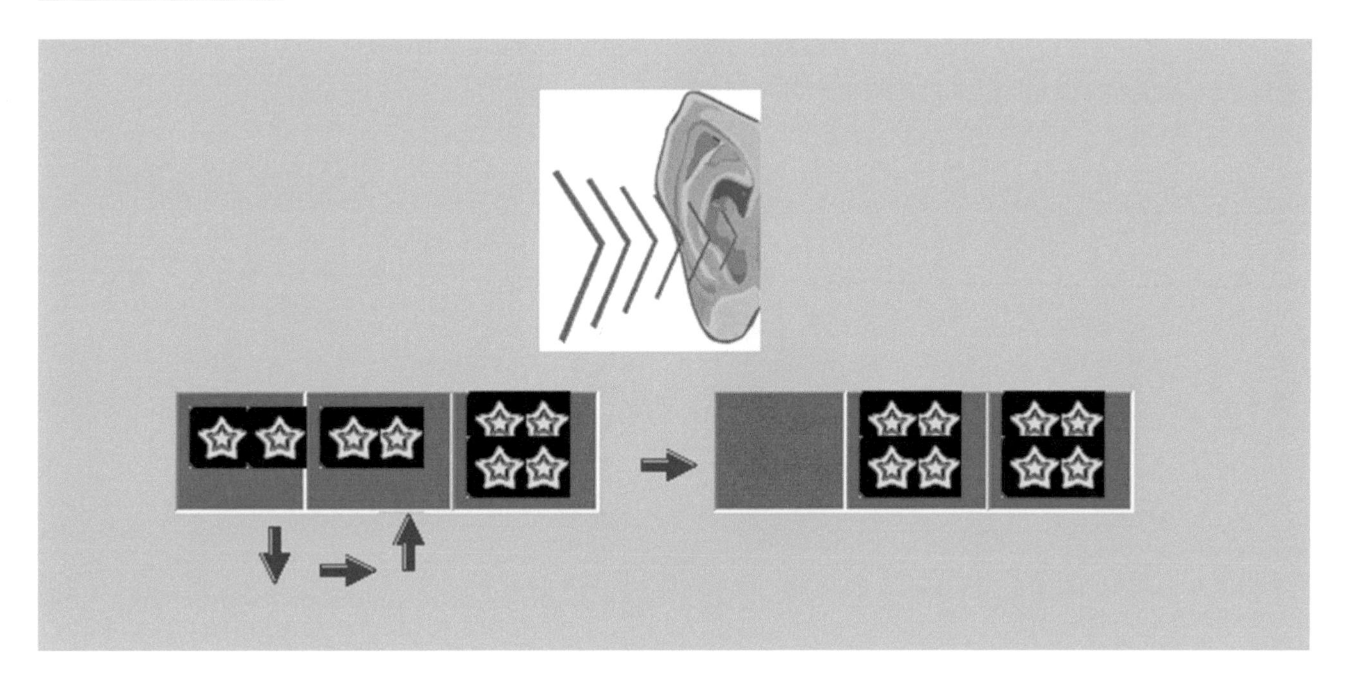

The ***kinesthetic*** representation of the filled up position is announced by clamping of the right hand. An empty position is represented by the clamping of the left hand. The kinesthetic representation for three pictures is the clamping of the left hand thrice and right hand.

MATHEMATICS

ADDITION, EXERCISE 28, STEP 28.3

In exercise 28, step 28.3 in ***visual*** display of addition we see two pictures of boxes. The first picture has a symbol in the second and third positions. The second picture has the third position filled with a symbol. For purpose of addition we are introducing the concept of moving. When a position passes its maximum capacity the next symbol is simultaneously moved in the ascending position. We are moving from the first to the second position. However the audio representation for the picture does not change because the numerical value does not change.

The ***audio*** representation for the three pictures consumes of a **double knock, double knock, double knock, and knock.**

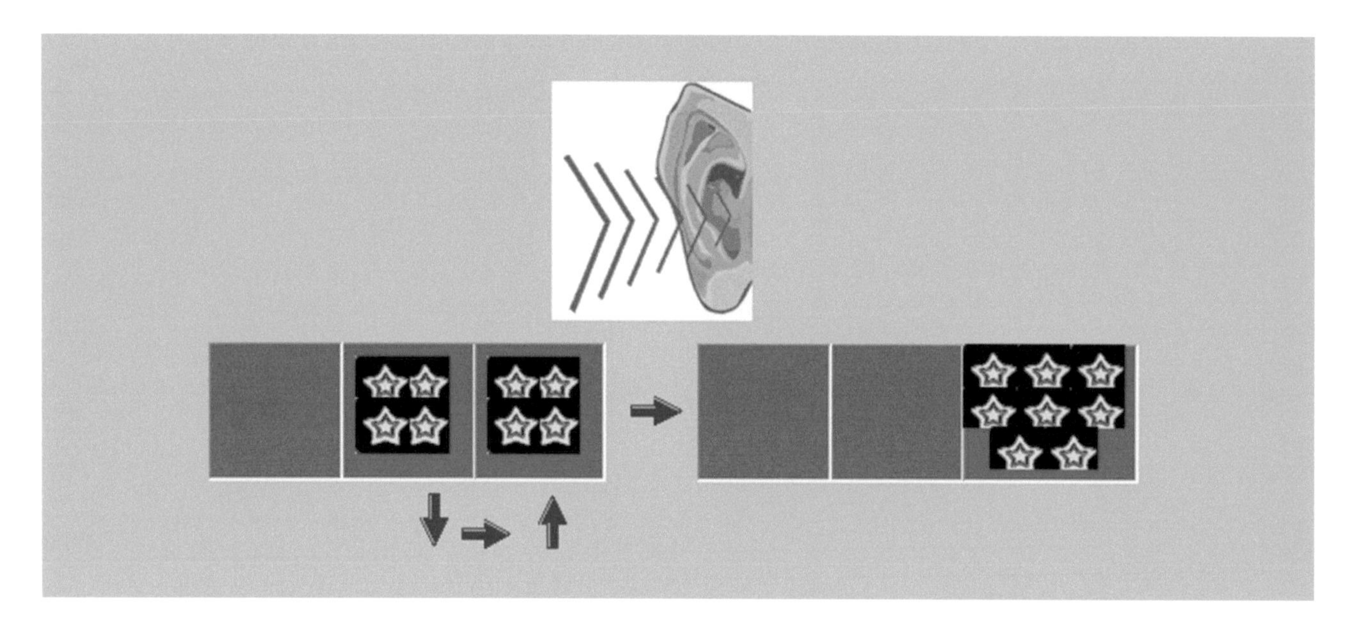

The ***kinesthetic*** representation of the filled up position is announced by clamping of the right hand. An empty position is represented by the clamping of the left hand. The kinesthetic representation for three pictures is the clamping of the left hand thrice and right hand.

ADDITION, EXERCISE 28, STEP 28.4

In exercise 28, step 28.4 in ***visual*** display of addition we see two pictures of boxes. The first picture has a symbol in the third position. The second picture has the fourth position filled with a symbol. For purpose of addition we are introducing the concept of moving. When a position passes its maximum capacity the next symbol is simultaneously moved in the ascending position. We are moving from the first to the second position. However the audio representation for the picture does not change because the numerical value does not change.

The ***audio*** representation for the three pictures consumes of a **double knock, double knock, double knock, and knock.**

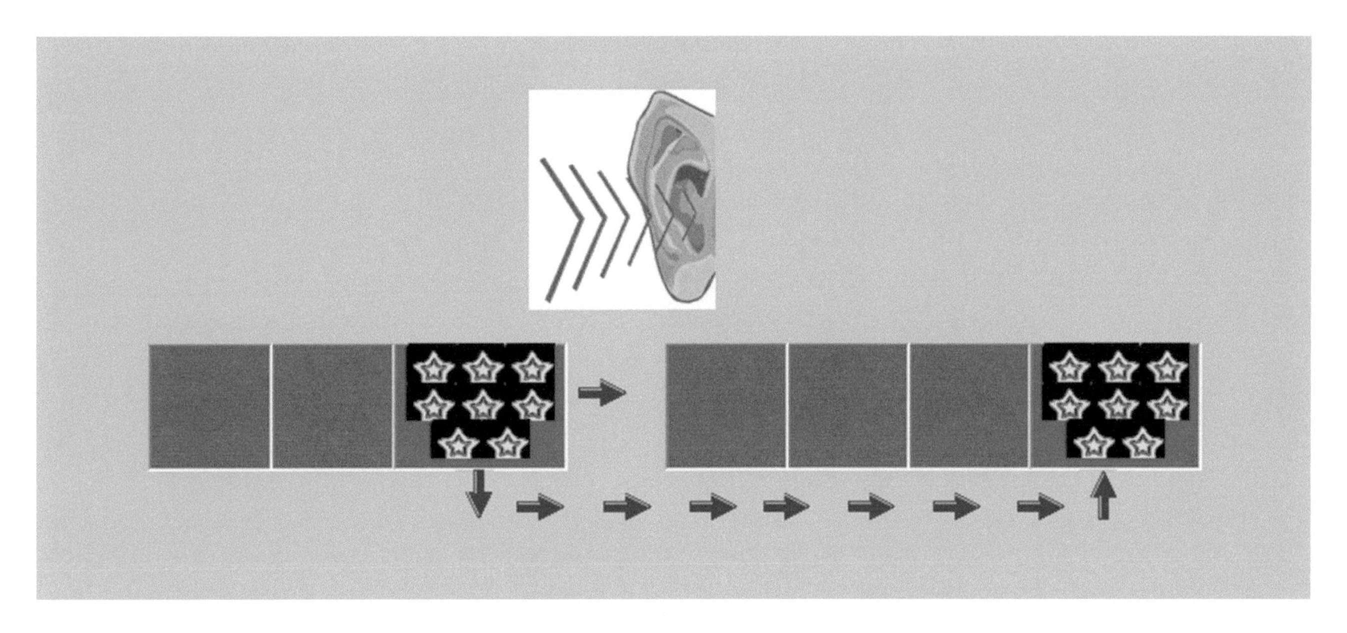

The ***kinesthetic*** representation of the filled up position is announced by clamping of the right hand. An empty position is represented by the clamping of the left hand. The kinesthetic representation for three pictures is the clamping of the left hand thrice and right hand.

MATHEMATICS ADDITION, EXERCISE 28, STEP 28.5

In exercise 28, step 28.5 in ***visual*** display of addition, we see two pictures of boxes with equal sign. Both pictures have the first three of their positions empty and the last position is filled with a symbol. The first picture has a yellow star in its positions but the second picture has a red ball in its positions. The form and shape of the symbols do not change mathematical value and the numerical value of both pictures is equal to eight.

The ***audio*** representation for this picture consumes of a **double knock, double knock, double knock, knock, and cling.**

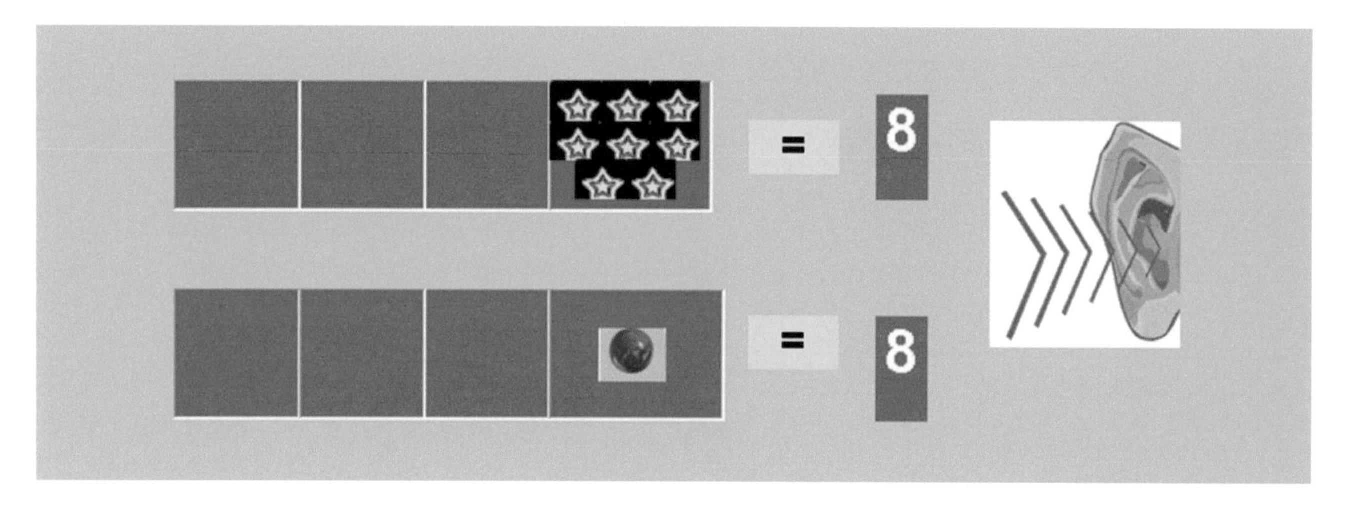

The ***kinesthetic*** representation of the filled up position is announced by clamping of the right hand. An empty position is represented by the clamping of the left hand. The equal sign is crossing both hands in front. The kinesthetic representation for this picture is clamping the left hand thrice, right hand, and crossing of hands.

ADDITION, EXERCISE 28

In exercise 28, in ***visual*** display of addition we see two pictures of boxes. In the bottom we see three pictures with probable answers. We want to specify, that between the first and second pictures we see the addition sign. Bellow the second picture we see the equal line, which separates the problem from the answer. The ***audio*** representation for two pictures consumes of a **knock, double knock, knock, tram, knock, knock, and cling.**

From the probable answers for addition we need to find the right one. The audio signals are: 1) **double knock, double knock, double knock, and knock.** 2) **knock, double knock, double knock, and double knock** 3) **double knock, knock, double knock, and double knock**.

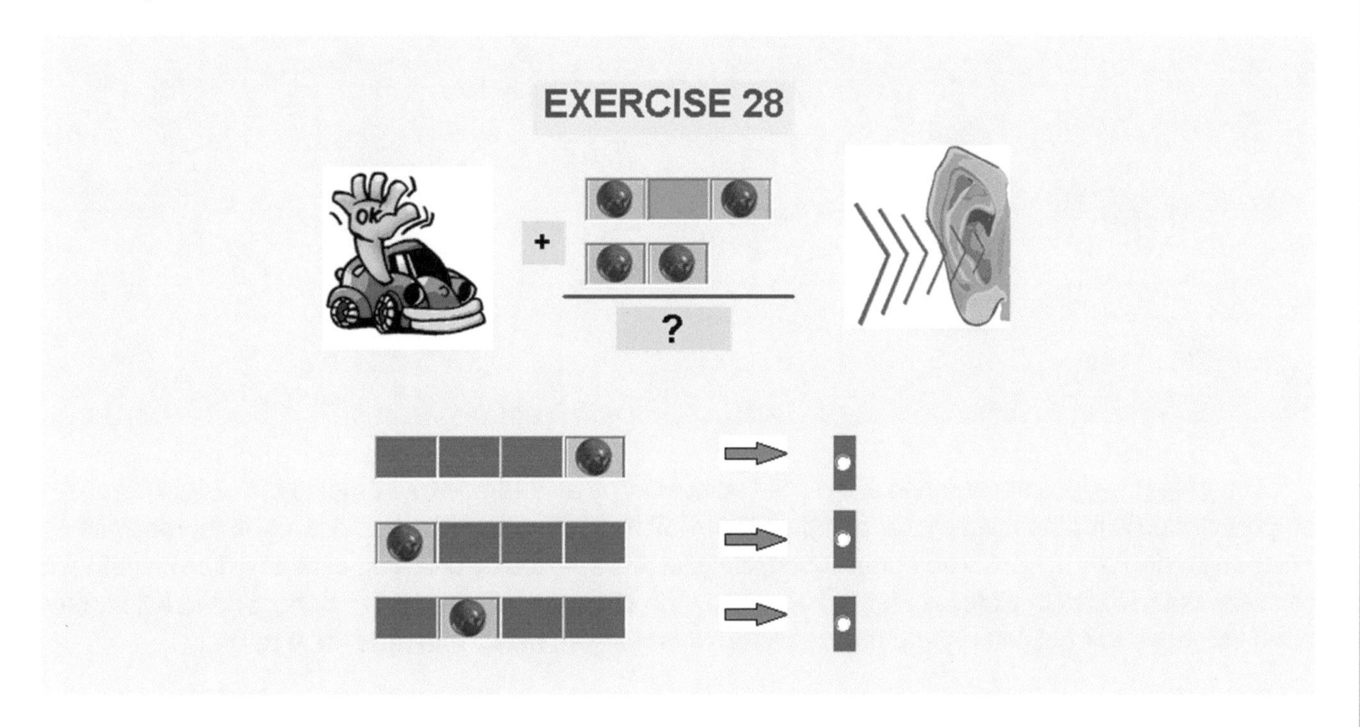

The ***kinesthetic*** representation of the filled up position is announced by clamping of the right hand. An empty position is represented by the clamping of the left hand. The addition sound, tram, is represented by one hand extended in front. The equal sound, cling, is represented by crossing of hands. The kinesthetic representation for both pictures and the answer is the clamping of the right hand, left hand, right hand, one hand extended in front, right hand twice, crossing of hands, left hand thrice, and right hand.

MATHEMATICS

ADDITION, EXERCISE 29

In exercise 29, in ***visual*** display of addition we see three pictures of boxes. We want to specify, that between the first and second pictures we see the addition sign. Below the second picture we see the equal line, which separates the problem from the answer. The ***audio*** representation for the three pictures consumes of a **double knock, knock, knock, tram, double knock, double knock, double knock, cling, double knock, knock, and knock.**

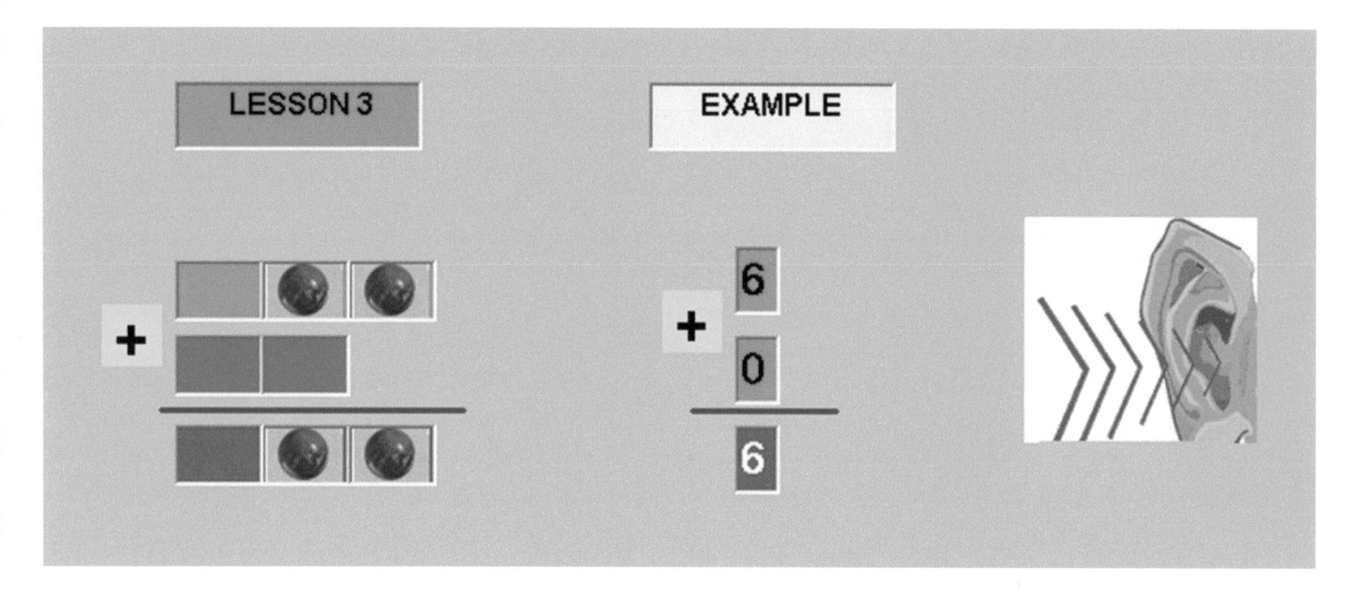

The ***kinesthetic*** representation of the filled up position is announced by clamping of the right hand. An empty position is represented by the clamping of the left hand. The addition sound, tram, is represented by one hand extended in front. The equal sound, cling, is represented by crossing of hands. The kinesthetic representation for both pictures and the answer is the clamping of the left hand, right hand twice, one hand extended in front, left hand thrice, crossing of hands, left hand, and right hand twice.

ADDITION, EXERCISE 29

In exercise 29, in ***visual*** display of addition we see two pictures of boxes. In the bottom we see three pictures with probable answers. We want to specify, that between the first and second pictures we see the addition sign. Bellow the second picture we see the equal line, which separates the problem from the answer. The ***audio*** representation for two pictures consumes of a **double knock, knock, knock, tram, double knock, double knock, double knock, and cling.**

From the probable answers for addition we need to find the right one. The audio signals are: 1) **knock, knock, and double knock.** 2) **knock, double knock, and knock** 3) **double knock, knock, and knock**.

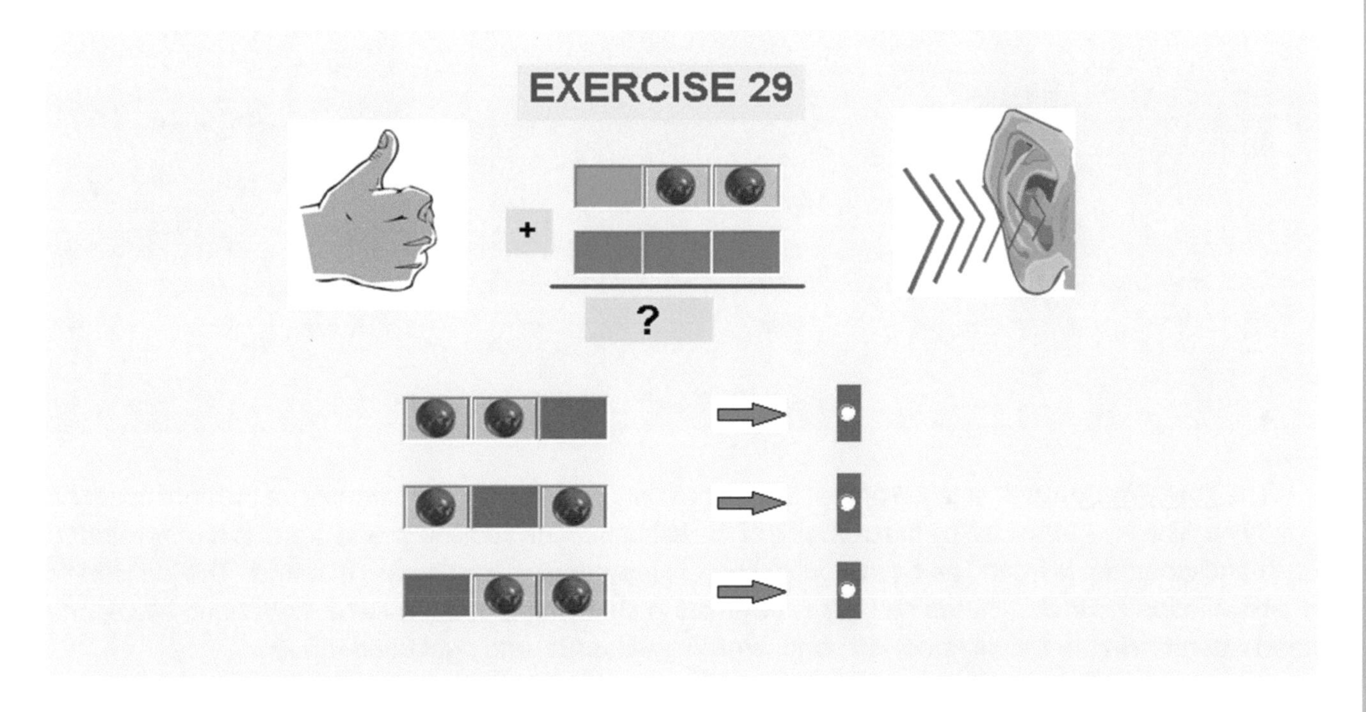

The ***kinesthetic*** representation of the filled up position is announced by clamping of the right hand. An empty position is represented by the clamping of the left hand. The addition sound, tram, is represented by one hand extended in front. The equal sound, cling, is represented by crossing of hands. The kinesthetic representation for both pictures and the answer is the clamping of the left hand, right hand twice, one hand extended in front, left hand thrice, crossing of hands, left hand, and right hand twice.

MATHEMATICS ADDITION, EXERCISE 30

In exercise 30, in ***visual*** display of addition we see three pictures of boxes. We want to specify, that between the first and second pictures we see the addition sign. Below the second picture we see the equal line, which separates the problem from the answer. The ***audio*** representation for the three pictures consumes of a **double knock, knock, knock, tram, knock, double knock, cling, knock, knock, and knock.**

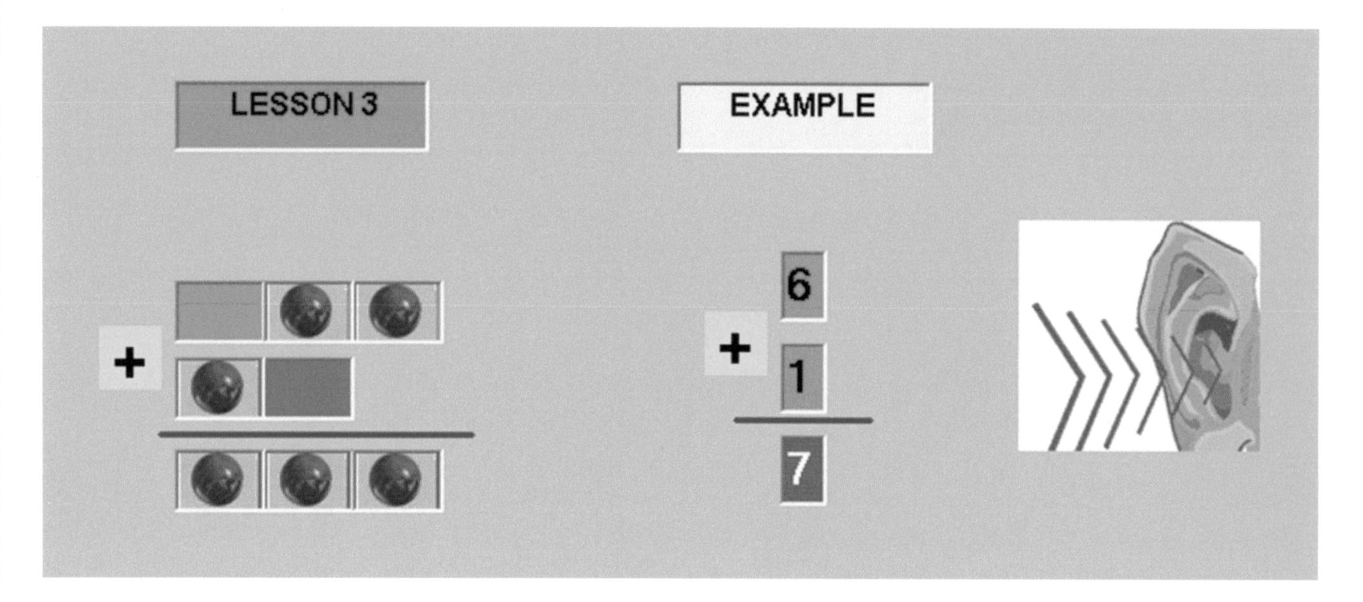

The ***kinesthetic*** representation of the filled up position is announced by clamping of the right hand. An empty position is represented by the clamping of the left hand. The addition sound, tram, is represented by one hand extended in front. The equal sound, cling, is represented by crossing of hands. The kinesthetic representation for both pictures and the answer is the clamping of the left hand, right hand twice, one hand extended in front, right hand, left hand, crossing of hands, and right hand thrice.

ADDITION, EXERCISE 30

In exercise 30, in ***visual*** display of addition we see three pictures of boxes. We want to specify, that between the first and second pictures we see the addition sign. Below the second picture we see the equal line, which separates the problem from the answer. When we are adding two numbers and the position on the first number is filled with a symbol and the corresponding position on the second number is also with a symbol, we move the two symbols to the following positions in the answer as one symbol. When two symbols form, two corresponding positions move forward to the following position of the answer then the two symbols become one symbol. The ***audio*** representation for the three pictures consumes of a **double knock, knock, knock, tram, knock, double knock, cling, knock, knock, and knock.**

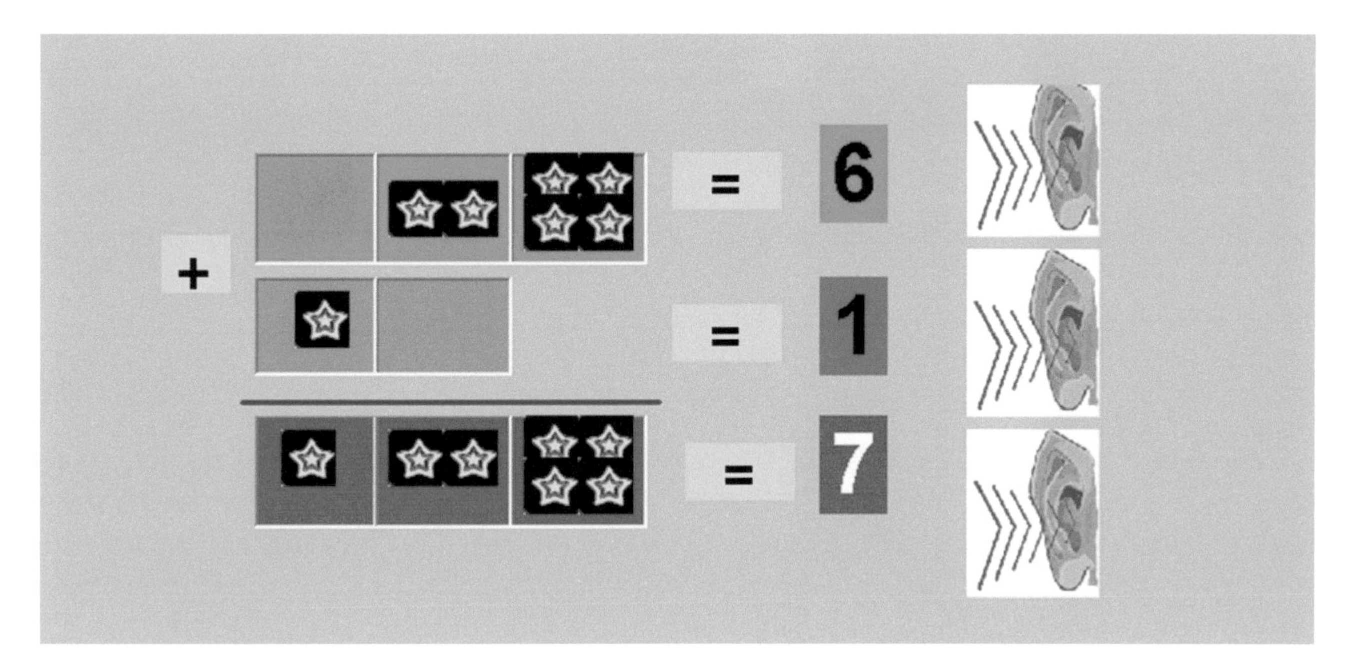

The ***kinesthetic*** representation of the filled up position is announced by clamping of the right hand. An empty position is represented by the clamping of the left hand. The addition sound, tram, is represented by one hand extended in front. The equal sound, cling, is represented by crossing of hands. The kinesthetic representation for both pictures and the answer is the clamping of the left hand, right hand twice, one hand extended in front, right hand, left hand, crossing of hands, and right hand thrice.

MATHEMATICS

ADDITION, EXERCISE 30, STEP 30.2

In exercise 30, step 30.2 in ***visual*** display of addition, we see two pictures of boxes with equal sign. Both pictures have all three of their positions filled with a symbol. The first picture has a yellow star in its positions but the second picture has a red ball in its positions. The form and shape of the symbols do not change mathematical value and the numerical value of both pictures is equal to seven.

The ***audio*** representation for this picture consumes of a **knock, knock, knock, and cling.**

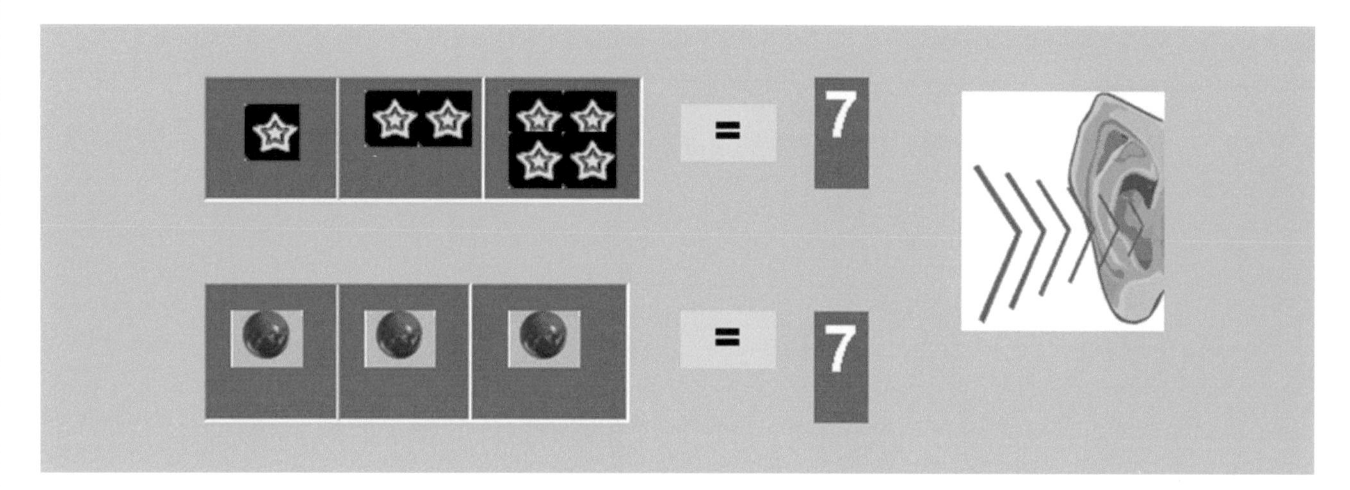

The ***kinesthetic*** representation of the filled up position is announced by clamping of the right hand. An empty position is represented by the clamping of the left hand. The equal sign is crossing both hands in front. The kinesthetic representation for this picture is clamping the right hand thrice and crossing of hands.

ADDITION, EXERCISE 30

In exercise 30, in ***visual*** display of addition we see two pictures of boxes. In the bottom we see three pictures with probable answers. We want to specify, that between the first and second pictures we see the addition sign. Bellow the second picture we see the equal line, which separates the problem from the answer. The ***audio*** representation for two pictures consumes of a **double knock, knock, knock, tram, knock, double knock, and cling.**

From the probable answers for addition we need to find the right one. The audio signals are: 1) **double knock, knock, and knock.** 2) **knock, knock, and knock** 3) **knock, double knock, and knock**.

The ***kinesthetic*** representation of the filled up position is announced by clamping of the right hand. An empty position is represented by the clamping of the left hand. The addition sound, tram, is represented by one hand extended in front. The equal sound, cling, is represented by crossing of hands. The kinesthetic representation for both pictures and the answer is the clamping of the left hand, right hand twice, one hand extended in front, right hand, left hand, crossing of hands, and right hand thrice.

ADDITION, EXERCISE 31

In exercise 31, in ***visual*** display of addition we see three pictures of boxes. We want to specify, that between the first and second pictures we see the addition sign. Below the second picture we see the equal line, which separates the problem from the answer. The ***audio*** representation for the three pictures consumes of a **double knock, knock, knock, tram, double knock, knock, cling, double knock, double knock, double knock, and knock.**

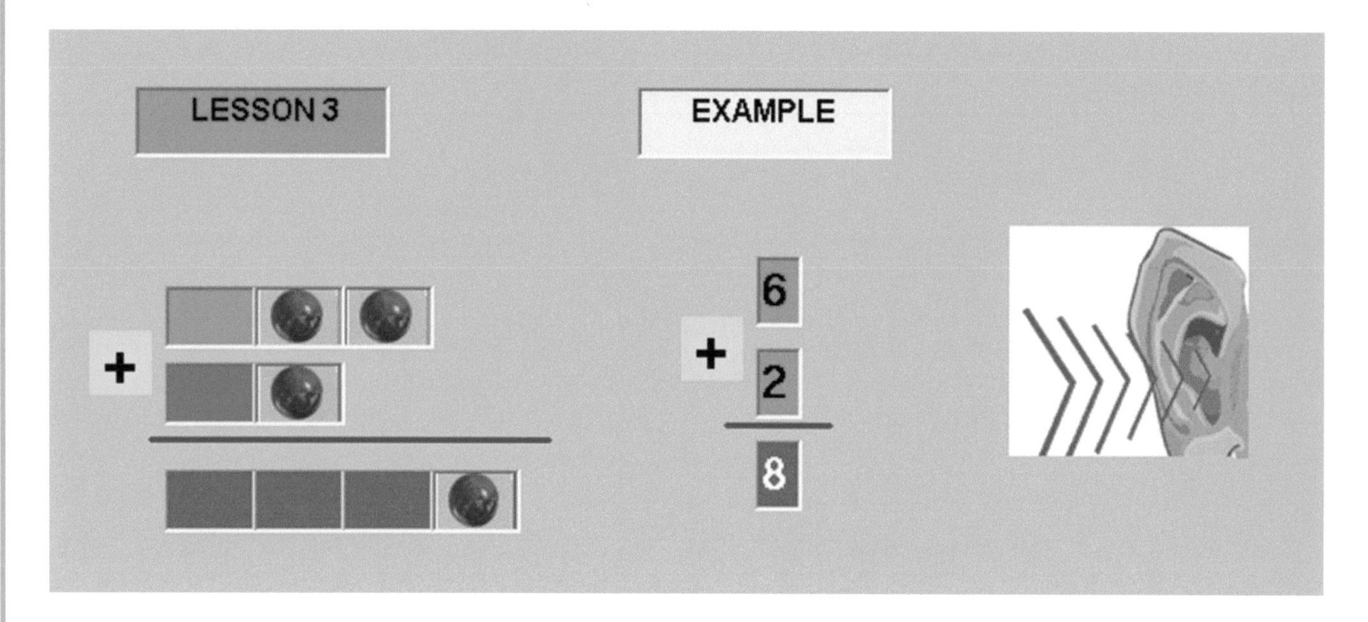

The ***kinesthetic*** representation of the filled up position is announced by clamping of the right hand. An empty position is represented by the clamping of the left hand. The addition sound, tram, is represented by one hand extended in front. The equal sound, cling, is represented by crossing of hands. The kinesthetic representation for both pictures and the answer is the clamping of the left hand, right hand twice, one hand extended in front, left hand, right hand, crossing of hands, left hand thrice, and right hand.

ADDITION, EXERCISE 31, STEP 31.1

In exercise 31, step 31.1 in ***visual*** display of addition we see three pictures of boxes. We want to specify, that between the first and second pictures we see the addition sign. Below the second picture we see the equal line, which separates the problem from the answer. When we are adding two numbers and the position on the first number is filled with a symbol and the corresponding position on the second number is also with a symbol, we move the two symbols to the following positions in the answer as one symbol. When two symbols form, two corresponding positions move forward to the following position of the answer then the two symbols become one symbol. The ***audio*** representation for the three pictures consumes of a **double knock, knock, knock, tram, double knock, knock, cling, double knock, double knock, double knock, and knock.**

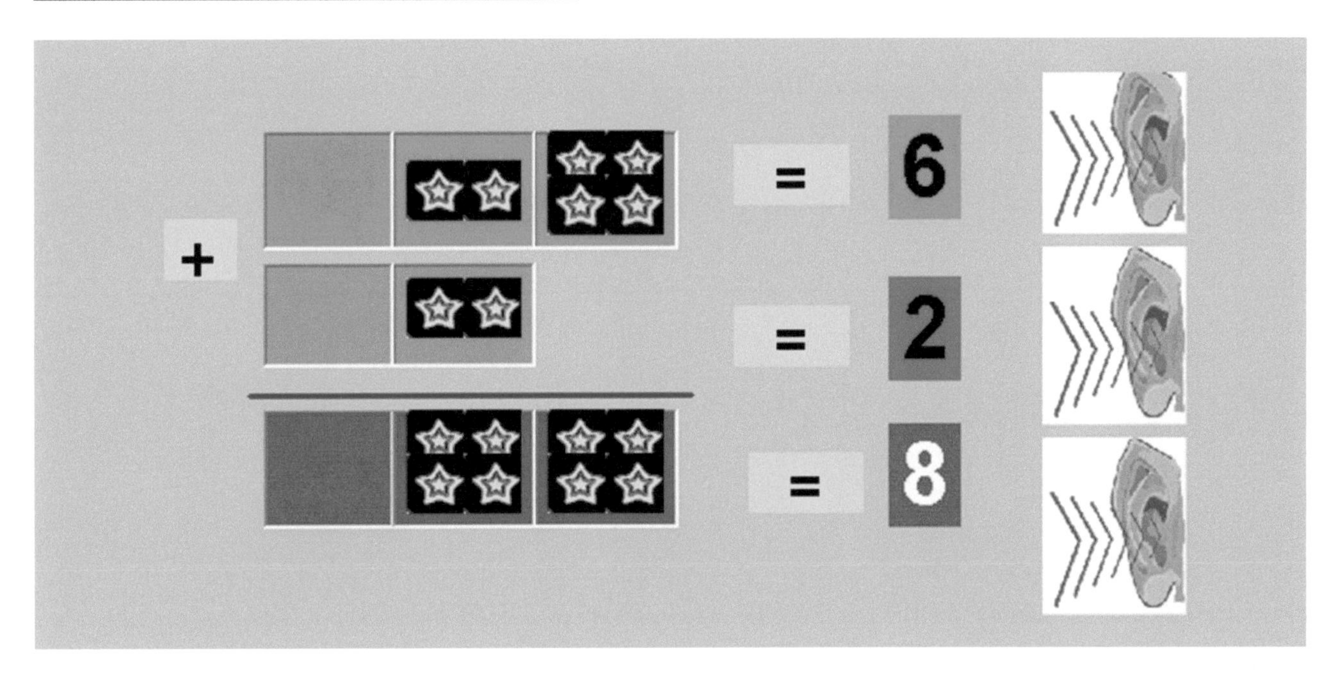

The ***kinesthetic*** representation of the filled up position is announced by clamping of the right hand. An empty position is represented by the clamping of the left hand. The addition sound, tram, is represented by one hand extended in front. The equal sound, cling, is represented by crossing of hands. The kinesthetic representation for both pictures and the answer is the clamping of the left hand, right hand twice, one hand extended in front, left hand, right hand, crossing of hands, left hand thrice, and right hand.

MATHEMATICS ADDITION, EXERCISE 31, STEP 31.2

In exercise 31, step 31.2 in ***visual*** display of addition we see two pictures of boxes. The first picture has a symbol in the second and third positions. The second picture has the third position filled with a symbol. For purpose of addition we are introducing the concept of moving. When a position passes its maximum capacity the next symbol is simultaneously moved in the ascending position. We are moving from the first to the second position. However the audio representation for the picture does not change because the numerical value does not change.

The ***audio*** representation for the three pictures consumes of a **double knock, double knock, double knock, and knock.**

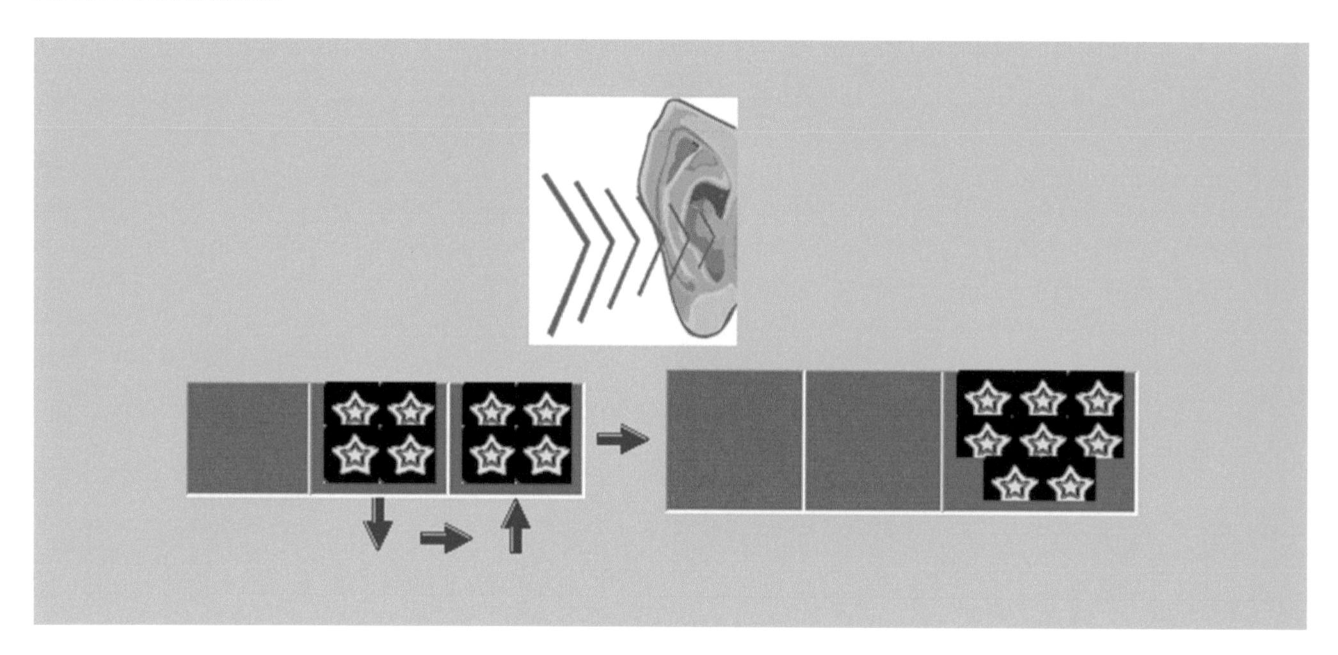

The ***kinesthetic*** representation of the filled up position is announced by clamping of the right hand. An empty position is represented by the clamping of the left hand. The kinesthetic representation for three pictures is the clamping of the left hand thrice and right hand.

ADDITION, EXERCISE 31, STEP 31.3

In exercise 31, step 31.3 in ***visual*** display of addition we see two pictures of boxes. The first picture has a symbol in the third position. The second picture has the fourth position filled with a symbol. For purpose of addition we are introducing the concept of moving. When a position passes its maximum capacity the next symbol is simultaneously moved in the ascending position. We are moving from the first to the second position. However the audio representation for the picture does not change because the numerical value does not change.

The ***audio*** representation for the three pictures consumes of a **double knock, double knock, double knock, and knock.**

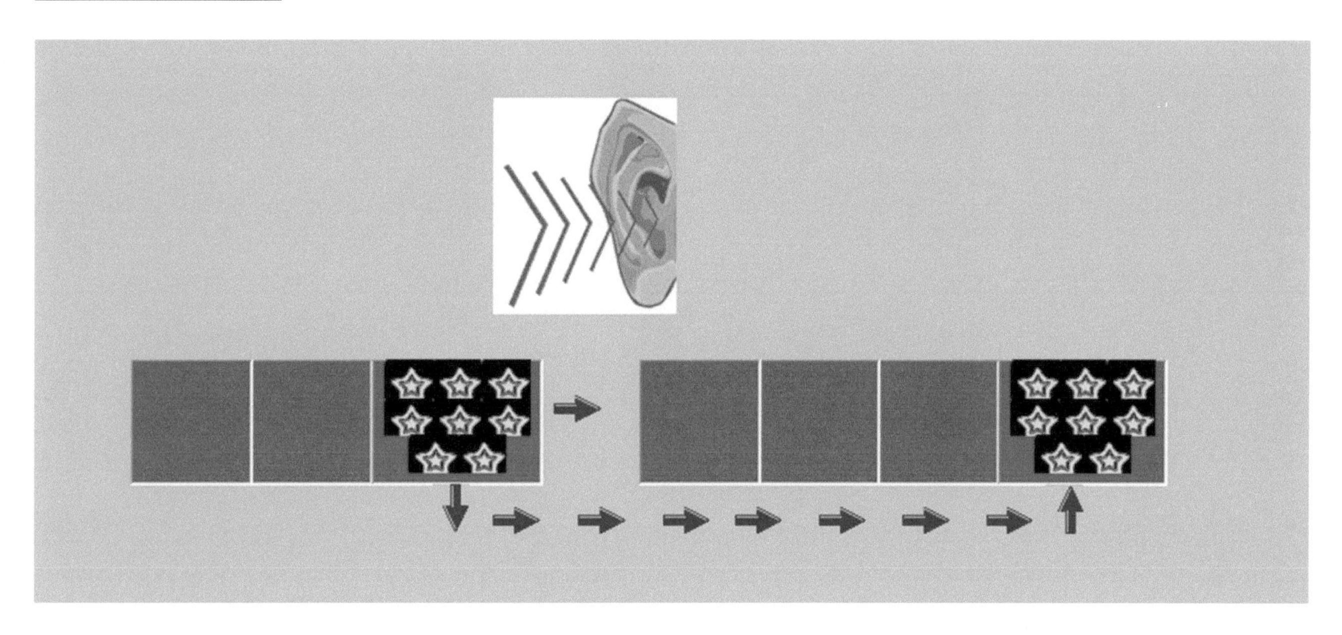

The ***kinesthetic*** representation of the filled up position is announced by clamping of the right hand. An empty position is represented by the clamping of the left hand. The kinesthetic representation for three pictures is the clamping of the left hand thrice and right hand.

MATHEMATICS

ADDITION, EXERCISE 31, STEP 31.4

In exercise 31, step 31.4 in ***visual*** display of addition, we see two pictures of boxes with equal sign. Both pictures have the first three of their positions empty and the last position is filled with a symbol. The first picture has a yellow star in its positions but the second picture has a red ball in its positions. The form and shape of the symbols do not change mathematical value and the numerical value of both pictures is equal to eight.

The ***audio*** representation for this picture consumes of a **double knock, double knock, double knock, knock, and cling.**

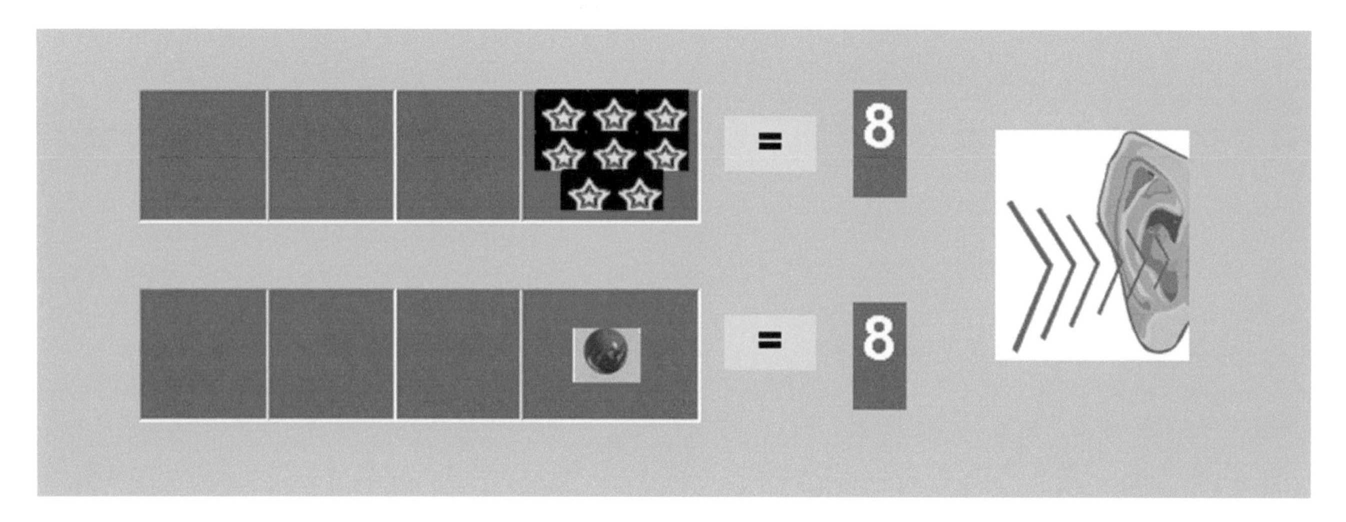

The ***kinesthetic*** representation of the filled up position is announced by clamping of the right hand. An empty position is represented by the clamping of the left hand. The equal sign is crossing both hands in front. The kinesthetic representation for this picture is clamping the left hand thrice, right hand, and crossing of hands.

ADDITION, EXERCISE 31

In exercise 31, in ***visual*** display of addition we see two pictures of boxes. In the bottom we see three pictures with probable answers. We want to specify, that between the first and second pictures we see the addition sign. Bellow the second picture we see the equal line, which separates the problem from the answer. The ***audio*** representation for two pictures consumes of a **double knock, knock, knock, tram, double knock, knock, and cling.**

From the probable answers for addition we need to find the right one. The audio signals are: 1) **double knock, double knock, double knock, and knock.** 2) **knock, double knock, double knock, and double knock** 3) **double knock, knock, double knock, and double knock**.

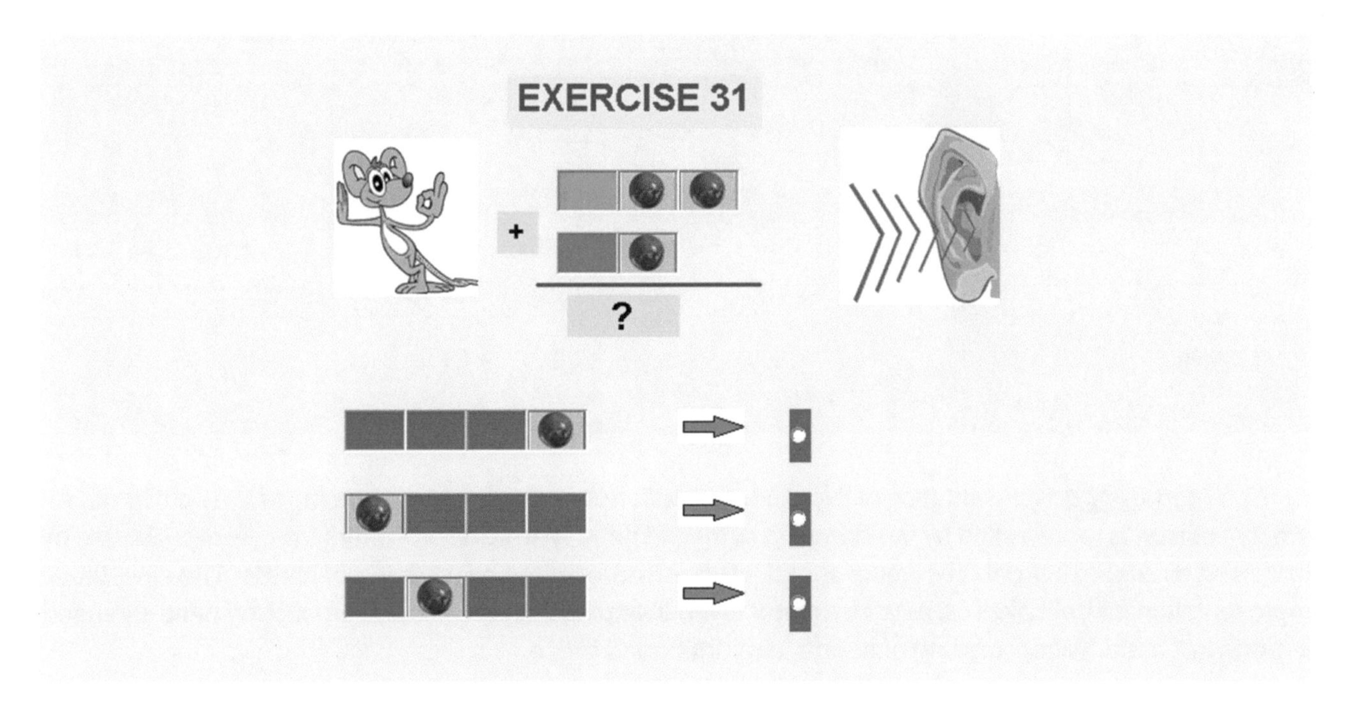

The ***kinesthetic*** representation of the filled up position is announced by clamping of the right hand. An empty position is represented by the clamping of the left hand. The addition sound, tram, is represented by one hand extended in front. The equal sound, cling, is represented by crossing of hands. The kinesthetic representation for both pictures and the answer is the clamping of the left hand, right hand twice, one hand extended in front, left hand, right hand, crossing of hands, left hand thrice, and right hand.

MATHEMATICS

ADDITION, EXERCISE 32

In exercise 32, in ***visual*** display of addition we see three pictures of boxes.. We want to specify, that between the first and second pictures we see the addition sign. Below the second picture we see the equal line, which separates the problem from the answer. The ***audio*** representation for two pictures consumes of a **knock, knock, knock, tram, double knock, double knock, cling, knock, knock, and knock.**

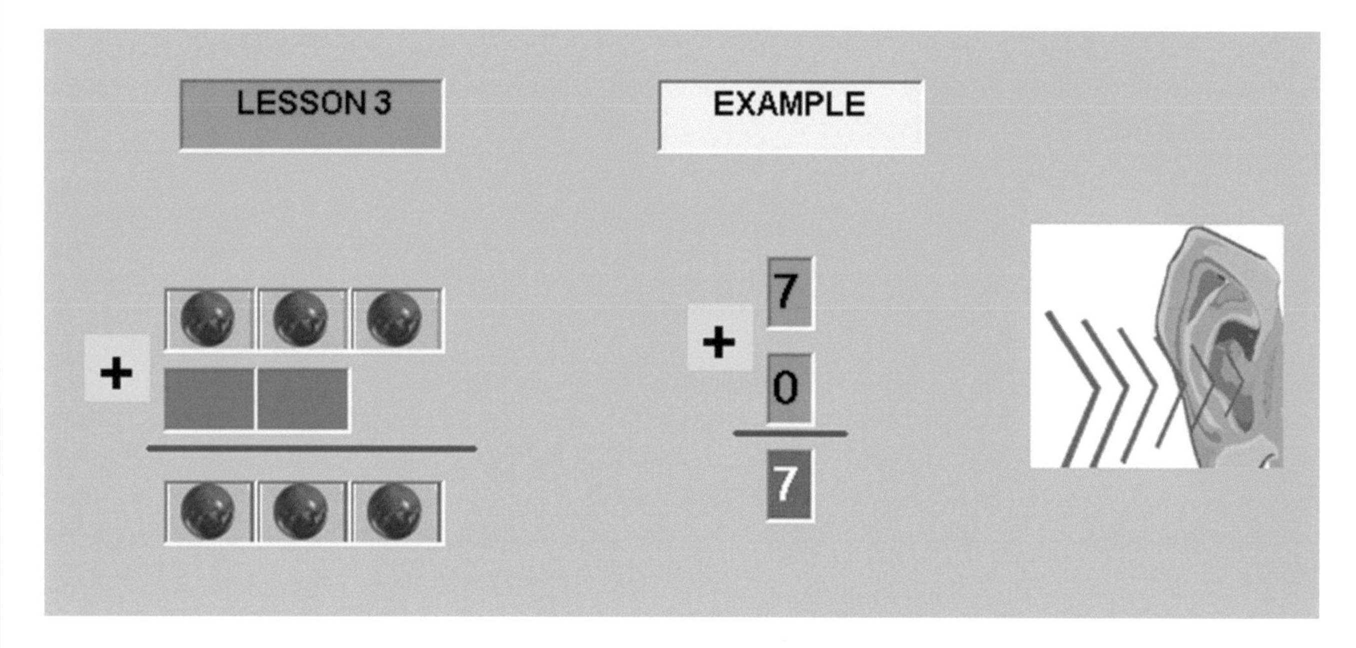

The ***kinesthetic*** representation of the filled up position is announced by clamping of the right hand. An empty position is represented by the clamping of the left hand. The addition sound, tram, is represented by one hand extended in front. The equal sound, cling, is represented by crossing of hands. The kinesthetic representation for both pictures and the answer is the clamping of the right hand thrice, one hand extended in front, left hand twice, crossing of hands, and right hand thrice.

ADDITION, EXERCISE 32

In exercise 32, in ***visual*** display of addition we see two pictures of boxes. In the bottom we see three pictures with probable answers. We want to specify, that between the first and second pictures we see the addition sign. Bellow the second picture we see the equal line, which separates the problem from the answer. The ***audio*** representation for two pictures consumes of a **knock, knock, knock, tram, double knock, double knock, and cling.**

From the probable answers for addition we need to find the right one. The audio signals are: 1) **double knock, knock, and knock.** 2) **knock, knock, and knock** 3) **knock, double knock, and knock**.

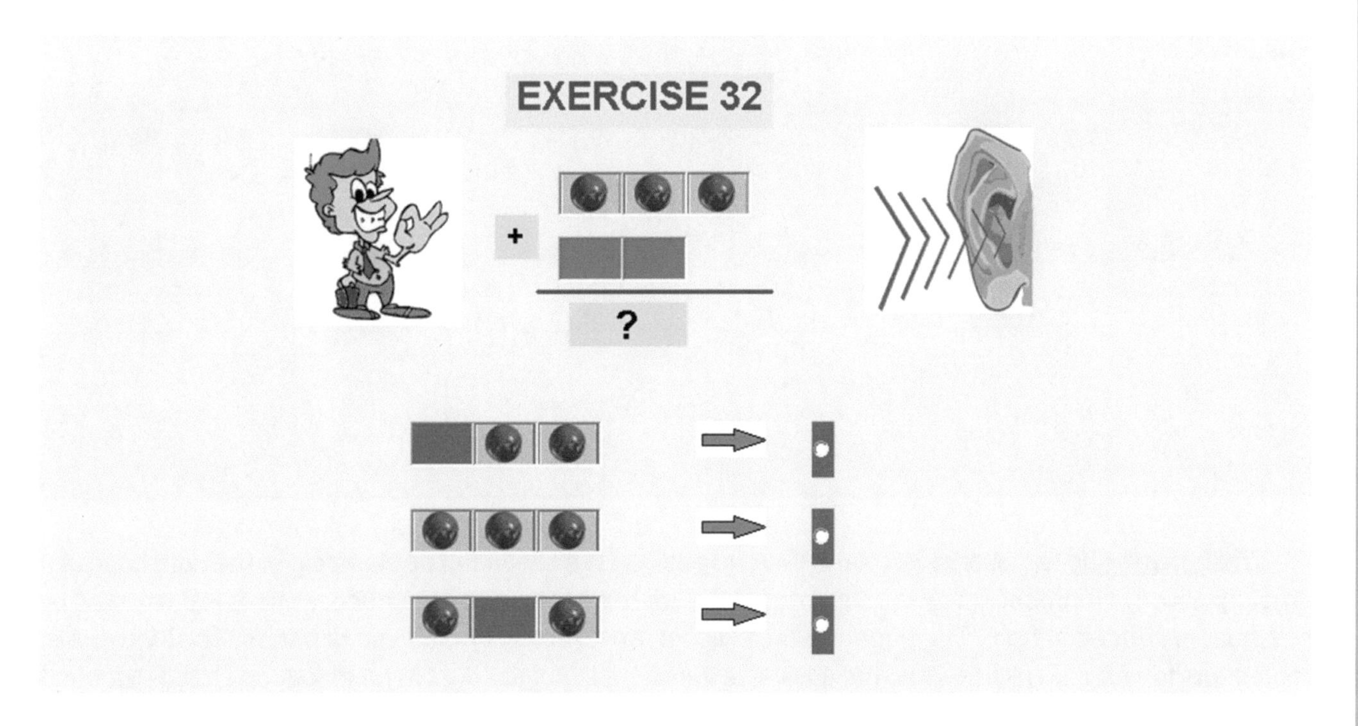

The ***kinesthetic*** representation of the filled up position is announced by clamping of the right hand. An empty position is represented by the clamping of the left hand. The addition sound, tram, is represented by one hand extended in front. The equal sound, cling, is represented by crossing of hands. The kinesthetic representation for both pictures and the answer is the clamping of the right hand thrice, one hand extended in front, left hand twice, crossing of hands, and right hand thrice.

MATHEMATICS

ADDITION, EXERCISE 33

In exercise 33, in ***visual*** display of addition we see three pictures of boxes. We want to specify, that between the first and second pictures we see the addition sign. Below the second picture we see the equal line, which separates the problem from the answer. The ***audio*** representation for the three pictures consumes of a **knock, knock, knock, tram, knock, double knock, cling, double knock, double knock, double knock, and knock.**

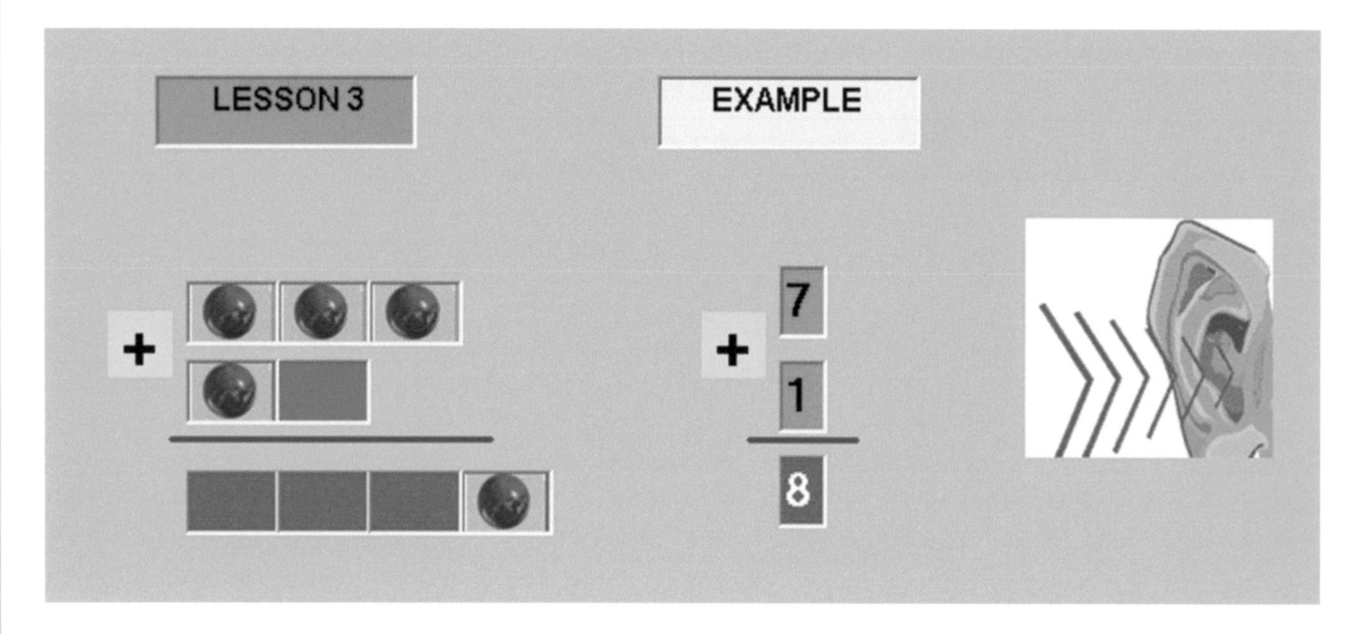

The ***kinesthetic*** representation of the filled up position is announced by clamping of the right hand. An empty position is represented by the clamping of the left hand. The addition sound, tram, is represented by one hand extended in front. The equal sound, cling, is represented by crossing of hands. The kinesthetic representation for both pictures and the answer is the clamping of the right hand thrice, one hand extended in front, right hand, left hand, crossing of hands, left hand thrice, and right hand.

ADDITION, EXERCISE 33, STEP 33.1

In exercise 33, step 33.1 in ***visual*** display of addition we see three pictures of boxes. We want to specify, that between the first and second pictures we see the addition sign. Below the second picture we see the equal line, which separates the problem from the answer. When we are adding two numbers and the position on the first number is filled with a symbol and the corresponding position on the second number is also with a symbol, we move the two symbols to the following positions in the answer as one symbol. When two symbols form, two corresponding positions move forward to the following position of the answer then the two symbols become one symbol. The ***audio*** representation for the three pictures consumes of a **knock, knock, knock, tram, knock, double knock, cling, double knock, double knock, double knock, and knock.**

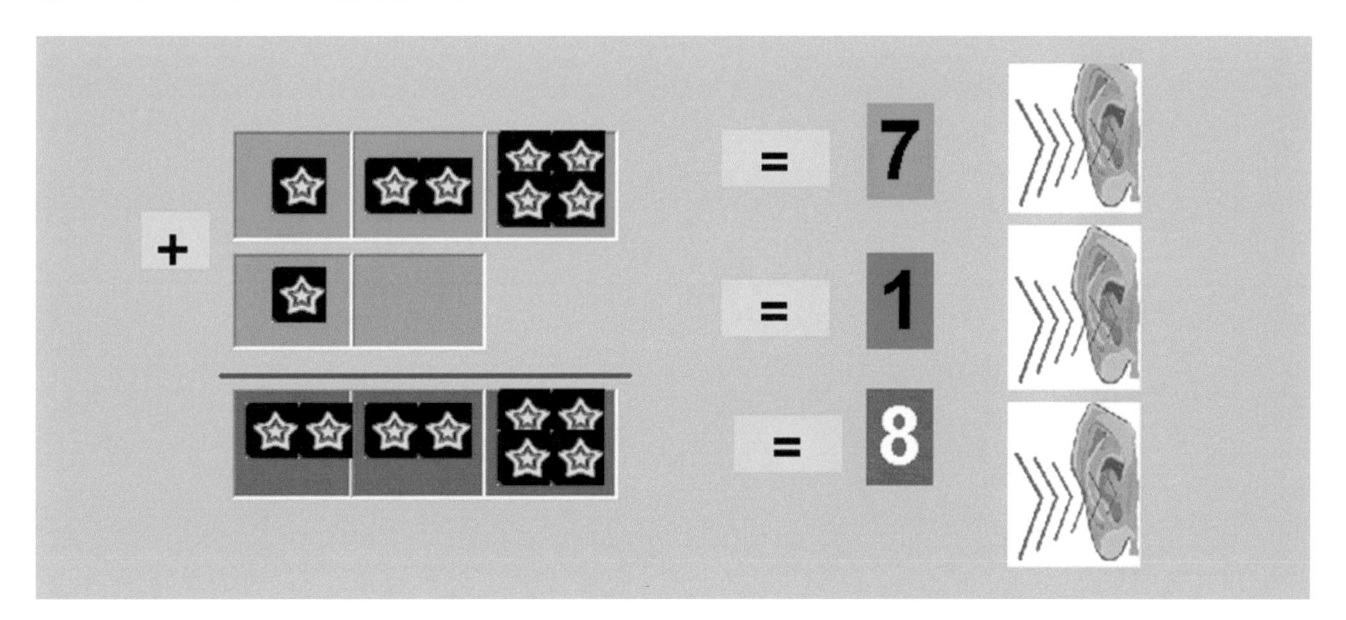

The ***kinesthetic*** representation of the filled up position is announced by clamping of the right hand. An empty position is represented by the clamping of the left hand. The addition sound, tram, is represented by one hand extended in front. The equal sound, cling, is represented by crossing of hands. The kinesthetic representation for both pictures and the answer is the clamping of the right hand thrice, one hand extended in front, right hand, left hand, crossing of hands, left hand thrice, and right hand.

MATHEMATICS ADDITION, EXERCISE 33, STEP 33.2

In exercise 33, step 33.2 in ***visual*** display of addition we see two pictures of boxes. The first picture has a symbol in all three positions. The second picture has the second and third positions filled with a symbol. For purpose of addition we are introducing the concept of moving. When a position passes its maximum capacity the next symbol is simultaneously moved in the ascending position. We are moving from the first to the second position. However the audio representation for the picture does not change because the numerical value does not change.

The ***audio*** representation for the three pictures consumes of a **double knock, double knock, double knock, and knock.**

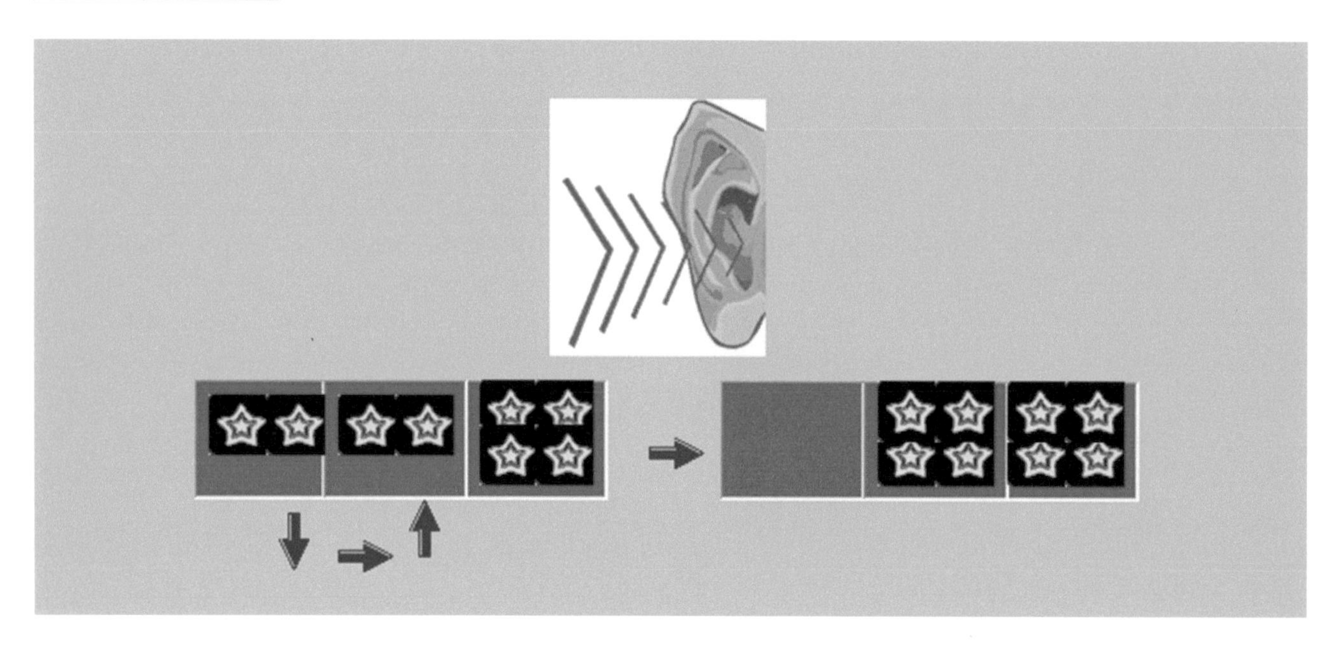

The ***kinesthetic*** representation of the filled up position is announced by clamping of the right hand. An empty position is represented by the clamping of the left hand. The kinesthetic representation for three pictures is the clamping of the left hand thrice and right hand.

ADDITION, EXERCISE 33, STEP 33.3

In exercise 33, step 33.3 in ***visual*** display of addition we see two pictures of boxes. The first picture has a symbol in the second and third positions. The second picture has the third position filled with a symbol. For purpose of addition we are introducing the concept of moving. When a position passes its maximum capacity the next symbol is simultaneously moved in the ascending position. We are moving from the first to the second position. However the audio representation for the picture does not change because the numerical value does not change.

The ***audio*** representation for the three pictures consumes of a **double knock, double knock, double knock, and knock.**

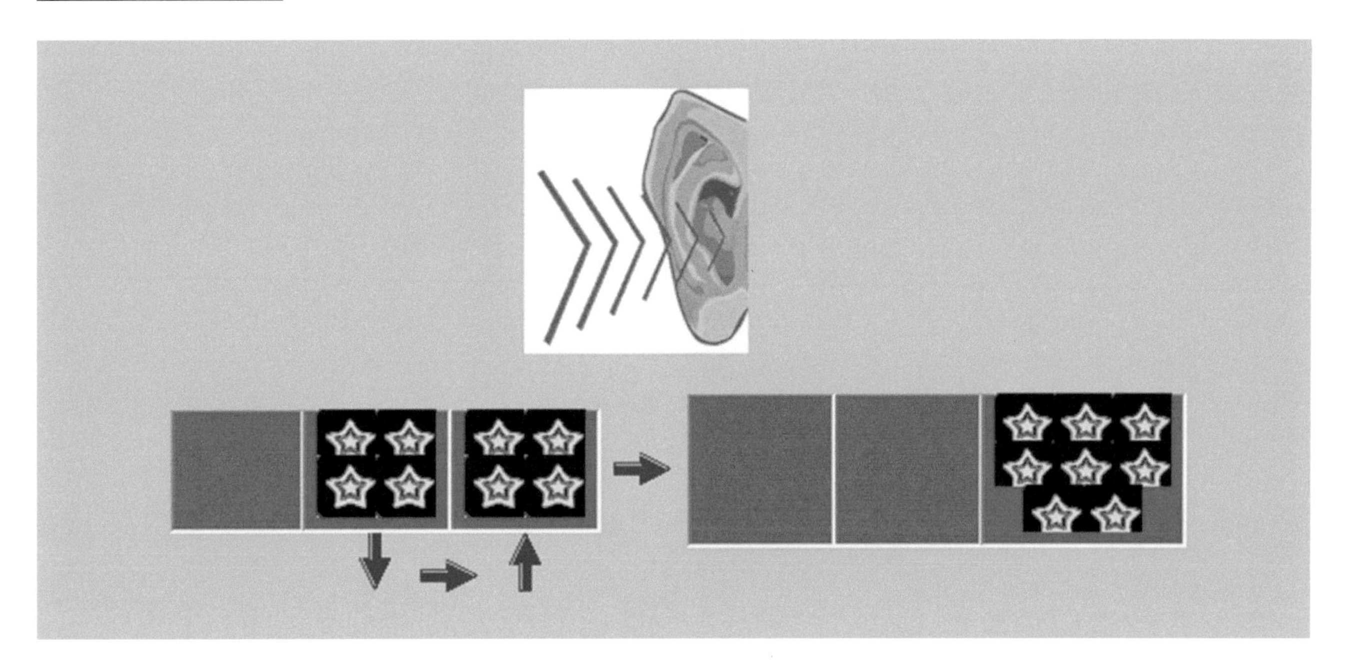

The ***kinesthetic*** representation of the filled up position is announced by clamping of the right hand. An empty position is represented by the clamping of the left hand. The kinesthetic representation for three pictures is the clamping of the left hand thrice and right hand.

MATHEMATICS

ADDITION, EXERCISE 34, STEP 34.4

In exercise 34, step 34.4 in ***visual*** display of addition we see two pictures of boxes. The first picture has a symbol in the third position. The second picture has the fourth position filled with a symbol. For purpose of addition we are introducing the concept of moving. When a position passes its maximum capacity the next symbol is simultaneously moved in the ascending position. We are moving from the first to the second position. However the audio representation for the picture does not change because the numerical value does not change.

The ***audio*** representation for the three pictures consumes of a **double knock, double knock, double knock, and knock.**

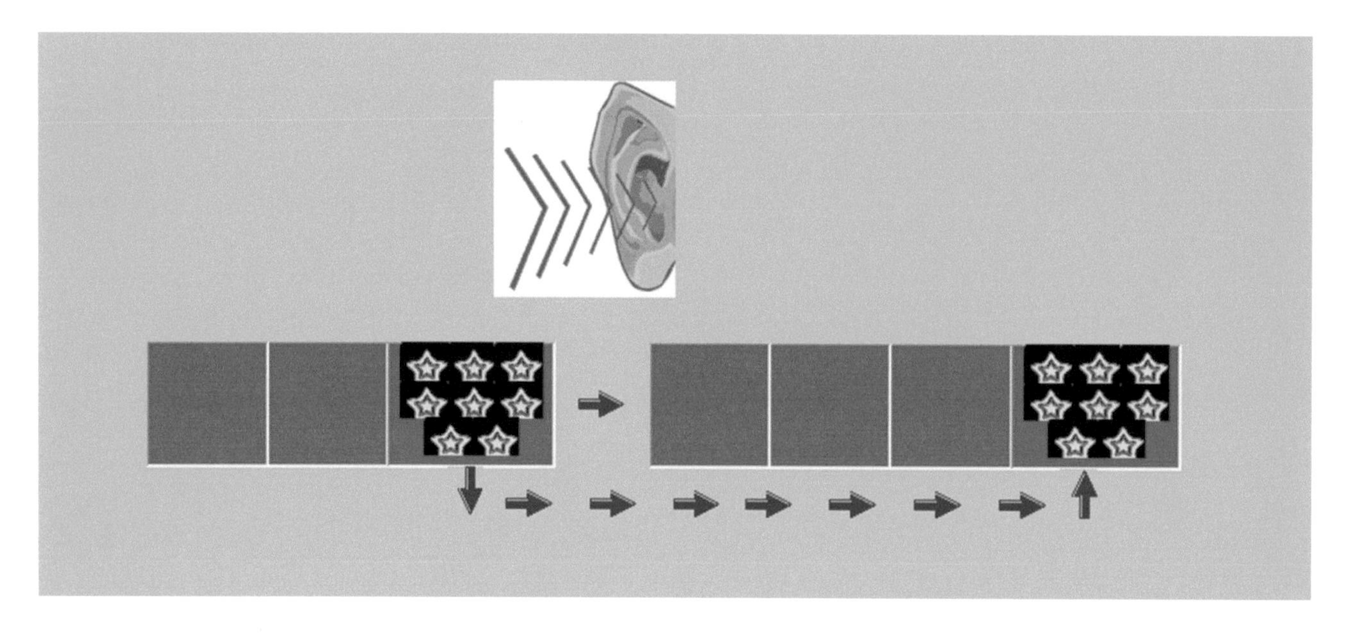

The ***kinesthetic*** representation of the filled up position is announced by clamping of the right hand. An empty position is represented by the clamping of the left hand. The kinesthetic representation for three pictures is the clamping of the left hand thrice and right hand.

ADDITION, EXERCISE 33, STEP 33.5

In exercise 33, step 33.5 in ***visual*** display of addition, we see two pictures of boxes with equal sign. Both pictures have the first three of their positions empty and the last position is filled with a symbol. The first picture has a yellow star in its positions but the second picture has a red ball in its positions. The form and shape of the symbols do not change mathematical value and the numerical value of both pictures is equal to eight.

The ***audio*** representation for this picture consumes of a **double knock, double knock, double knock, knock, and cling.**

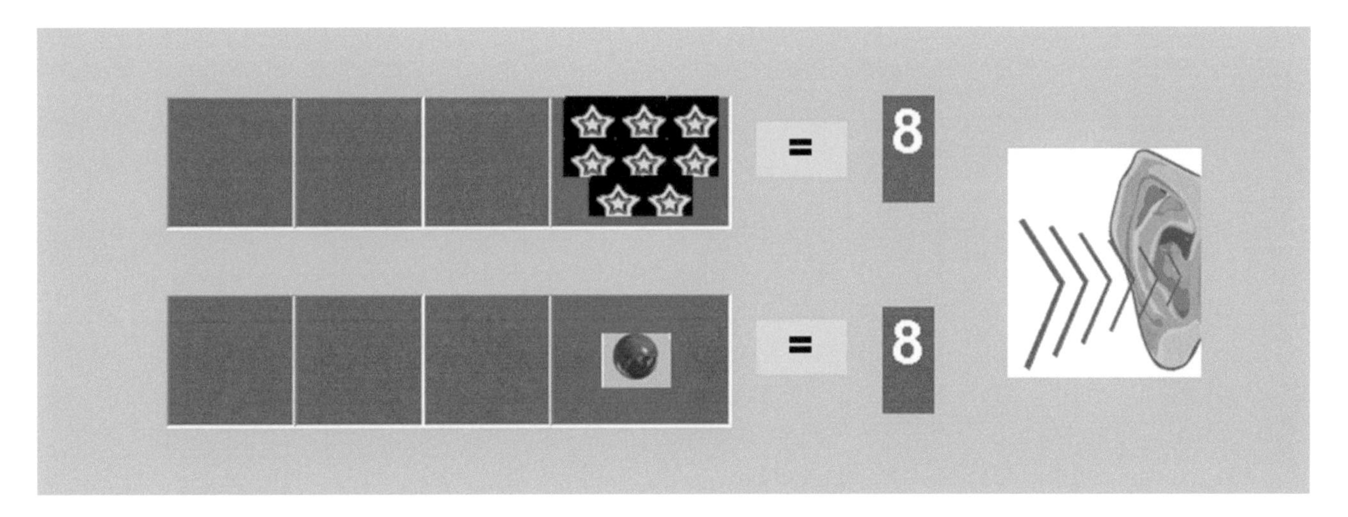

The ***kinesthetic*** representation of the filled up position is announced by clamping of the right hand. An empty position is represented by the clamping of the left hand. The equal sign is crossing both hands in front. The kinesthetic representation for this picture is clamping the left hand thrice, right hand, and crossing of hands.

MATHEMATICS

ADDITION, EXERCISE 33

In exercise 33, in ***visual*** display of addition we see two pictures of boxes. In the bottom we see three pictures with probable answers. We want to specify, that between the first and second pictures we see the addition sign. Bellow the second picture we see the equal line, which separates the problem from the answer. The ***audio*** representation for two pictures consumes of a **knock, knock, knock, tram, knock, double knock, and cling.**

From the probable answers for addition we need to find the right one. The audio signals are: 1) **double knock,double knock, double knock, and knock.** 2) **double knock, knock, double knock, and double knock** 3) **double knock, double knock, knock, and double knock**.

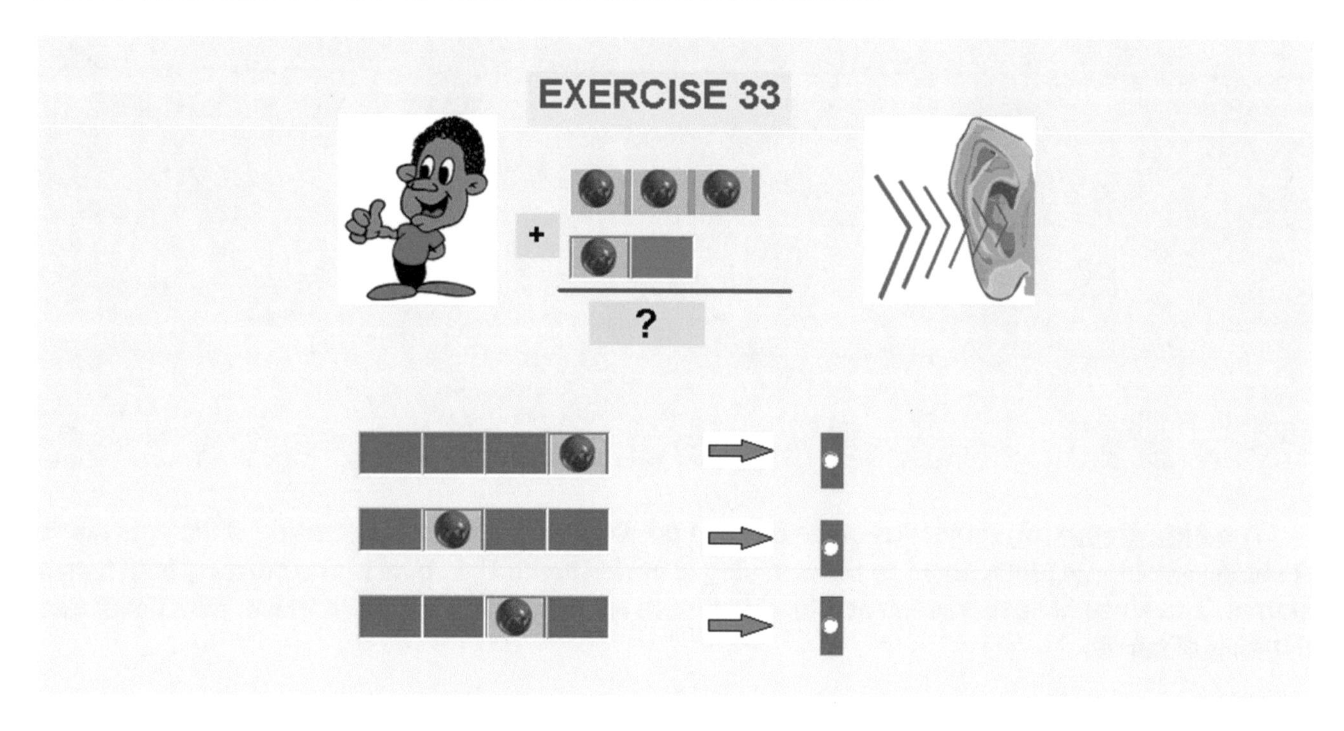

The ***kinesthetic*** representation of the filled up position is announced by clamping of the right hand. An empty position is represented by the clamping of the left hand. The addition sound, tram, is represented by one hand extended in front. The equal sound, cling, is represented by crossing of hands. The kinesthetic representation for both pictures and the answer is the clamping of the right hand thrice, one hand extended in front, right hand, left hand, crossing of hands, left hand thrice, and right hand.

SUBTRACTION

To subtract one number from another, Litvin's Code uses symbols (dots) in assigned positions, instead of digits.

Rules:

- When subtracting two numbers, where a given position of the first number is empty but the corresponding position of the second number is filled with a symbol, we borrow a symbol from the following ascending positions of the first number, which becomes two symbols in the prior position.

- When we borrow a symbol from a following position, this symbol becomes two symbols in the previous position. We continue borrowing from each higher position to fulfill the lower position with a symbol (dot), because it is impossible subtract from an empty position.

- When doing a subtraction, we borrow available symbols from the higher position. It is important to remember that all symbols borrowed from a higher position are twice as large as the one in the prior position.

- Once the symbol is borrowed, the position that the symbol is borrowed from becomes empty and two symbols appear in the prior position. If needed, we leave one symbol in the prior position and take one symbol to the next descending position which is empty. Once again, this one symbol contains two symbols in the next descending position.

- When a symbol is borrowed, then this one symbol becomes turns into two symbols in the next descending position.

Example 3.1

In Example 3.1, we could subtract 1 from 5.

First number is 5 "x 0 x"

	•		•	5
−	•			− 1
			•	4

Second number is 1 "x"

Answer is 4 "0 0 x"

In Example 3.1, we have two numbers with three positions each. In the first number, the first and third positions are filled with symbols. In the second number, only the first position is filled with a symbol.

In Summary:

- When we have two symbols (dots) in the first positions of the numbers being subtracted, then after the subtraction, the first position in the answer becomes empty or zero.

- When there is a symbol present in the third position of the first number and the same position in the second number is empty, we move the symbol from the third position of the first number to the third position of the answer.

- Generally, when subtracting two numbers and one of the positions of the first number has a symbol and the second number does not have a symbol in the same position, we just move the symbol to the same position of the answer.

End.

Example 3.2

First number is 6 "0 x x"

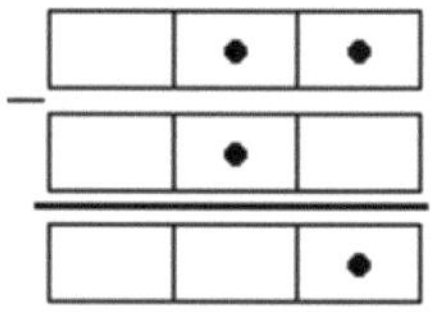

	•	•
	•	
		•

$$6 - 2 = 4$$

Second number is 2 " 0 x 0 " Answer is 4 " 0 0 x "

In Example 3.2, we have two numbers with three positions each. In the first number, positions two and three filled with symbols. In the second number, the second position is filled with a symbol.

Explanation

During the subtraction, we eliminate the two symbols in the corresponding second positions of the numbers in the subtraction. The two symbols (dots) in the parallel positions are replaced in the answer by an empty space or "0". We move the single symbol left in the third position of the first number to the third position of the answer.

In Summary:

- During the subtraction of the numbers above, the first number has a symbol in the third position and the second number has empty space in the same position. Therefore, the answer has a symbol in the third position.

- When there is one symbol in parallel positions of the numbers in the subtraction, then it is moved to the answer.

End.

Example 3.3

First number is 5 "x 0 x "

	•		•	5
−		•		− 2
	•	•		3

Second number is 2 " 0 x 0"

Answer is 3 "x x 0"

In Example 3.3, we have two numbers with three positions each. The first number has symbols in its first and third positions. The second number has a symbol in the second position.

Explanation

In Example 3.3, in the first number, we borrow a symbol from the third position to the prior second position. In the third position of the first number there is a symbol (dot), but the second number has empty space in the same position. We transfer the symbol from the first position of the first number to the answer. Because it is impossible to subtract from empty space, we borrow a symbol from the third position of the first number. The one symbol from the third position becomes two symbols in second position, leaving the third position of the first number empty. We subtract one symbol from the second position of the first number, because the first number has two symbols in the second position and the second number in the second position has one symbol. Now the answer has symbols in the first and second positions.

In Summary:

- When borrowing a symbol from the third position, we need to remember that the binary number in the third position of the first number is twice as big as the prior one. In our example, the prior descending position is position two in the first number. Now we have two symbols in the second position. The binary number which is expressed with one symbol in the third position is expressed with two symbols in the second position. Now we subtract two symbols (dots) from the second position of the first number and one symbol from second position of the second number or (2-1). This leaves us with one symbol for the answer, which is placed in the second position of the answer.

- When subtracting one symbol from the second position in the second number, the symbols from the first and second positions of the first number are moved to the answer.

End.

Example 3.4

First number is 6 “0 x x”

Second number is 3 “x x 0”

Answer is 3 “x x 0“

		•	•	6
−	•	•		− 3
	•	•		3

In Example 3.4, we have two numbers with the three positions each. In the first number, the first position is empty, while the second and third positions are filled with symbols (dot). In the second number, the first and second positions are filled with symbols (dot). As mentioned before, it is impossible to subtract form an empty position. In this case, the first position of the first number is empty. Therefore, we borrow from the second position of the first number and one symbol becomes two in the prior descending position of the first number, leaving the second position of the first number empty.

Explanation

After borrowing two symbols from the second position of the first number, we subtract one symbol in the first position of the second number from two symbols in the first position of the first number. We place the one leftover symbol from the subtraction in the first position of the answer. After the first borrowing, we have the second position of the first number empty. Now we borrow a symbol from the third position of the first number. One symbol in the third position becomes two symbols in the second position of the first number and the third position of the first number is becomes empty. We subtract two symbols in the second position of the first number and one symbol in second position of the second number and place the symbol in the second position of the answer. In the end, we have the first and second positions of the answer filled with symbols (dot).

In Summary:

- During the subtraction, the first number has empty positions over the second number that has the same position filled up. We borrow an available symbol from the next position of the first number and the position from which the symbol was borrowed becomes empty.

- In a situation when the borrowed position becomes empty and is over the second number where there a position is filled, we need to borrow again from the next available position and the symbol from the next position becomes two symbols in prior positions.

Example 3.5

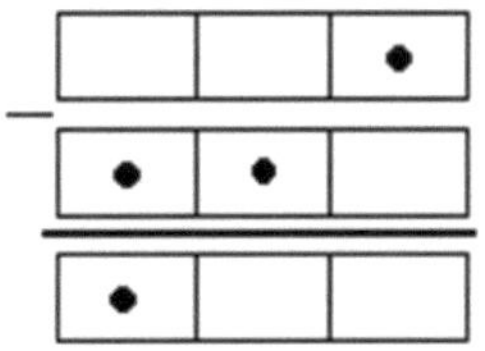

First number is 4 “0 0 x

Second number is 3 “x x 0”

Answer is 1 “x 0 0”

−			•	4
	•	•		− 3
	•			1

In Example 3.5, we have two numbers with three positions each. The third position in the first number is filled with a symbol. The first and second positions in the second number are filled with symbols as well. If the higher position of the first number (the third position) is filled with a symbol and the lowest two positions of the second number (positions one & two) are filled with symbols, the answer has only one position filled with a symbol (position one). Since the first number has two empty spaces in the positions under which there are symbols to be subtracted from, we need to begin borrowing.

Explanation

In Example 3.5, we have the third position of the first number filled with a symbol (dot), while the first two positions are empty. The second number is filled with symbols (dot) in its first two positions, while the third position is empty. As mentioned in previous examples, we cannot subtract from positions that have empty spaces. Therefore, we borrow a symbol from the third position of the first number to the second position of the first number. Now the second position of the first number has two symbols and the third position of the first number is empty. We subtract the symbol in the second position of the second number from the symbol in the second position of the first number. As a result, we are left with one symbol in the second position of the first number. We continue the same process of borrowing a symbol from the second position of the first number into the first position of the first number. In the end, we are left with one symbol in the first position of the answer.

In Summary:

- When the second position in the first number has two symbols, then we borrow a symbol from the next position, which is equal to two symbols in the prior position. The next position becomes empty.

- When subtracting one symbol in the second position of the second number from two symbols on the second position of the first number, one symbol is left in the second position of the first number. We borrow this symbol from the second position of the first number to the first position of the first number.

End.

SUBTRACTION, EXERCISE 1, STEP 1

In exercise 1, steps 1 in ***visual*** display of subtraction we see three pictures of boxes. Between the first and second pictures we see the subtraction sign. Under the second picture we see the equal line, which separates the problem from the answer. Below the equal line is the third picture, which is the answer of subtraction. The ***audio*** representation for subtraction consumes of a **knock, double nock, double tram, knock, double knock, cling, double knock, double knock.** Step 1

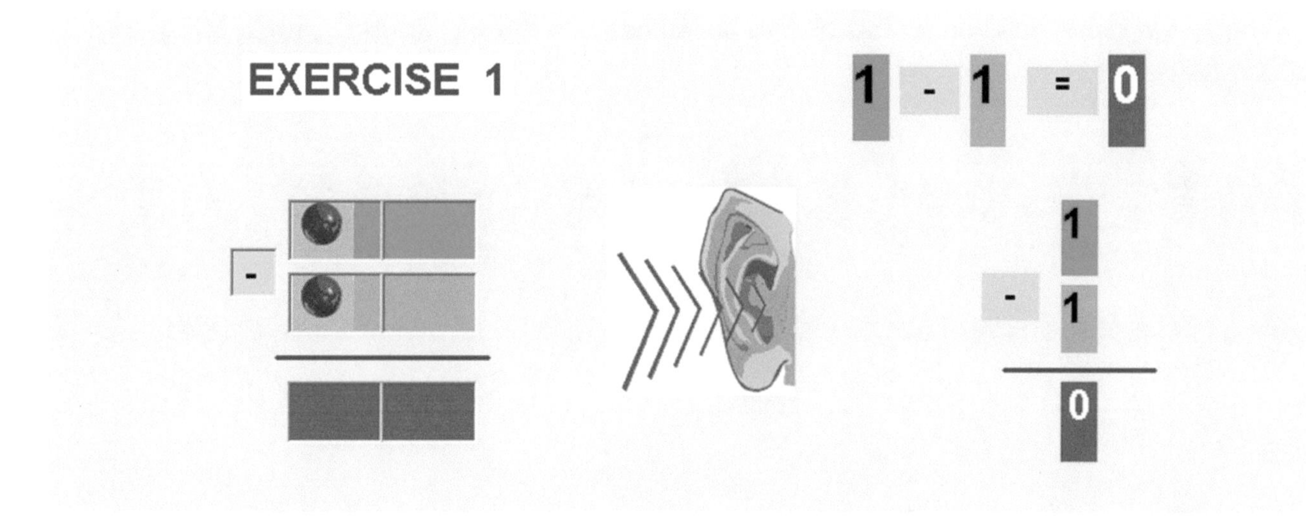

The ***kinesthetic*** representation of the filled up position is announced by clamping of the right hand. An empty position is represented by the clamping of the left hand. The subtraction sound, double tram, is represented by both hands extended in front. The equal sound, cling, is represented by crossing both hands. The kinesthetic representation for three pictures is the clamping of the right hand, left, both hands extended in front, right hand, left hand, crossing of both hands, and the right hand twice.

MATHEMATICS SUBTRACTION, EXERCISE 1, STEP 1.2

In exercise 1, step 1.2 in ***visual*** display of the last step of subtraction, we see two pictures of boxes with equal sign between them. On both pictures we see that the first position is full but the second position is empty. The first picture has a red ball in the first position but the second picture has a yellow star in the first position. The form and shape of the symbols do not change mathematical value and the numerical value of both pictures is equal to one.

The ***audio*** representation for this picture consumes of a **knock, double knock, cling, knock, double knock**.

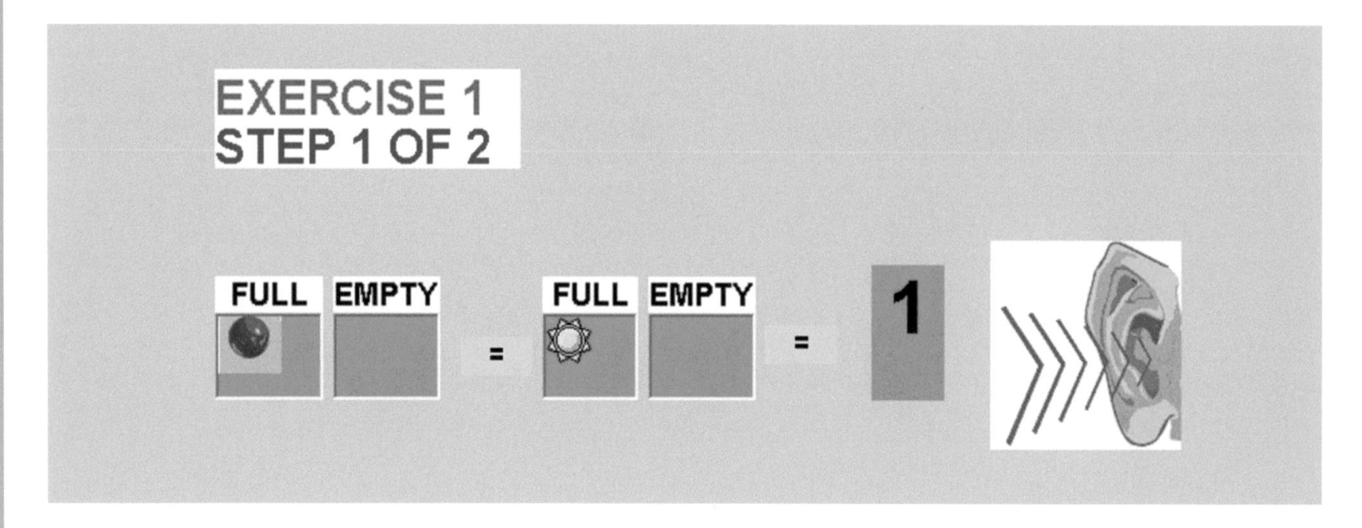

The ***kinesthetic*** representation of the filled up position is announced by clamping of the right hand. An empty position is represented by the clamping of the left hand. The equal sign is crossing both hands in front. The kinesthetic representation for this picture is clamping the right hand, left hand, crossing hands, right hand and left hand.

SUBTRACTION, EXERCISE 1, STEP 2

In exercise 1, steps 2 in ***visual*** display of subtraction we see three pictures of boxes. Between the first and second pictures we see the subtraction sign. Under the second picture we see the equal line, which separates the problem from the answer. Below the equal line is the third picture, which is the answer of subtraction. The ***audio*** representation for subtraction consumes of a **knock, double knock, double tram, knock, double knock, cling, double knock, and double knock.**

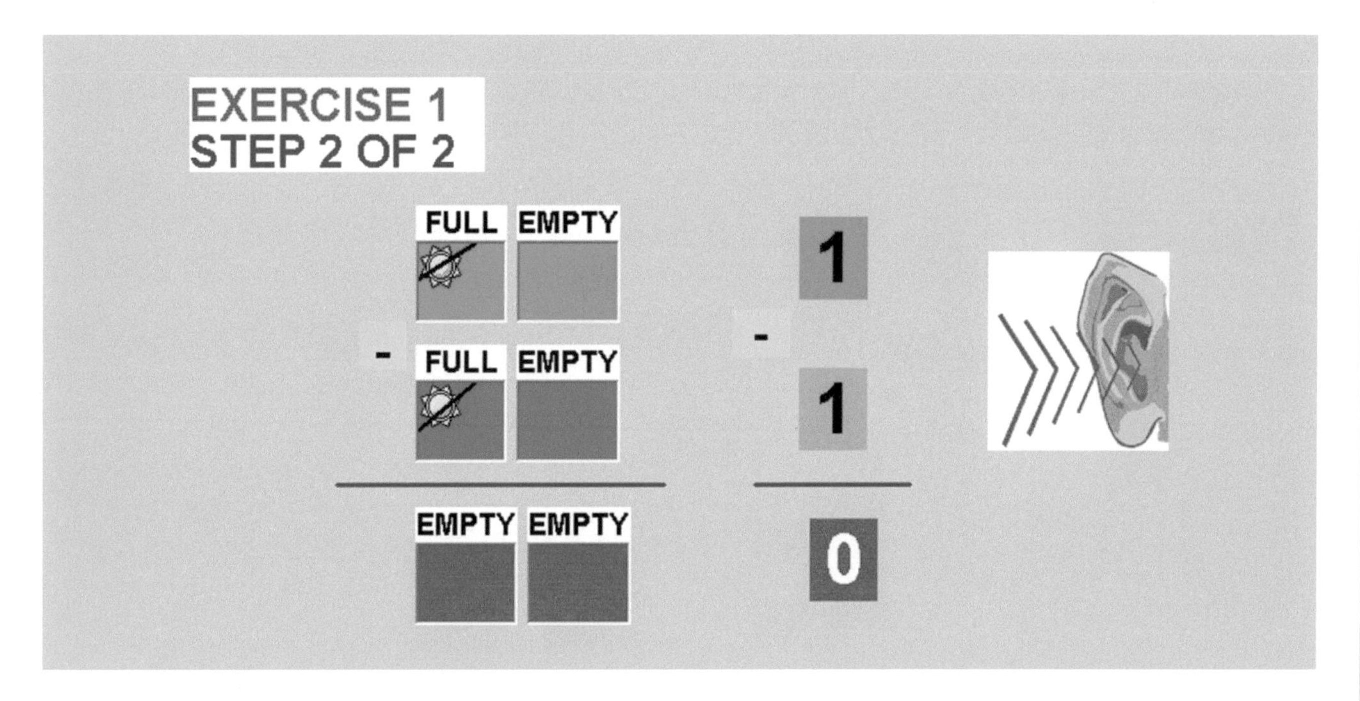

The ***kinesthetic*** representation of the filled up position is announced by clamping of the right hand. An empty position is represented by the clamping of the left hand. The subtraction sound, double tram, is represented by both hands extended in front. The equal sound, cling, is represented by crossing both hands. The kinesthetic representation for three pictures is the clamping of the right hand, left, both hands extended in front, right hand, left hand, crossing of hands, and the left hand twice.

MATHEMATICS

SUBTRACTION, FIND AN ANSWER, EXERCISE 1

In exercise 1, in ***visual*** display of subtraction we see two pictures of boxes. In the bottom we see three pictures with probable answers. Both pictures have a symbol in the first position and the second position is empty. We want to specify, that between the first and second pictures we see the subtraction sign. Bellow the second picture we see the equal line, which separates the problem from the answer. The ***audio*** representation for two pictures consumes of a **knock, double knock, double tram, knock, double knock, and cling.**

From the probable answers for addition we need to find the right one. The audio signals are: 1) **double knock and double knock.** 2) **knock and double knock** 3) **double knock and knock**.

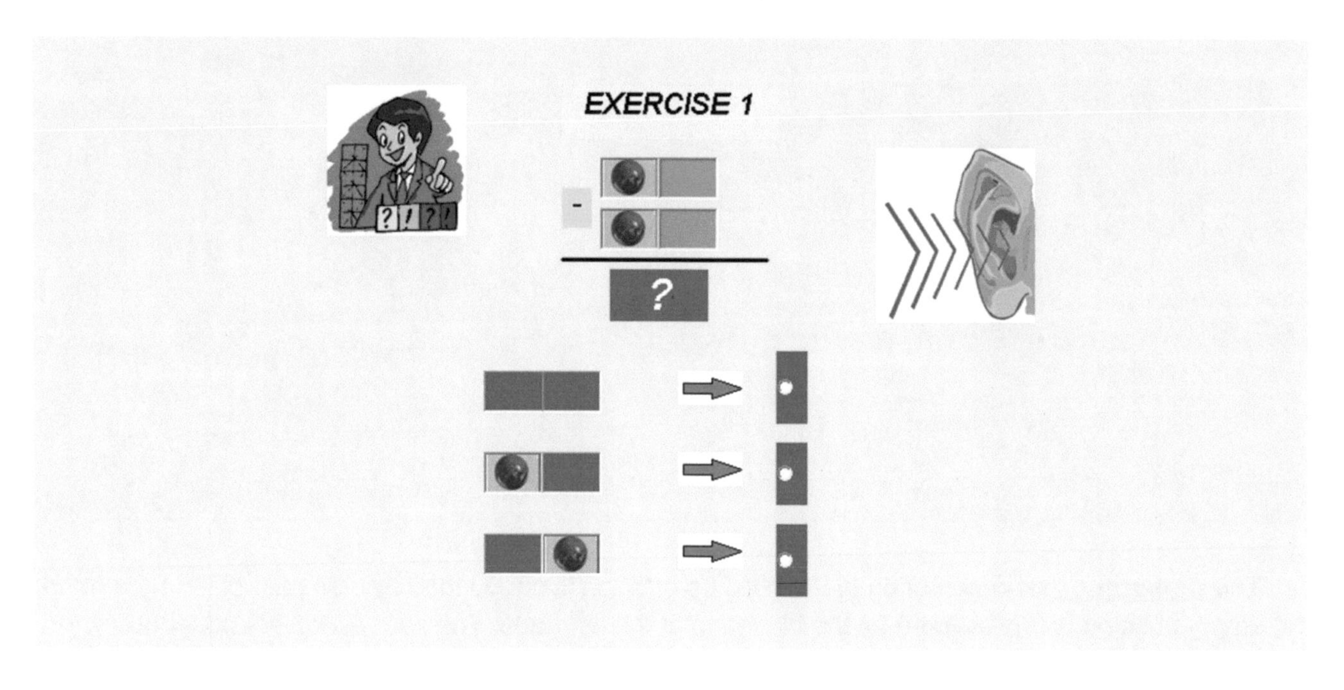

The ***kinesthetic*** representation of the filled up position is announced by clamping of the right hand. An empty position is represented by the clamping of the left hand. The subtraction sound, double tram, is represented by both hands extended in front. The equal sound, cling, is represented by crossing of hands. The kinesthetic representation for both pictures and the answer is the clamping of the right hand, left hand, both hands extended in front, clamping the right hand, left hand, crossing of hands, left hand twice.

SUBTRACTION, EXERCISE 2

In exercise 2, in ***visual*** display of subtraction we see three pictures of boxes. Between the first and second pictures we see the subtraction sign. Under the second picture we see the equal line, which separates the problem from the answer. Below the equal line is the third picture, which is the answer of subtraction. The ***audio*** representation for subtraction consumes of a **double knock, knock, double tram, knock, double knock, cling, knock, double knock.**

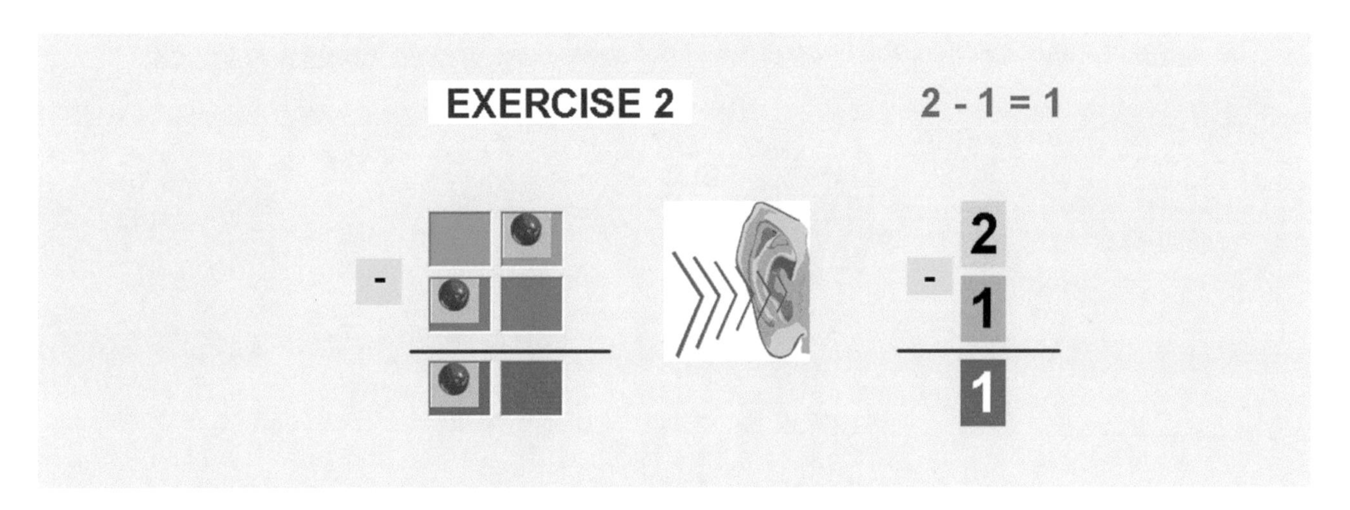

The ***kinesthetic*** representation of the filled up position is announced by clamping of the right hand. An empty position is represented by the clamping of the left hand. The subtraction sound, double tram, is represented by both hands extended in front. The equal sound, cling, is represented by crossing both hands. The kinesthetic representation for three pictures is the clamping of the left hand, right hand, both hands extended in front, right hand, left hand, crossing of both hands, right hand, and left hand.

MATHEMATICS SUBTRACTION, EXERCISE 2, STEP 2.1

In exercise 2, step 2.1 in ***visual*** display of subtraction we see three pictures of boxes. In the first two pictures the first position is empty while the second position is filled with a symbol. The third picture has a symbol in the first position and is empty in the second position. For purpose of subtraction we are introducing the concept of borrowing. We are borrowing from the second to the first position. However the audio representation for the picture does not change because the numerical value does not change.

The ***audio*** representation for the three pictures consumes of a **double knock and knock.**

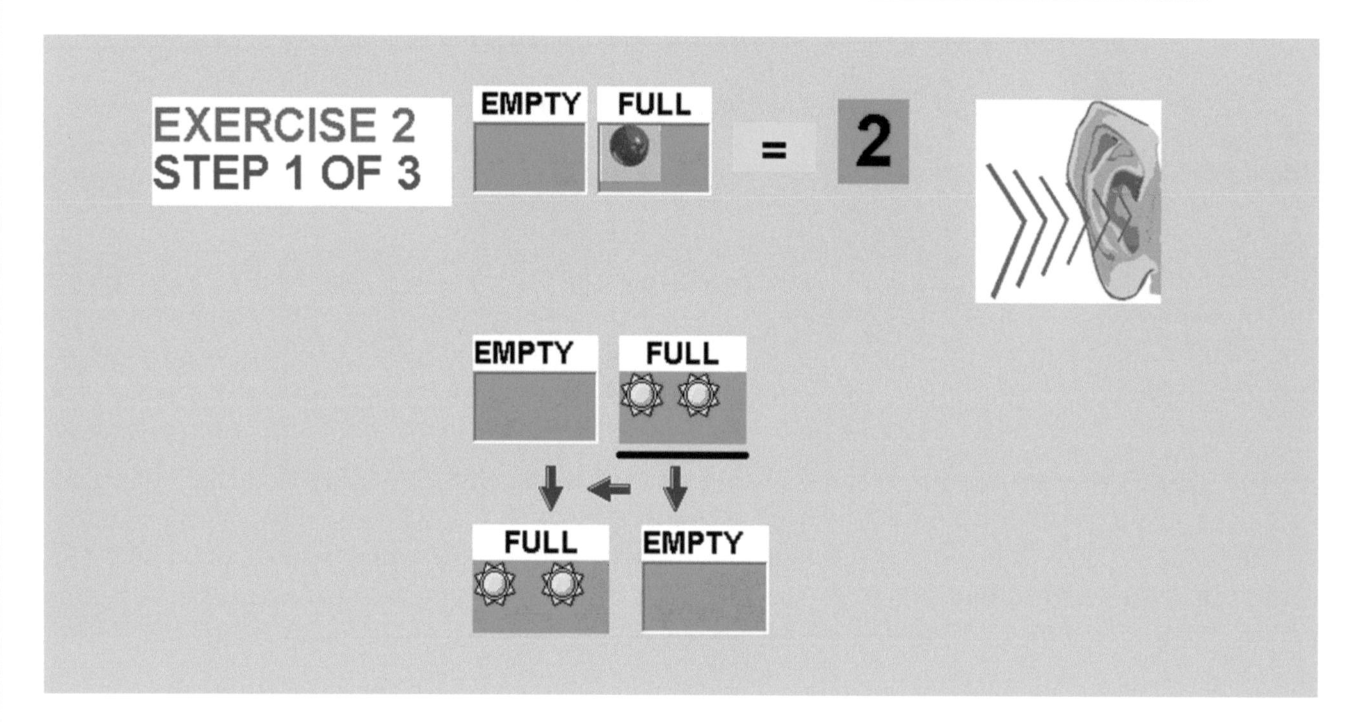

The ***kinesthetic*** representation of the filled up position is announced by clamping of the right hand. An empty position is represented by the clamping of the left hand. The equal sound, cling, is represented by crossing both hands. The kinesthetic representation for three pictures is the clamping of the left hand, right hand, and crossing of hands.

SUBTRACTION, EXERCISE 2.2

In exercise 2, step 2.2 in ***visual*** display of subtraction we see six pictures of boxes. There are three pictures on the right side and three pictures on the left side. Both sides represent the same equation; however the left side has used the concept of borrowing to further show the problem. When we do subtraction we are borrowing available symbols from the higher position. We need to remember that all symbols borrowed from the higher position are twice larger than the one in the prior descending position. Once the symbol is borrowed the position borrowed becomes empty and the two symbols appear in the prior position. The ***audio*** representation for subtraction consumes of a **double knock, knock, double tram, knock, double knock, cling, knock, and double knock.**

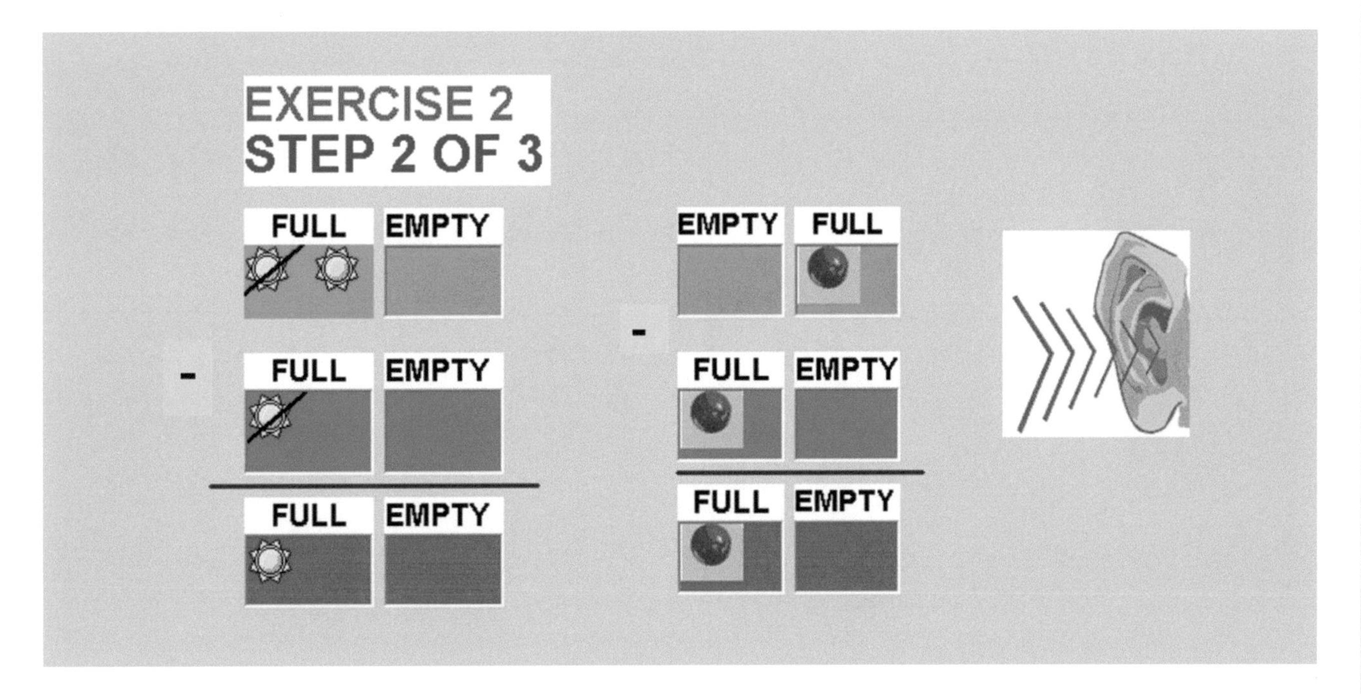

The ***kinesthetic*** representation of the filled up position is announced by clamping of the right hand. An empty position is represented by the clamping of the left hand. The subtraction sound, double tram, is represented by both hands extended in front. The equal sound, cling, is represented by crossing both hands. The kinesthetic representation for the subtraction is the clamping of the left hand, right hand, hands extended in front, right hand, left hand, crossing of hands, right hand, and left hand.

MATHEMATICS

SUBTRACTION, EXERCISE 2, STEP 2.3

In exercise 2, step 2.3 in ***visual*** display of the last step of subtraction, we see two pictures of boxes with equal sign between them. On both pictures we see that the first position is full but the second position is empty. The first picture has a yellow star in the first position but the second picture has a red ball in the first position. The form and shape of the symbols do not change mathematical value and the numerical value of both pictures is equal to one.

The ***audio*** representation for this picture consumes of a **knock, double knock, cling, knock, and double knock.**

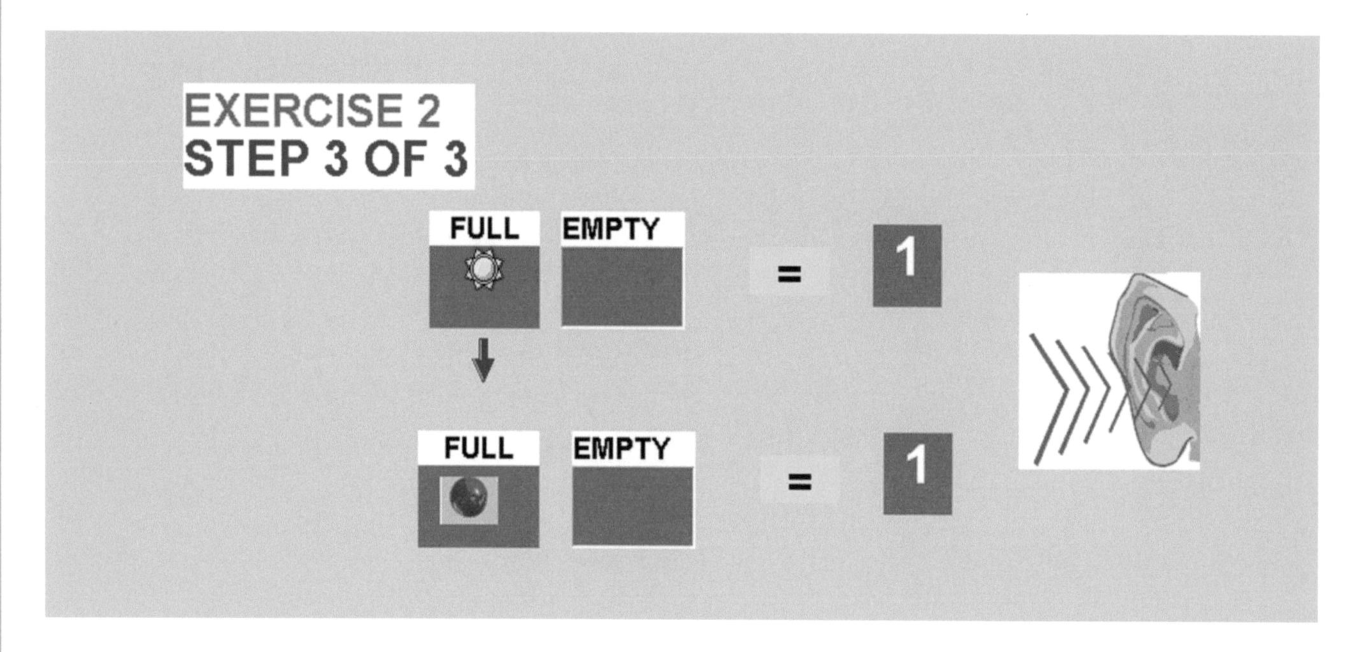

The ***kinesthetic*** representation of the filled up position is announced by clamping of the right hand. An empty position is represented by the clamping of the left hand. The equal sign is crossing both hands in front. The kinesthetic representation for this picture is clamping the right hand, left hand, crossing of hands, right hand, and left hand.

SUBTRACTION, FIND AN ANSWER, EXERCISE 2

In exercise 2, in ***visual*** display of subtraction we see two pictures of boxes. In the bottom we see three pictures with probable answers. The first picture has the first position empty and the second position is filled up with a symbol. The second picture has a symbol in the first position and the second position is empty. We want to specify, that between the first and second pictures we see the subtraction sign. Bellow the second picture we see the equal line, which separates the problem from the answer. The ***audio*** representation for two pictures consumes of a **double knock, knock, double tram, knock, double knock, and cling.**

From the probable answers for addition we need to find the right one. The audio signals are: 1) **double knock and double knock.** 2) **double knock and knock** 3) **knock and double knock**.

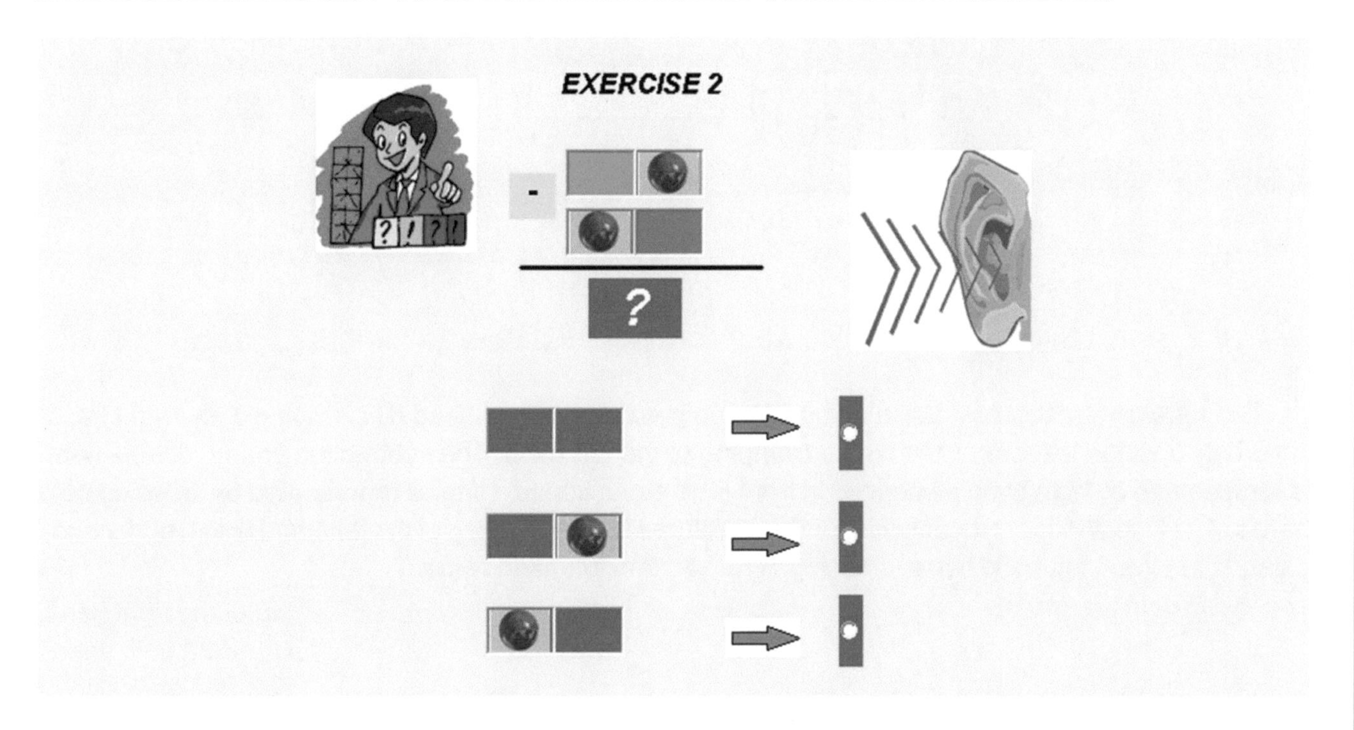

The ***kinesthetic*** representation of the filled up position is announced by clamping of the right hand. An empty position is represented by the clamping of the left hand. The subtraction sound, double tram, is represented by both hands extended in front. The equal sound, cling, is represented by crossing of hands. The kinesthetic representation for both pictures and the answer is the clamping of the left hand, right hand, hands extended in front, right hand, left hand, crossing of hands, left hand, and right hand.

MATHEMATICS SUBTRACTION, EXERCISE 3

In exercise 3, in ***visual*** display of subtraction we see three pictures of boxes. Between the first and second pictures we see the subtraction sign. Under the second picture we see the equal line, which separates the problem from the answer. Below the equal line is the third picture, which is the answer of subtraction. The ***audio*** representation for subtraction consumes of a **double knock, double knock, double tram, double knock, double knock, cling, double knock, double knock.**

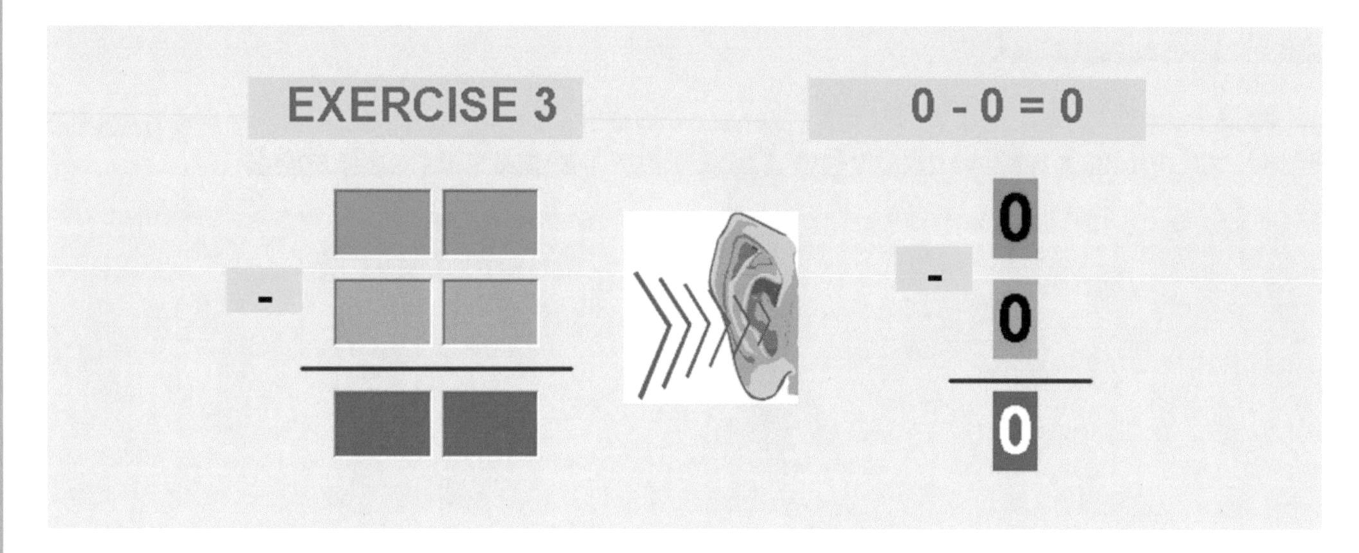

The ***kinesthetic*** representation of the filled up position is announced by clamping of the right hand. An empty position is represented by the clamping of the left hand. The subtraction sound, double tram, is represented by both hands extended in front. The equal sound, cling, is represented by crossing both hands. The kinesthetic representation for three pictures is the clamping of the left hand twice, both hands extended in front, left hand twice, crossing of hands, and left hand twice.

SUBTRACTION, FIND AN ANSWER, EXERCISE 3

In exercise 3, in ***visual*** display of subtraction we see two pictures of boxes. In the bottom we see three pictures with probable answers. Both pictures have a symbol in the first position and the second position is empty. We want to specify, that between the first and second pictures we see the subtraction sign. Bellow the second picture we see the equal line, which separates the problem from the answer. The ***audio*** representation for two pictures consumes of a **double knock, double knock, double tram, double knock, double knock, and cling.**

From the probable answers for addition we need to find the right one. The audio signals are: 1) **knock and double knock.** 2) **double knock and double knock** 3) **double knock and knock**.

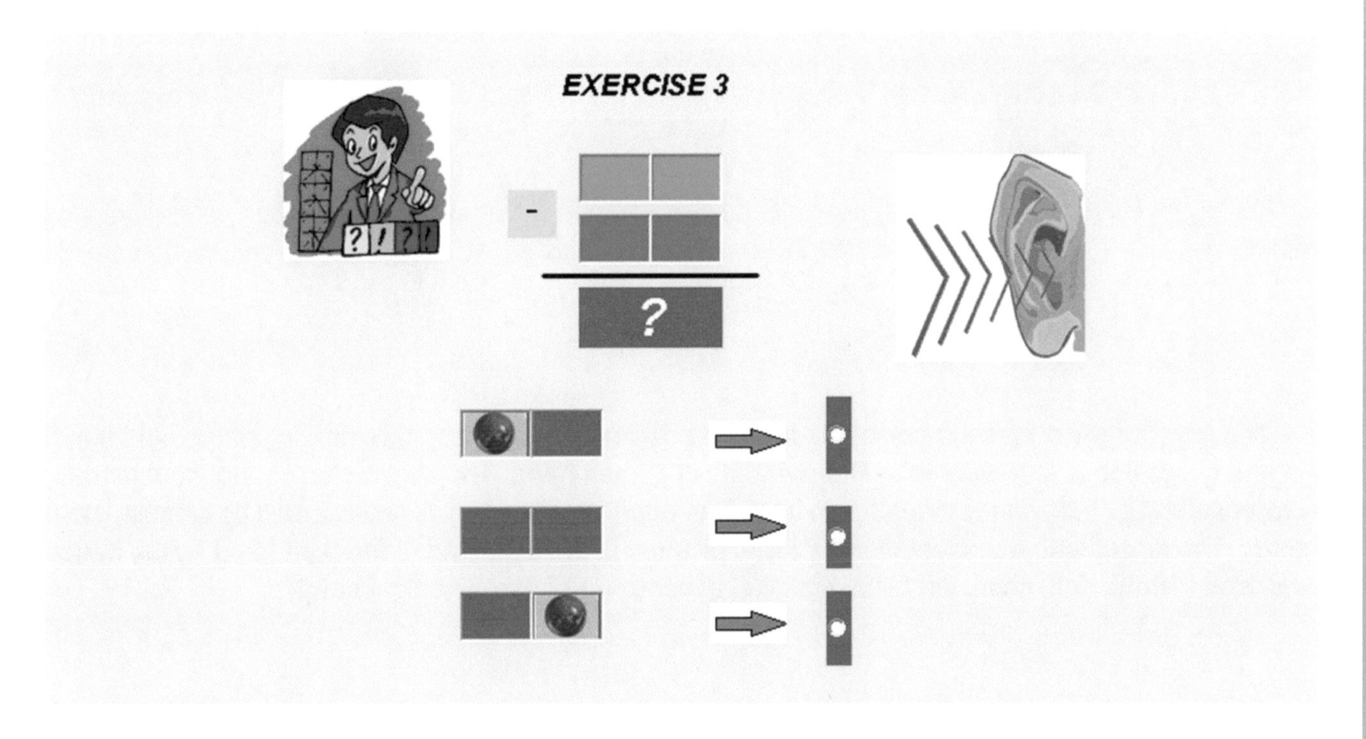

The ***kinesthetic*** representation of the filled up position is announced by clamping of the right hand. An empty position is represented by the clamping of the left hand. The subtraction sound, double tram, is represented by both hands extended in front. The equal sound, cling, is represented by crossing of hands. The kinesthetic representation for both pictures and the answer is the clamping of the left hand twice, hands extended in front, left hand twice, crossing of hands, left hand twice.

MATHEMATICS

SUBTRACTION, EXERCISE 5

In exercise 5, in ***visual*** display of subtraction we see three pictures of boxes. Between the first and second pictures we see the subtraction sign. Under the second picture we see the equal line, which separates the problem from the answer. Below the equal line is the third picture, which is the answer of subtraction. The ***audio*** representation for subtraction consumes of a **knock, knock, double tram, knock, double knock, cling, double knock, and knock.**

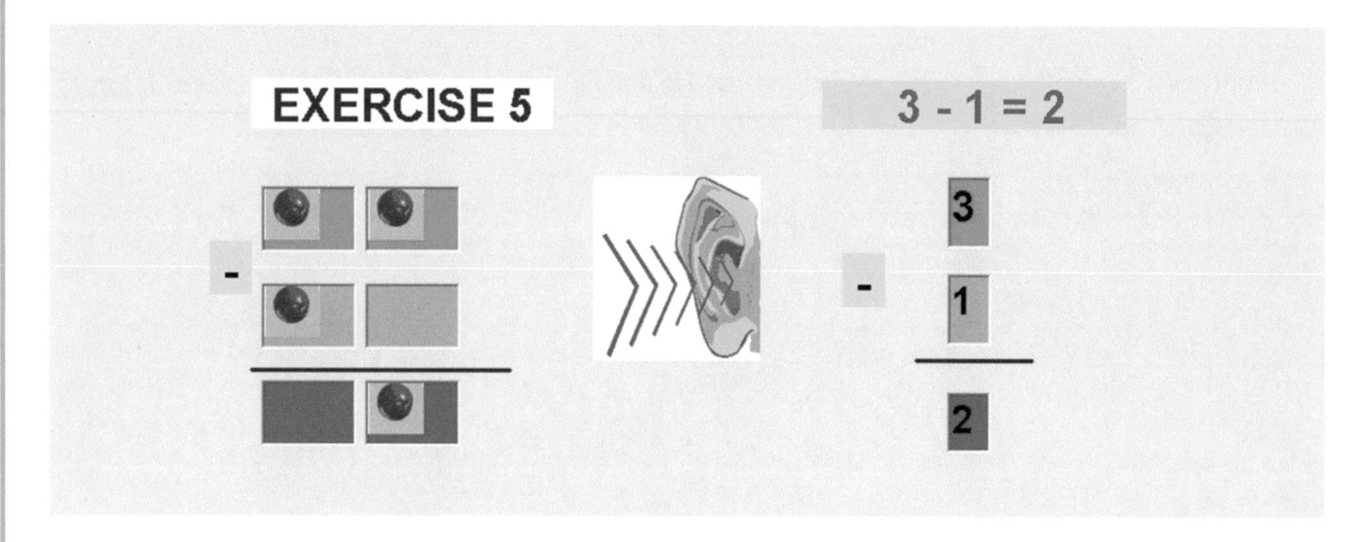

The ***kinesthetic*** representation of the filled up position is announced by clamping of the right hand. An empty position is represented by the clamping of the left hand. The subtraction sound, double tram, is represented by both hands extended in front. The equal sound, cling, is represented by crossing both hands. The kinesthetic representation for three pictures is the clamping of the right hand twice, hands extended in front, right hand, left hand, crossing of hands, left hand, and right hand.

SUBTRACTION EXERCISE 5, STEP 5.1

In exercise 5, step 5.1 in ***visual*** display of subtraction we see three pictures of boxes. The first picture has symbols in both positions. The second picture has a symbol in the first position while the second position is empty. The answer of the equation has the first position empty and the second position is filled with a symbol. We want to specify, that between the first and second pictures we see the subtraction sign. Bellow the second picture we see the equal line, which separates the problem from the answer. The ***audio*** representation for the three pictures consumes of a **knock, knock, double tram, knock, double knock, cling, double knock, and knock.**

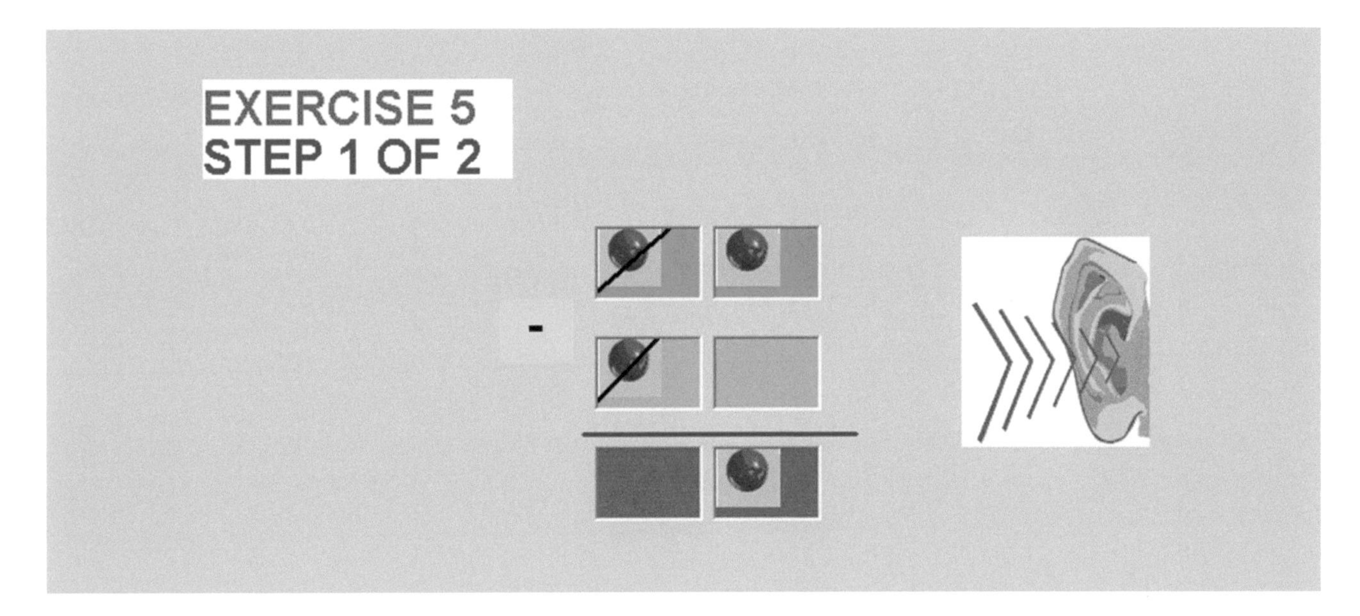

The ***kinesthetic*** representation of the filled up position is announced by clamping of the right hand. An empty position is represented by the clamping of the left hand. The subtraction sound, double tram, is represented by both hands extended in front. The equal sound, cling, is represented by crossing of hands. The kinesthetic representation for both pictures and the answer is the clamping of the right hand twice, hands extended in front, right hand, left hand, crossing of hands, left hand, and right hand.

MATHEMATICS ADDITION, EXERCISE 5, STEP 2

In exercise 5, step 5.2 in ***visual*** display of the last step of subtraction, we see one picture of boxes, which is the answer of the problem. We see that the first position is empty while the second position is filled up with a symbol. The answer has a value of two.

The ***audio*** representation for this picture consumes of a **double knock, knock, and cling.**

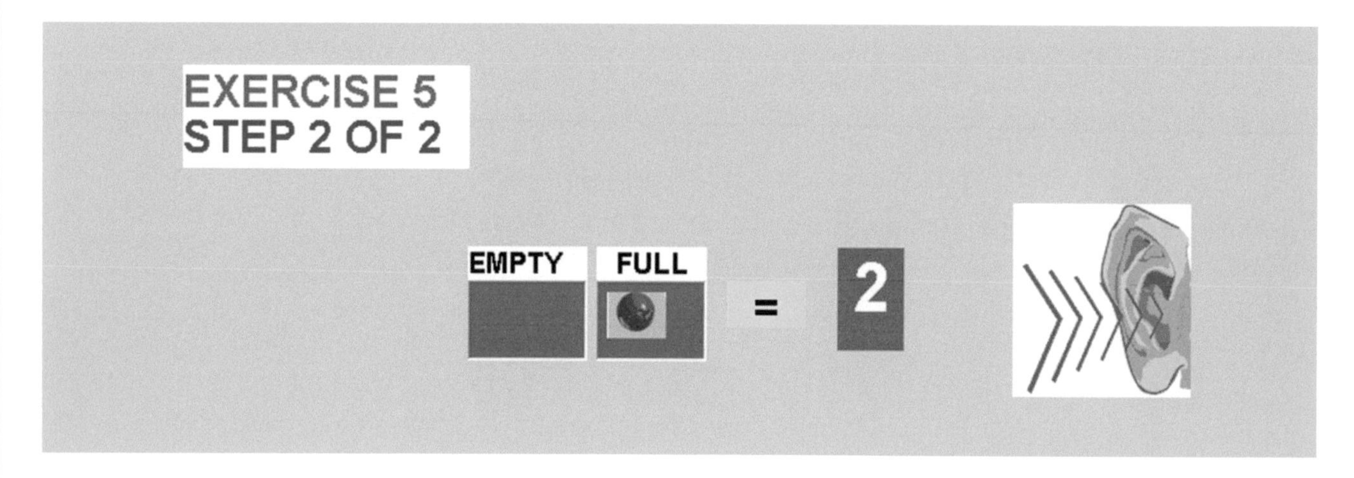

The ***kinesthetic*** representation of the filled up position is announced by clamping of the right hand. An empty position is represented by the clamping of the left hand. The kinesthetic representation for this picture is clamping the left hand and right hand.

SUBTRACTION, FIND AN ANSWER, EXERCISE 5

In exercise 5, in ***visual*** display of subtraction we see two pictures of boxes. In the bottom we see three pictures with probable answers. Both pictures have a symbol in the first position and the second position is empty. We want to specify, that between the first and second pictures we see the subtraction sign. Bellow the second picture we see the equal line, which separates the problem from the answer. The ***audio*** representation for two pictures consumes of a **knock, knock, double tram, knock, double knock, and cling.**

From the probable answers for addition we need to find the right one. The audio signals are: 1) **knock and double knock.** 2) **double knock and knock** 3) **knock and knock**.

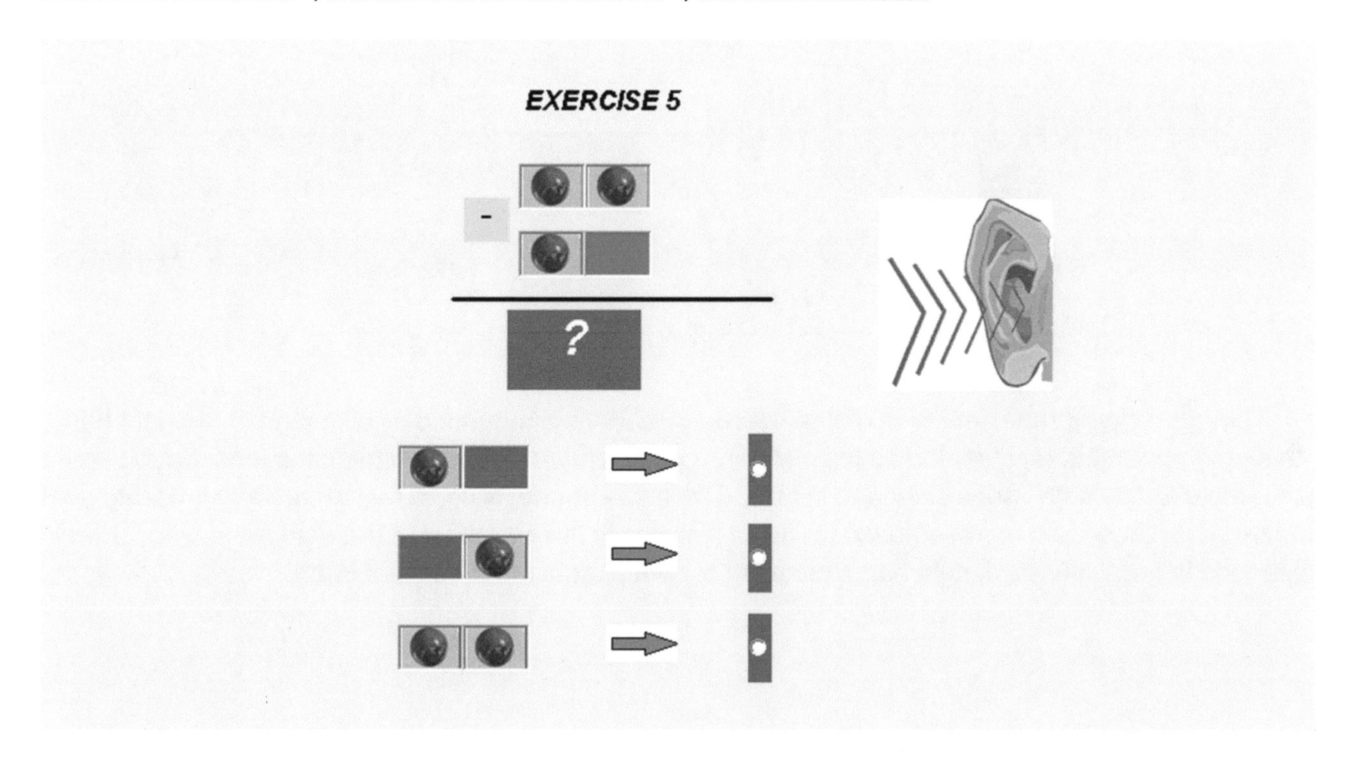

The ***kinesthetic*** representation of the filled up position is announced by clamping of the right hand. An empty position is represented by the clamping of the left hand. The subtraction sound, double tram, is represented by both hands extended in front. The equal sound, cling, is represented by crossing of hands. The kinesthetic representation for both pictures and the answer is the clamping of the right hand twice, hands extended in front, clamping the right hand, left hand, crossing of hands, left hand and right hand.

MATHEMATICS

SUBTRACTION, EXERCISE 6

In exercise 6, in ***visual*** display of subtraction we see three pictures of boxes. Between the first and second pictures we see the subtraction sign. Under the second picture we see the equal line, which separates the problem from the answer. Below the equal line is the third picture, which is the answer of subtraction. The ***audio*** representation for subtraction consumes of a **knock, knock, double tram, knock, double knock, cling, knock, and double knock.**

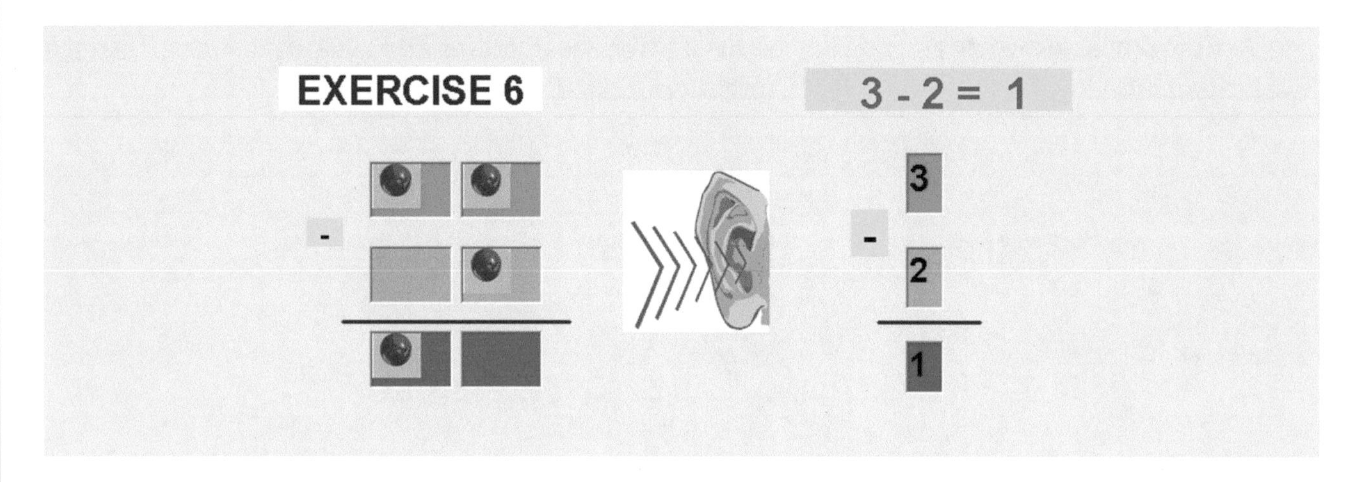

The ***kinesthetic*** representation of the filled up position is announced by clamping of the right hand. An empty position is represented by the clamping of the left hand. The subtraction sound, double tram, is represented by both hands extended in front. The equal sound, cling, is represented by crossing both hands. The kinesthetic representation for three pictures is the clamping of the right hand twice, hands extended in front, left hand, right hand, crossing of hands, right hand, and left hand.

SUBTRACTION EXERCISE 6, STEP 6.1

In exercise 6, step 6.1in ***visual*** display of subtraction we see three pictures of boxes. The first picture has symbols in both positions. The second picture has a symbol in the second position while the first position is empty. The answer of the equation has the first position filled with a symbol and the second position is empty. We want to specify, that between the first and second pictures we see the subtraction sign. Bellow the second picture we see the equal line, which separates the problem from the answer. The ***audio*** representation for the three pictures consumes of a **knock, knock, double tram, double knock, knock, cling, knock, and double knock.**

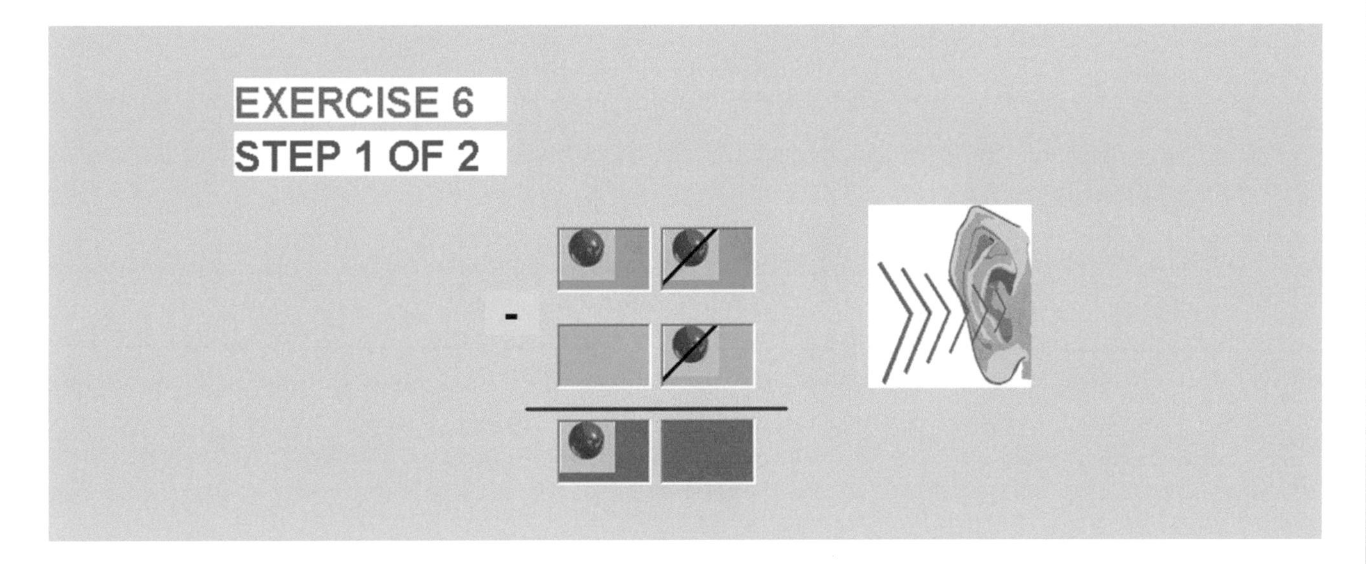

The ***kinesthetic*** representation of the filled up position is announced by clamping of the right hand. An empty position is represented by the clamping of the left hand. The subtraction sound, double tram, is represented by both hands extended in front. The equal sound, cling, is represented by crossing of hands. The kinesthetic representation for both pictures and the answer is the clamping of the right hand twice, hands extended in front, left hand, right hand, crossing of hands, right hand, and left hand.

ADDITION, EXERCISE 6, STEP 2

In exercise 6, step 6.2 in ***visual*** display of the last step of subtraction, we see one picture of boxes, which is the answer of the problem. We see that the first position is empty while the second position is filled up with a symbol. The answer has a value of one.

The ***audio*** representation for this picture consumes of a **double knock, knock, and cling.**

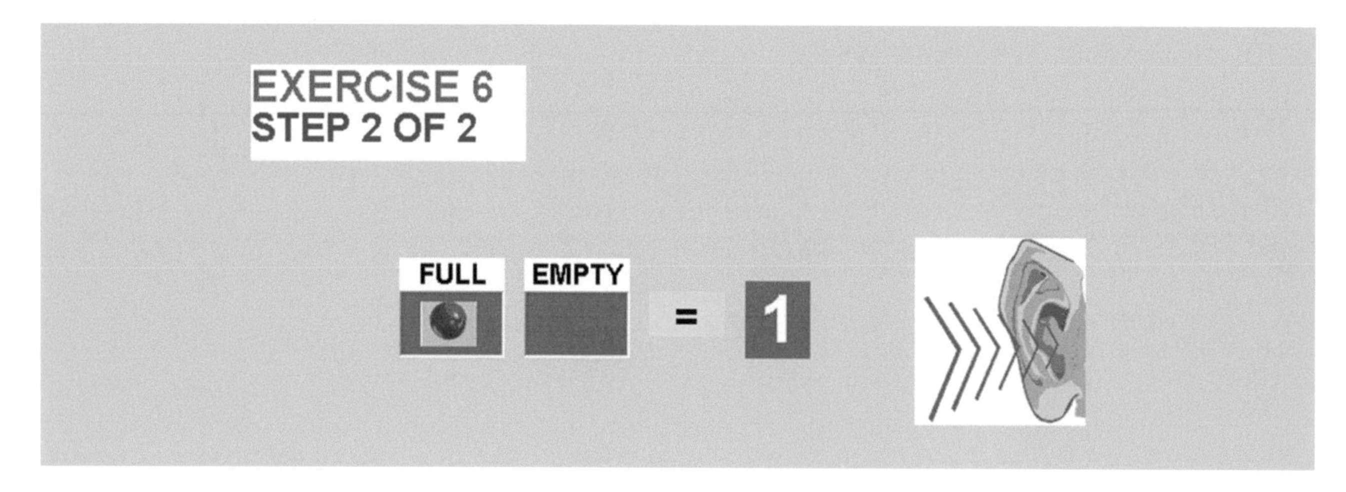

The ***kinesthetic*** representation of the filled up position is announced by clamping of the right hand. An empty position is represented by the clamping of the left hand. The kinesthetic representation for this picture is clamping the right hand and left hand.

SUBTRACTION, FIND AN ANSWER, EXERCISE 6

In exercise 6, in ***visual*** display of subtraction we see two pictures of boxes. In the bottom we see three pictures with probable answers. Both pictures have a symbol in the first position and the second position is empty. We want to specify, that between the first and second pictures we see the subtraction sign. Bellow the second picture we see the equal line, which separates the problem from the answer. The ***audio*** representation for two pictures consumes of a **knock, knock, double tram, double knock, knock, and cling.**

From the probable answers for addition we need to find the right one. The audio signals are: 1) **double knock and knock.** 2) **knock and knock** 3) **knock and double knock**.

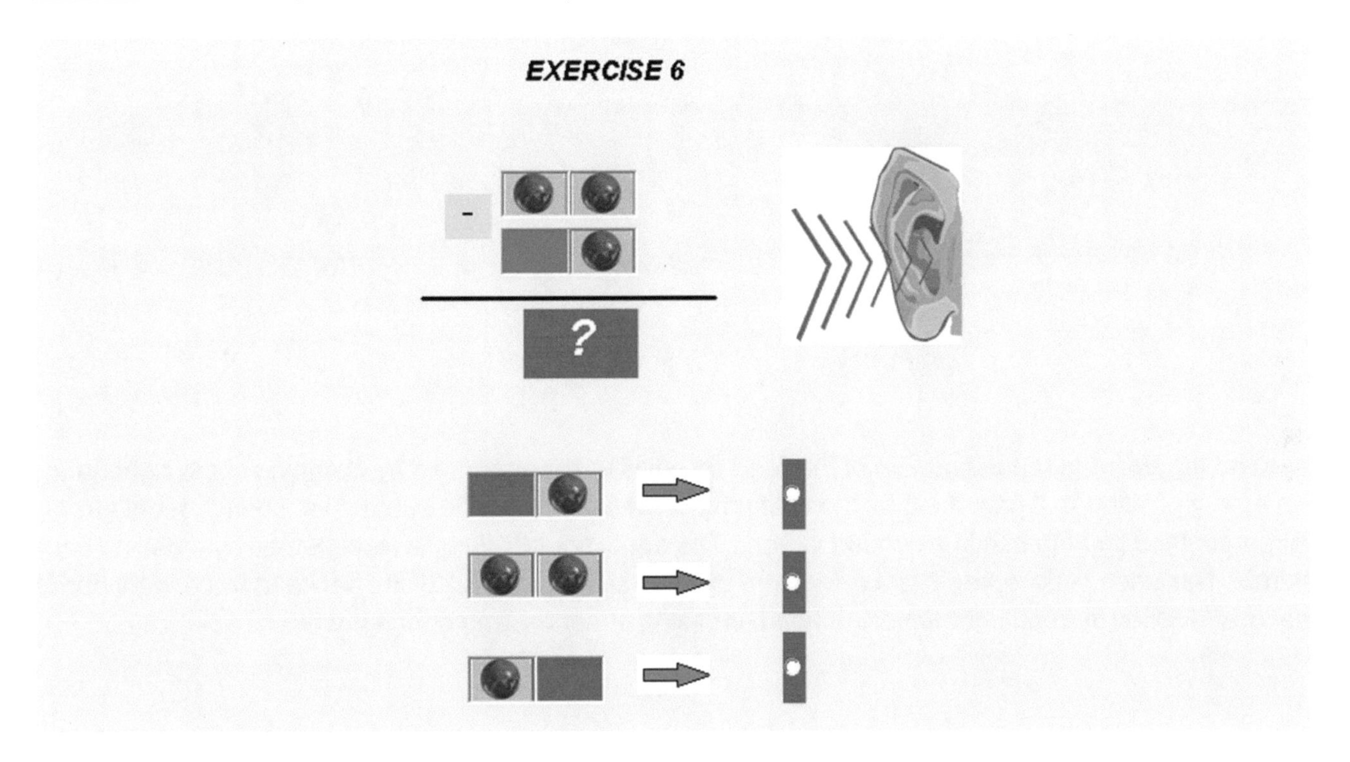

The ***kinesthetic*** representation of the filled up position is announced by clamping of the right hand. An empty position is represented by the clamping of the left hand. The subtraction sound, double tram, is represented by both hands extended in front. The equal sound, cling, is represented by crossing of hands. The kinesthetic representation for both pictures and the answer is the clamping of the right hand twice, hands extended in front, left hand, right hand, crossing of hands, right hand, and left hand.

MATHEMATICS

SUBTRACTION, EXERCISE 7

In exercise 7, in ***visual*** display of subtraction we see three pictures of boxes. Between the first and second pictures we see the subtraction sign. Under the second picture we see the equal line, which separates the problem from the answer. Below the equal line is the third picture, which is the answer of subtraction. The ***audio*** representation for subtraction consumes of a **double knock, double knock, knock, double tram, knock, double knock, cling, knock, and knock.**

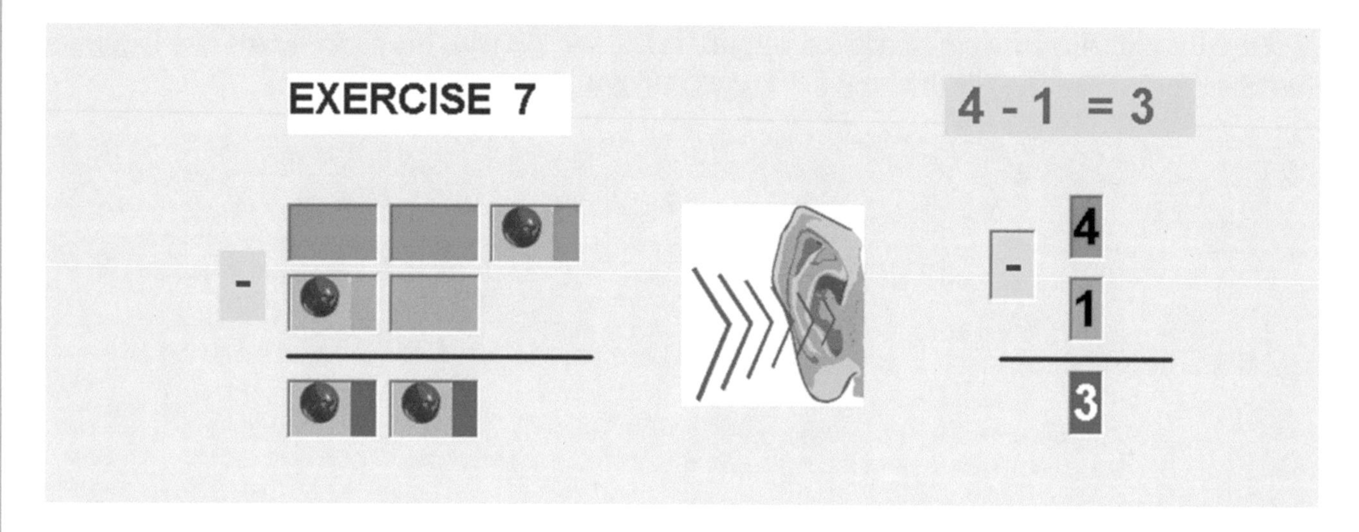

The ***kinesthetic*** representation of the filled up position is announced by clamping of the right hand. An empty position is represented by the clamping of the left hand. The subtraction sound, double tram, is represented by both hands extended in front. The equal sound, cling, is represented by crossing both hands. The kinesthetic representation for three pictures is the clamping of the left hand twice, right hand, hands extended in front, right hand, left hand, crossing of hands, and right hand twice.

SUBTRACTION, EXERCISE 7, STEP 7.1

In exercise 7, step 7.1 in ***visual*** display we see two pictures of boxes. In the first picture we see that the first and second positions are empty and the third position is filled with a symbol. This pictures numerical value is 4.

The **audio** representation for the first picture is **double knock, double knock, and knock.**

The second picture has the first position filled with a symbol and the second position is empty.

The **audio** representation for the second picture is **knock and double knock.**

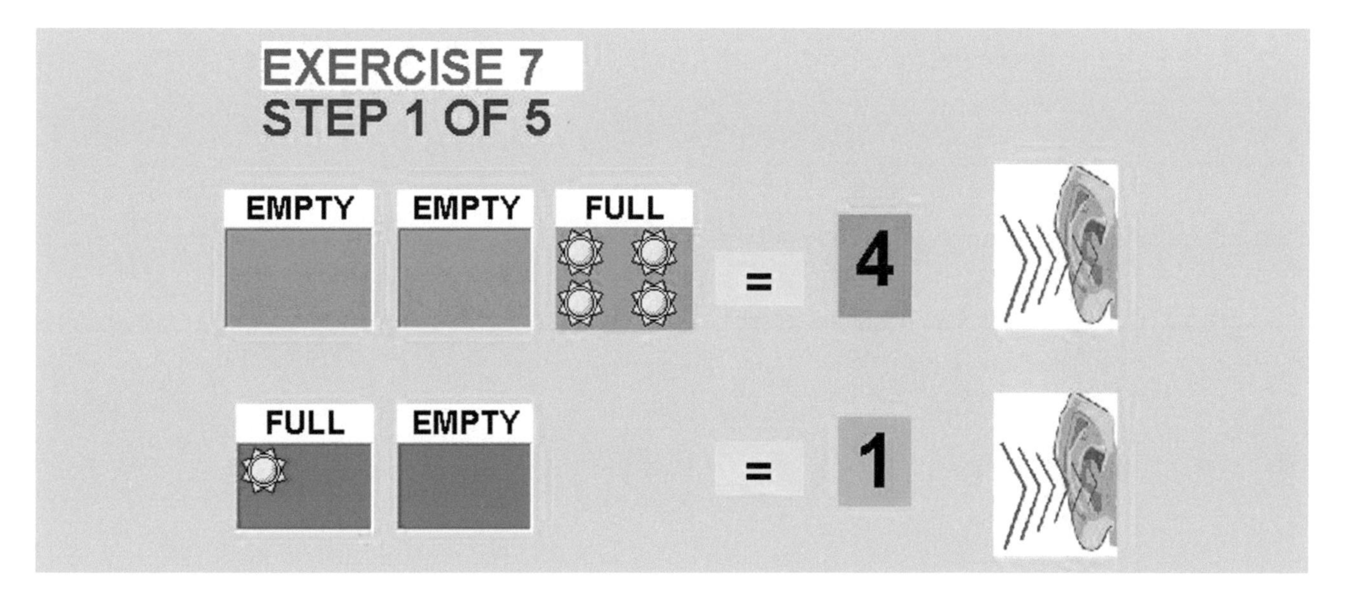

The ***kinesthetic*** representation of the filled up position is announced by clamping of the right hand. An empty position is represented by the clamping of the left hand. The equal sound, cling, is represented by crossing both hands. The kinesthetic representation for the first picture is the clamping of the left hand twice and right hand. The kinesthetic representation for the second picture is clamping of right hand and left hand.

MATHEMATICS SUBTRACTION, EXERCISE 7, STEP 7.2

In exercise 7, step 7.2 in ***visual*** display of subtraction we see three pictures of boxes. In the first two pictures the first two positions are empty while the third position is filled with a symbol. The third picture has a symbol in the second position and has the first and third positions empty. For purpose of subtraction we are introducing the concept of borrowing. We are borrowing from the third to the second position. However the audio representation for the picture does not change because the numerical value does not change.

The ***audio*** representation for the three pictures consumes of a **double knock, double knock, and knock.**

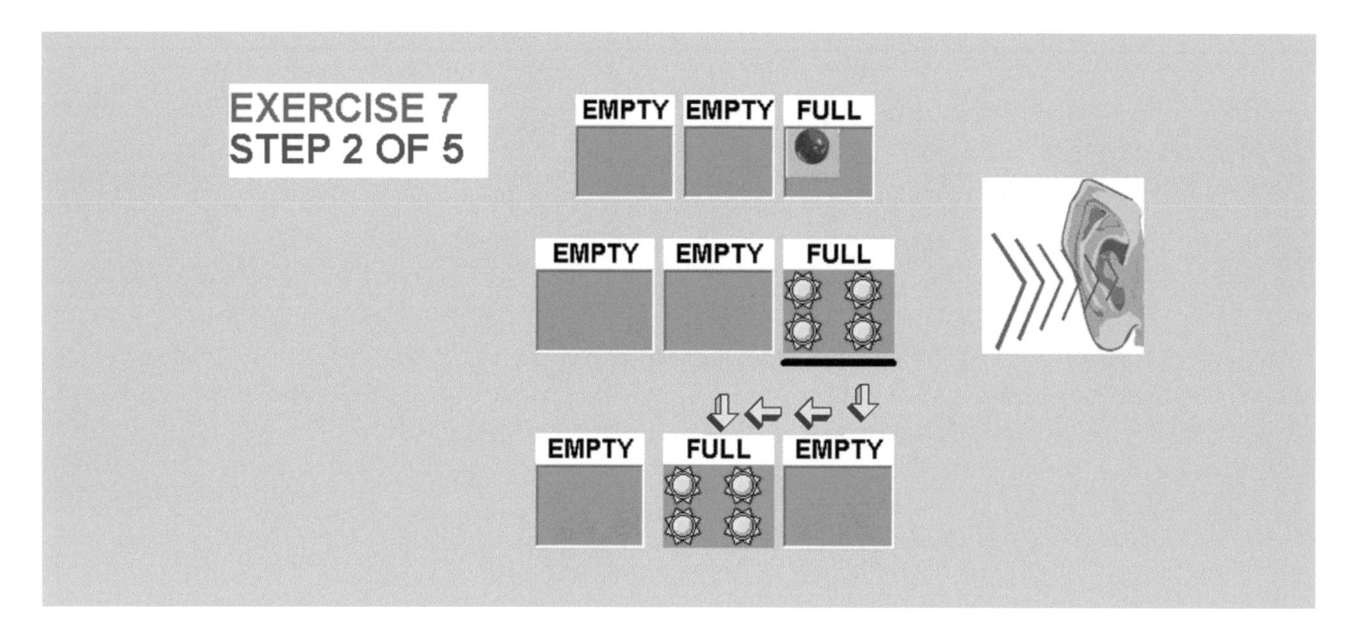

The ***kinesthetic*** representation of the filled up position is announced by clamping of the right hand. An empty position is represented by the clamping of the left hand. The kinesthetic representation for three pictures is the clamping of the left hand twice and right hand.

SUBTRACTION, EXERCISE 7, STEP 7.3

In exercise 7, step 7.3 in ***visual*** display of subtraction we see three pictures of boxes. In the first picture the first two positions are empty while the third position is filled with a symbol. The second picture has a symbol in the second position and has the first and third positions empty. The third picture has symbols in the first and second positions while the third position is empty. For purpose of subtraction we are introducing the concept of borrowing. We are borrowing from the second to the first position. However the audio representation for the picture does not change because the numerical value does not change.

The ***audio*** representation for the three pictures consumes of a **double knock, double knock, and knock.**

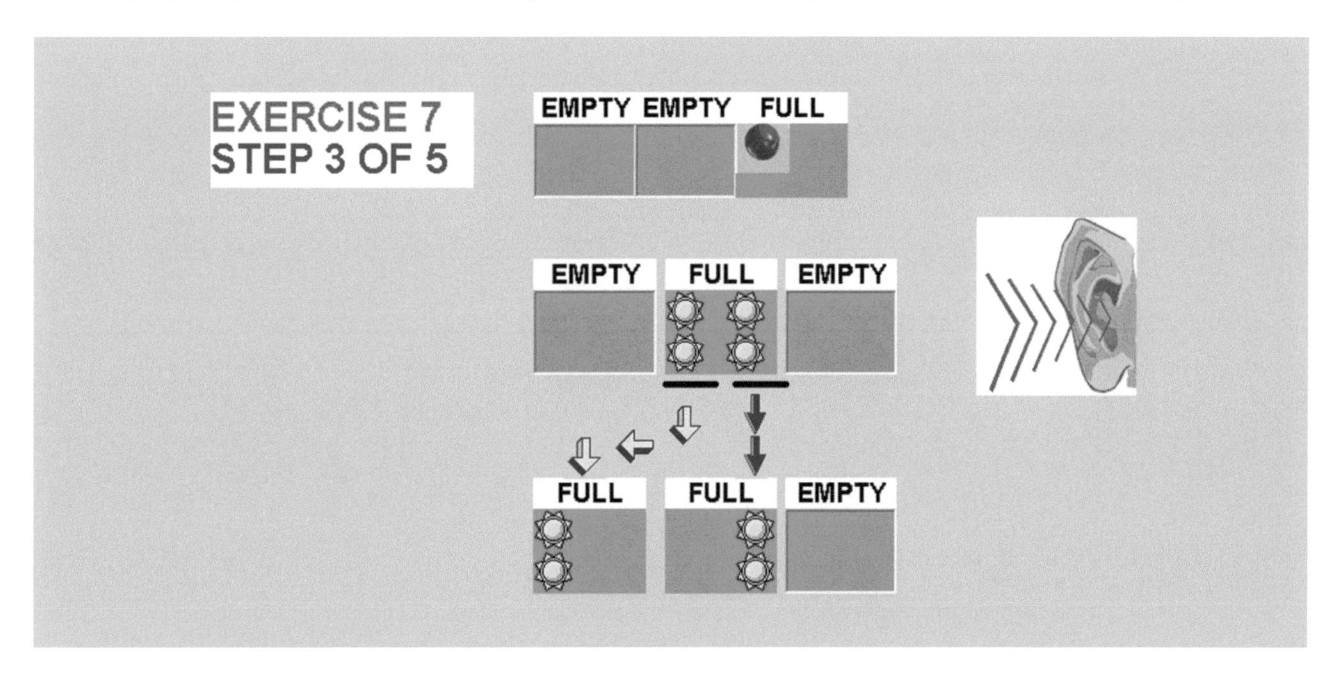

The ***kinesthetic*** representation of the filled up position is announced by clamping of the right hand. An empty position is represented by the clamping of the left hand. The kinesthetic representation for three pictures is the clamping of the left hand twice and right hand.

MATHEMATICS

SUBTRACTION, EXERCISE 7.4

In exercise 7, step 7.4 in ***visual*** display of subtraction we see six pictures of boxes. There are three pictures on the right side and three pictures on the left side. Both sides represent the same equation; however the left side has used the concept of borrowing to further show the problem. When we do subtraction we are borrowing available symbols from the higher position. We need to remember that all symbols borrowed from the higher position are twice larger than the one in the prior descending position. Once the symbol is borrowed the position borrowed becomes empty and the two symbols appear in the prior position. If needed we leave one symbol to the next descending position which is empty and once again this one symbol becomes two symbols in the next descending position. The ***audio*** representation for subtraction consumes of a **double knock, double knock, knock, double tram, knock, double knock, cling, knock, and knock.**

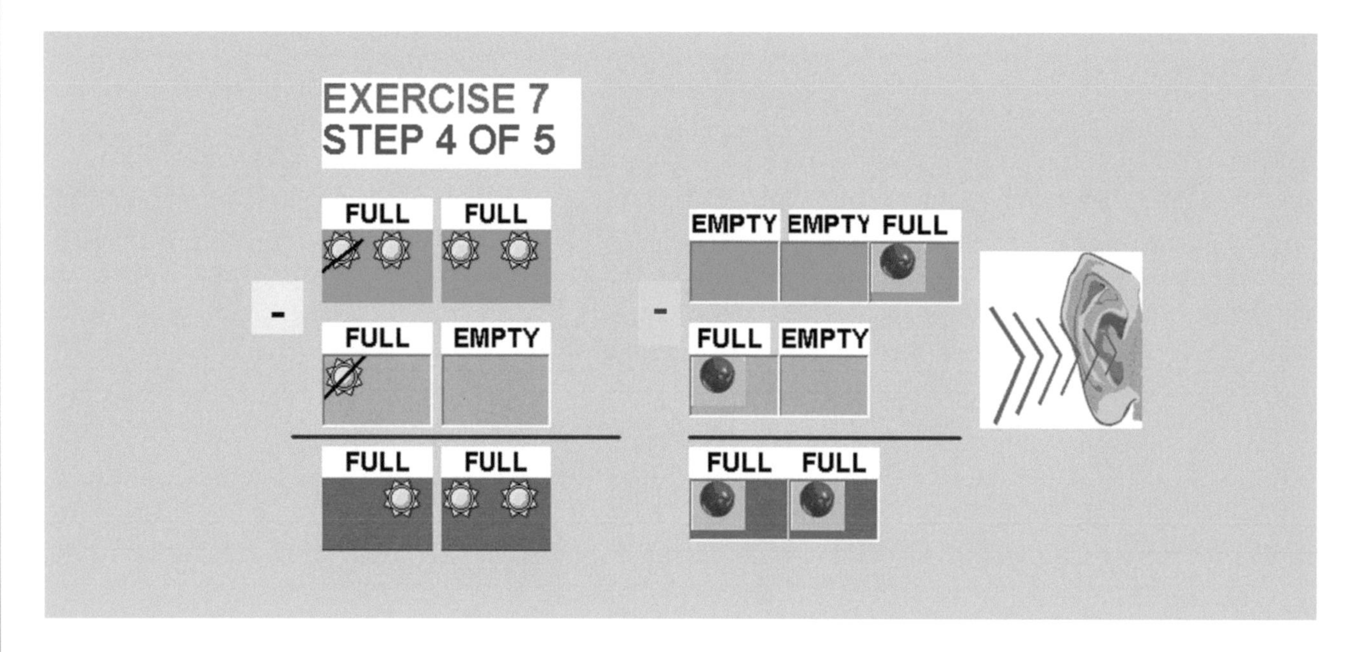

The ***kinesthetic*** representation of the filled up position is announced by clamping of the right hand. An empty position is represented by the clamping of the left hand. The subtraction sound, double tram, is represented by both hands extended in front. The equal sound, cling, is represented by crossing both hands. The kinesthetic representation for the subtraction is the clamping of the left hand twice, right hand, hands extended in front, right hand, left hand, crossing of hands, and right hand twice.

ADDITION, EXERCISE 7, STEP 7.5

In exercise 7, step 7.5 in ***visual*** display of the last step of subtraction, we see two pictures of boxes, which is the answer of the problem. We see that the both positions are filled up with a symbol. The answer has a value of three.

The ***audio*** representation for the pictures consumes of a **knock, knock, and cling.**

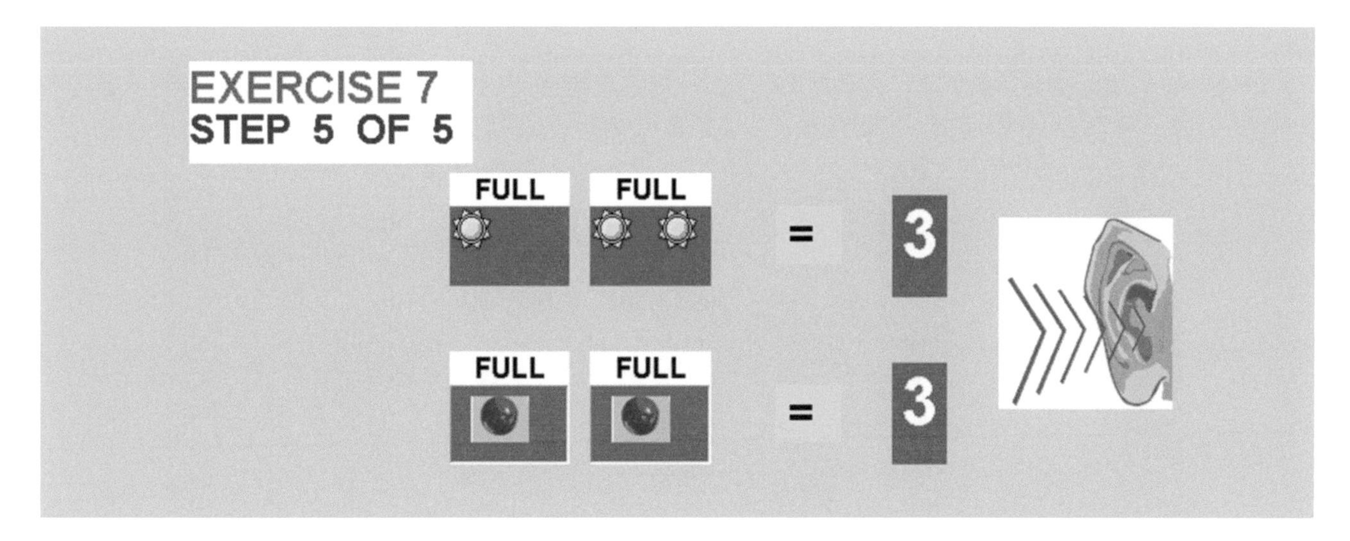

The ***kinesthetic*** representation of the filled up position is announced by clamping of the right hand. An empty position is represented by the clamping of the left hand. The kinesthetic representation for this picture is clamping the right hand twice.

MATHEMATICS

SUBTRACTION, FIND AN ANSWER, EXERCISE 7

In exercise 7, in ***visual*** display of subtraction we see two pictures of boxes. In the bottom we see three pictures with probable answers. Both pictures have a symbol in the first position and the second position is empty. We want to specify, that between the first and second pictures we see the subtraction sign. Bellow the second picture we see the equal line, which separates the problem from the answer. The ***audio*** representation for two pictures consumes of a **double knock, double knock, knock, double tram, knock, double knock, and cling.**

From the probable answers for addition we need to find the right one. The audio signals are: 1) **knock and knock.** 2) **knock and double knock** 3) **double knock and knock**.

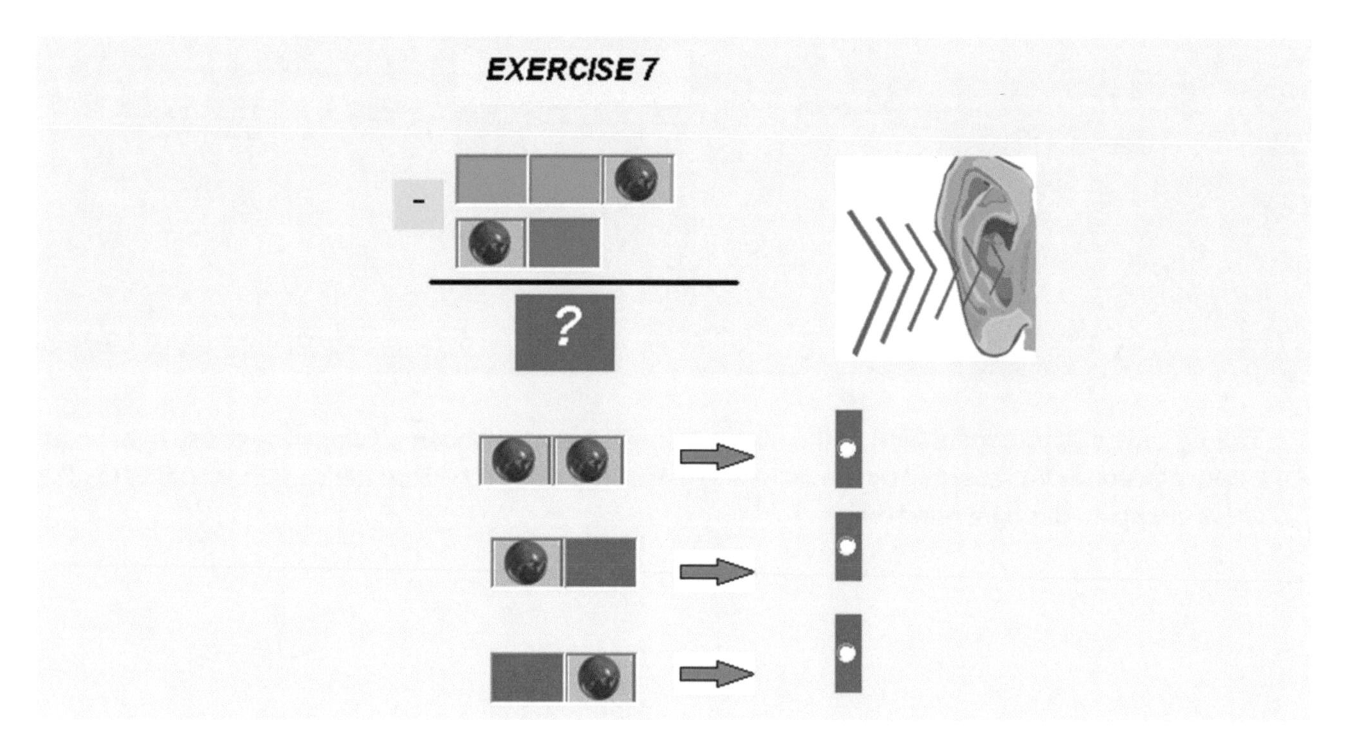

The ***kinesthetic*** representation of the filled up position is announced by clamping of the right hand. An empty position is represented by the clamping of the left hand. The subtraction sound, double tram, is represented by both hands extended in front. The equal sound, cling, is represented by crossing of hands. The kinesthetic representation for both pictures and the answer is the clamping of the left hand twice, right hand, hands extended in front, right hand, left hand, crossing of hands, and right hand twice.

SUBTRACTION, EXERCISE 8

In exercise 8, in ***visual*** display of subtraction we see three pictures of boxes. Between the first and second pictures we see the subtraction sign. Under the second picture we see the equal line, which separates the problem from the answer. Below the equal line is the third picture, which is the answer of subtraction. The ***audio*** representation for subtraction consumes of a **double knock, double knock, knock, double tram, knock, knock, cling, knock, and double knock.**

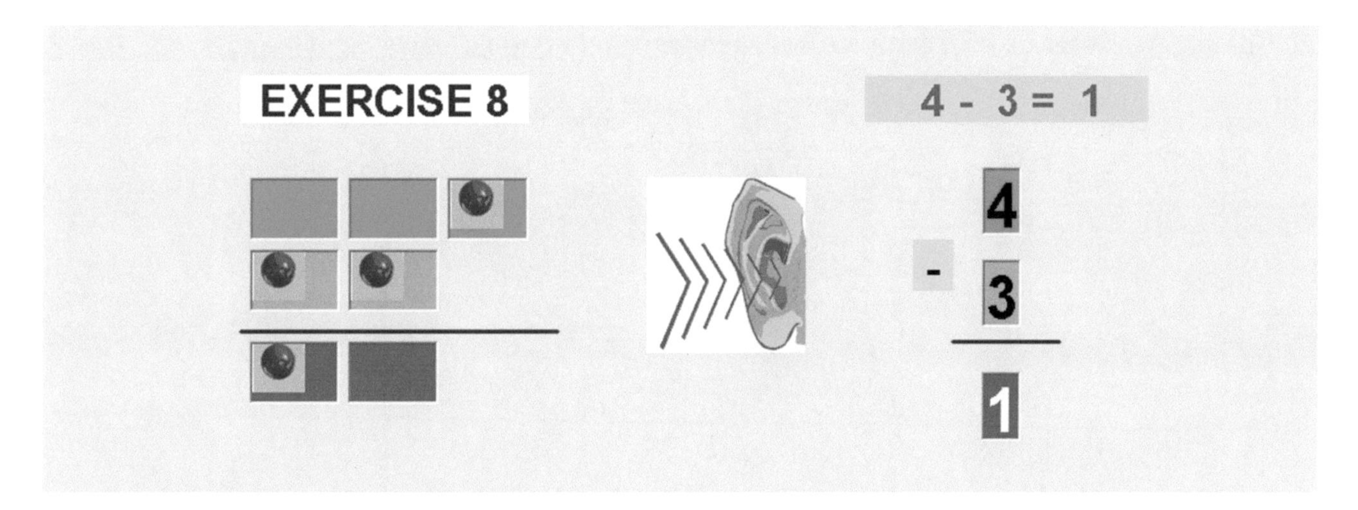

The ***kinesthetic*** representation of the filled up position is announced by clamping of the right hand. An empty position is represented by the clamping of the left hand. The subtraction sound, double tram, is represented by both hands extended in front. The equal sound, cling, is represented by crossing both hands. The kinesthetic representation for three pictures is the clamping of the left hand twice, right hand, hands extended in front, right hand twice, crossing of hands, right hand, and left hand.

MATHEMATICS SUBTRACTION, EXERCISE 8, STEP 8.1

In exercise 8, step 8.1 in ***visual*** display of subtraction we see three pictures of boxes. In the first two pictures the first two positions are empty while the third position is filled with a symbol. The third picture has a symbol in the second position and has the first and third positions empty. For purpose of subtraction we are introducing the concept of borrowing. We are borrowing from the third to the second position. However the audio representation for the picture does not change because the numerical value does not change.

The ***audio*** representation for the three pictures consumes of a **double knock, double knock, and knock.**

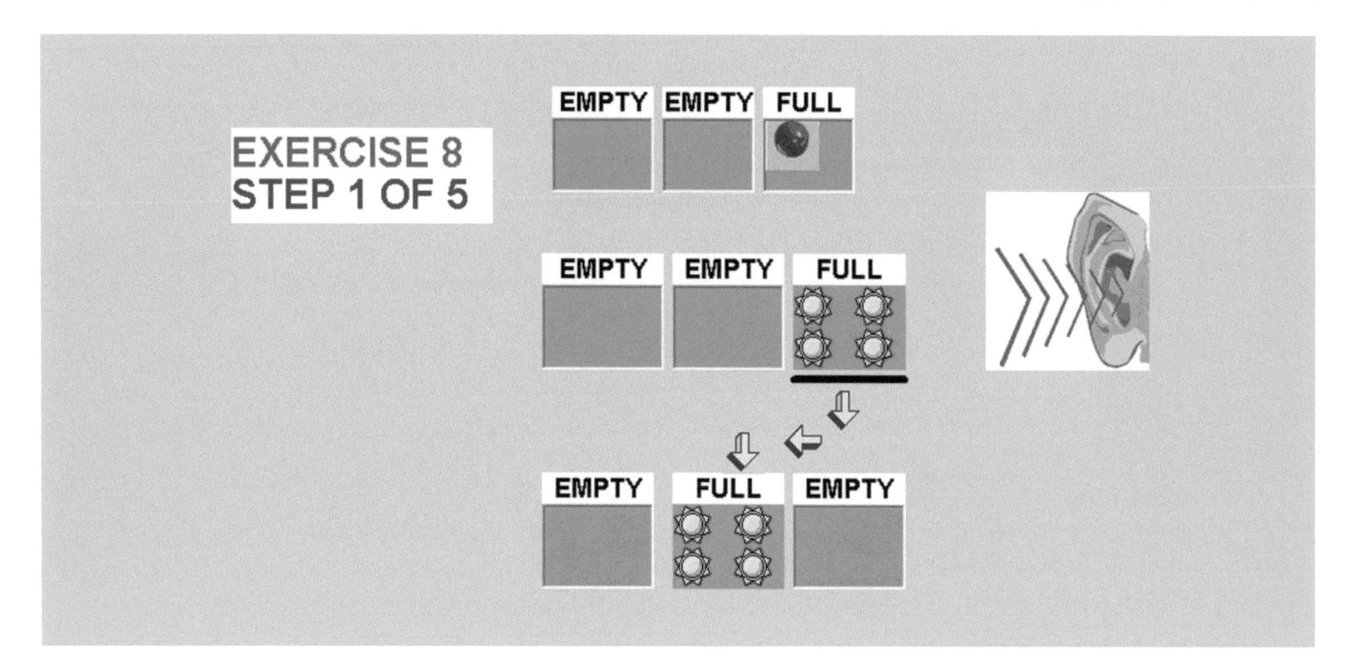

The ***kinesthetic*** representation of the filled up position is announced by clamping of the right hand. An empty position is represented by the clamping of the left hand. The kinesthetic representation for three pictures is the clamping of the left hand twice and right hand.

SUBTRACTION, EXERCISE 8, STEP 8.2

In exercise 8, step 8.2 in ***visual*** display of subtraction we see three pictures of boxes. In the first picture the first two positions are empty while the third position is filled with a symbol. The second picture has a symbol in the second position and has the first and third positions empty. The third picture has symbols in the first and second positions while the third position is empty. For purpose of subtraction we are introducing the concept of borrowing. We are borrowing from the second to the first position. However the audio representation for the picture does not change because the numerical value does not change.

The ***audio*** representation for the three pictures consumes of a **double knock, double knock, and knock.**

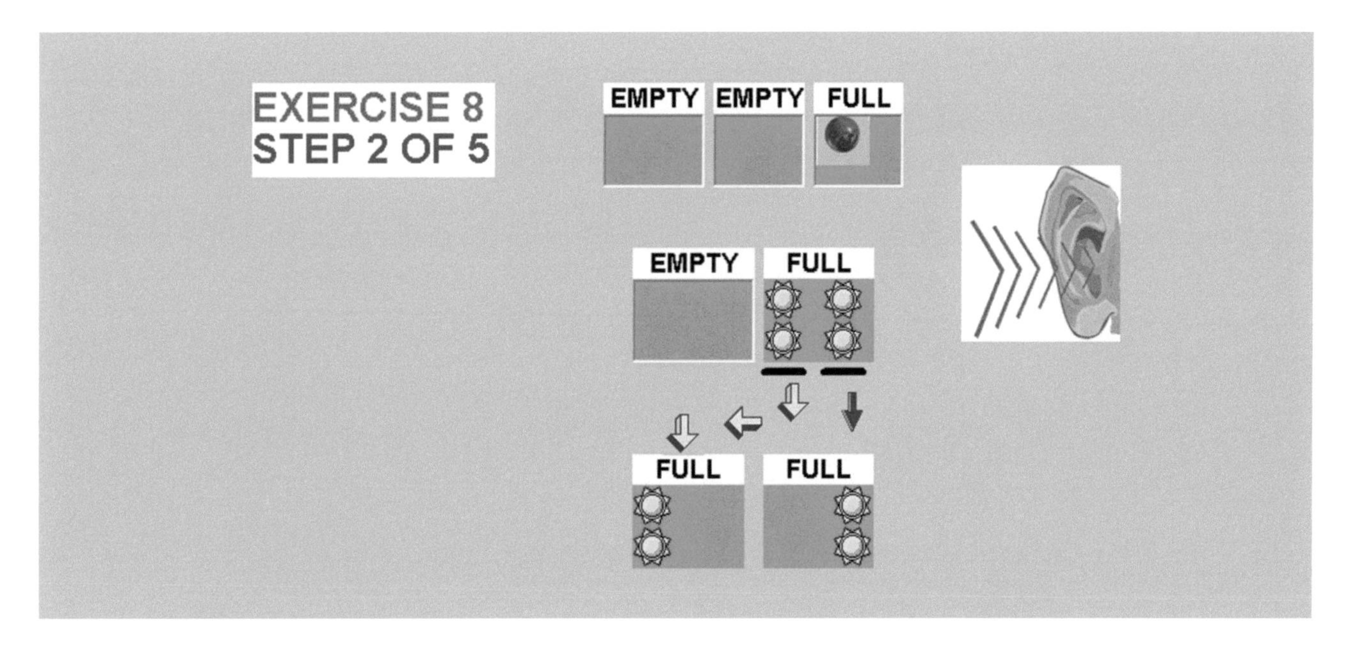

The ***kinesthetic*** representation of the filled up position is announced by clamping of the right hand. An empty position is represented by the clamping of the left hand. The kinesthetic representation for three pictures is the clamping of the left hand twice and right hand.

MATHEMATICS

SUBTRACTION, EXERCISE 8.3

In exercise 8, step 8.3 in ***visual*** display of subtraction we see six pictures of boxes. There are three pictures on the right side and three pictures on the left side. Both sides represent the same equation; however the left side has used the concept of borrowing to further show the problem. When we do subtraction we are borrowing available symbols from the higher position. We need to remember that all symbols borrowed from the higher position are twice larger than the one in the prior descending position. Once the symbol is borrowed the position borrowed becomes empty and the two symbols appear in the prior position. If needed we leave one symbol to the next descending position which is empty and once again this one symbol becomes two symbols in the next descending position. The ***audio*** representation for subtraction consumes of a **double knock, double knock, knock, double tram, knock, knock, cling, knock, and double knock.**

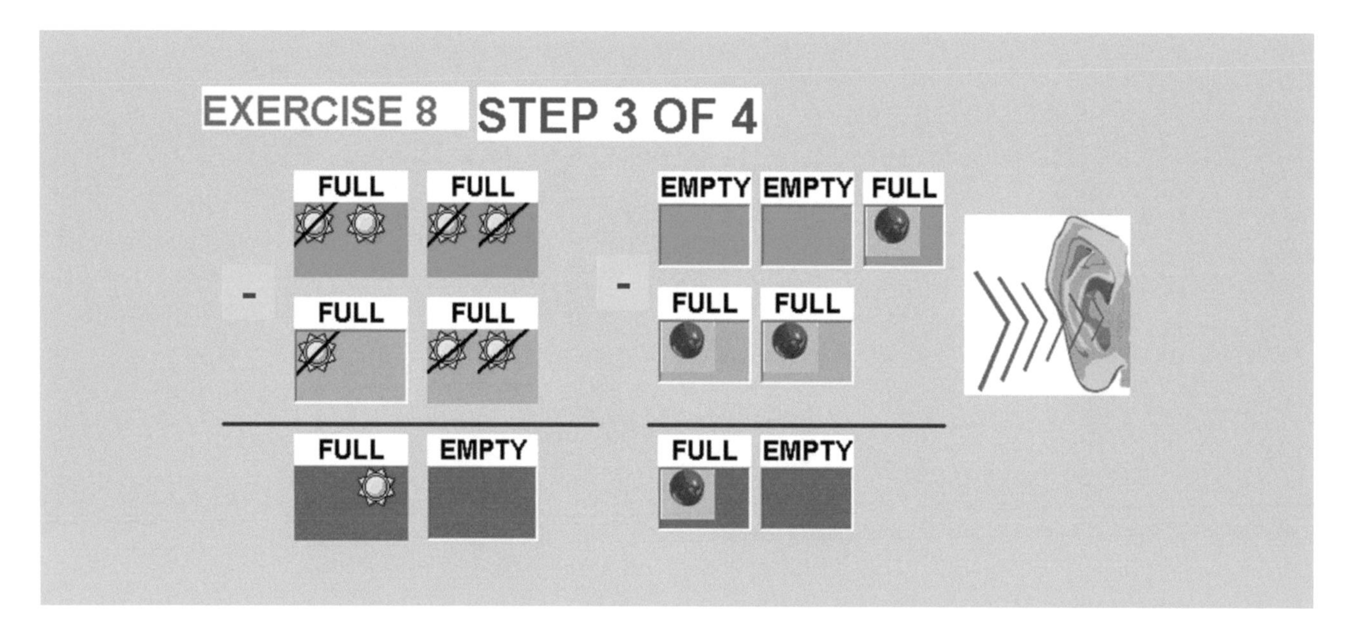

The ***kinesthetic*** representation of the filled up position is announced by clamping of the right hand. An empty position is represented by the clamping of the left hand. The subtraction sound, double tram, is represented by both hands extended in front. The equal sound, cling, is represented by crossing both hands. The kinesthetic representation for the subtraction is the clamping of the left hand twice, right hand, hands extended in front, right hand twice, crossing of hands, right hand, and left hand.

ADDITION, EXERCISE 8, STEP 8.4

In exercise 8, step 8.4 in ***visual*** display of the last step of subtraction, we see two pictures of boxes, which is the answer of the problem. We see that the first position is filled up with a symbol while the second position is empty. The answer has a value of one.

The ***audio*** representation for the pictures consumes of a **knock, double knock, and cling.**

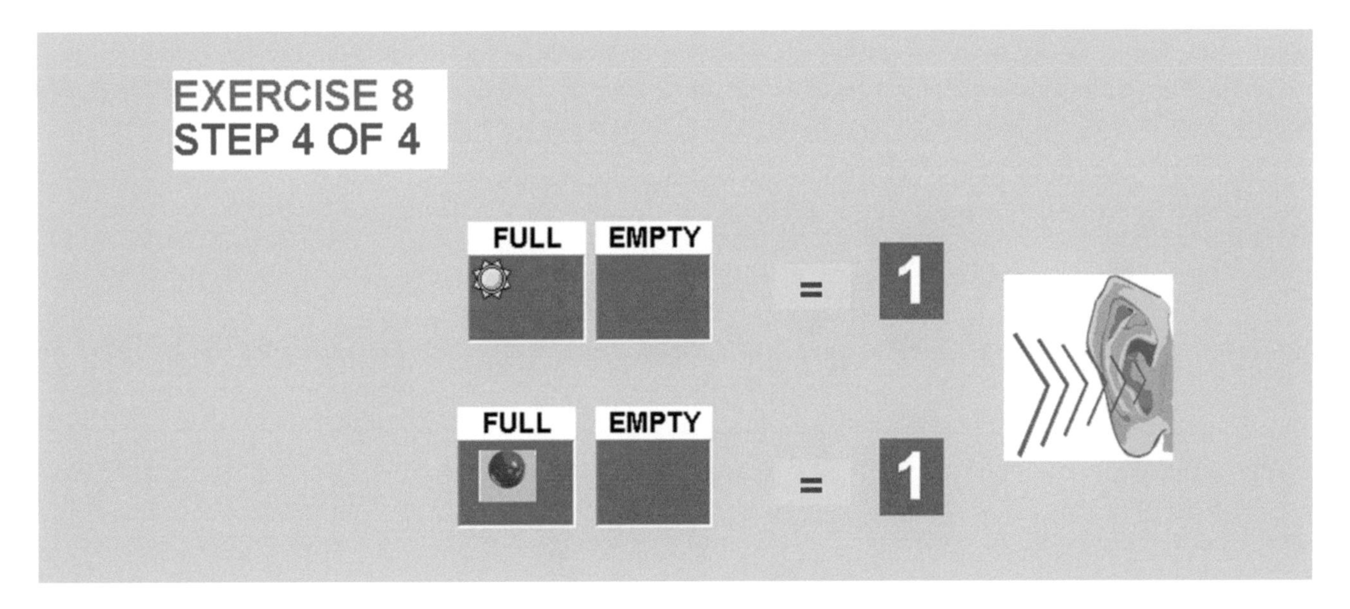

The ***kinesthetic*** representation of the filled up position is announced by clamping of the right hand. An empty position is represented by the clamping of the left hand. The kinesthetic representation for this picture is clamping the right hand and left hand.

MATHEMATICS

SUBTRACTION, FIND AN ANSWER, EXERCISE 8

In exercise 8, in ***visual*** display of subtraction we see two pictures of boxes. In the bottom we see three pictures with probable answers. Both pictures have a symbol in the first position and the second position is empty. We want to specify, that between the first and second pictures we see the subtraction sign. Bellow the second picture we see the equal line, which separates the problem from the answer. The ***audio*** representation for two pictures consumes of a **double knock, double knock, knock, double tram, knock, knock, and cling.**

From the probable answers for addition we need to find the right one. The audio signals are: 1) **double knock and knock.** 2) **knock and double knock** 3) **knock and knock**.

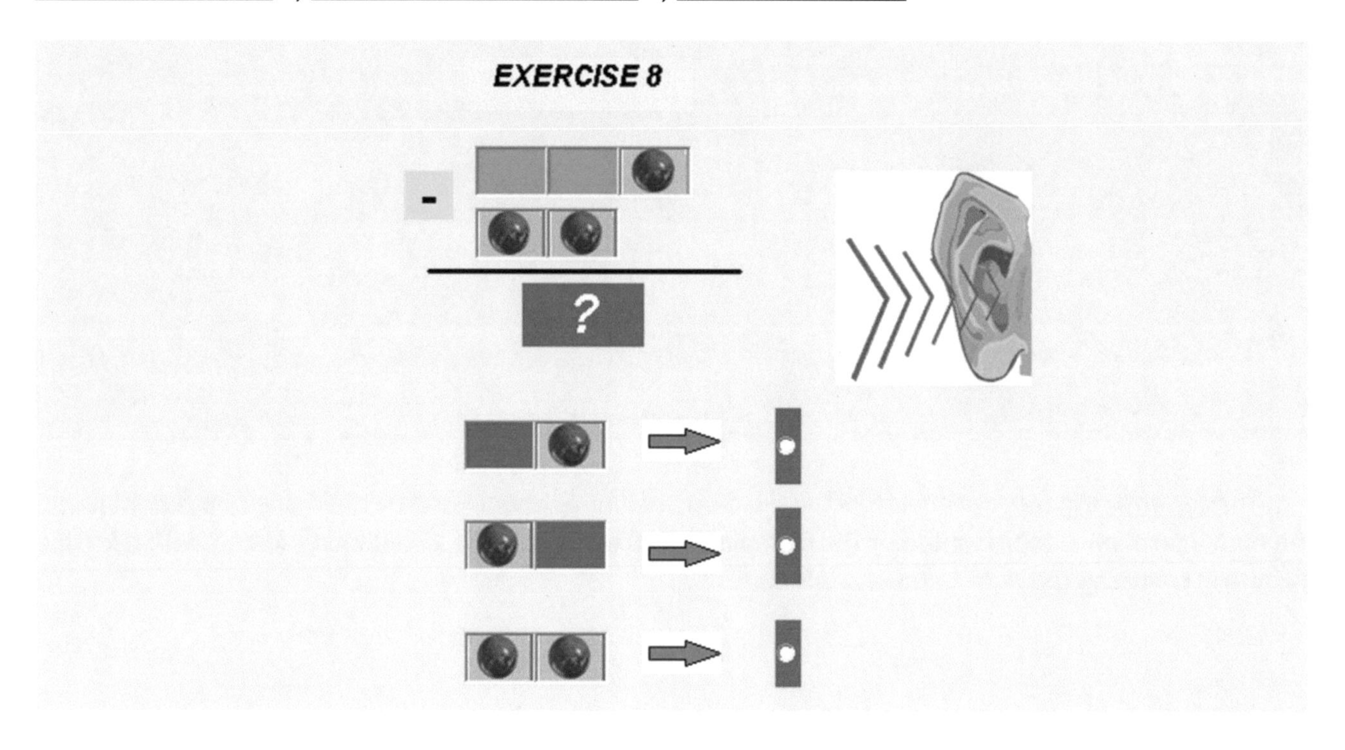

The ***kinesthetic*** representation of the filled up position is announced by clamping of the right hand. An empty position is represented by the clamping of the left hand. The subtraction sound, double tram, is represented by both hands extended in front. The equal sound, cling, is represented by crossing of hands. The kinesthetic representation for both pictures and the answer is the clamping of the left hand twice, right hand, hands extended in front, right hand twice, crossing of hands, right hand, and left hand.

SUBTRACTION, FIND AN ANSWER, EXERCISE 9

In exercise 9, in ***visual*** display of subtraction we see two pictures of boxes. In the bottom we see three pictures with probable answers. Both pictures have a symbol in the first position and the second position is empty. We want to specify, that between the first and second pictures we see the subtraction sign. Bellow the second picture we see the equal line, which separates the problem from the answer. The ***audio*** representation for two pictures consumes of a **knock, double knock, knock, double tram, double knock, knock, and cling.**

From the probable answers for addition we need to find the right one. The audio signals are: 1) **double, knock, and knock.** 2) **knock, double knock, and knock** 3) **knock, knock, and double knock**.

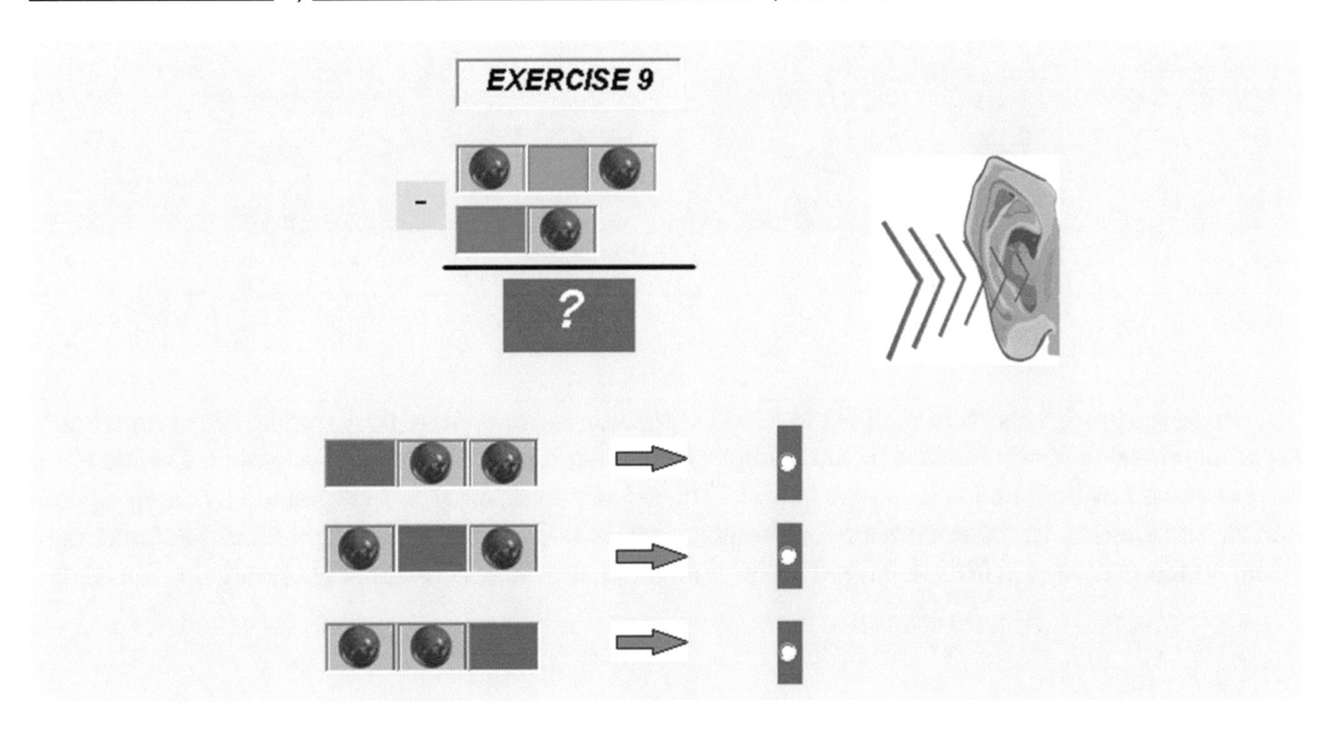

The ***kinesthetic*** representation of the filled up position is announced by clamping of the right hand. An empty position is represented by the clamping of the left hand. The subtraction sound, double tram, is represented by both hands extended in front. The equal sound, cling, is represented by crossing of hands. The kinesthetic representation for both pictures and the answer is the clamping of the right hand, left hand, right hand, hands extended in front, left hand, right hand, crossing of hands, right hand twice, and left hand.

MATHEMATICS SUBTRACTION, EXERCISE 10

In exercise 10, in ***visual*** display of subtraction we see three pictures of boxes. Between the first and second pictures we see the subtraction sign. Under the second picture we see the equal line, which separates the problem from the answer. Below the equal line is the third picture, which is the answer of subtraction. The ***audio*** representation for subtraction consumes of a **knock, double knock, knock, double tram, double knock, double knock, knock, cling, knock, and double knock.**

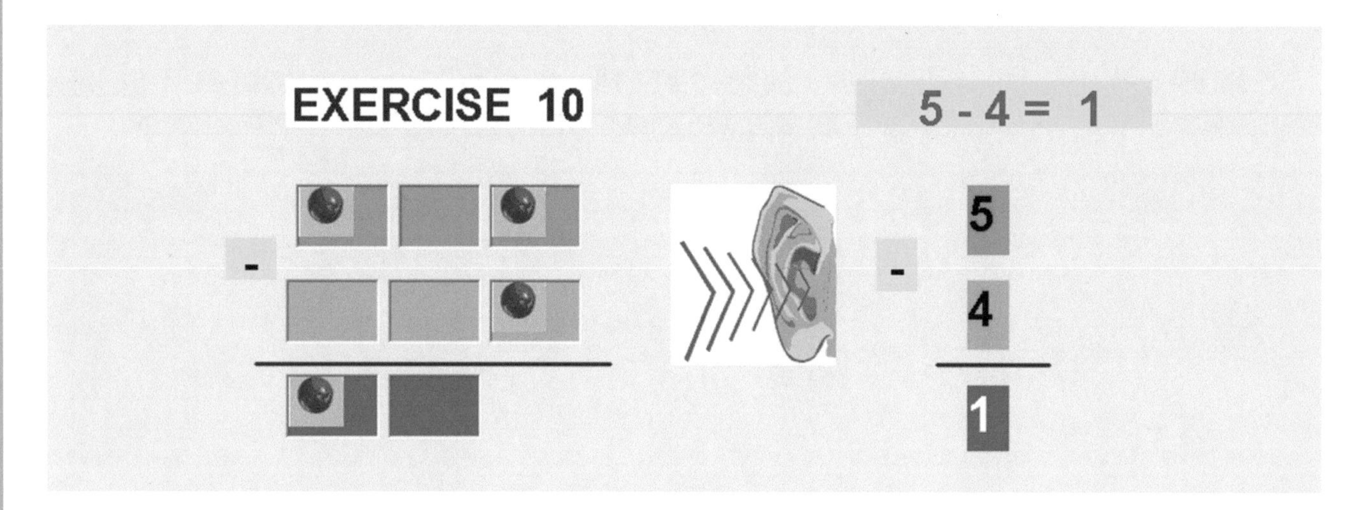

The ***kinesthetic*** representation of the filled up position is announced by clamping of the right hand. An empty position is represented by the clamping of the left hand. The subtraction sound, double tram, is represented by both hands extended in front. The equal sound, cling, is represented by crossing both hands. The kinesthetic representation for three pictures is the clamping of the right hand, left hand, right hand, hands extended in front, left hand twice, right hand, crossing of hands, right hand, and left hand.

SUBTRACTION, EXERCISE 10, STEP 10.1

In exercise 10, step 10.1 in ***visual*** display we see two pictures of boxes. In the first picture we see that the first and third positions are filled with a symbol while the second position is empty. This pictures numerical value is 5.

The **audio** representation for the first picture is **knock, double knock, and knock.**

The second picture has the first two positions empty and the last position is filled with a symbol.

The **audio** representation for the second picture is **double knock, double knock, and knock.**

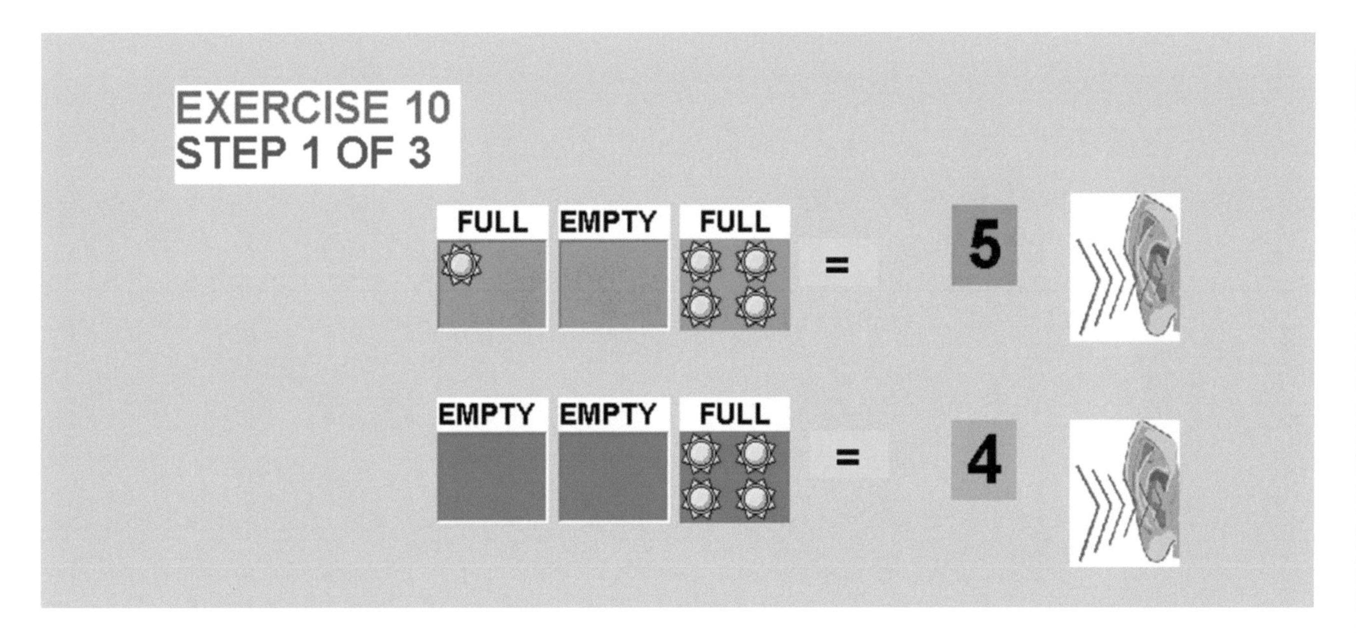

The ***kinesthetic*** representation of the filled up position is announced by clamping of the right hand. An empty position is represented by the clamping of the left hand. The equal sound, cling, is represented by crossing both hands. The kinesthetic representation for the first picture is the clamping of the right hand, left hand, and right hand. The kinesthetic representation for the second picture is clamping of left hand twice and right hand.

MATHEMATICS

SUBTRACTION, EXERCISE 10.2

In exercise 10, step 10.2 in ***visual*** display of subtraction we see six pictures of boxes. There are three pictures on the right side and three pictures on the left side. Both sides represent the same equation; however the left side has used the concept of borrowing to further show the problem. When we do subtraction we are borrowing available symbols from the higher position. We need to remember that all symbols borrowed from the higher position are twice larger than the one in the prior descending position. Once the symbol is borrowed the position borrowed becomes empty and the two symbols appear in the prior position. If needed we leave one symbol to the next descending position which is empty and once again this one symbol becomes two symbols in the next descending position. The ***audio*** representation for subtraction consumes of a **knock, double knock, knock, double tram, double knock, double knock, knock, cling, knock, and double knock.**

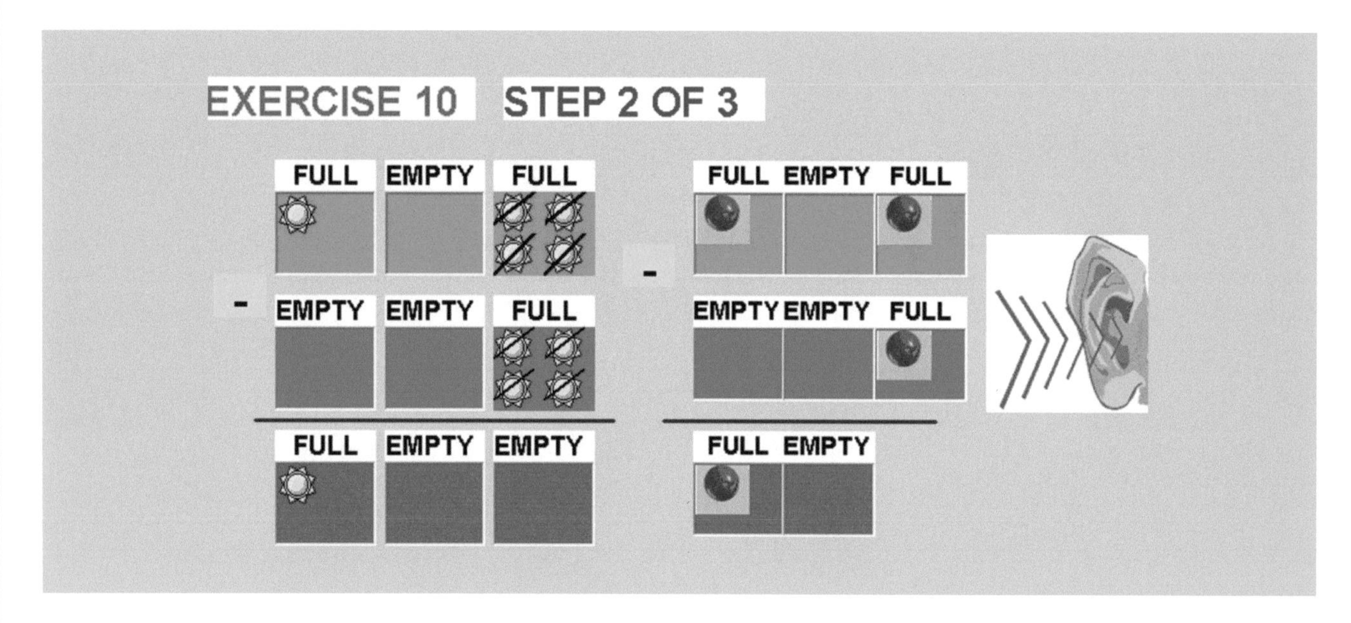

The ***kinesthetic*** representation of the filled up position is announced by clamping of the right hand. An empty position is represented by the clamping of the left hand. The subtraction sound, double tram, is represented by both hands extended in front. The equal sound, cling, is represented by crossing both hands. The kinesthetic representation for the subtraction is the clamping of the right hand, left hand, right hand, hands extended in front, left hand twice, right hand, crossing of hands, right hand, and left hand.

ADDITION, EXERCISE 10, STEP 10.3

In exercise 10, step 10.3 in ***visual*** display of the last step of subtraction, we see two pictures of boxes, which is the answer of the problem. We see that the first position is filled up with a symbol while the second position is empty. The answer has a value of one.

The ***audio*** representation for the pictures consumes of a **knock, double knock, and cling.**

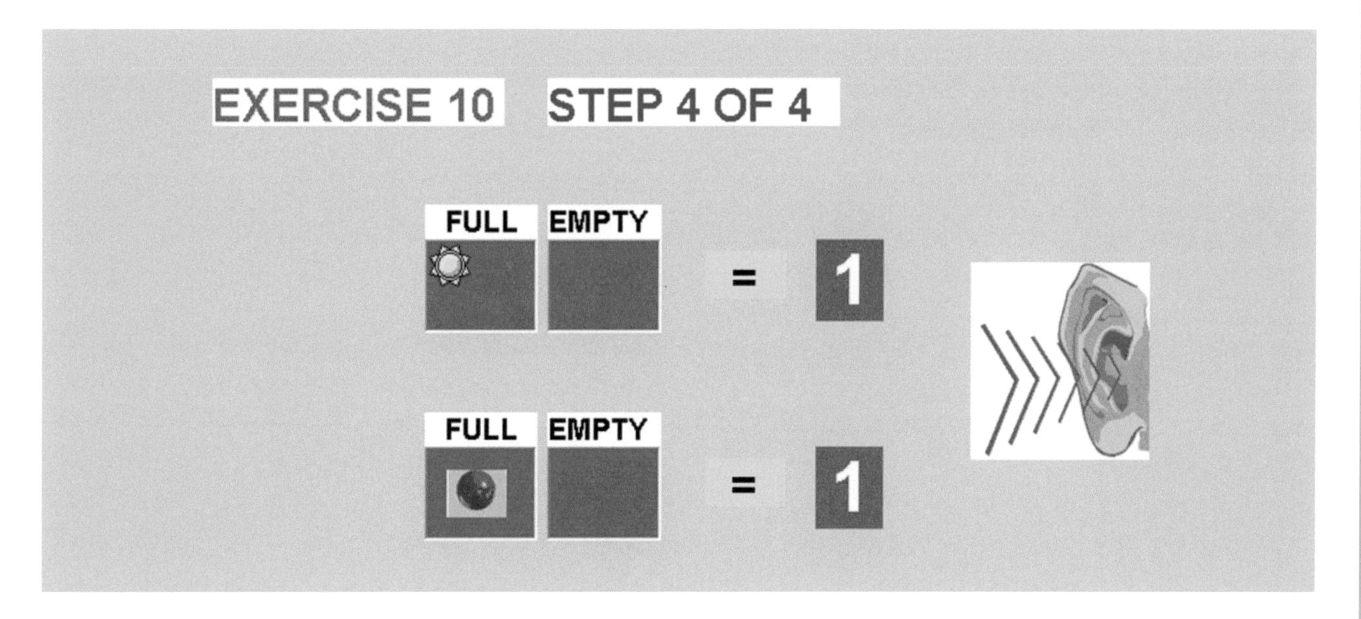

The ***kinesthetic*** representation of the filled up position is announced by clamping of the right hand. An empty position is represented by the clamping of the left hand. The kinesthetic representation for this picture is clamping the right hand and left hand.

MATHEMATICS

SUBTRACTION, FIND AN ANSWER, EXERCISE 10

In exercise 10, in ***visual*** display of subtraction we see two pictures of boxes. In the bottom we see three pictures with probable answers. Both pictures have a symbol in the first position and the second position is empty. We want to specify, that between the first and second pictures we see the subtraction sign. Bellow the second picture we see the equal line, which separates the problem from the answer. The ***audio*** representation for two pictures consumes of a **knock, double knock, knock, double tram, double knock, double knock, knock, and cling.**

From the probable answers for addition we need to find the right one. The audio signals are: 1) **knock and knock.** 2) **knock and double knock** 3) **double knock and knock**.

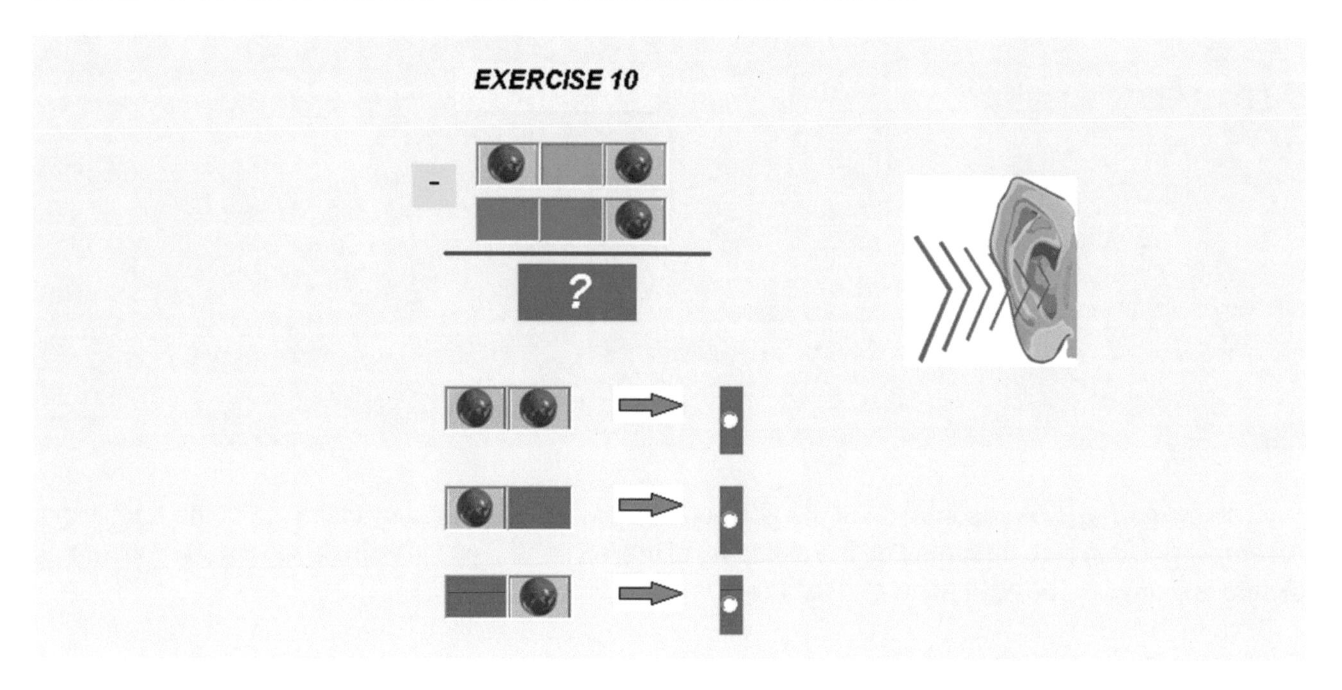

The ***kinesthetic*** representation of the filled up position is announced by clamping of the right hand. An empty position is represented by the clamping of the left hand. The subtraction sound, double tram, is represented by both hands extended in front. The equal sound, cling, is represented by crossing of hands. The kinesthetic representation for both pictures and the answer is the clamping of the right hand, left hand, right hand, hands extended in front, left hand twice, right hand, crossing of hands, right hand, and left hand.

SUBTRACTION, EXERCISE 11

In exercise 11, in ***visual*** display of subtraction we see three pictures of boxes. Between the first and second pictures we see the subtraction sign. Under the second picture we see the equal line, which separates the problem from the answer. Below the equal line is the third picture, which is the answer of subtraction. The ***audio*** representation for subtraction consumes of a **knock, double knock, knock, double tram, knock, knock, cling, double knock, and knock.**

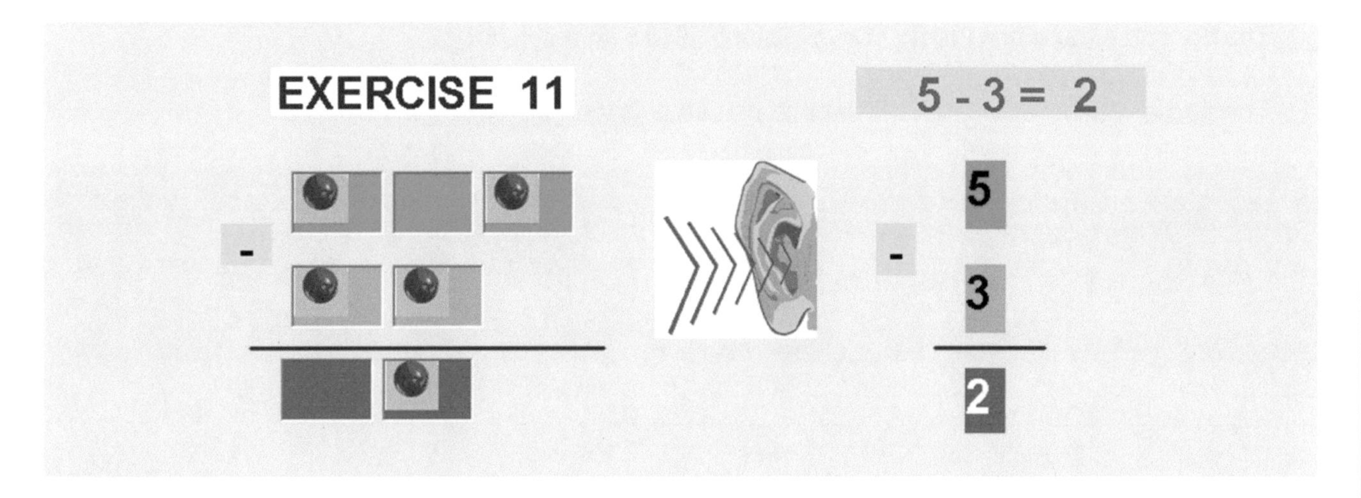

The ***kinesthetic*** representation of the filled up position is announced by clamping of the right hand. An empty position is represented by the clamping of the left hand. The subtraction sound, double tram, is represented by both hands extended in front. The equal sound, cling, is represented by crossing both hands. The kinesthetic representation for three pictures is the clamping of the right hand, left hand, right hand, hands extended in front, right hand twice, crossing of hands, left hand, and right hand.

MATHEMATICS

SUBTRACTION, EXERCISE 11, STEP 11.1

In exercise 11, step 11.1 in ***visual*** display we see two pictures of boxes. In the first picture we see that the first and third positions are filled with a symbol while the second position is empty. This pictures numerical value is 5.

The **audio** representation for the first picture is **knock, double knock, and knock.**

The second picture has the first two positions filled with a symbol.

The **audio** representation for the second picture is **knock and knock.**

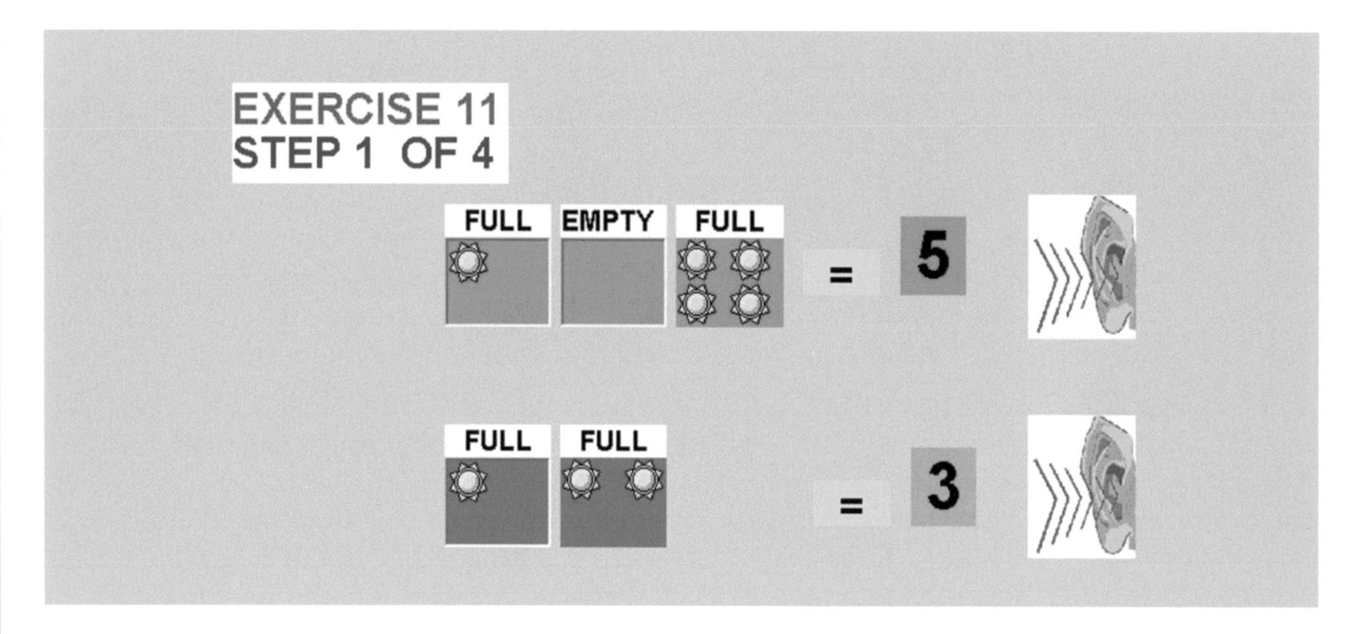

The ***kinesthetic*** representation of the filled up position is announced by clamping of the right hand. An empty position is represented by the clamping of the left hand. The equal sound, cling, is represented by crossing both hands. The kinesthetic representation for the first picture is the clamping of the right hand, left hand, and right hand. The kinesthetic representation for the second picture is clamping of right hand twice.

SUBTRACTION, EXERCISE 11, STEP 11.2

In exercise 11, step 11.2 in ***visual*** display of subtraction we see three pictures of boxes. In the first two pictures the first position is filled with a symbol, the second position is empty, and the third position is filled with a symbol. The third picture has a symbol in the first and second positions while the third position is empty. For purpose of subtraction we are introducing the concept of borrowing. We are borrowing from the third to the second position. However the audio representation for the picture does not change because the numerical value does not change.

The ***audio*** representation for the three pictures consumes of a **knock, double knock, and knock.**

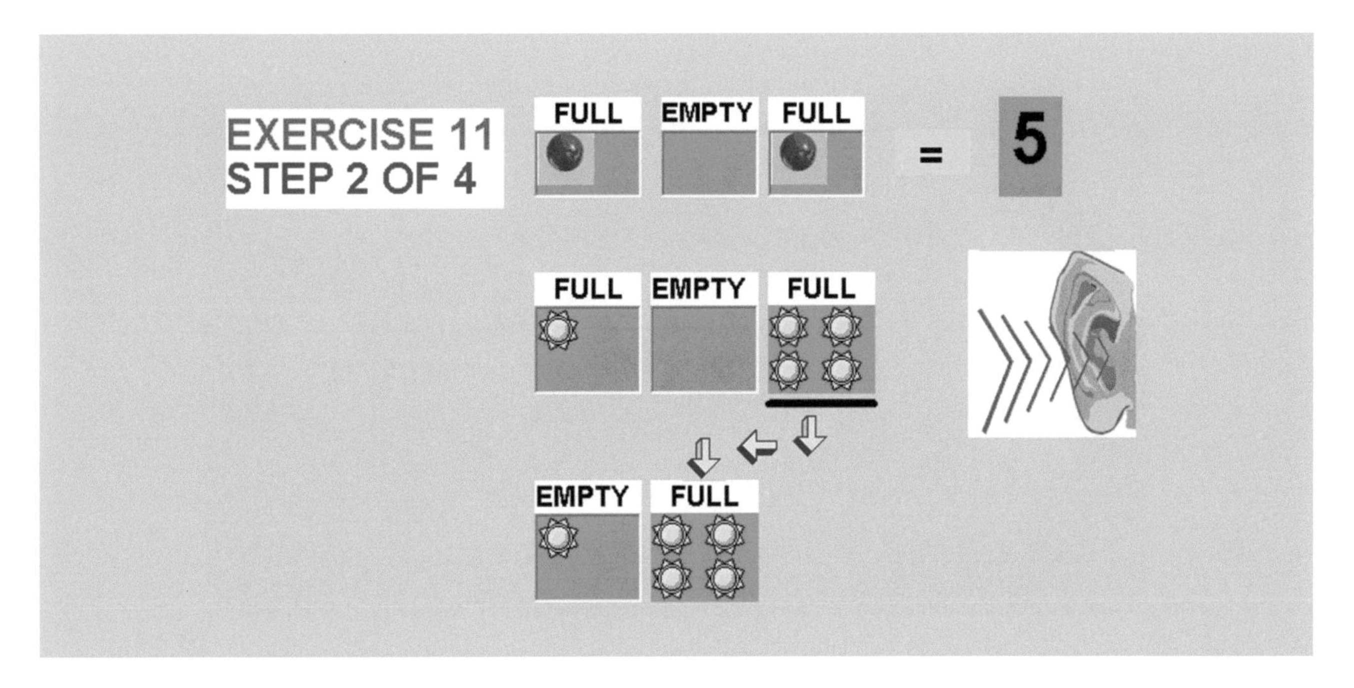

The ***kinesthetic*** representation of the filled up position is announced by clamping of the right hand. An empty position is represented by the clamping of the left hand. The kinesthetic representation for three pictures is the clamping of the right hand, left hand, and right hand.

MATHEMATICS

SUBTRACTION, EXERCISE 11.3

In exercise 11, step 11.3 in ***visual*** display of subtraction we see six pictures of boxes. There are three pictures on the right side and three pictures on the left side. Both sides represent the same equation; however the left side has used the concept of borrowing to further show the problem. When we do subtraction we are borrowing available symbols from the higher position. We need to remember that all symbols borrowed from the higher position are twice larger than the one in the prior descending position. Once the symbol is borrowed the position borrowed becomes empty and the two symbols appear in the prior position. If needed we leave one symbol to the next descending position which is empty and once again this one symbol becomes two symbols in the next descending position. The ***audio*** representation for subtraction consumes of a **knock, double knock, knock, double tram, knock, knock, cling, double knock, and knock.**

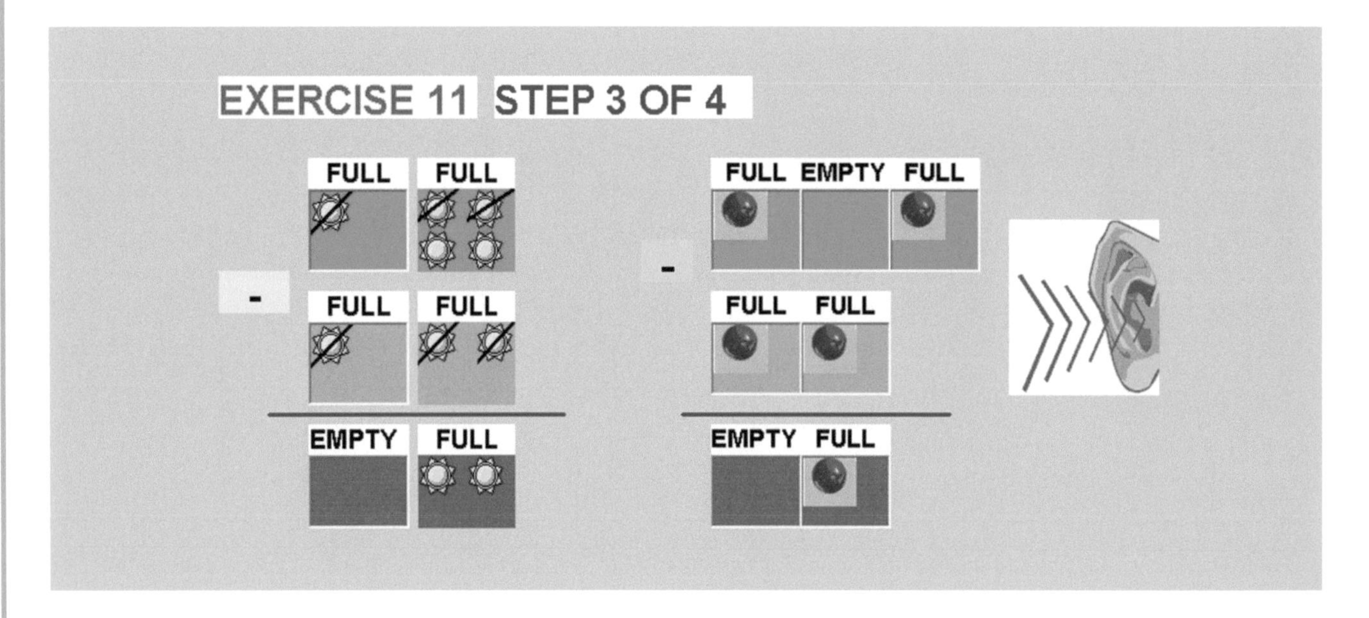

The ***kinesthetic*** representation of the filled up position is announced by clamping of the right hand. An empty position is represented by the clamping of the left hand. The subtraction sound, double tram, is represented by both hands extended in front. The equal sound, cling, is represented by crossing both hands. The kinesthetic representation for the subtraction is the clamping of the right hand, left hand, right hand, hands extended in front, right hand twice, crossing of hands, left hand, and right hand.

ADDITION, EXERCISE 11, STEP 11.4

In exercise 11, step 11.4 in ***visual*** display of the last step of subtraction, we see two pictures of boxes, which is the answer of the problem. We see that the first position is empty while the second position is filled up with a symbol. The answer has a value of two.

The ***audio*** representation for the pictures consumes of a **double knock, knock, and cling.**

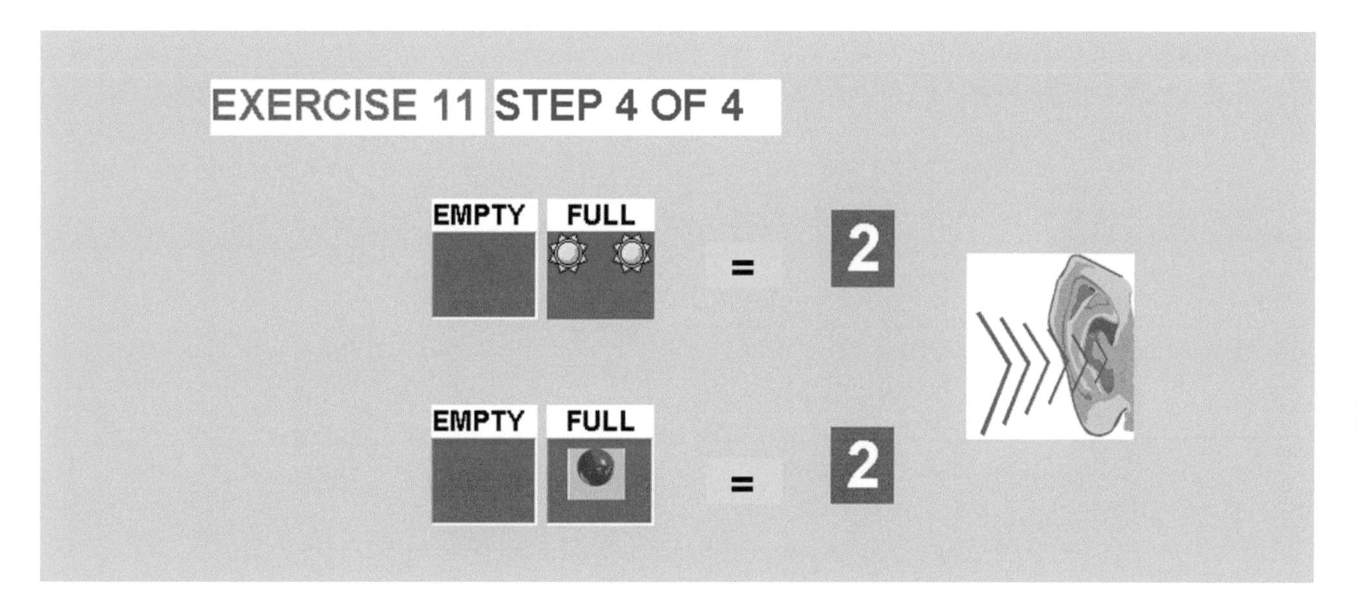

The ***kinesthetic*** representation of the filled up position is announced by clamping of the right hand. An empty position is represented by the clamping of the left hand. The kinesthetic representation for this picture is clamping the left hand and right hand.

MATHEMATICS

SUBTRACTION, FIND AN ANSWER, EXERCISE 11

In exercise 11, in ***visual*** display of subtraction we see two pictures of boxes. In the bottom we see three pictures with probable answers. Both pictures have a symbol in the first position and the second position is empty. We want to specify, that between the first and second pictures we see the subtraction sign. Bellow the second picture we see the equal line, which separates the problem from the answer. The ***audio*** representation for two pictures consumes of a **knock, double knock, knock, double tram, knock, knock, and cling.**

From the probable answers for addition we need to find the right one. The audio signals are: 1) **knock, knock, and double knock.** 2) **knock, double knock, and knock** 3) **double knock, knock, and double knock**.

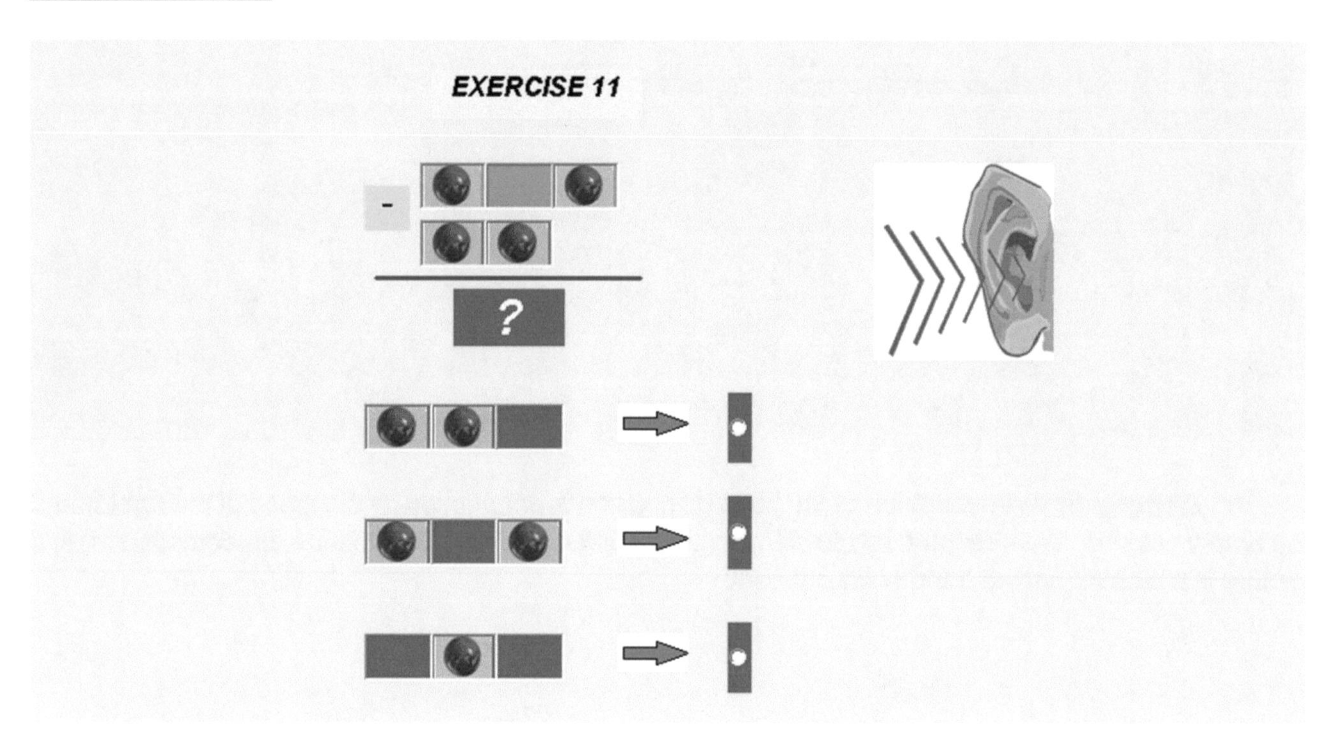

The ***kinesthetic*** representation of the filled up position is announced by clamping of the right hand. An empty position is represented by the clamping of the left hand. The subtraction sound, double tram, is represented by both hands extended in front. The equal sound, cling, is represented by crossing of hands. The kinesthetic representation for both pictures and the answer is the clamping of the right hand, left hand, right hand, hands extended in front, right hand twice, crossing of hands, left hand, right hand, and left hand.

Dedicated to my nephew, David Gimelfarb,
Lost in Costa Rico in 2009.

Advance brain stimulation by psychoconduction

Chester Litvin, Ph.D., Clinical Psychologist

CONTENTS

LESSON ONE INTRODUCTION TO SIGNS

There are five different signs. There is an addition sign, division sign, subtraction sign, multiplication sign, and an equal sign. Addition is represented by the sound, **tram**. Division is represented by the sound, **double click**. Subtraction is represented by the sound, **double tram**. Multiplication is represented by the sound, **blick**. The equal sign is represented by the sound, **cling**.

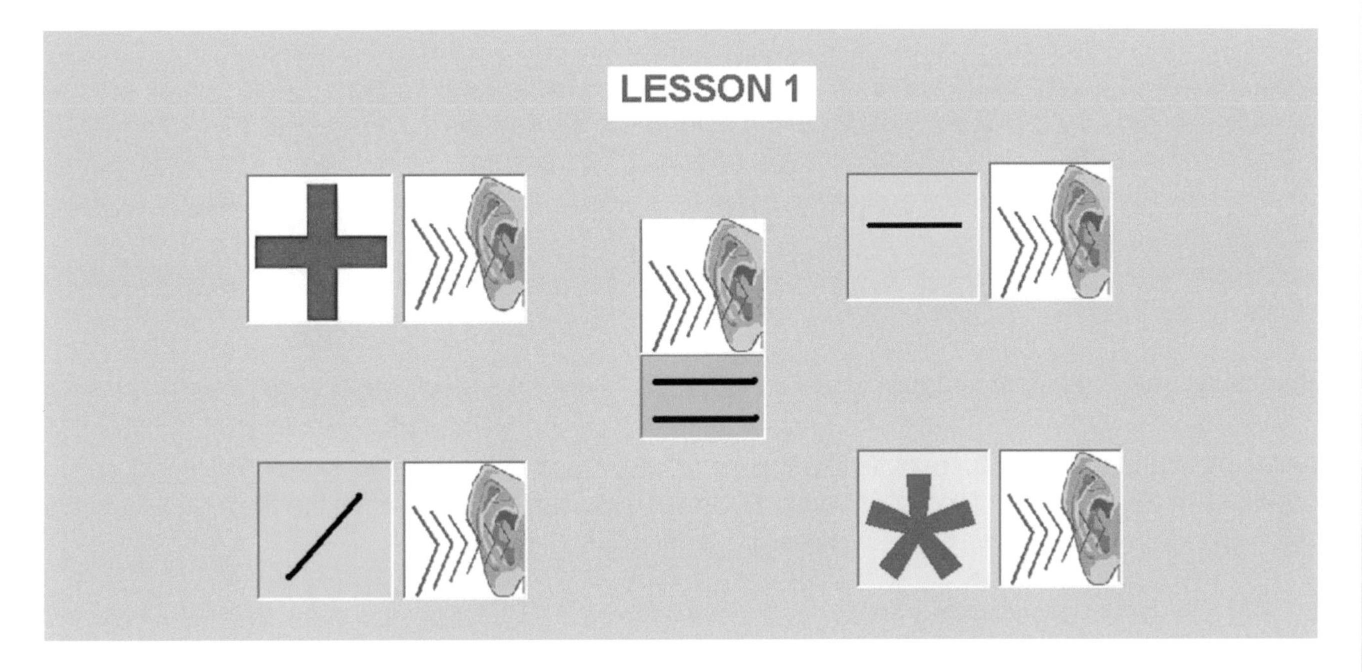

The addition sign is represented by one hand extended in front. The division sign is represented by both hands extended up. The subtraction sign is represented by both hands extended in front. The multiplication sign is represented by one hand raised up. The equal sign is represented by crossing both hands.

MULTIPLICATION

When we use Litvin's Code for multiplication, the multiplication table is not required. Instead, we use a series of simple steps. The second number in the multiplication is used as an indicator for moving the first multiplicand and shows how many steps the multiplication process has. In the end, we sum up the steps to receive the answer. In the first step, we move the first multiplicand to the position prior to the first occupied position of the second multiplicand. If the first position of the second number is filled with a symbol, then the result of the first step is equal to the first number. As mentioned before, the first position indicates the binary number 2^0 which is equal to 1. So any number multiplied by 1 is equal to the first multiplicand. On the other hand, if the first position of the second multiplicand is empty, we move the first multiplicand to the right of the number of empty positions, which are located before the first filled position. By this movement, we get the first number to add for the final sum, which represents the answer of the multiplication. If the second multiplicand has the second position filled with symbol, then we move the first multiplicand to the right of the number of positions, which are before the second position occupied with a symbol in the second multiplicand. By this movement, we get the second number to add to get the sum, which represents the answer of the multiplication. If the second multiplicand has its third position filled with symbol, we move the first multiplicand to the right of the number corresponding to the positions before the third filled with the symbol in the second multiplicand. By this movement, we get the third number to add to get the sum, which represents the answer of the multiplication. The answer of the multiplication depends on the number of filled positions in the second multiplicand. We move the first multiplicand to the number of positions before the filled position in the second multiplicand.

We add the results of all steps and arrive at the answer. It is important to understand the difference between the powers of binary numbers in different positions and the sequential number of a position. For example, if the power of the binary number is 2, then the position of this binary number is 3. This happens because the first power of a binary number is 0 but is located in position 1. The power is one unit behind the position. The power of the binary numbers in the sequential positions is as follows: 0, 1, 2, 3, 4, …, n-1. The sequential positions are as follows: 1, 2, 3, 4, 5, …, n. When the power of a binary number is equal to 0, it corresponds to position 1. In other words, when the power of each position is equal to (n –1), the corresponding position is equal to n.

Rules:

- When multiplying two numbers, we need to complete a number of steps which are equal to the number of positions filled with symbols in the second multiplicand. Then we add up the results of the steps. The sum of the addition represents the answer for the multiplication.

- When multiplying two numbers, we move the first multiplicand in the ascending direction by the number, which corresponds to the amount of empty positions in the second multiplicand. We count the number of empty positions prior the ones filled with symbols in the second multiplicand. We move the first multiplicand to the right, by the counted number of positions.

- When the first position of the multiplicand is filled with a symbol, we do not need to move the first multiplicand and the result of this step is equal to the first number. When the second multiplicand has two or more positions filled with symbols, then in each step we consequently move the first multiplicand to the right of the numbers of positions. The number of positions filled with symbols in the second multiplicand indicates how many steps we have in the multiplication.

- When adding up the results of each step, we arrive at the final answer of the multiplication.

Example 4.1

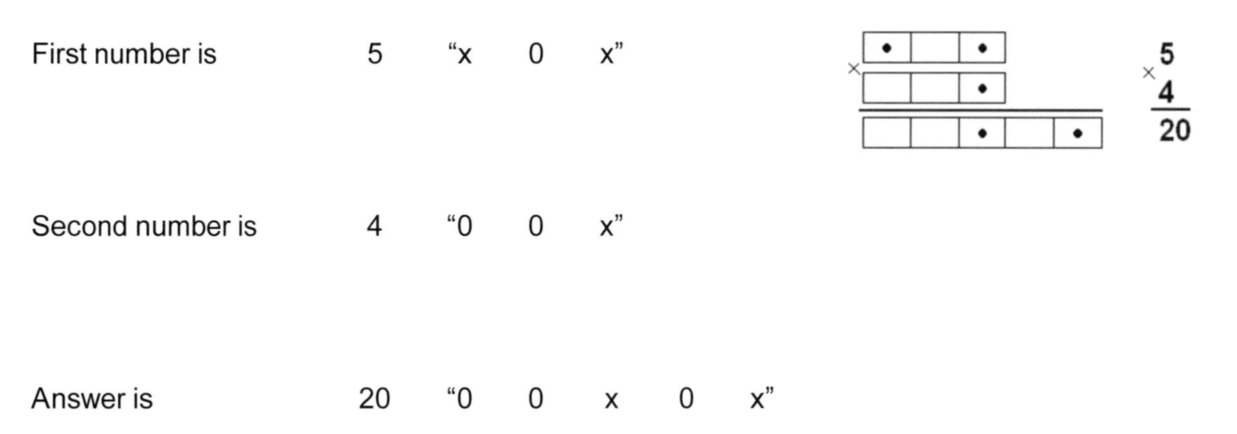

In Example 4.1, the two numbers to be multiplied have three positions each. To have a decimal equivalent of the number from the binary representation, we sum up the numbers of positions with symbols. Each position contains 2 to the power of (n-1), where n represents a filled position in the number. Since the second number has only the third position filled with a symbol, then the second number is 2 to the power of 2 and is equal to 4. To complete this multiplication, we move the first multiplicand in the ascending direction by two positions. The number of the empty positions is equal to (n-1), where n corresponds to a sequential number of a filled position in the second multiplicand, and (n – 1) is the power of a filled up position in the second multiplicand. In Example 4.1, the number n corresponds to position 3, (n – 1) is equal to 2. The second multiplicand is equal to number 4, which is equal to number 2 to the power of 2.

Explanation

In Example 4.1, the second multiplicand is an even number and has only one symbol in the third position - n. Therefore, we move the first multiplicand in position (n-1) to the right. The second number has the third position filled with a symbol. The power of the binary number in this position is also 2 and the number, itself, is 4. We move the first number in two positions to the ascending direction to get the result of this step. This method of calculation requires moving the first multiplicand to the right by the number of positions, specified by the second multiplicand. The number of moving positions corresponds to (n – 1), which is the power of the binary number of the position that is filled with a symbol in the second multiplicand. We move the binary number 5 two positions to the right and the answer is twenty.

In Summary:

- When the second number has two empty positions prior to the position filled with a symbol, we move the first number two positions in the ascending direction.

Example 4.2

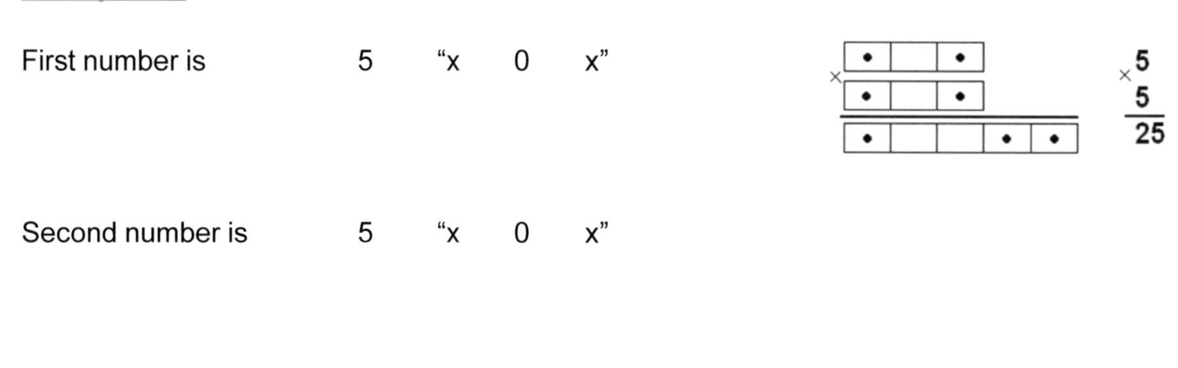

First number is 5 “x 0 x”

Second number is 5 “x 0 x”

Answer is 25 “x 0 0 x x”

In Example 4.2, we have two numbers with three positions each. The first and the second numbers are odd and equal to one another. They are represented in Litvin’s Code by three positions where the first and third are filled with symbols and the second is empty. In the first step, we move the first multiplicand according to the first filled position in the second multiplicand. The first position of the second number is filled with a symbol and since its decimal value is one, we do not move the first number, and the result for the first step is equal to the first multiplicand. If the second symbol of the second multiplicand is in the

nth position, we move the first multiplicand in the (***n***-1) position in the ascending direction. The result of the second step is the first multiplicand moved two positions to the right. In the end, we add the results of the steps and arrive at the answer of the multiplication.

Explanation

In the first step of Example 4.2, in the second number the first position is filled with a symbol and is equal to one. Therefore, we eave the first number as it is. If we multiply any number by 1, we have the same number in the result of this step. When we use the formula for moving (n-1) position, then we subtract 1 from 1, and have 0 as the result. We move the first number 0 positions, which means that we do not move the first number at all.

Result for the first step is 5 "x 0 x"

In the second step of Example 4.2, in the second number, the next filled position is the third one and corresponds to number four. The third position is represented by 2 to the power of 2 and is equal to 4. As discussed before, we move the first number (***n***-1) positions, which is two positions in the ascending direction. (***n*** - 1) is a formula, which is equal to the power of the binary number in the position in the second multiplicand, filled with a symbol and equal to 2.

The power of the binary number in this position is the sequence number of this position minus one. To make things simple, the first multiplicand is moved two positions in the ascending direction in the result of the second step.

Result for step two is 20 "0 0 x 0 x"

In the end, we add the results of the first and se cond steps, and arrive at the answer of the multiplication, which is equal to the number 25 and is represented in Litvin's code by five positions, out of which the first, fourth and fifth are filled with symbols.

First step is 5 "x 0 x"

Second step is 20 “0 0 x 0 x”

Answer is 25 “x 0 0 x x”

In Summary:

- When the second number has positions 1 and 3 filled with symbols, then the powers of the binary numbers in those positions are 0 and 2 and the decimal numbers are 1 and 4.

- When any number is multiplied by one, then the result is the same number. In this case, since the power of binary number in position 1 is 0, the number two to the power of 0 is equal to one. Any number to the power of 0 is equal to 1.

- When the first position in the second number is filled with a symbol, then we do not move the first number in the first step and the answer for that step is equal to the first number.

- During the second step, we move the first number two positions in the ascending direction. In the end, we add the results of the two steps.

- When using this method of calculation, we move the numbers in specified positions that correspond to the power of the binary number in this position in the second multiplicand. The power of the binary number in the third position is equal to 2 and we move the first number two positions.

End.

Example 4.3

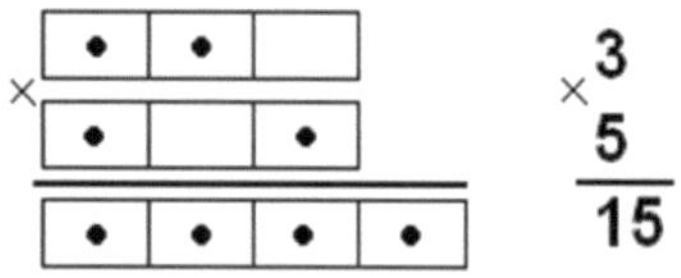

First number is 3 “x x 0”

$$\begin{array}{r} 3 \\ \times\ 5 \\ \hline 15 \end{array}$$

Second number is 5 “x 0 x”

Answer is “x x x x”

In Example 4.3, we have two numbers with three positions each. The first and second numbers are odd. The first number has symbols in its first and second positions. The second number has its first and third positions filled with symbols. By looking at the positions filled with symbols in the second number, we move the first number in the ascending direction. The symbols in the second number are in positions 1 and 3, which represent binary number in the power of 0 and 2. To receive the answer, we need to move the first number twice. The first time, we move it 0 positions and the second time, 2 positions. Then we add the results of the steps.

Explanation

In Step One for Example 4.3, the second number has the first position filled with a symbol, so the power of the binary number in this position is 0. When the power of the binary number two is equal to 0, then number is equal to 1. When we multiply it by 1, we do not move the first multiplicand and the result of the first step is equal to the first multiplicand.

In Step two for Example 4.3, the second number has the third position filled with a symbol, which means that the power of the binary number in this position is 2. When power of binary number is equal to 2, we move the first number two positions, and the result of the second step is the first number moved two positions in the ascending direction.

Result for step two is 12 “0 0 x x”

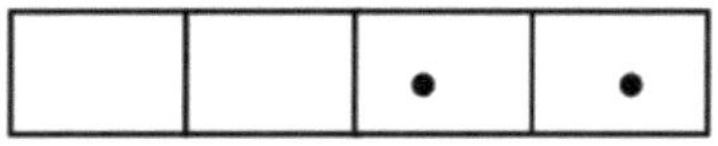

In the end, we add the results of the first and the second steps and arrive at the result of the multiplication.

The first number is 3 “x x”

The second number is 5 “x 0 x”

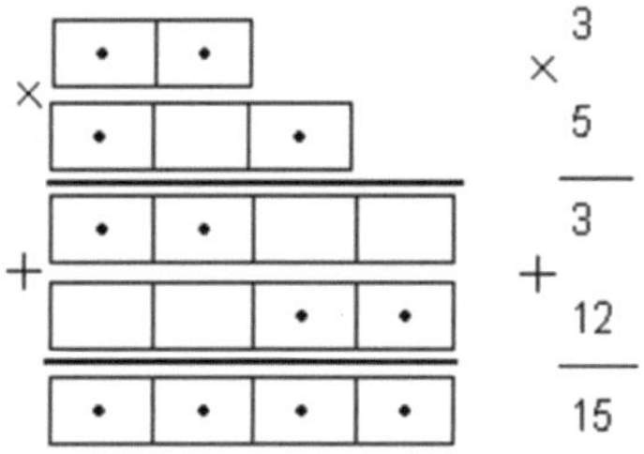

	3
×	5
	3
+	12
	15

Result of step one is 3 “x x”

Result of step two is 12 “0 0 x x”

Answer is “x x x x”

In Summary:

- When multiplying the numbers above, where the first position of the second number is filled with a symbol, the result of the first step is equal to the first number.

- Then we move the first number two positions in the ascending direction and then add the results of the two steps.

End.

Example 4.4

First number is 14 “0 x x x”

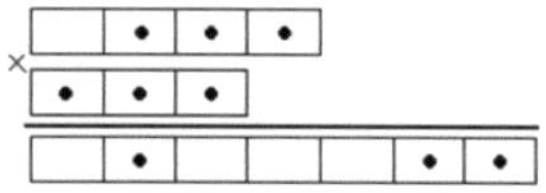

× 14
7
98

Second number is 7 “x x x”

Answer is 98 “0 x 0 0 0 x x ”

In Example 4.4, we multiply two numbers with four positions in the first multiplicand and three positions in the second. The second multiplicand has all positions occupied with symbols. In this example, positions 1, 2, and 3 are filled with symbols and the power of binary number (***n***-1) is equal to ***0, 1, and 2***. The filled positions in the second multiplicand are the first, second, and third and their sum is 7.

In the first step, we do not move the first multiplicand in any position, because the power of binary number in the first position of the second multiplicand is 0. If we move the number on 0 positions, then the result will equal the first number. In the second step, we move the first multiplicand one position in the ascending direction, because the power of the binary number in the second position of the second multiplicand is equal to 1. In the third step, we move the first multiplicand the two positions in the ascending direction, because the power of the binary number in the third position of the second multiplicand is equal to 2.

Explanation

In the first step of Example 4.4, in the second multiplicand, the first position is equal to 1, when (n - 1) = 0 and n=1, (n - 1) is the power, or $2^{n-1} = 2^0 = 1$.

In the first position of the second multiplicand, we have two to the power of 0. Any number to the power of 0 is 1. We do not move the first multiplicand, because if we multiply it by 1, the answer is the same.

The result for the first step is equal to the first multiplicand or 14.

"0 x x x"

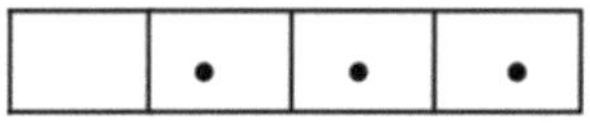

	•	•	•

In the second step, the second occupied position in the second multiplicand is n=2 and the power of the binary number is (n - 1)=1. The power of the binary number in the second position of the second multiplicand is equal to 1. We need to move the first multiplicand one position in the ascending direction and the result of step two is 28.

"0 0 x x x"

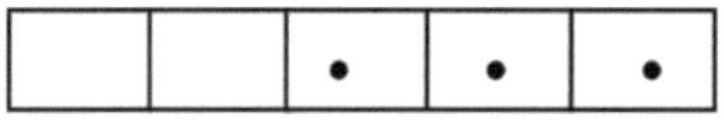

		•	•	•

In the third step, the third occupied position in the second multiplicand is 3. We move the first multiplicand (n-1) positions. (n-1) corresponds to the power of the binary number in this position and is equal to 2; (n-1) = 2, n=3.

We move the first multiplicand two positions in the ascending direction and the result of this step is 56.

"0 0 0 x x x"

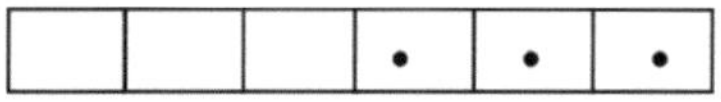

			•	•	•

Below is represented the whole process of the steps. We add all binary numbers in the filled positions of the answer.

First step 14 “0 x x x”

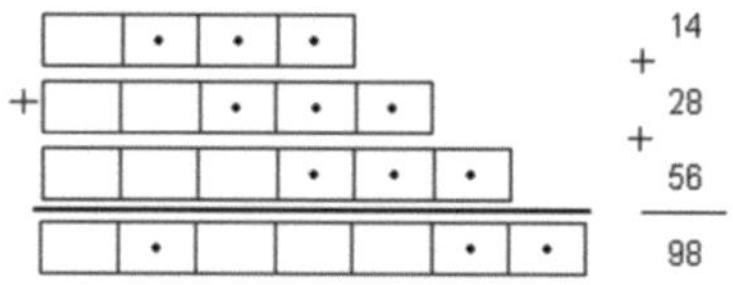

14
\+
28
\+
56
98

Second step 28 “0 0 x x x”

Third step 56 “0 0 0 x x x”

Answer 98 “0 x 0 0 0 x x”

In Summary:

- When multiplying using Litvin’s Code, we move the symbols in the ascending direction.

- When the first position in the second number is filled with a symbol, the result for the first step is equal to the first number, which is 14.

- Then we move the first number one position in the ascending direction and the result of the second step is 28.

- After this, we move the first number two positions in the ascending direction and the result of the third step is 56. We add the results of all steps and arrive at the final answer 98.

End.

MULTIPLICATION, EXERCISE 1

In Exercise 1, in a ***visual*** display of multiplication we see three pictures of boxes. The first picture has its first two positions empty while its third position is filled with a symbol. In the second picture, both positions are filled with symbols. The third picture, which is the answer for the multiplication, has its first two positions empty and the next two filled with symbols. The multiplication sign is between the first and second pictures. Below the second picture is the equal line, which separates the problem from the answer. The ***audio*** representation for three pictures consumes of a **double knock, double knock, knock, blick, knock, knock, cling, double knock, double knock, knock and knock.**

Figure 1

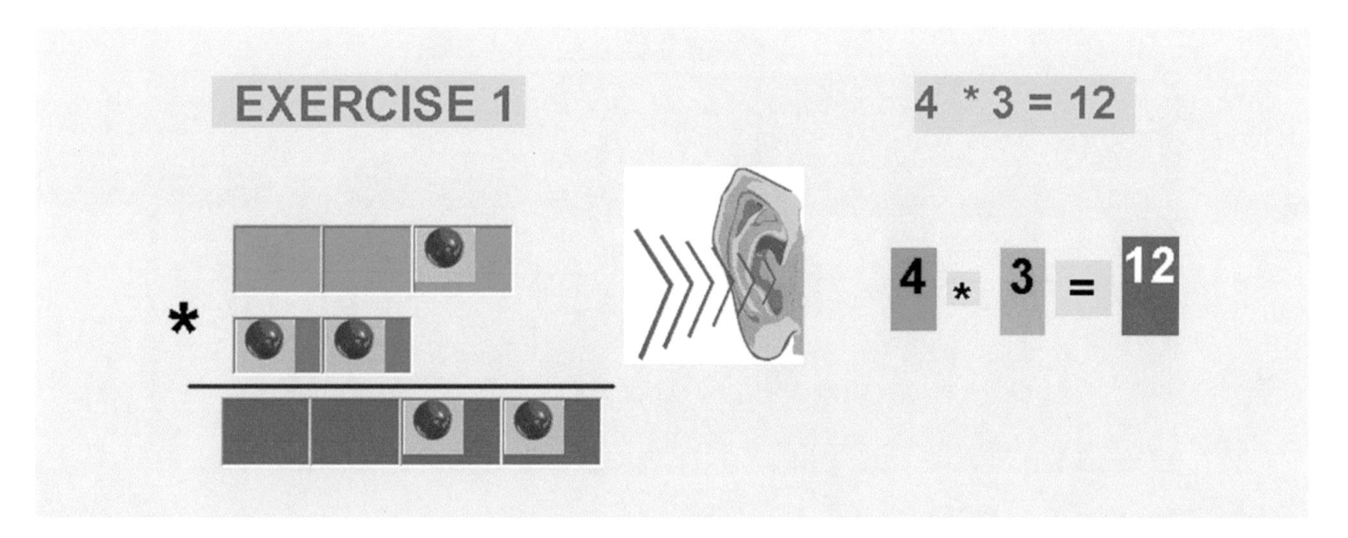

The ***kinesthetic*** representation of a filled up position is represented by the clenching of the right hand. An empty position is represented by the clenching of the left hand. The multiplication sign, blick, is represented by one hand raised up. The equal sign, cling, is represented by crossing both hands. The kinesthetic representation for both pictures is *the clenching of the left hand twice, right, extending both hands towards the front, the clenching of the right twice, crossing both hands, the clenching of the left twice and right hand twice.*

MATHEMATICS

MULTIPLICATION, EXERCISE 1, STEP 1.1

In Exercise 1 Steps 1.1, in a ***visual*** display of multiplication we see three pictures of boxes. In the first picture, which has a value of four, the first two positions are empty while the third is filled with a symbol. In the second picture, which has a value of one, the first position is filled with a symbol and the second is empty. Between the first and second pictures is the multiplication sign. Below the second picture is the equal line, which separates the problem from the answer. In the third picture, which is the answer for the first step of the multiplication, the first and second positions are empty, while the third is filled with a symbol.

The ***audio*** representation for the multiplication consists of a **double knock, double knock, knock, blick, knock, double knock, cling, double, knock, double knock, and knock**.

Exercise 1, Step 1.1

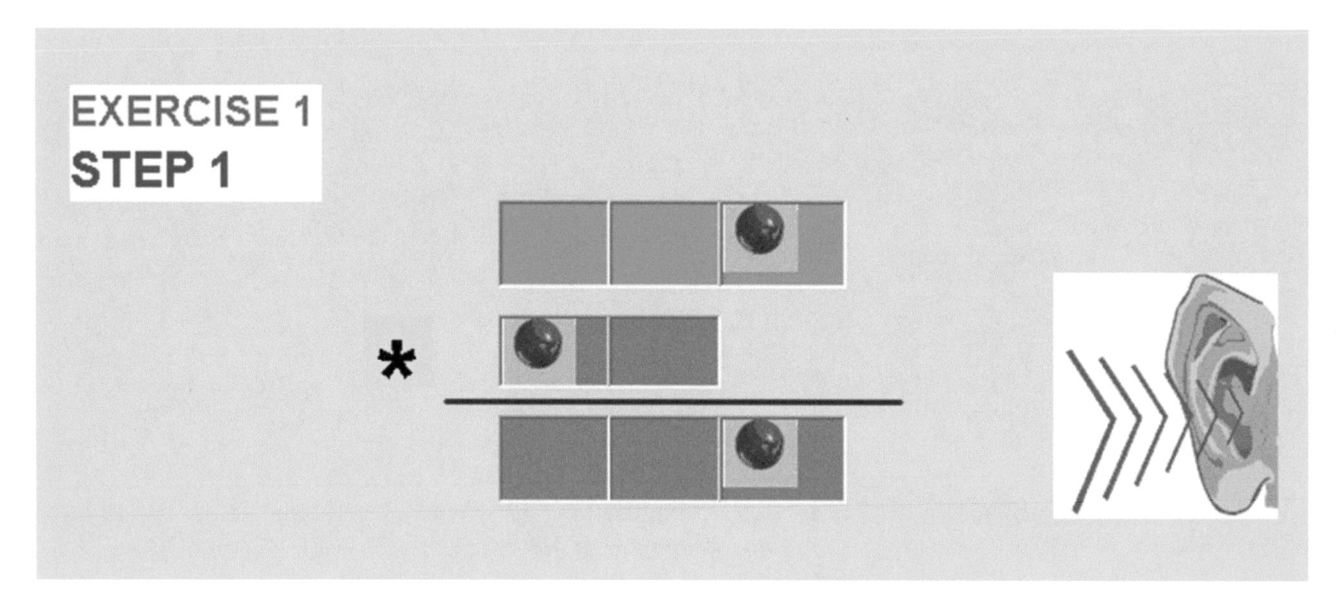

The ***kinesthetic*** representation of a filled position is represented by the clenching of the right hand. An empty position is represented by the clenching of the left hand. The multiplication sound, blick, is represented by one hand raised up. The equal sound, cling, is represented by crossing both hands. The kinesthetic representation for the multiplication is *the clenching of the left twice, right, one hand raised up, right hand, left hand, crossing of hands, the clenching of the left hand twice and the right hand.*

MULTIPLICATION, EXERCISE 1, STEP 2

In Exercise 1 Step 1.2, in a ***visual*** display for the next step of multiplication, we see three pictures of boxes. We move the first picture one position to the right. In the first picture, which has a value of four, the third position is filled with a symbol. In the second picture, which has a value of two, the second position is filled with a symbol. The multiplication sign is between the first and second pictures. Below the second picture is the equal line, which separates the problem from the answer. In the third picture, which is the answer for the multiplication and has a value of eight, the first three positions are empty, while the fourth is filled with a symbol.

The ***audio*** representation for the multiplication consists of a **double knock, double knock, knock, blick, double knock, knock, cling, double knock, double knock, double knock and knock.**

Exercise 1, Step 1.2

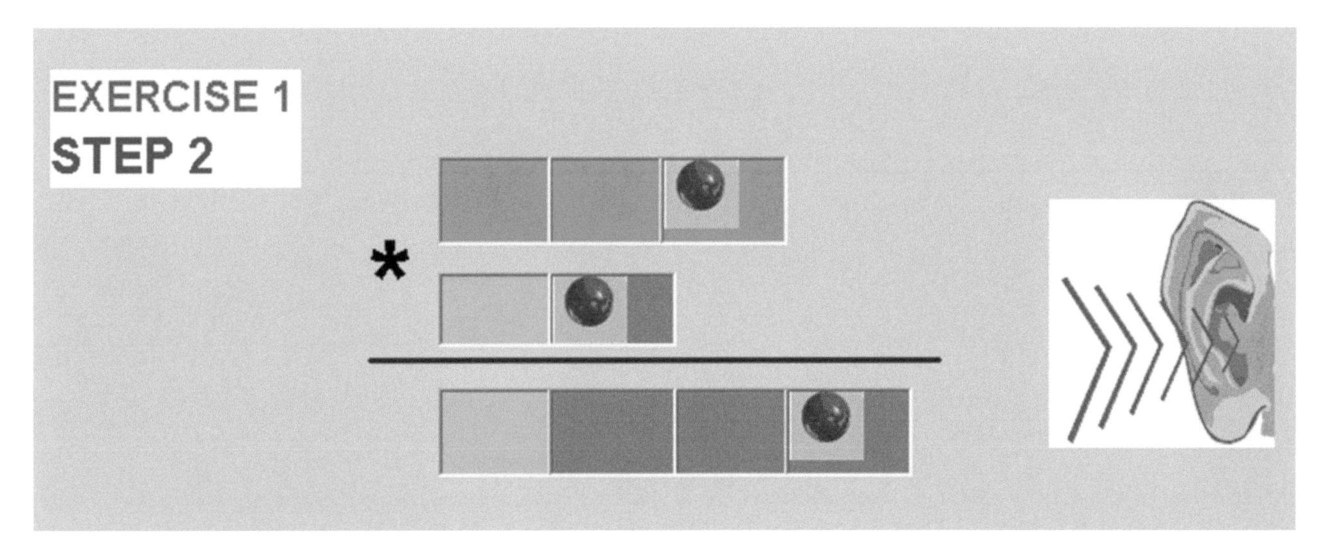

The ***kinesthetic*** representation of a filled position is represented by the clenching of the right hand. An empty position is represented by the clenching of the left hand. The multiplication sign, blick, is represented by one hand raised up. The equal sign, cling, is represented by crossing both hands. The kinesthetic representation for the multiplication is *the clenching of the left hand twice, right hand once, one hand raised up, the clenching of the left hand once, right hand once, crossing the both hands, the clenching of the left hand three times and the right hand once.*

MATHEMATICS

MULTIPLICATION BY ADDITION, EXERCISE 1, STEP 3

In Exercise 1 Step 1.3, in a ***visual*** display of the third step of the multiplication, we are adding together the results of the first and second steps. In the first picture, which has a value of four, the third position is filled with a symbol. In the second picture, which has a value of eight, the fourth position is filled with a symbol. The addition sign is between the first and second pictures. Below the second picture is the equal line, which separates the problem from the answer. In the third picture, which is the answer for the third step of the multiplication, positions three and four are filled with symbols, while the first two positions are empty. The answer has a value of twelve.

The ***audio*** representation for the three pictures consists of a **double knock, double knock, knock, tram, double knock, double knock, double knock, knock, cling, double knock, double knock, knock and knock**.

Exercise 1, Step 1.3

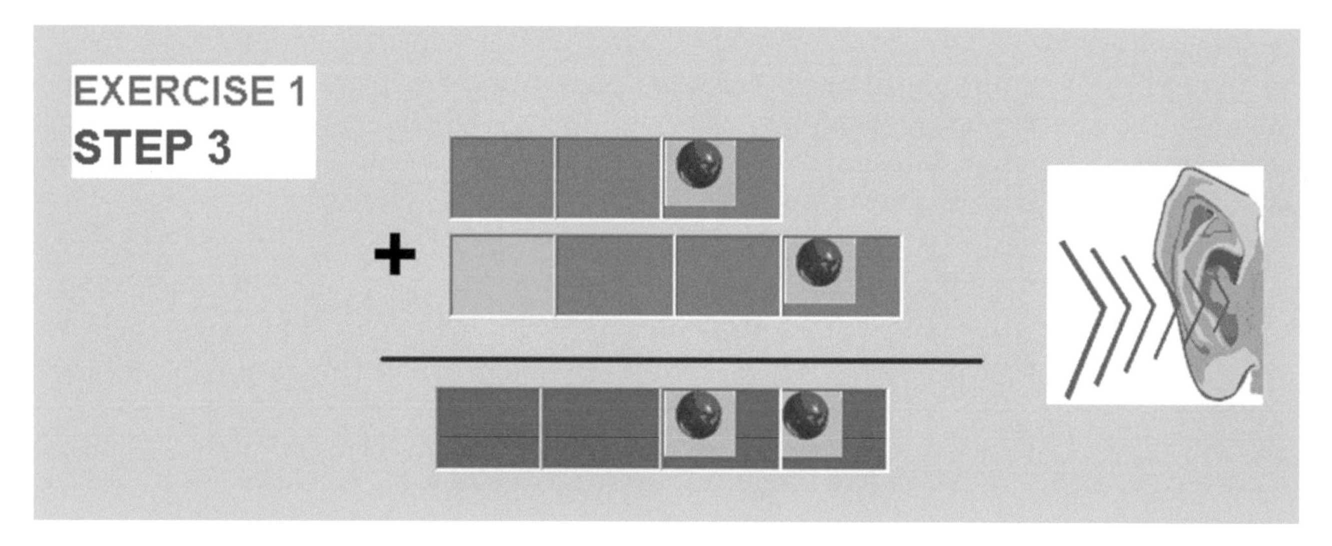

The ***kinesthetic*** representation of a filled position is represented by the clenching of the right hand. An empty position is represented by the clenching of the left hand. The addition sound, tram, is represented by extending the left hand in front. The equal sound, cling, is represented by crossing both hands. The kinesthetic representation for the three pictures is *the clenching of the left hand twice, right hand once, one hand raised up, the clenching of the left hand three times, right hand once, crossing both hands, the clenching of the left hand twice and the right hand twice.*

MULTIPLICATION, EXERCISE 1, STEP 4

In Exercise 1 Step 1.4, in a ***visual*** display of the last step of the multiplication, we see one picture of boxes, which is the answer of the problem. The first and second positions are empty, while the third and fourth three are filled with symbols. The answer has a value of twelve.

The ***audio*** representation for this picture consists of a **double knock, double knock, knock and knock**.

Figure 1.4

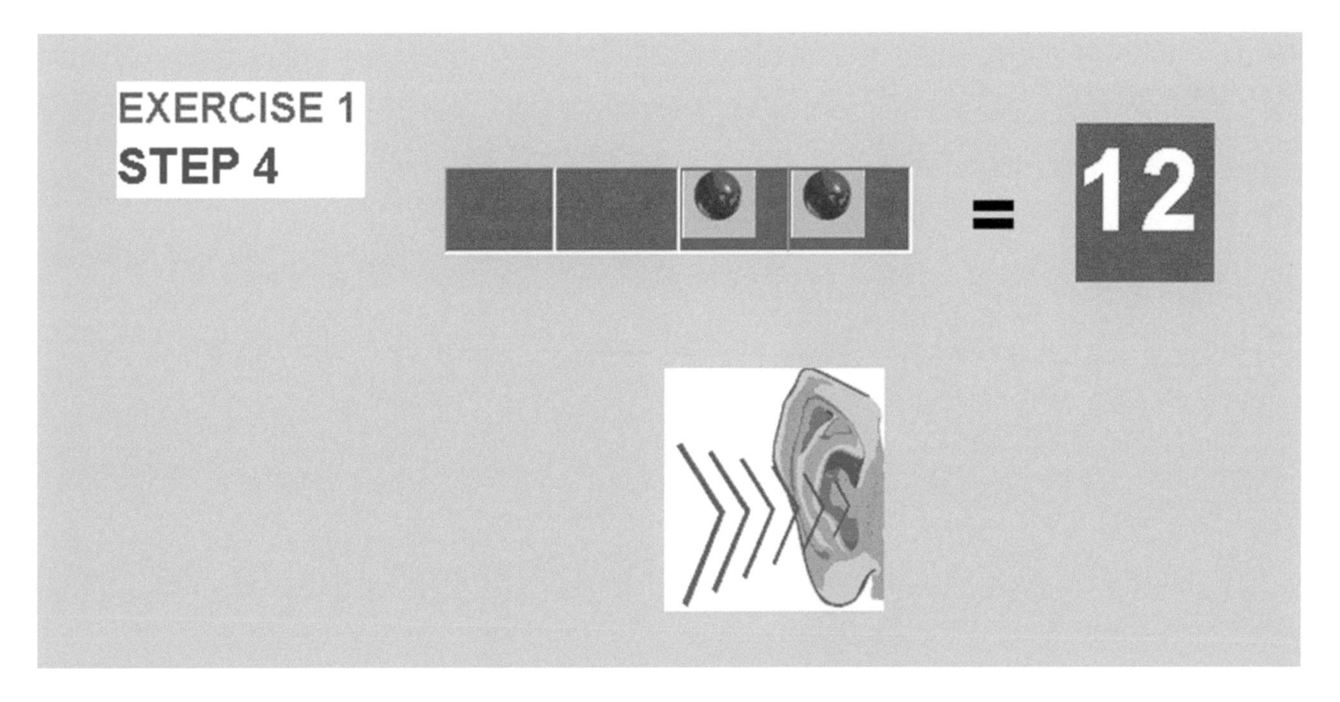

The ***kinesthetic*** representation of a filled position is represented by the clenching of the right hand. An empty position is represented by the clenching of the left hand. The kinesthetic representation for this picture is *the clenching of the left hand twice and the right hand twice.*

MATHEMATICS

MULTIPLICATION, EXERCISE 1

In Exercise 1, in a ***visual*** display of multiplication we see two pictures of boxes. At the bottom are three pictures with probable answers. In the first multiplicand, the first two positions are empty, while the third position is filled with a symbol. In the second multiplicand, both positions are filled with symbols. The multiplication sign is between the first and second numbers. Below the second number is the equal line, which separates the problem from the answer. The ***audio*** representation for the multiplication consists of a **double knock, double knock, knock, blick, knock, knock, and cling.**

We need to choose the answer from the given options. The audio signals are: 1) **double knock, knock, double knock, and knock**, 2) **double knock, double knock, knock, and knock**, 3) **double knock, knock, knock, and double knock**.

Exercise 1

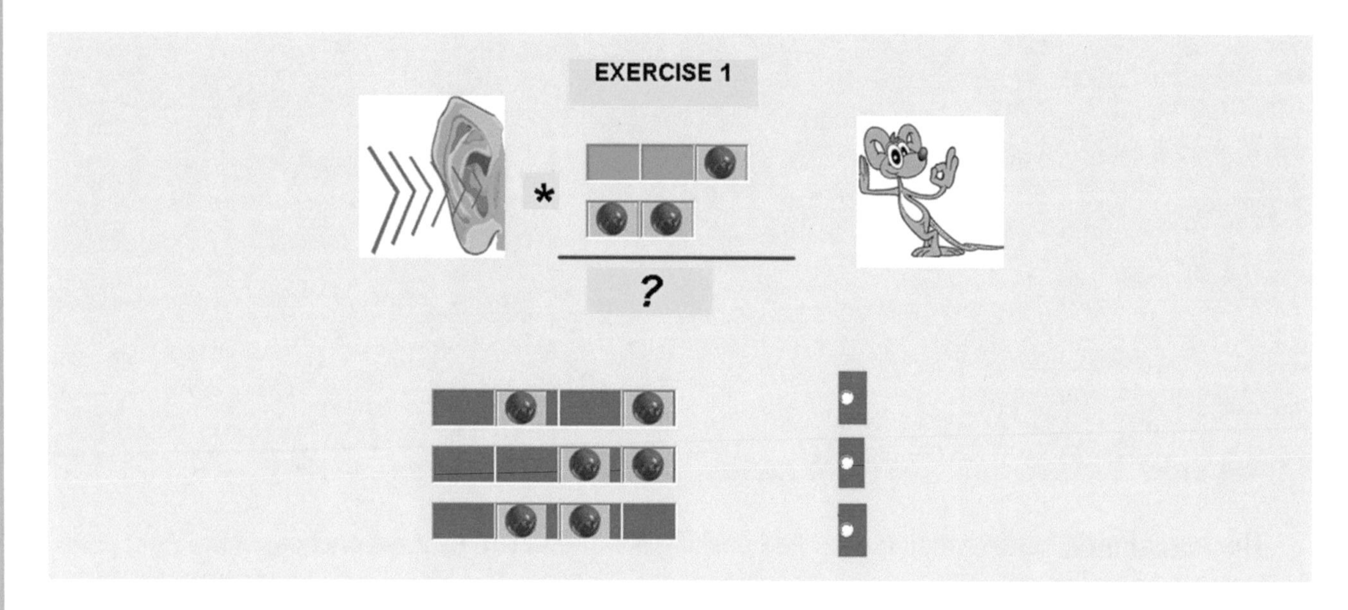

The ***kinesthetic*** representation of a filled position is represented by the clenching of the right hand. An empty position is represented by the clenching of the left hand. The multiplication sound, blick, is represented by one hand raised up. The equal sound, cling, is represented by crossing both hands. The kinesthetic representation for both pictures and the answer is the *clenching of the left hand twice, right hand, one hand raised up, the clenching of the right hand twice, crossing both hands, the clenching of the left hand twice and the right hand twice.*

MULTIPLICATION, EXERCISE 2

In Exercise 2, in a ***visual*** display of multiplication we see three pictures of boxes. The first picture, which has a value of five, has its first and third positions filled with symbols, while position two is empty. In the second picture, which has a value of four, the first two positions are empty, while the third is filled with a symbol. The multiplication sign is between the first and second pictures. Below the second picture is the equal line, which separates the problem from the answer. In the third picture, which is the answer for the multiplication and has a value of twenty, the first two positions are empty, the third is filled with a symbol, the fourth is empty, and the fifth is filled with a symbol.

The ***audio*** representation for the multiplication consists of a **knock, double knock, knock, blick, double knock, double knock, knock, cling, double knock, double knock, knock, double knock and knock.**

Figure 2

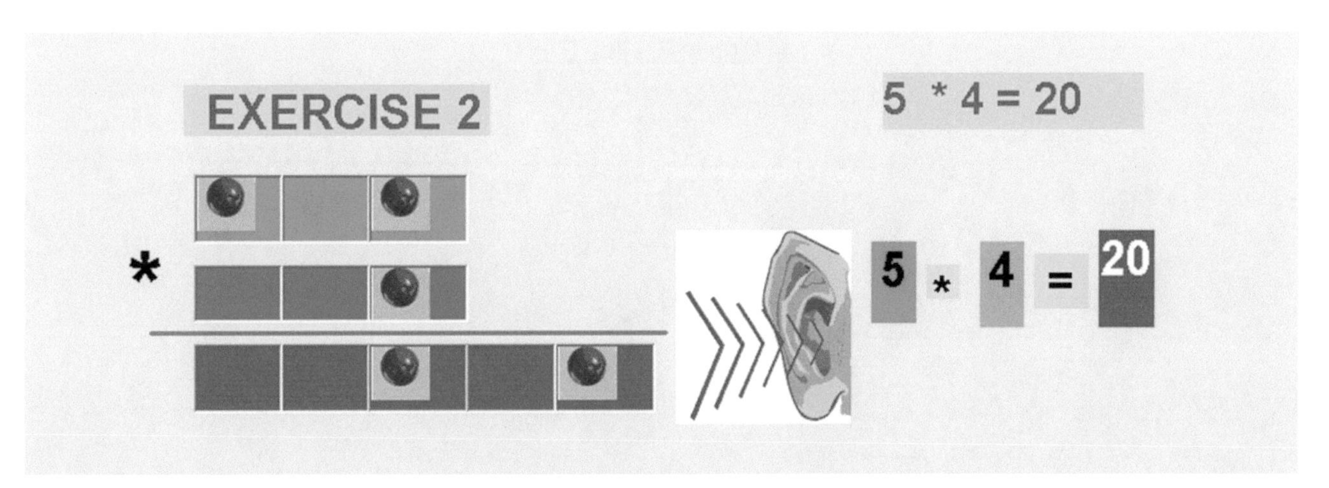

The ***kinesthetic*** representation of a filled position is represented by the clenching of the right hand. An empty position is represented by the clenching of the left hand. The multiplication sign, blick, is represented by one hand raised up. The equal sign, cling, is represented by crossing both hands. The kinesthetic representation for the multiplication is the clenching of the right hand, left hand, right hand, one hand raised up, the clenching of the left hand twice, right hand, crossing both hands, the clenching of the left twice, right, left and then the right hand once.

MATHEMATICS

MULTIPLICATION, EXERCISE 2, STEP 2.1

In Exercise 2 Step 2.1, in a ***visual*** display of multiplication we see three pictures of boxes. We move the first picture two positions to the right.

In the first picture, which has a value of five, positions one and three are filled with symbols. In the second picture, which has a value of three, the first two positions are empty, while the third is filled with a symbol. The multiplication sign is between the first and second pictures. Below the second picture is the equal line, which separates the problem from the answer. In the third picture, which is the answer for this step one of the multiplication, positions one and two are empty, position three is filled with a symbol, position four is empty, and position five is filled with a symbol.

The ***audio*** representation for the multiplication consists of a **knock, double knock, knock, blick, double knock, double knock, knock, cling, double knock, double knock, knock, double knock and knock**.

Exercise 2, Step 2.1

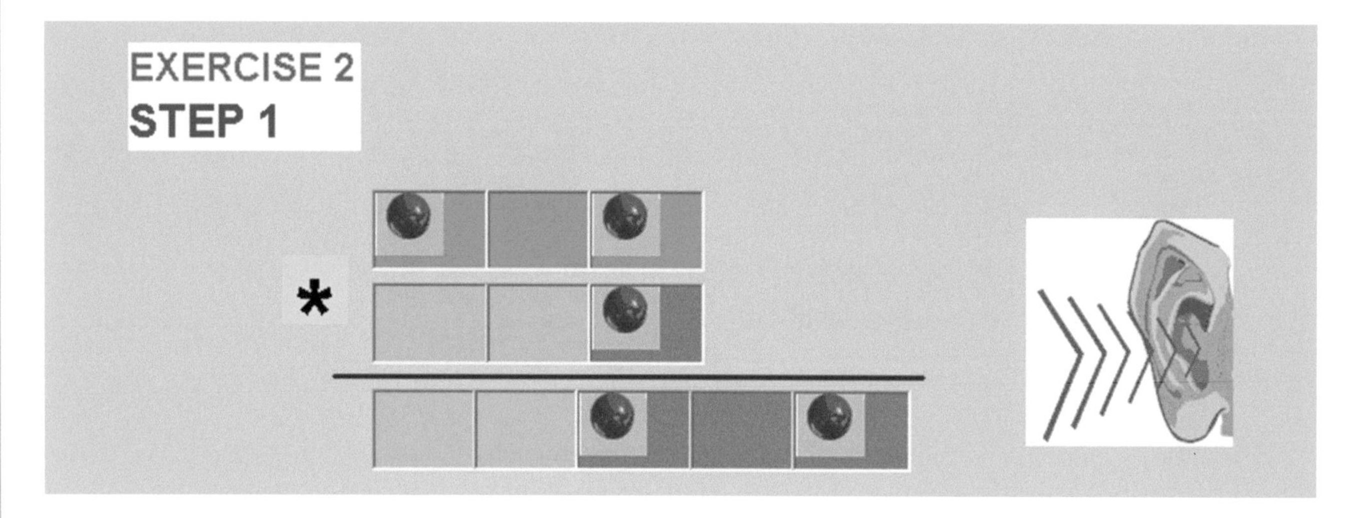

The ***kinesthetic*** representation of a filled position is represented by the clenching of the right hand. An empty position is represented by the clenching of the left hand. The multiplication sound, blick, is represented by one hand raised up. The equal sound, cling, is represented by crossing both hands. The kinesthetic representation for the multiplication is the clenching of the right hand, left, right, one hand raised up, the clenching of the left hand twice, right hand once, crossing of both hands, the clenching of the left hand twice, right once, left once and the right hand once.

MULTIPLICATION, EXERCISE 2, STEP 2.2

In Exercise 2 Step 2.2, in a ***visual*** display of the last step of the multiplication, we see one picture of boxes, which is the answer of the problem. The first two positions are empty, the third is filled with a symbol, the fourth is empty, and the fifth is filled with a symbol. The value of the pictures is twenty.

The ***audio*** representation for this picture consists of a **double knock, double knock, knock, double knock and knock**.

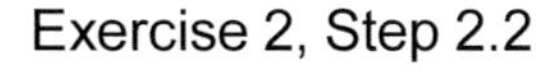

Exercise 2, Step 2.2

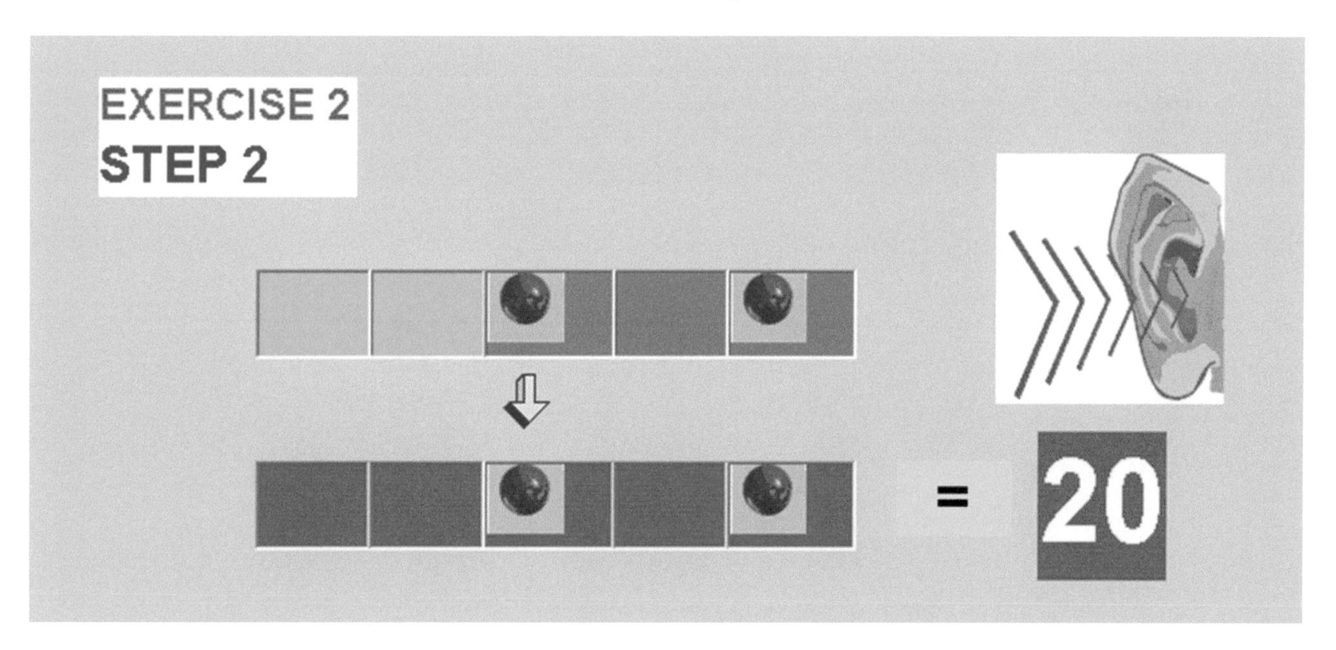

The ***kinesthetic*** representation of a filled position is represented by the clenching of the right hand. An empty position is represented by the clenching of the left hand. The kinesthetic representation for this picture is clenching of the left hand twice, right once, left once and the right hand once.

MATHEMATICS

MULTIPLICATION, EXERCISE 2

In Exercise 2, in a ***visual*** display of multiplication we see two pictures of boxes. At the bottom, there are three pictures with probable answers. In the first picture, which is the first multiplicand, the first and third positions are filled with symbols, while the second position is empty. In the second picture, the first two positions are empty and position three is filled with a symbol. The multiplication sign is between the first and second numbers. Below the second picture is the equal line, which separates the problem from the answer. The ***audio*** representation for the multiplication consists of a **knock, double knock, and knock, blick, double knock, double knock, knock, and cling.**

We need to find the right answer from the given choices below. The audio signals are: 1) **knock, double knock, knock, double knock, and double knock**, 2) **double knock, knock, knock, double knock, and knock**, 3) **double knock, double knock, knock, double knock, and knock**.

Exercise 2

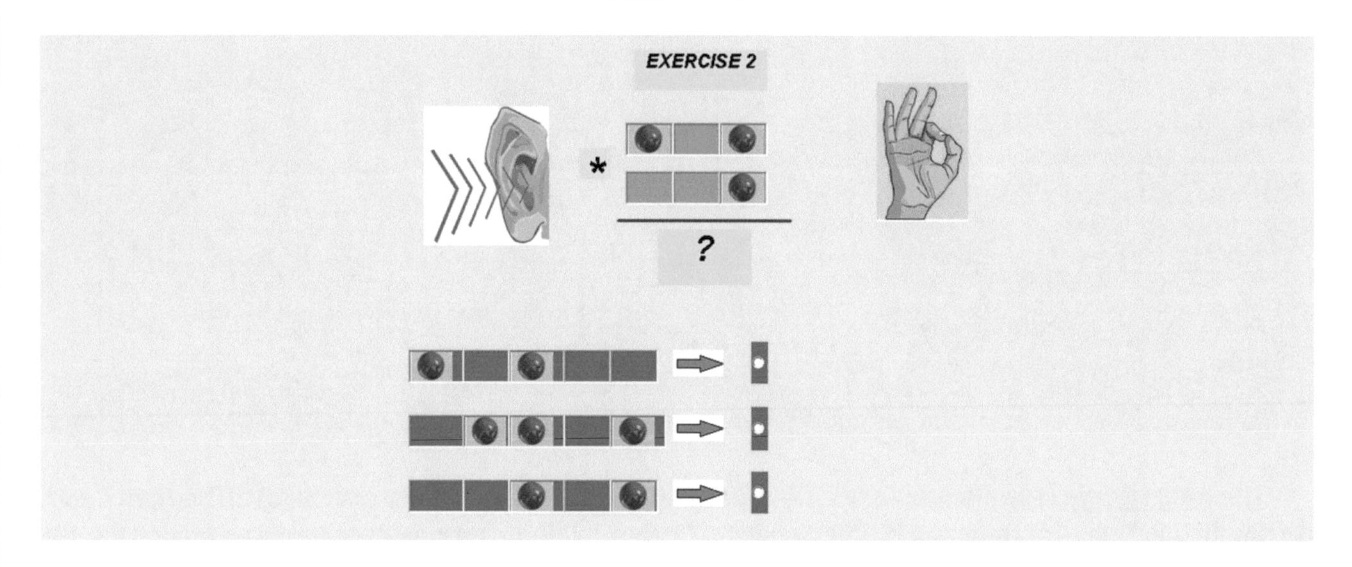

The ***kinesthetic*** representation of a filled position is represented by the clenching of the right hand. An empty position is represented by the clenching of the left hand. The multiplication sound, blick, is represented by one hand raised up. The equal sound, cling, is represented by crossing both hands. The kinesthetic representation for both pictures and the answer is *the clenching of the right hand, left, right, one hand raised up, the clenching of the left hand twice, right hand, crossing the both hands, the clenching of the left hand twice and then the right hand, left and right again.*

MULTIPLICATION, EXERCISE 3

In Exercise 3, in a ***visual*** display of multiplication we see three pictures of boxes. The first two pictures have a value of five and have their first and third positions filled with symbols, while their second positions are empty. The multiplication sign is between the first and second numbers. Below the second picture is the equal line, which separates the problem from the answer. In the third picture, which is the answer for the multiplication and has a value of twenty five, the second and third positions are empty, while positions one, four and five are filled with symbols. The ***audio*** representation for the multiplication consists of a **knock, double knock, knock, blick, knock, double knock, knock, cling, knock, double knock, double knock, knock and knock.**

Exercise 3

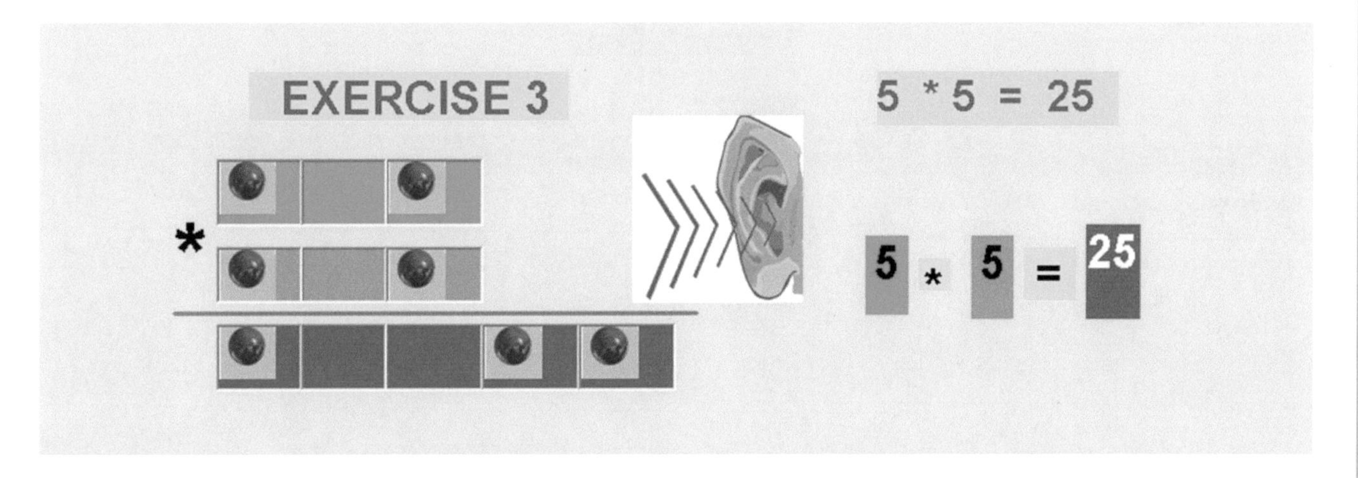

The ***kinesthetic*** representation of a filled position is represented by the clenching of the right hand. An empty position is represented by the clenching of the left hand. The multiplication sign, blick, is represented by one hand raised up. The equal sign, cling, is represented by crossing both hands. The kinesthetic representation for the multiplication is *the clenching of the right hand, left, right, one hand raised up, left twice, right, crossing both hands, left twice, right, left and then the right hand once.*

MATHEMATICS

MULTIPLICATION, EXERCISE 3, STEP 3.1

In Exercise 3 Steps 3.1, in a ***visual*** display of multiplication we see three pictures of boxes. In the first picture, which has a value of five, the first and third positions are filled with symbols, while the second position is empty. In the second picture, position one is filled with a symbol and position two is empty. The multiplication sign is between the first and second numbers. Below the second picture is the equal line, which separates the problem from the answer. In the third picture, which is the answer for step 3.1 of multiplication, position two is empty, while positions one and three are filled with symbols. The answer has a value of five.

The ***audio*** representation for the picture below consists of a **knock, double knock, knock, blick, knock, double knock, cling, knock, double knock, knock and knock**.

Exercise 3, Step 1

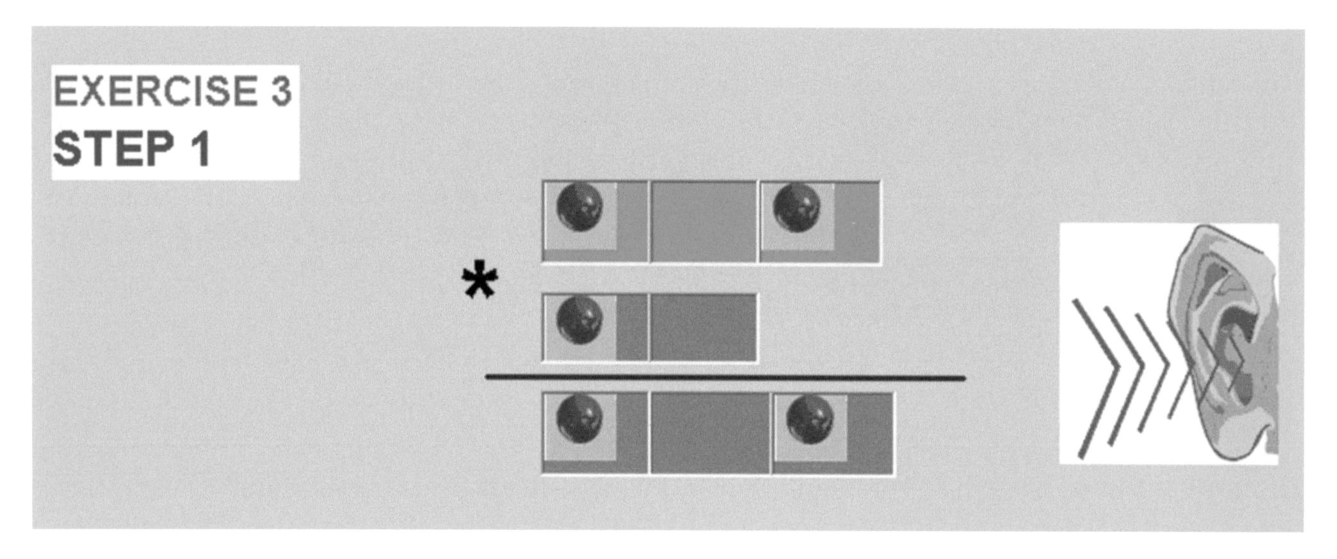

The ***kinesthetic*** representation of a filled position is represented by the clenching of the right hand. An empty position is represented by the clenching of the left hand. The multiplication sound, blick, is represented by one hand raised up. The equal sound, cling, is represented by crossing both hands. The kinesthetic representation for the multiplication is *the clenching of the right hand, left, right, one hand raised up, the clenching of the right, left, crossing of both hands, the clenching of the right once, left once and the right hand once.*

MULTIPLICATION, EXERCISE 3, STEP 2

In Exercise 3 Step 1.2, in a ***visual*** display for this step of the multiplication we see three pictures of boxes. We move the first picture two positions to the right. In the first picture, which has a value of five, the first and third positions are filled with symbols. In the second picture, which has a value of four, the third position is filled with a symbol. The multiplication sign is between the first and second pictures. Below the second picture is the equal line, which separates the problem from the answer. In the third picture, which is the answer for the multiplication and has a value of eight, positions one, two and four are empty and positions three and five are filled with symbols.

The ***audio*** representation for the multiplication consists of a **knock, double knock, knock, blick, double knock, double knock, knock, cling, double knock, double knock, knock, double knock and knock**.

Exercise 3, Step 2

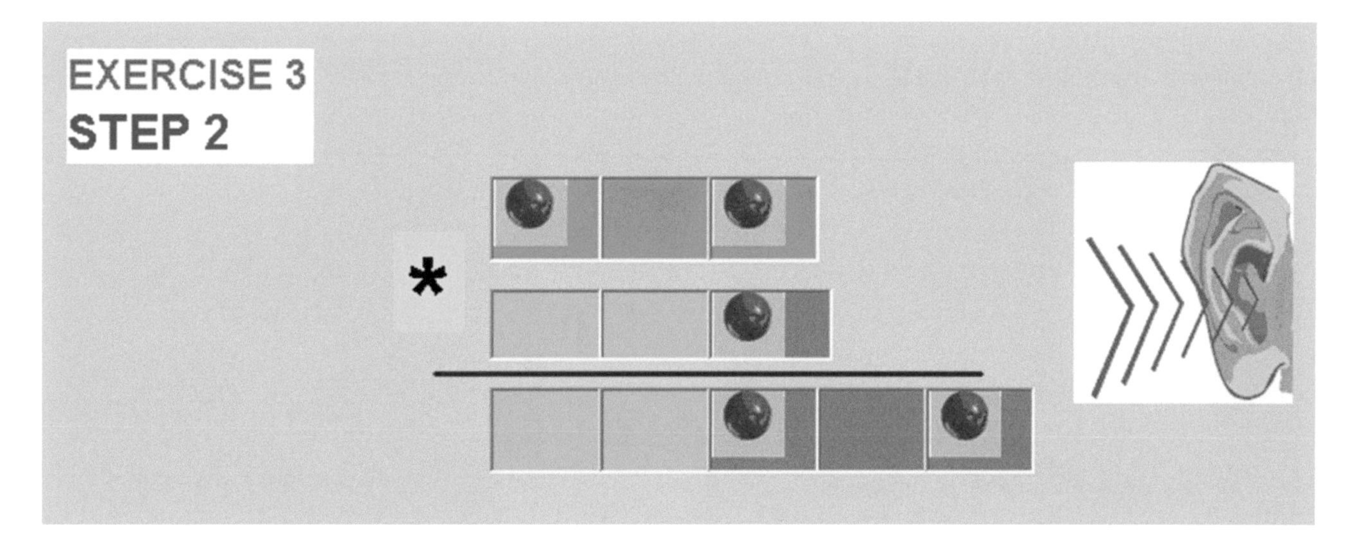

The ***kinesthetic*** representation of a filled position is representation by the clenching of the right hand. An empty position is represented by the clenching of the left hand. The multiplication sign, blick, is represented by one hand raised up. The equal sign, cling, is represented by crossing both hands. The kinesthetic representation for the pictures is *the clenching of the right hand, left, right hand, one hand raised up, the clenching of the left hand twice, right hand once, crossing both hands, the clenching of the left hand twice, right hand, left, and right.*

MATHEMATICS

MULTIPLICATION BY ADDITION, EXERCISE 3, STEP 3

In Exercise 3 Step 3.3, in a ***visual*** display of the third step of the multiplication, we are adding together the results of the first and second steps. In the first picture, which has a value of five, the first and third positions are filled with symbols. In the second picture, which has a value of twenty, the third and fifth positions filled with symbols. The addition sign is between the first and second pictures. Below the second picture is the equal line, which separates the problem from the answer. In the third picture, which is the answer for the third step of the multiplication, positions one, four and five are filled with symbols. The answer has a value of twenty five.

The ***audio*** representation for the pictures consists of a **knock, double knock, knock, tram, double knock, double knock, knock, double knock, knock, cling, knock, double knock, double knock, knock and knock**.

Exercise 3, Step 3

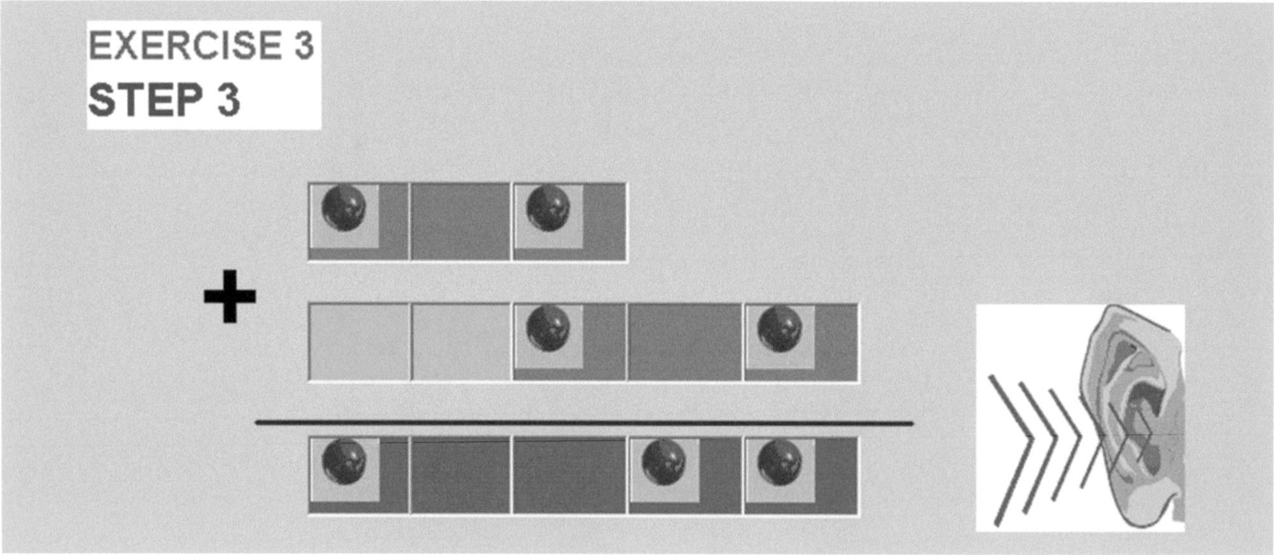

EXERCISE 3
STEP 3

The ***kinesthetic*** representation of a filled position is represented by the clenching of the right hand. An empty position is represented by the clenching of the left hand. The addition sound, tram, is represented by extending the left hand in front. The equal sound, cling, is represented by crossing both hands. The kinesthetic representation for the multiplication is the clenching of the right hand, left hand, right hand once, one hand raised up, the clenching of the left hand twice, right hand once, left and right, crossing both hands, the clenching of the right hand, left hand twice and the right hand twice.

MULTIPLICATION, EXERCISE 3, STEP 4

In Exercise 3, Step 3.4, in a ***visual*** display of the last step of the multiplication, we see one picture of boxes, which is the answer of the problem. The first, fourth, and fifth positions are filled with symbols, while positions two and three are empty.

The ***audio*** representation for this picture consists of a **knock, double knock, double knock, knock and knock**.

Exercise 3, Step 4

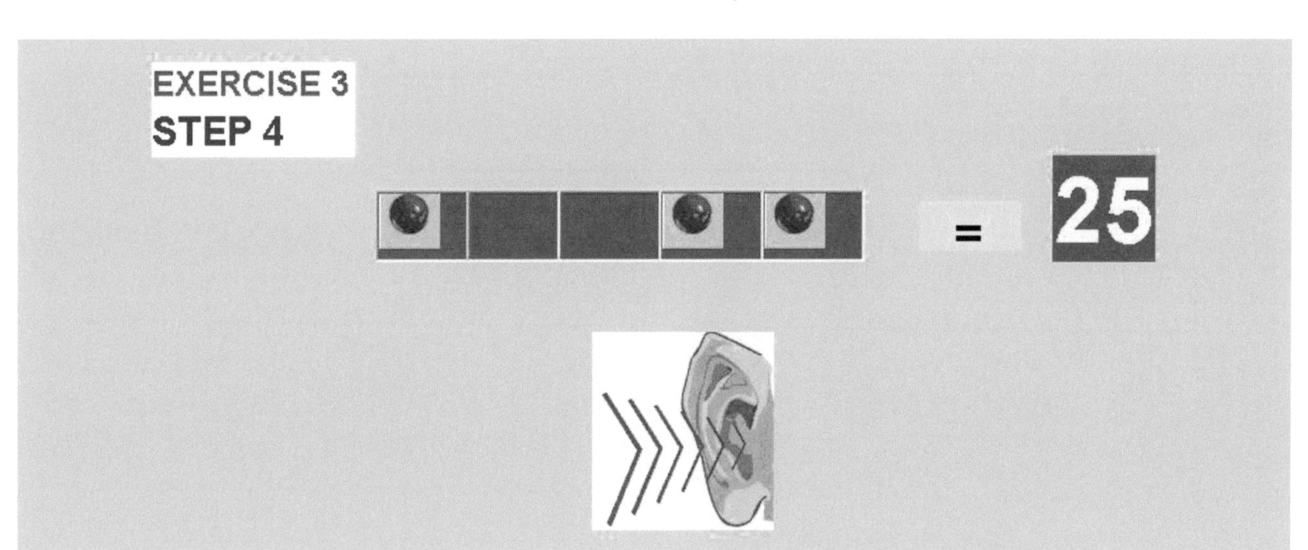

The ***kinesthetic*** representation of a filled position is represented by the clenching of the right hand. An empty position is represented by the clenching of the left hand. The kinesthetic representation for this picture is *the clenching of the right hand, left hand twice and the right hand twice.*

MATHEMATICS MULTIPLICATION, EXERCISE 3

In Exercise 3, in a ***visual*** display of multiplication we see two pictures of boxes. At the bottom are three pictures with probable answers. In the first multiplicand, the first and third positions are filled with symbols, while the second position is empty. The same is true for the second multiplicand. The multiplication sign is between the first and second numbers. Below the second picture is the equal line, which separates the problem from the answer. The ***audio*** representation for the multiplication consists of a **knock, double knock, and knock, blick, knock, double knock, knock, and cling.**

We need to choose the right answer from the options below. The audio signals are: 1) **double knock, knock, double knock, double knock, and knock**, 2) **double knock, knock, double knock, knock, and double knock**, 3) **knock, double knock, double knock, knock, and knock**.

Exercise 3

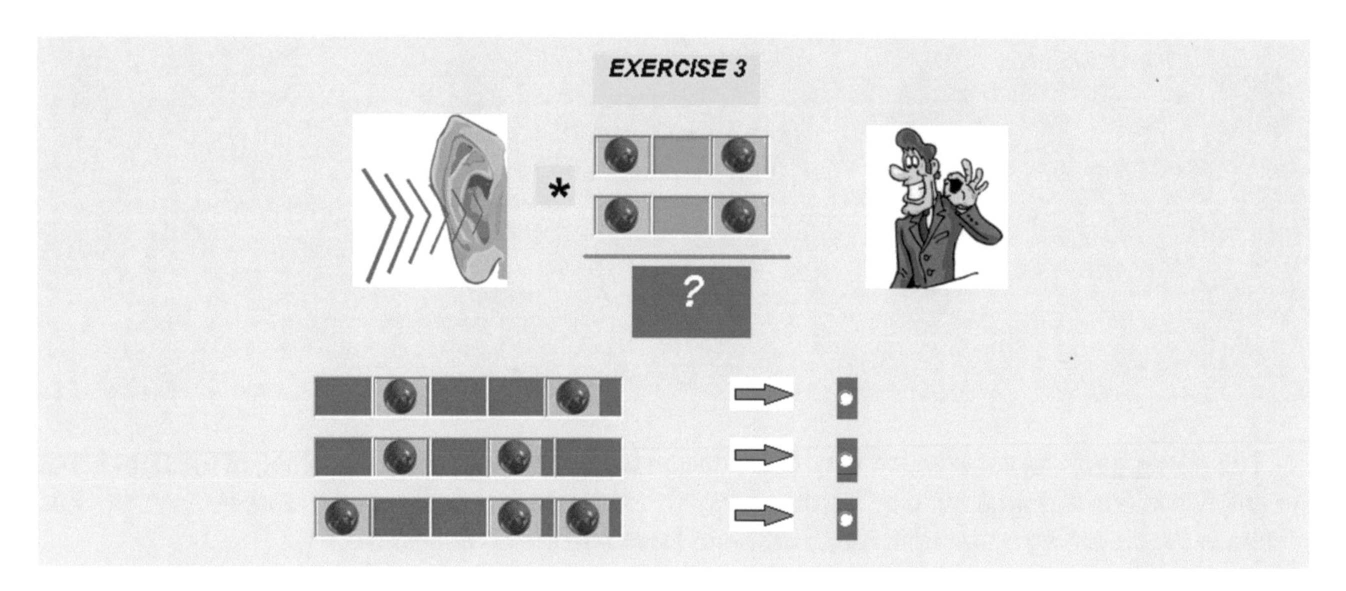

The ***kinesthetic*** representation of a filled position is represented by the clenching of the right hand. An empty position is represented by the clenching of the left hand. The multiplication sound, blick, is represented by one hand raised up. The equal sound, cling, is represented by crossing both hands. The kinesthetic representation for both pictures and the answer is *the clenching of the right hand, left hand, right hand, one hand raised up, the clenching of the right, left hand and right hand, crossing the both hands, the clenching of the right, left hand twice and then the right hand twice.*

MULTIPLICATION, EXERCISE 4

In Exercise 4, in a ***visual*** display of multiplication we see three pictures of boxes. In the first picture, positions one is empty and the next two positions are filled with symbols. In picture two, the first two positions are empty and the third position is filled with a symbol. The multiplication sign is between the first and second pictures. Below the second picture is the equal line, which separates the problem from the answer. In the third picture, which is the answer for the multiplication and has a value of twenty four, the first, second and third positions are empty but positions four and five are filled with symbols.

The ***audio*** representation for the pictures consists of a **double knock, knock, knock, blick, double knock, double knock, knock, cling, double knock, double knock, double knock, knock and knock.**

Figure 4

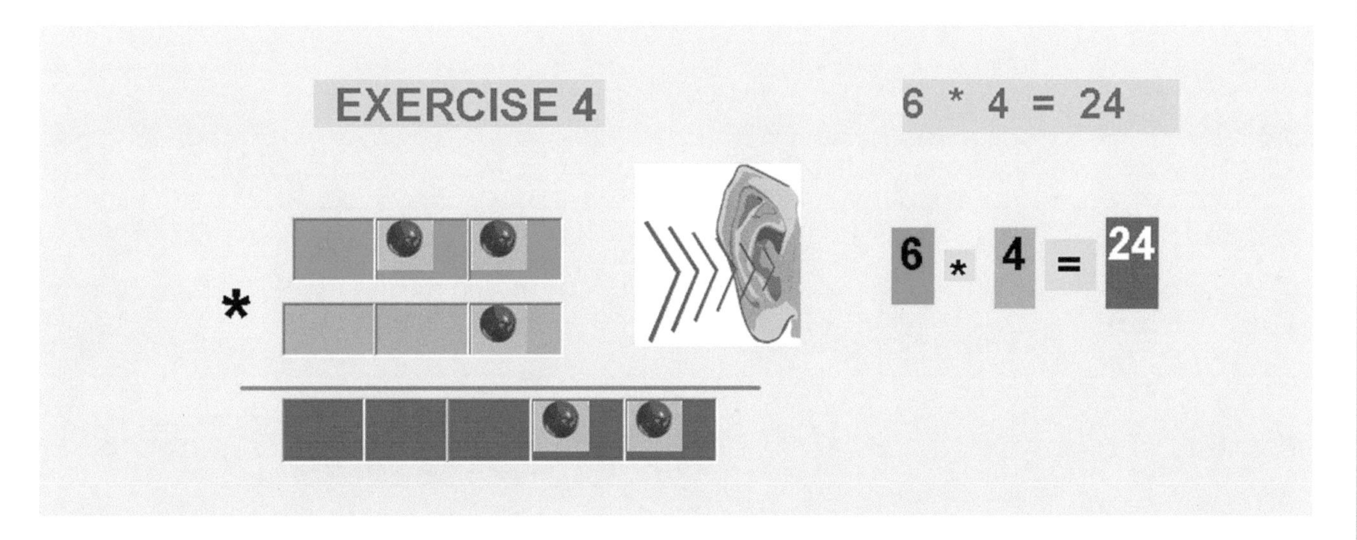

The ***kinesthetic*** representation of a filled position is represented by the clenching of the right hand. An empty position is represented by the clenching of the left hand. The multiplication sign, blick, is represented by one hand raised up. The equal sign, cling, is represented by crossing both hands. The kinesthetic representation for the pictures above is *the clenching of the left hand, right hand twice, one hand raised up, the clenching of the left hand twice, right once, crossing both hands, the clenching of the left hand three times and the right hand twice.*

MATHEMATICS

MULTIPLICATION, EXERCISE 4, STEP 4.1

In Exercise 4 Steps 4.1, in a ***visual*** display of multiplication we see three pictures of boxes. In the first picture, which has a value of six, position one is empty and positions two and three are filled with symbols. In the second picture, which has a value of four, position four is filled with a symbol. Positions one and two are empty in the second picture. The multiplication sign is between the first and second pictures. Below the second picture is the equal line, which separates the problem from the answer. In the third picture, which is the answer for Step 4.1 of the multiplication, positions one, two and three are empty while positions four and five are filled with symbols. The answer has a value of twenty four.

The ***audio*** representation for the pictures consists of a **double knock, knock, knock, blick, double knock, double knock, knock, cling, double knock, double knock, double knock, knock and knock**.

Exercise 4, Step 1

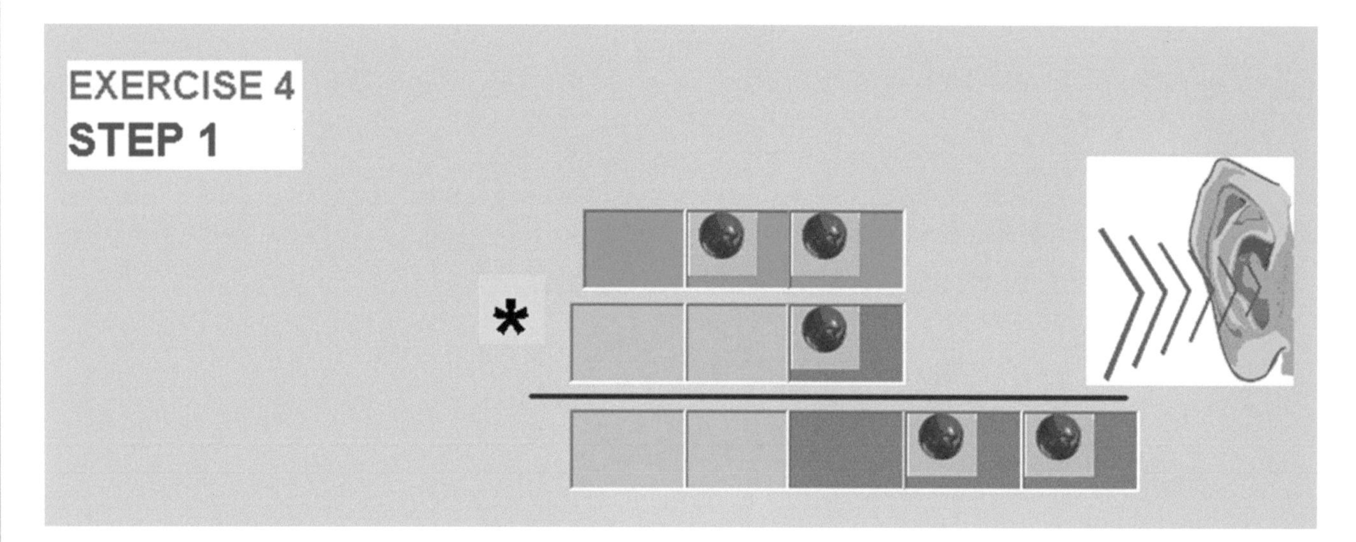

The ***kinesthetic*** representation of a filled position is represented by the clenching of the right hand. An empty position is represented by the clenching of the left hand. The multiplication sound, blick, is represented by one hand raised up. The equal sound, cling, is represented by crossing both hands. The kinesthetic representation for the pictures above is *the clenching of the left hand once, right hand twice, one hand raised up, the clenching of the left hand twice, right once, crossing of both hands, the clenching of the left hand three times, and the right hand twice.*

MULTIPLICATION, EXERCISE 2, STEP 4.2

In Exercise 4 Step 4.2, in a ***visual*** display of the last step of the multiplication, we see one picture of boxes, which is the answer of the problem. The first three positions are empty, while the fourth and fifth positions are filled with symbols. The value of this number is twenty four.

The ***audio*** representation for this picture consists of a **double knock, double knock, double knock, knock and knock**.

Figure 4.2

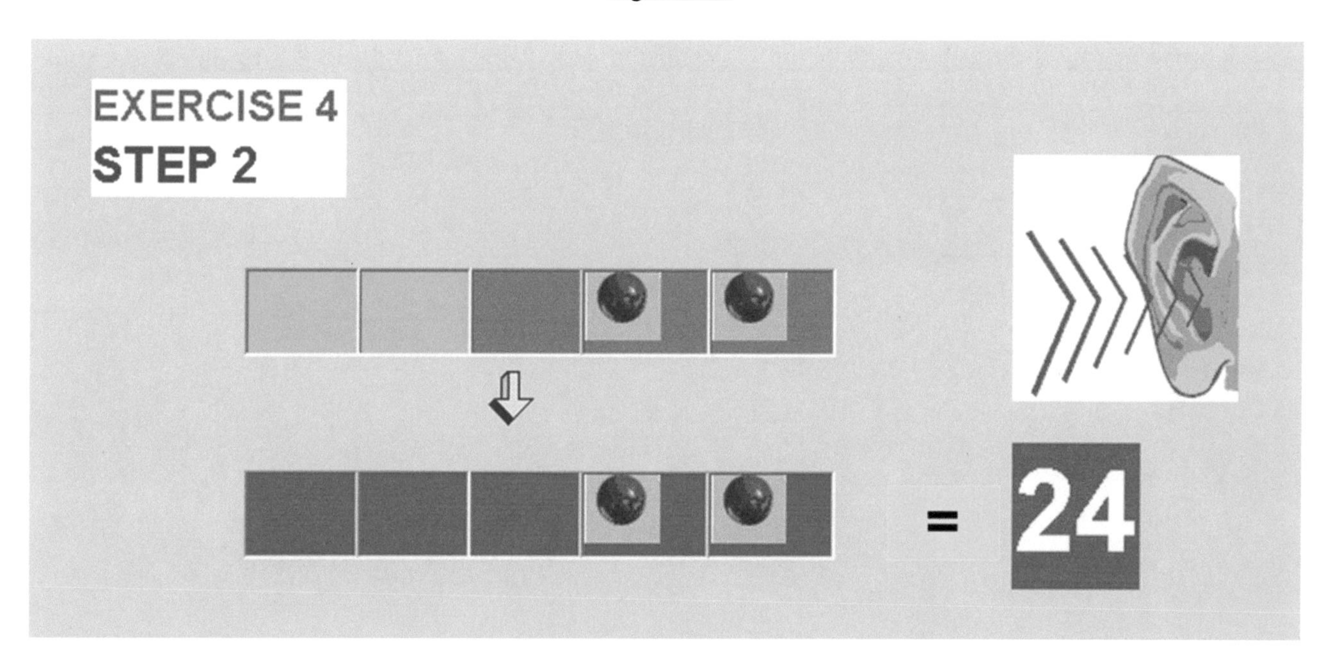

The ***kinesthetic*** representation of a filled position is represented by the clenching of the right hand. An empty position is represented by the clenching of the left hand. The kinesthetic representation for this picture is *the clenching of the left hand three times and the right hand twice.*

MATHEMATICS

MULTIPLICATION, EXERCISE 4

In Exercise 4, in a ***visual*** display of multiplication we see two pictures of boxes. At the bottom are three pictures with probable answers. In the first multiplicand, the first position is empty and the next two positions are filled with symbols. In the second multiplicand, the first two positions are empty, while the third is filled with a symbol. The multiplication sign is between the first and second pictures. Below the second picture is the equal line, which separates the problem from the answer. The ***audio*** representation for the multiplication consists of a **double knock, knock, and knock, blick, double knock, double knock, knock, and cling.**

We need to choose the right answer from the options below. The audio signals are: 1) **double knock, double knock, knock, double knock and knock**, 2) **double knock, double knock, double knock, knock and knock**, 3) **double knock, knock, double knock, knock and double knock**.

Exercise 4

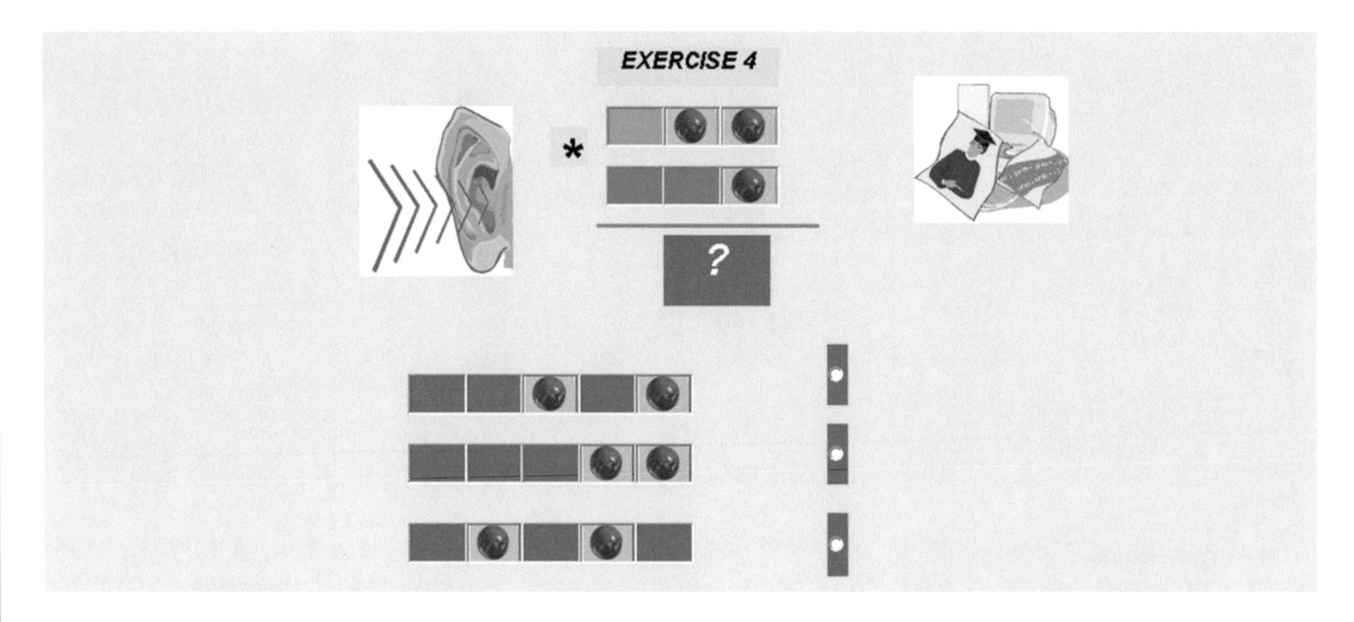

The ***kinesthetic*** representation of a filled position is represented by the clenching of the right hand. An empty position is represented by the clenching of the left hand. The multiplication sound, blick, is represented by one hand raised up. The equal sound, cling, is represented by crossing both hands. The kinesthetic representation for both pictures and the answer is *the clenching of the left hand, right hand twice, one hand raised up, the clenching of the left hand twice and right hand once, crossing both hands, the clenching of the left hand three times and then the right hand twice.*

MULTIPLICATION, EXERCISE 5

In Exercise 5, in a ***visual*** display of multiplication we see three pictures of boxes. The first picture has its all three positions filled with symbols, which has a value of seven. The second picture has its two positions filled with symbols, which has a value of three. The multiplication sign is between the first and second numbers. Below the second picture is the equal line, which separates the problem from the answer. In the third picture, which is the answer for the multiplication and has a value of twenty one, the second and fourth positions are empty while positions one, four and five are filled with symbols.

The ***audio*** representation for the pictures consists of a **knock, knock, knock, blick, knock, knock, cling, knock, double knock, knock, double knock and knock.**

Figure 5

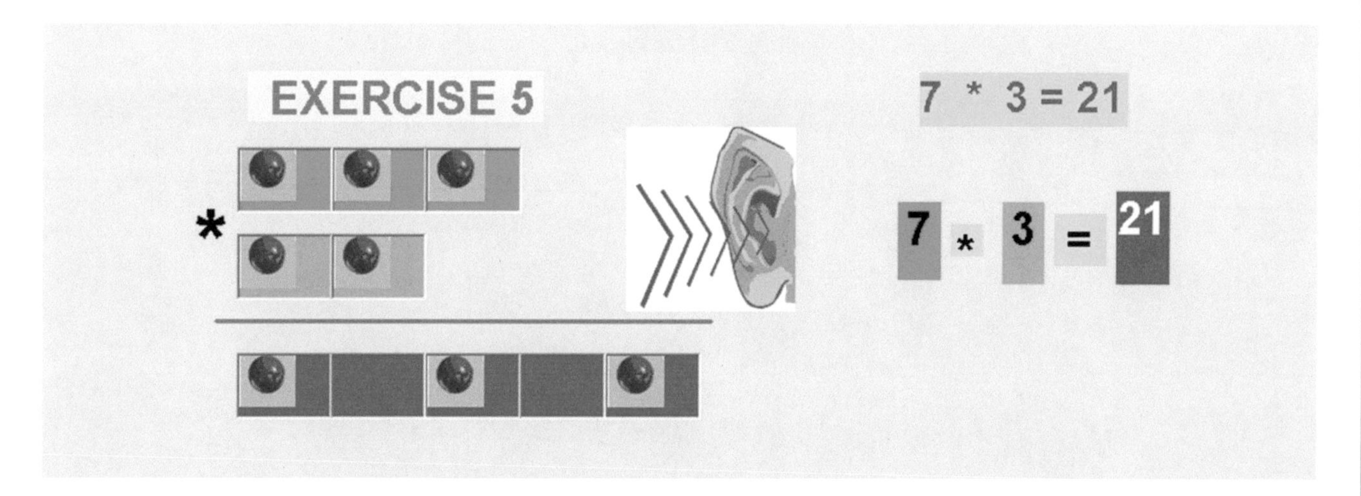

The ***kinesthetic*** representation of a filled position is represented by the clenching of the right hand. An empty position is represented by the clenching of the left hand. The multiplication sign, blick, is represented by one hand raised up. The equal sign, cling, is represented by crossing both hands. The kinesthetic representation for the pictures above is *the clenching of the right hand three times, one hand raised up, the clenching of the right hand twice, crossing both hands, the clenching of the right, left, right, left and right hand once.*

MULTIPLICATION, EXERCISE 5, STEP 5.1

In Exercise 5 Step 5.1, in a ***visual*** display of multiplication we see three pictures of boxes. We move the first picture two positions to the right.

In the first picture, which has a value of seven, all three positions are filled with red balls. In the second picture, which has a value of one, position one is filled with a symbol. The multiplication sign is between the first and second pictures. Below the second picture is the equal line, which separates the problem from the answer. In the third picture, which is the answer for this step one of the multiplication, all positions are filled with symbols and have a value of seven.

The ***audio*** representation for the pictures consists of a **knock, knock, knock, blick, knock, double knock, cling, knock, knock, and knock**.

Exercise 5, Step 1

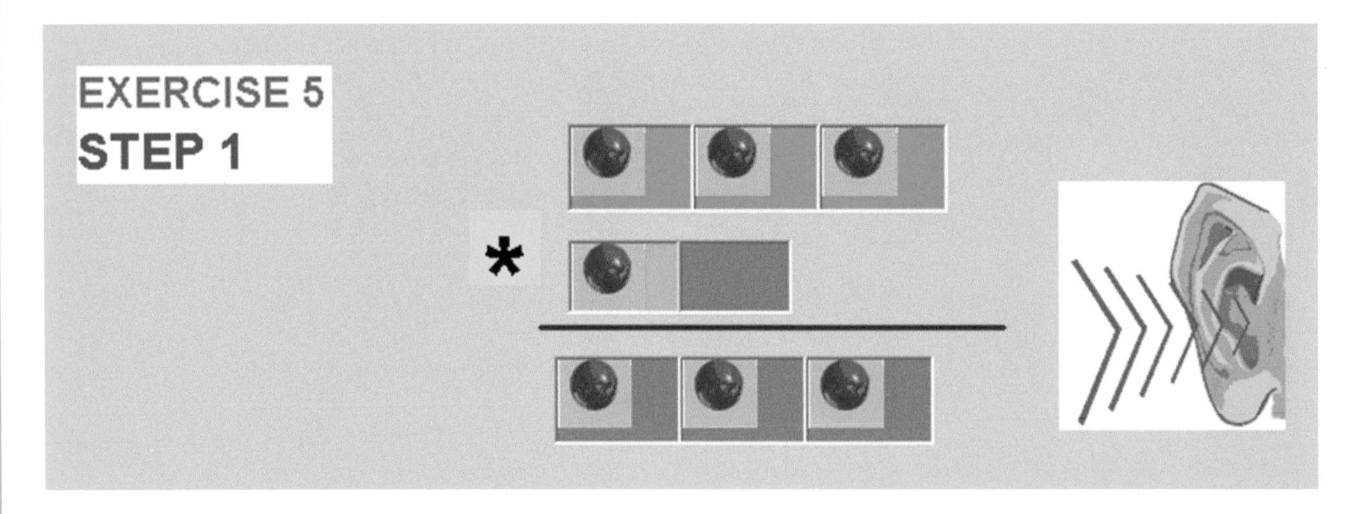

The ***kinesthetic*** representation of a filled position is represented by the clenching of the right hand. An empty position is represented by the clenching of the left hand. The multiplication sound, blick, is represented by one hand raised up. The equal sound, cling, is represented by crossing both hands. The kinesthetic representation for both pictures is the clenching of the right hand three times, one hand raised up, the clenching of the right hand once, left once, the crossing of both hands and the clenching of the right hand three times.

MULTIPLICATION, EXERCISE 5, STEP 5.2

In Exercise 5 Step 5.2, in a ***visual*** display for this step of multiplication we see three pictures of boxes. We move the first picture two positions to the right. In the first picture, which has a value of seven, all positions are filled with red balls. In the second picture, which has a value of two, the second position is filled with a symbol. The multiplication sign is between the first and second pictures. Below the second picture is the equal line, which separates the problem from the answer. In the third picture, which is the answer for the multiplication and has a value of fourteen, positions two, three and four are filled with symbols.

The ***audio*** representation for the pictures consists of a **knock, knock, knock, blick, double knock, knock, cling, double knock, knock, knock, knock and knock**.

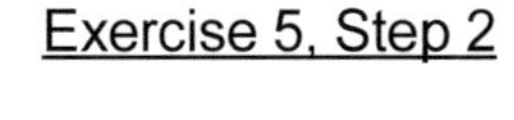

Exercise 5, Step 2

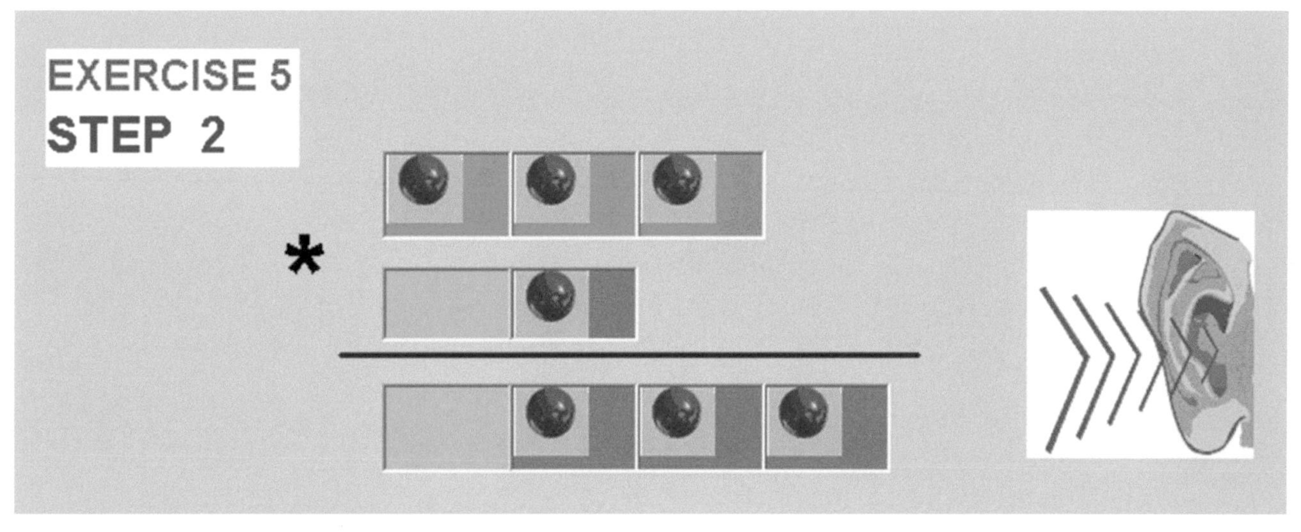

The ***kinesthetic*** representation of a filled position is represented by the clenching of the right hand. An empty position is represented by the clenching of the left hand. The multiplication sign, blick, is represented by one hand raised up. The equal sign, cling, is represented by crossing both hands. The kinesthetic representation for the pictures is *the clenching of the right hand three times, one hand raised up, the clenching of the left hand, then right, the crossing of both hands, the clenching of the left hand once and the right hand three times.*

MATHEMATICS MULTIPLICATION BY ADDITION, EXERCISE 5, STEP 5.3

In Exercise 5 Step 5.3, in a ***visual*** display of the third step of the multiplication we are adding together the results of the first and second steps. In the first picture, which has a value of seven, all positions are filled with red balls. In the second picture, which has a value of fourteen, the second, third and fourth positions are filled with red balls. The addition sign is between the first and second pictures. Below the second picture is the equal line, which separates the problem from the answer. In the third picture, which is the answer for the third step of the multiplication, positions one, three and five are filled with symbols. The answer has a value of twenty one.

The ***audio*** representation for the pictures consists of a **knock, knock, knock, tram, double knock, knock, knock, knock, cling, knock, double knock, knock, double knock and knock**.

Step 5.3

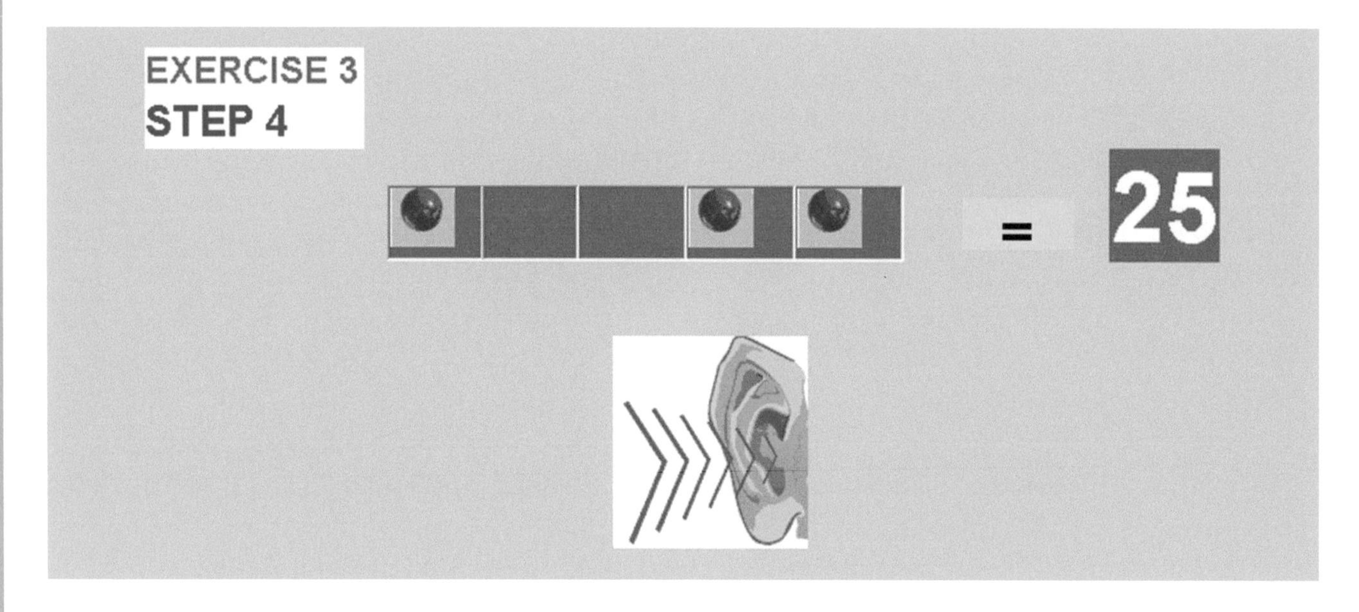

The ***kinesthetic*** representation of a filled position is represented by the clenching of the right hand. An empty position is represented by the clenching of the left hand. The addition sound, tram, is represented by extending the left hand in front. The equal sound, cling, is represented by crossing both hands. The kinesthetic representation for the pictures is *the clenching of the right hand three times, left hand extended in front, the clenching of the left hand once, the clenching of the right hand three times, crossing of both hands, the clenching of the right hand, left, right, left and the right hand once.*

MULTIPLICATION, EXERCISE 5, STEP 5.4

In Exercise 5 Step 5.4, in a ***visual*** display of the last step of the multiplication, we see one picture of boxes, which is the answer of the problem. The first, third and fifth positions are empty, while positions two and four are filled with red balls. The answer has a value of twenty one.

The ***audio*** representation for this picture consists of a **knock, double knock, knock, double knock and knock**.

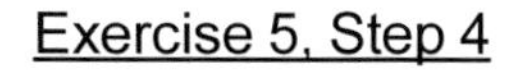

Exercise 5, Step 4

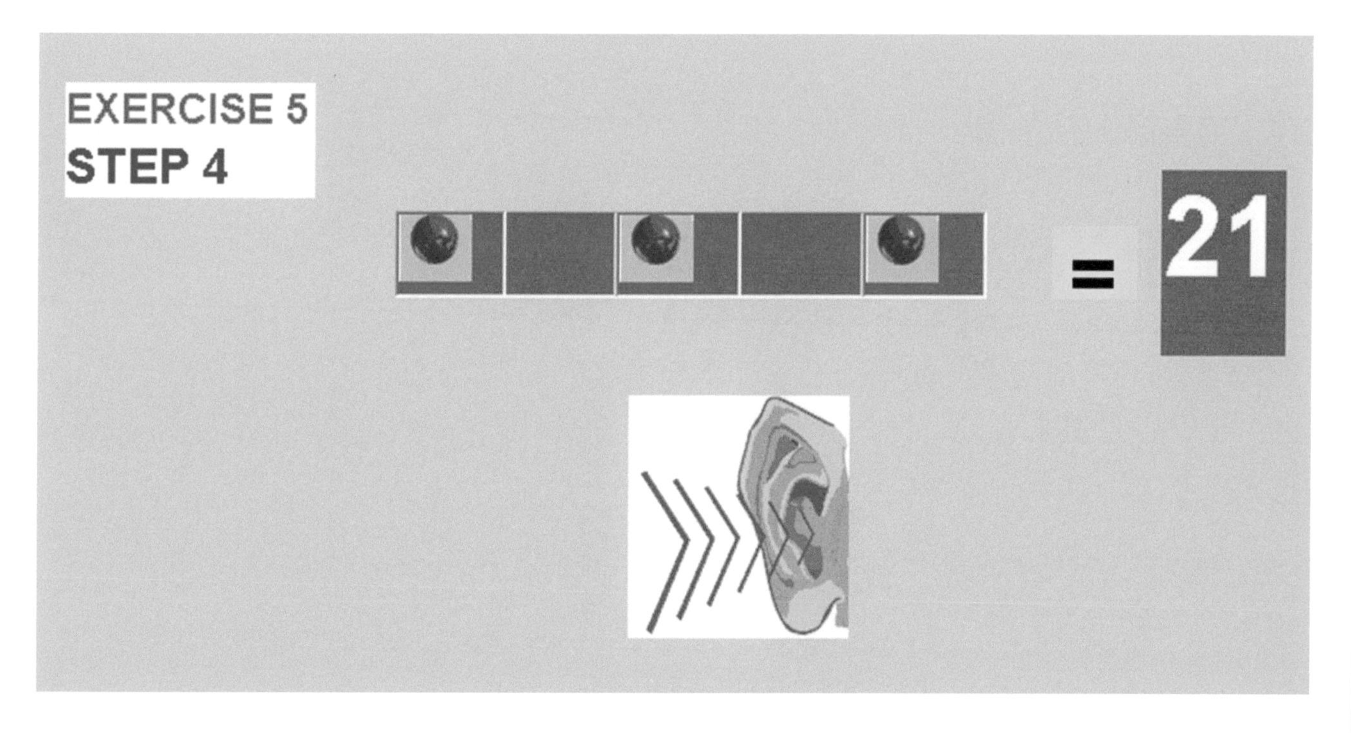

The ***kinesthetic*** representation of a filled position is represented by the clenching of the right hand. An empty position is represented by the clenching of the left hand. The kinesthetic representation for this picture is *the clenching of the right hand, left, right, left and the right hand once.*

MATHEMATICS

MULTIPLICATION, EXERCISE 5

In Exercise 5, in a ***visual*** display of multiplication we see two pictures of boxes. At the bottom are three pictures with probable answers. In the first multiplicand, all three positions are filled with symbols. In the second multiplicand, both positions are filled with symbols. The multiplication sign is between the first and second numbers. Below the second picture is the equal line, which separates the problem from the answer. The ***audio*** representation for the multiplication consists of a **knock, knock knock, blick, knock, knock, and cling.**

We need to choose the right answer from the options below. The audio signals are: 1)**knock, double knock, knock, double knock and knock**, 2) **double knock, knock, knock, double knock and knock**, 3) **double knock, knock, double knock, double knock and knock**.

Exercise 5

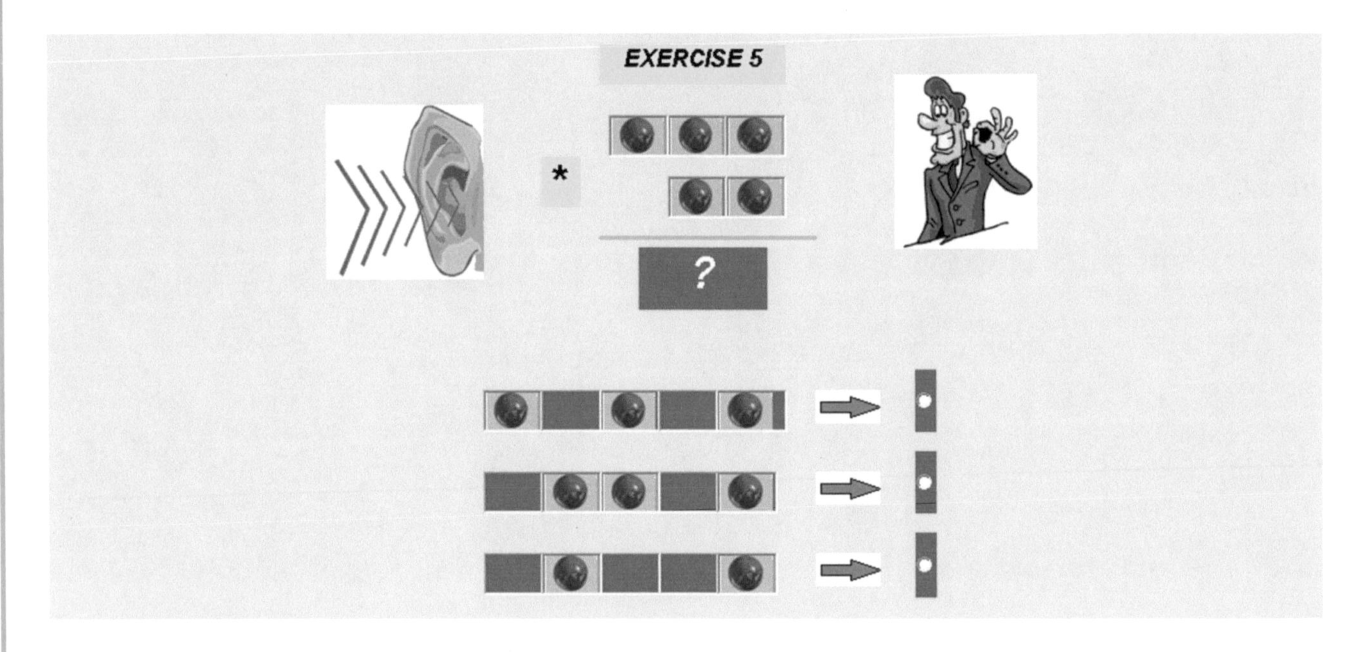

The ***kinesthetic*** representation of a filled position is represented by the clenching of the right hand. An empty position is represented by the clenching of the left hand. The multiplication sound, blick, is represented by one hand raised up. The equal sound, cling, is represented by crossing of both hands. The kinesthetic representation for both pictures and the answer is *the clenching of the right hand three times, one hand raised up, the clenching of the right hand twice, crossing of both hands, right, left, right, left and the right hand once.*

MULTIPLICATION, EXERCISE 6

In Exercise 6, in a ***visual*** display of multiplication, we see three pictures of boxes. The first picture has all of its three positions filled with symbols and has a value of seven. The second picture has its first and third positions filled with symbols, while its second position is empty. The value of this number is five. The multiplication sign is between the first and second pictures. Below the second picture is the equal line, which separates the problem from the answer. In the third picture, which is the answer for the multiplication and has a value of thirty five, the third fourth and fifth positions are empty while positions one, two and six are filled with symbols.

The ***audio*** representation for the pictures consists of a **knock, knock, knock, blick, knock, double knock, knock, cling, knock, knock, double knock, double knock, double knock and knock.**

Figure 6

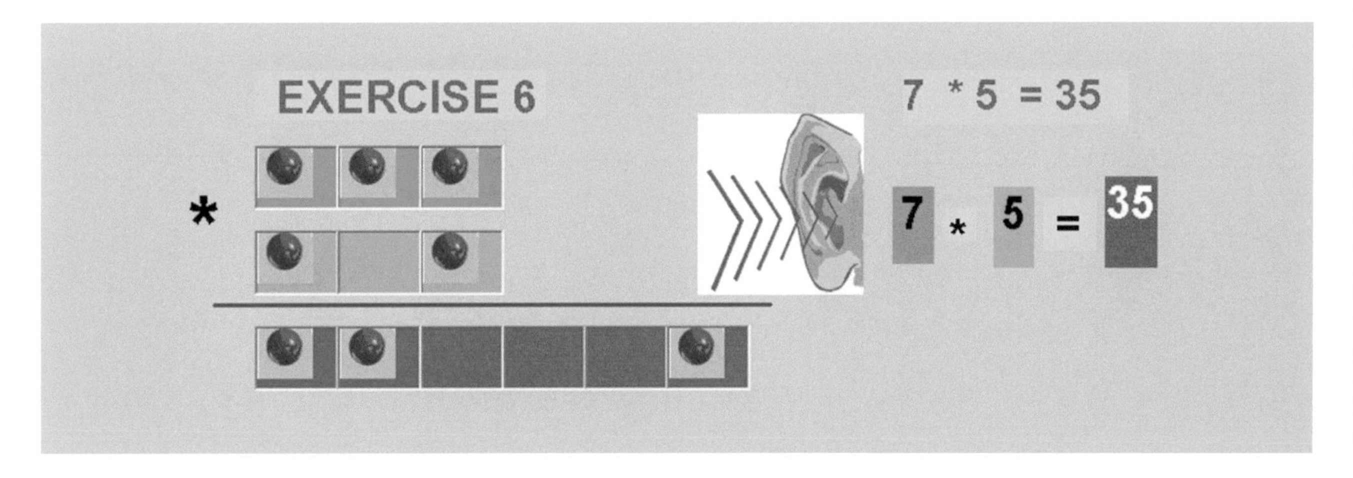

The ***kinesthetic*** representation of a filled position is represented by clenching of the right hand. An empty position is represented by the clenching of the left hand. The multiplication sign, blick, is represented by one hand raised up. The equal sign, cling, is represented by crossing both hands. The kinesthetic representation for both pictures is *the clenching of the right hand three times, one hand raised up, the clenching of the right hand, left, right, crossing both hands, the clenching of the right hand twice, left three times and right hand once.*

MULTIPLICATION, EXERCISE 6, STEP 6.1

In Exercise 6 Step 6.1, in a ***visual*** display of multiplication we see three pictures of boxes. We move the first picture two positions to the right.

In the first picture, which has a value of seven, all three positions are filled with red balls. In the second picture, which has a value of one, position one is filled with a symbol. The multiplication sign is between the first and second pictures. Below the second picture is the equal line, which separates the problem from the answer. In the third picture, which is the answer for step one of the multiplication, all positions are filled with symbols, which has a value of seven.

The ***audio*** representation for the pictures consists of a **knock, knock, knock, blick, knock, double knock, cling, knock, knock, and knock**.

Exercise 6, Step 1

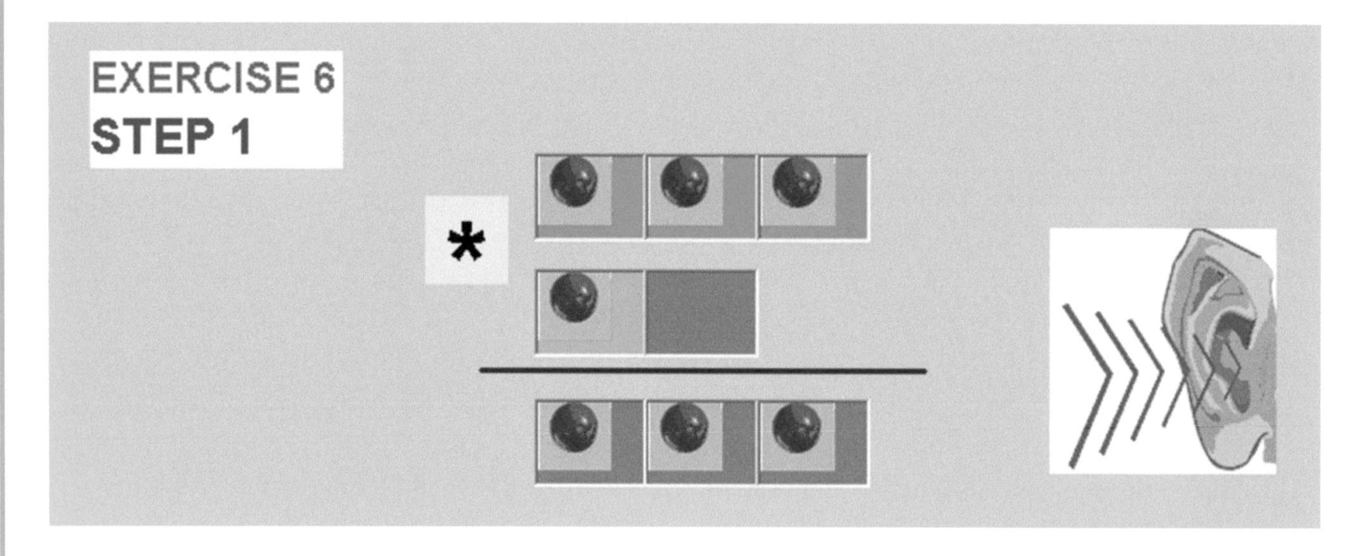

The ***kinesthetic*** representation of a filled position is represented by the clenching of the right hand. An empty position is represented by the clenching of the left hand. The multiplication sound, blick, is represented by one hand raised up. The equal sound; cling, is represented by crossing the hands. The kinesthetic representation for both pictures is *the clenching of the right three times, one hand raised up, the clenching of the right once, left once, crossing of both hands and the clenching of the right hand three times.*

MULTIPLICATION, EXERCISE 6, STEP 6.2

In Exercise 6 Step 6.2, in a ***visual*** display for this step of multiplication we see three pictures of boxes. We move the first picture two positions to the right. In the first picture, which has a value of seven, all positions are filled with red balls. In the second picture, which has a value of four, the third position is filled with a symbol. The multiplication sign is between the first and second pictures. Below the second picture is the equal line, which separates the problem from the answer. In the third picture, which is the answer for the multiplication and has a value of twenty four, positions three, four and five are filled with symbols.

The ***audio*** representation for the pictures consists of a **knock, knock, knock, blick, double knock, double knock, knock, cling, double knock, double knock, knock, knock and knock**.

Exercise 6, Step 2

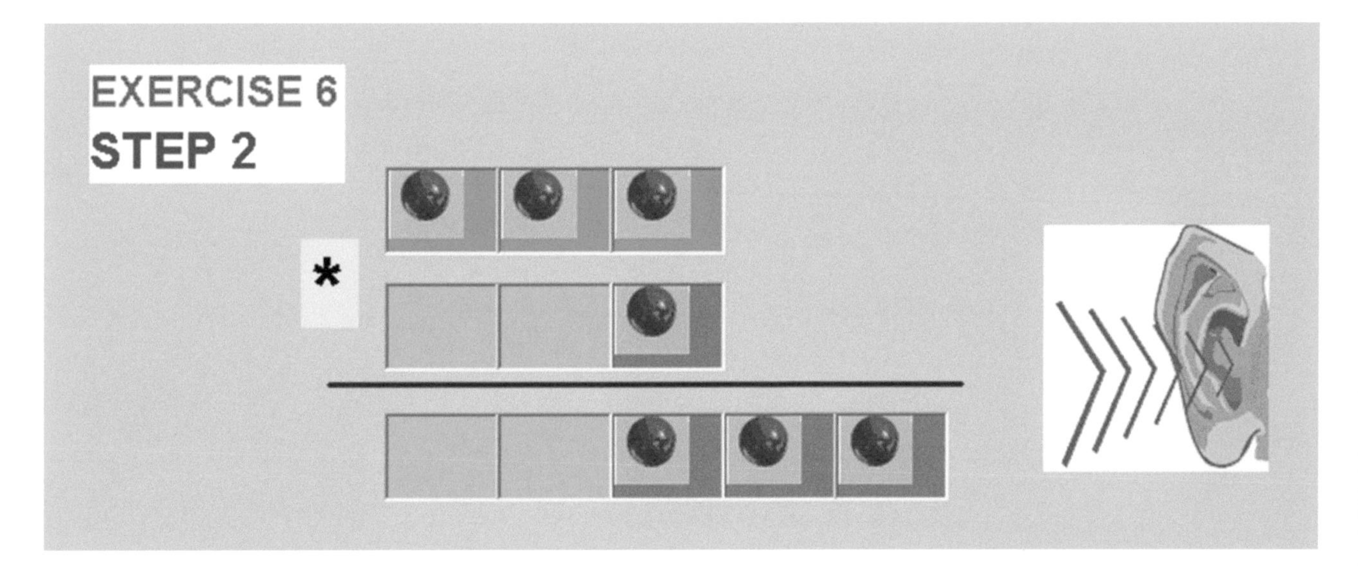

The ***kinesthetic*** representation of a filled position is represented by the clenching of the right hand. An empty position is represented by the clenching of the left hand. The multiplication sign, blick, is represented by one hand raised up. The equal sign, cling, is represented by crossing both hands. The kinesthetic representation for the pictures is the clenching of the right hand three times, one hand raised up, the clenching of the left hand twice, right once, crossing both hands, the clenching of the left hand twice and the right hand three times.

MATHEMATICS

MULTIPLICATION BY ADDITION, EXERCISE 6, STEP 6.3

In Exercise 6 Step 6.3, in a ***visual*** display of the third step of the multiplication, we add together the result of the first and second steps. In the first picture, which has a value of seven, all positions are filled with red balls. In the second picture, which has a value of twenty four, the third, fourth and fifth positions filled with red balls. The addition sign is between the first and second pictures. Below the second picture is the equal line, which separates the problem from the answer. In the third picture, which is the answer for the third step of the multiplication, positions one, two and six are filled with symbols. The answer has a value of thirty five.

The ***audio*** representation for the pictures consists of a **knock, knock, knock, tram, double knock, double knock, knock, knock, knock, cling, knock, knock, double knock, double knock, double knock and knock**.

Exercise 6, Step 3

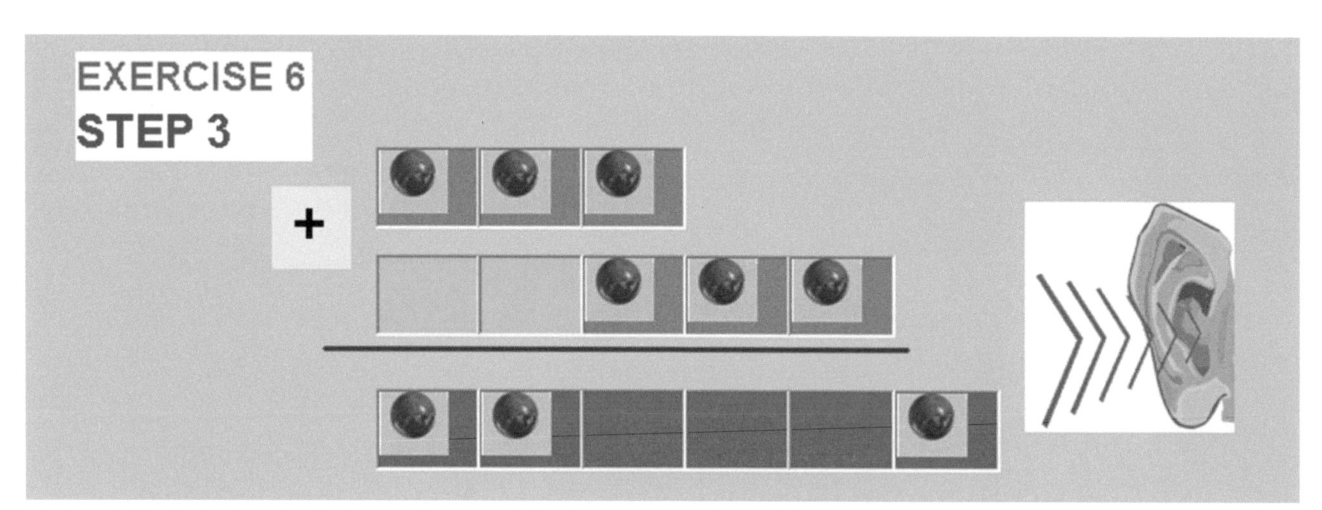

The ***kinesthetic*** representation of a filled position is represented by the clenching of the right hand. An empty position is represented by the clenching of the left hand. The addition sound, tram, is represented by extending the left hand in front. The equal sound, cling, is represented by crossing both hands. The kinesthetic representation for the pictures is *the clenching of the right hand three times, left hand extended in front, the clenching of the left hand twice, right hand three times, the crossing of both hands, the clenching of the right twice, left three times and the right hand once.*

MULTIPLICATION, EXERCISE 6, STEP 6.4

In Exercise 6 Step 6.4, in a ***visual*** display of the last step of the multiplication, we see one picture of boxes, which is the answer of the problem. The third, fourth and fifth positions are empty but positions one, two and six are filled with red balls. The answer has a value of thirty five.

The ***audio*** representation for this picture consists of a **knock, knock, double knock, double knock, double knock and knock**.

Exercise 6, Step 4

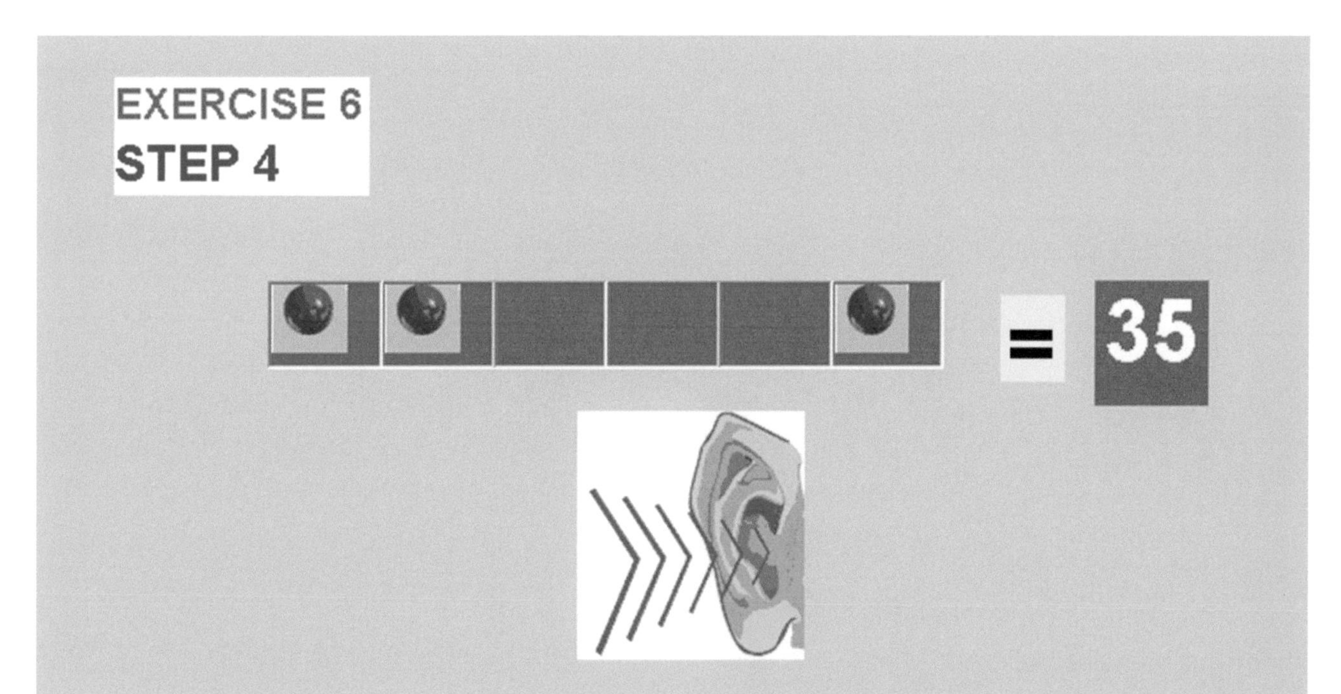

The ***kinesthetic*** representation of a filled position is represented by the clenching of the right hand. An empty position is represented by the clenching of the left hand. The kinesthetic representation for this picture is *the clenching of the right twice, left hand three times and the right hand once.*

MATHEMATICS MULTIPLICATION, EXERCISE 6

In Exercise 6, in a ***visual*** display of multiplication we see two pictures of boxes. At the bottom are three pictures with probable answers. In the first multiplicand, all three positions are filled with symbols. In the second multiplicand, the first and third positions are filled with symbols and the second is empty. The multiplication sign is between the first and second numbers. Below the second picture is the equal line, which separates the problem from the answer. The ***audio*** representation for the multiplication consists of a **knock, knock knock, blick, knock, double knock, knock, and cling.**

From the probable answers for the multiplication we need to find the right one. The audio signals are: 1)**knock, double knock, knock, double knock, knock, and double knock**, 2) **double knock, knock, double knock, knock, knock and double knock**, 3) **knock, knock, double knock, double knock, double knock, and knock**.

Exercise 6

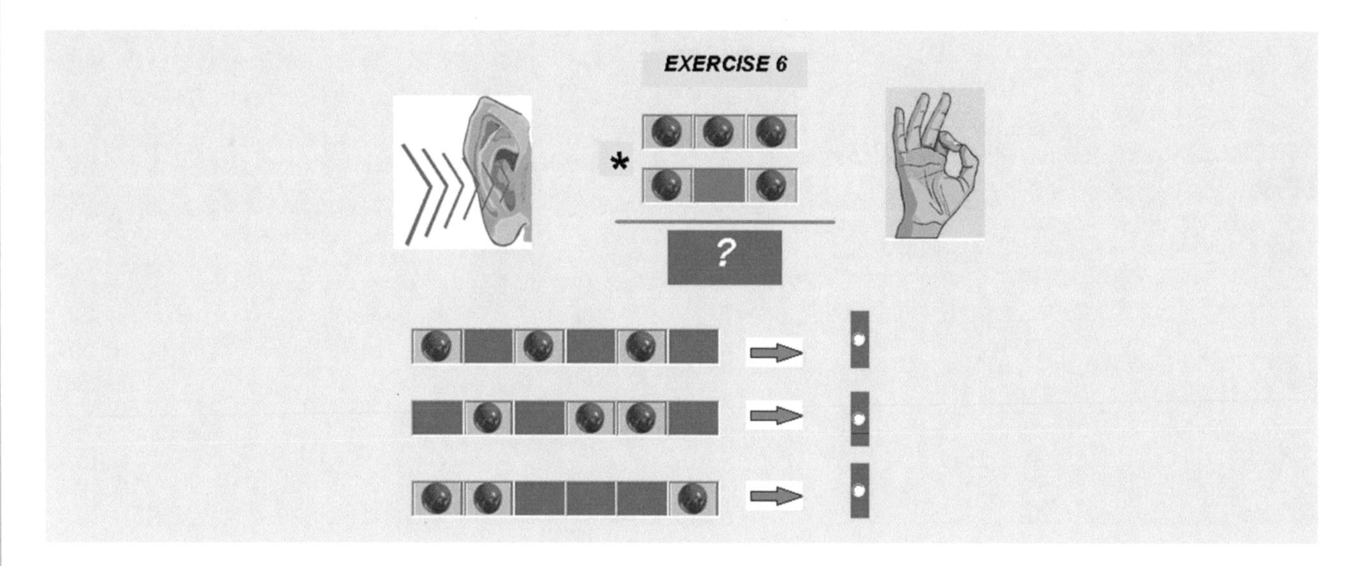

The ***kinesthetic*** representation of a filled position is represented by the clenching of the right hand. An empty position is represented by the clenching of the left hand. The multiplication sound, blick, is represented by one hand raised up. The equal sound, cling, is represented by crossing of both hands. The kinesthetic representation for both pictures and the answer is *the clenching of the right hand three times, one hand raised up, the clenching of the right, left, right, crossing of both hands, the clenching of the right hand twice, left hand three times and the right hand once.*

MULTIPLICATION, EXERCISE 7

In Exercise 7, in a ***visual*** display of multiplication we see three pictures of boxes. The first picture, which has a value of seven, all three positions are filled with symbols. In the second picture, only position four is filled with a symbol. The multiplication sign is between the first and second pictures. Below the second picture is the equal line, which separates the problem from the answer. In the third picture, which is the answer for the multiplication and has a value of fifty six, the first three positions are empty while positions four, five and six are filled with symbols.

The ***audio*** representation for the pictures consists of a **knock, knock, knock, blick, double knock, double knock, double knock, knock, cling, double knock, double knock, double knock, knock, knock and knock.**

Exercise 7

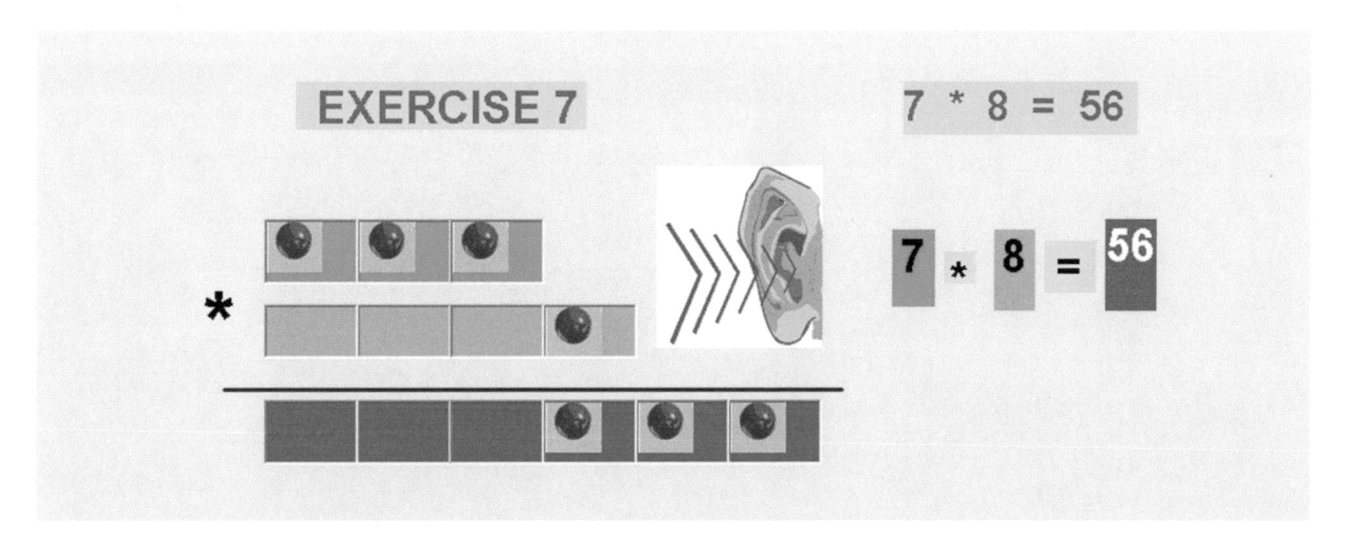

The ***kinesthetic*** representation of a filled position is represented by the clenching of the right hand. An empty position is represented by the clenching of the left hand. The multiplication sign, blick, is represented by both hands extended in front. The equal sign, cling, is represented by crossing both hands. The kinesthetic representation for both pictures is the clenching of the right hand three times, extending both hands forward, the clenching of the left hand three times, right once, crossing both hands, the clenching of the left hand three times and the right hand three times.

MULTIPLICATION, EXERCISE 7, STEP 7.1

In Exercise 7 Steps 7.1, in a ***visual*** display of multiplication we see three pictures of boxes. In the first picture, which has a value of seven, all three positions are filled with symbols. In the second picture, which has a value of eight, only position four is filled with a symbol. Positions one, two and three are empty in the second picture. The multiplication sign is between the first and second pictures. Below the second picture is the equal line, which separates the problem from the answer. In the third picture, which is the answer for Step 7.1 of the multiplication, positions one, two and three are empty while positions four, five and six are filled with symbols. The answer has a value of fifty six.

The ***audio*** representation for the pictures consists of a **knock, knock, knock, blick, double knock, double knock, double knock, knock, cling, double knock, double knock, double knock, knock, knock and knock**.

Exercise 7, Step 1

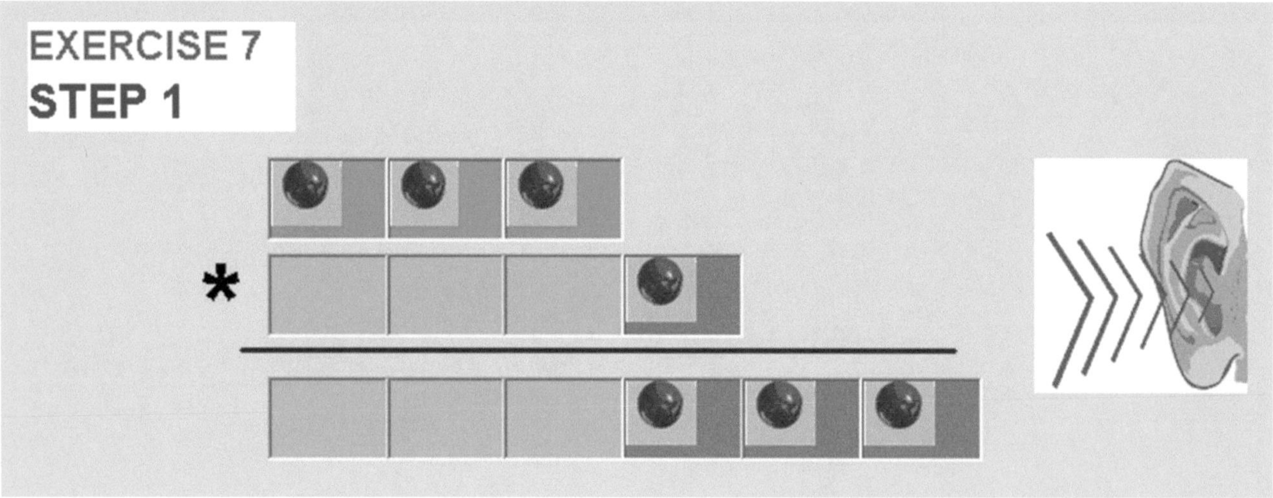

EXERCISE 7
STEP 1

*

The ***kinesthetic*** representation of a filled position is represented by the clenching of the right hand. An empty position is represented by the clenching of the left hand. The multiplication sound, blick, is represented by both hands extended forward. The equal sound, cling, is represented by crossing the hands. The kinesthetic representation for the pictures is *the clenching of the right hand three times, extending both hands forward, the clenching of the left hand three times, right hand once, the crossing of both hands, the clenching of the left hand three times, and the right hand three times.*

MULTIPLICATION, EXERCISE 7, STEP 7.2

In exercise 7, step 7.2 in ***visual*** display of the last step of multiplication, we see one picture of boxes, which is the answer of the problem. We see that the first, second and third positions are empty but position four, five and six are filled up with a red ball and the answer has a value of fifth six.

The ***audio*** representation for this picture consumes of a **double knock, double knock, double knock, knock, knock and knock**.

Figure 7.2

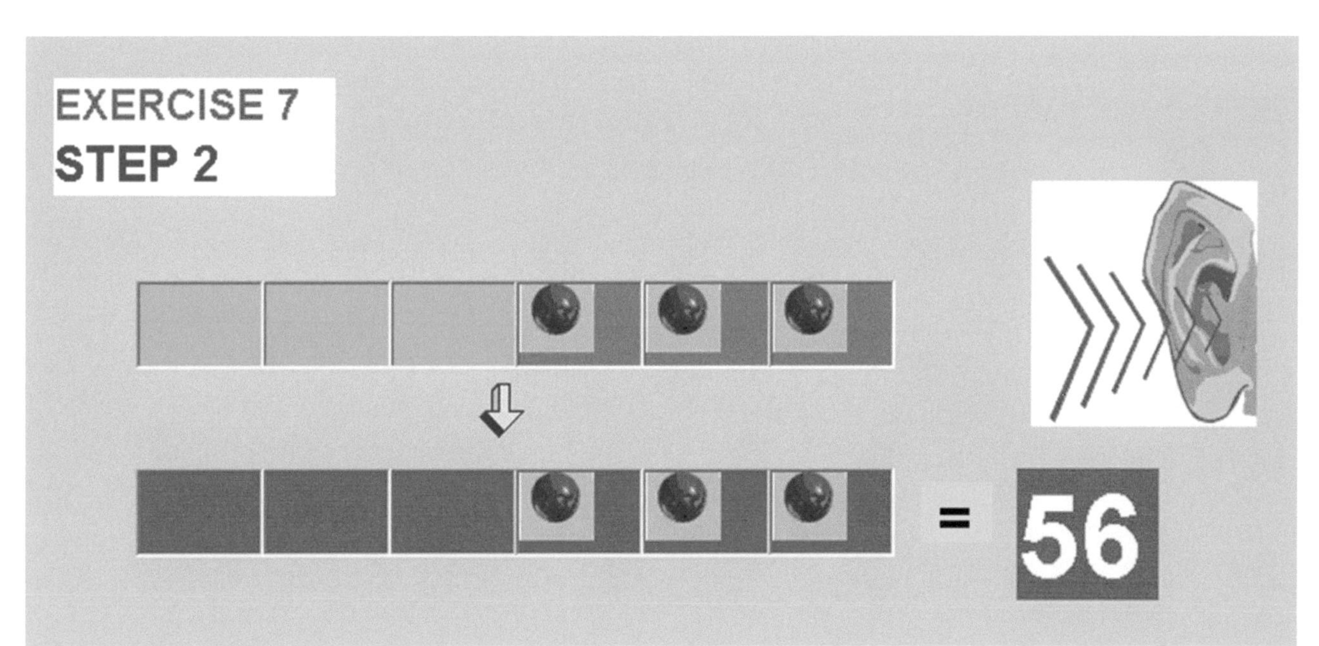

The ***kinesthetic*** representation of the filled up position is announced by clamping of the right hand. An empty position is represented by the clamping of the left hand. The kinesthetic representation for this picture is clamping the left hand three times and the right hand also three times.

MATHEMATICS FIND AN ANSWER, EXERCISE 7

In Exercise 7, in a ***visual*** display of multiplication we see two pictures of boxes. The probable answers are at the bottom. In the first picture, all three positions are filled with symbols. In the second picture, position one, two and three are empty, while position four is filled with a symbol. The multiplication sign is between the first and second pictures. Below the second picture is the equal line, which separates the problem from the answer. The ***audio*** representation for the pictures consists of a **knock, knock, knock, blick, knock, double knock, knock, and cling.**

From the probable answers for the multiplication we need to find the right one. The audio signals are: 1)**knock, double knock, knock, double knock, knock, and double knock**, 2) **double knock, knock, double knock, knock, knock and double knock**, 3) **knock, knock, double knock, double knock, double knock, and knock**.

Exercise 7

The ***kinesthetic*** representation of a filled position is represented by the clenching of the right hand. An empty position is represented by the clenching of the left hand. The multiplication sound, blick, is represented by both hands extended in front. The equal sound, cling, is represented by crossing both hands. The kinesthetic representation for both pictures and the answer is *the clenching of the right hand three times, both hands extended in front, the clenching of the left hand three times, right once, the crossing of both hands, the clenching of the left hand three times and the right hand three times.*

MULTIPLICATION, EXERCISE 8

In exercise 8 in ***visual*** display of multiplication we see three pictures of boxes, where in the first picture, which has value of eight; we only have the fourth position filled up with a red ball. In the second picture, we have position three filled up with a red ball. We want to specify, that between the first and second picture we see the multiplication sign. Bellow the second picture we see the equal line, which separates the problem from the answer. In the third picture, which is the answer for the multiplication and has a value of third two, we see that the first five positions are empty but position six is filled up with a symbol.

The ***audio*** representation for three pictures consumes of a **double knock, double knock, double knock, knock, blick, double knock, double knock, knock, cling, double knock, double knock, double knock, double knock, double knock and knock.**

Figure 8

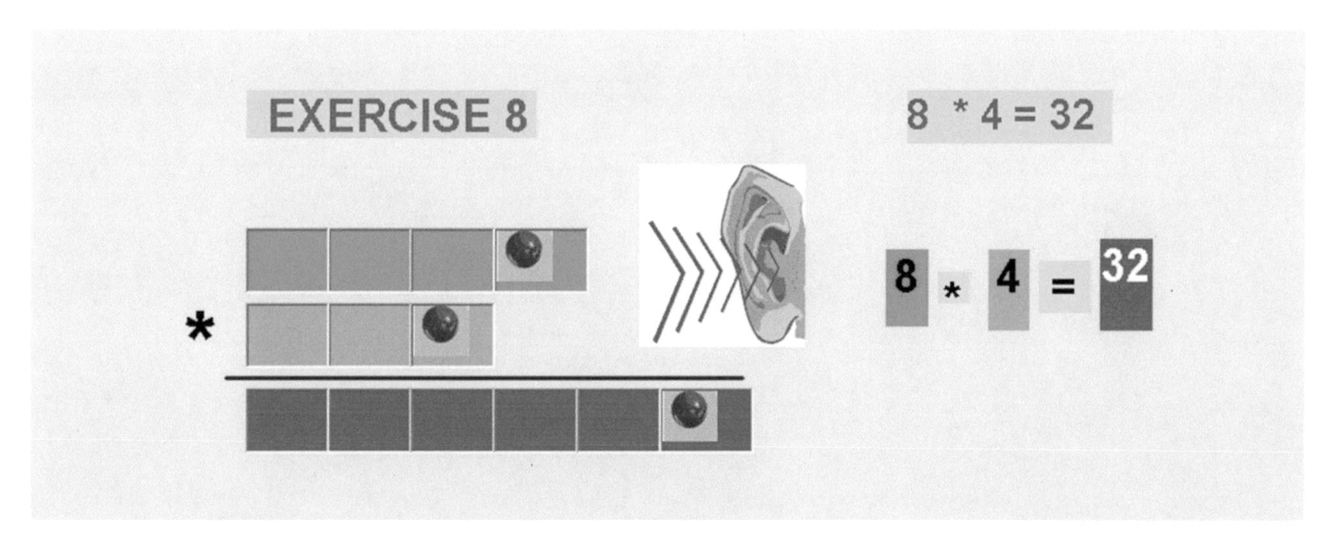

The ***kinesthetic*** representation of the filled up position is announced by clamping of the right hand. An empty position is represented by the clamping of the left hand. The multiplication sign; blick, is represented by both hands extended in front. The equal sign, cling, is represented by crossing both hands. The kinesthetic representation for both pictures are the clamping of the left hand three times, right once, extending both hands forward, left hand twice, right hand once, crossing both hands, left hand five times and the right hand once.

MATHEMATICS

MULTIPLICATION, EXERCISE 8, STEP 8.1

In exercise 8, steps 8.1 in ***visual*** display of multiplication we see three pictures of boxes. In the first picture, which has value of eight; we only have the fourth position filled up with a red ball. In the second picture, which has the value of four, we see only position three filled up with a symbol. Position one and two are empty in the second picture. We want to specify, that between the first and second pictures we see the multiplication sign. Bellow the second picture we see the equal line, which separates the problem from the answer. In the third picture, which is the answer for step 8.1 of multiplication, we see that positions one through five are empty but position six is filled up with a red ball, which has a value of third two.

The ***audio*** representation for three pictures consumes of a **double knock, double knock, double knock, knock, blick, double knock, double knock, knock, cling, double knock, double knock, double knock, double knock, double knock and knock.**

Step 8.1

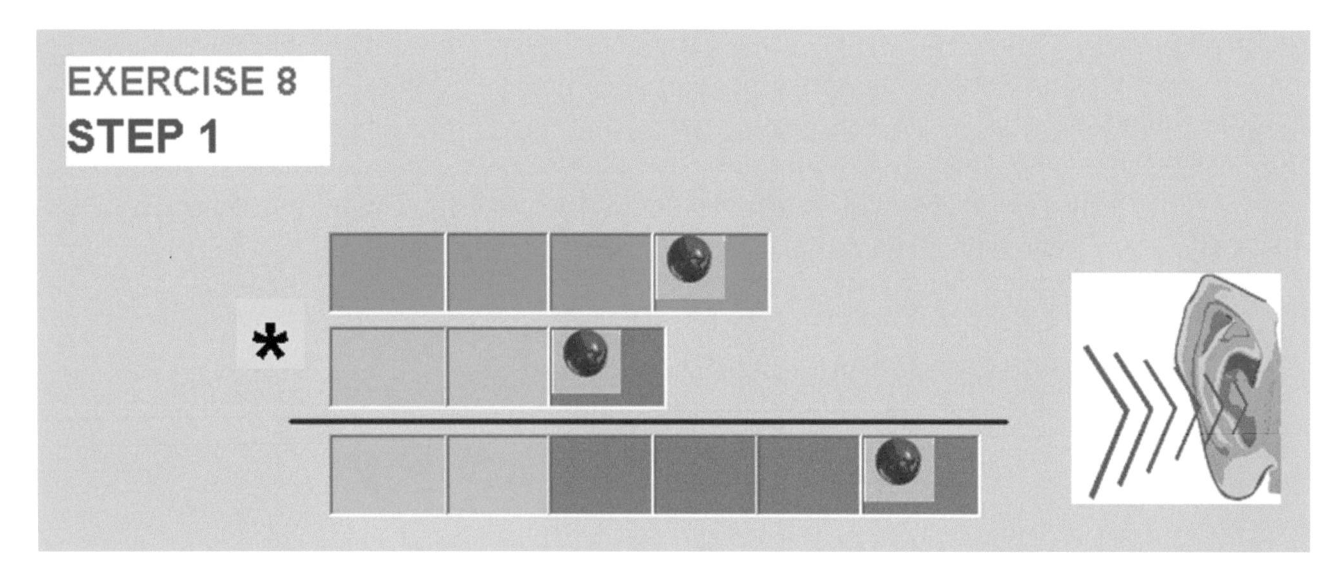

The ***kinesthetic*** representation of the filled up position is announced by clamping of the right hand. An empty position is represented by the clamping of the left hand. The multiplication sound; blick, is represented by both hands extended forward. The equal sound; cling, is represented by crossing hands. The kinesthetic representation for both pictures are the clamping of the left hand three times, right once, extending both hands forward, left hand twice, right hand once, crossing both hands, left hand five times and the right hand once.

MULTIPLICATION, EXERCISE 8, STEP 8.2

In exercise 8, step 8.2 in ***visual*** display of the last step of multiplication, we see one picture of boxes, which is the answer of the problem. We see that the first, second and third positions are empty but position four, five and six are filled up with a red ball and the answer has a value of fifth six.

The ***audio*** representation for this picture consumes of a **double knock, double knock, double knock, knock, knock and knock**.

Figure 8.2

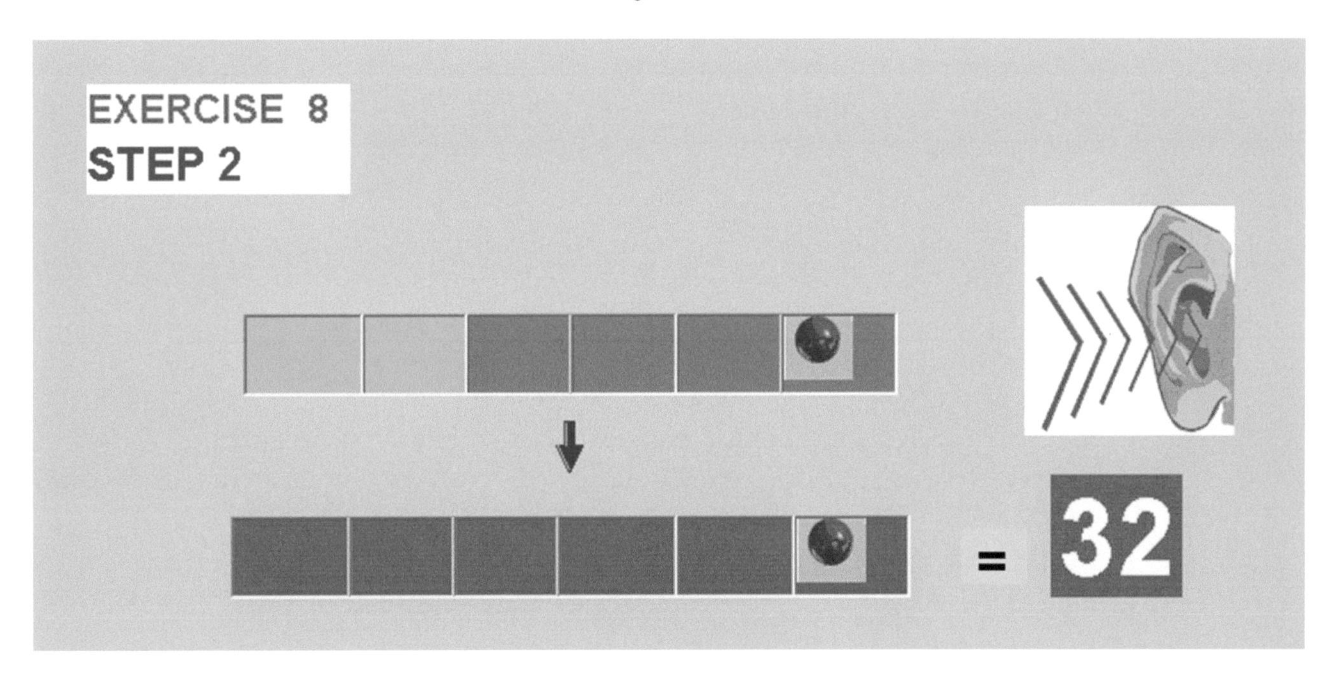

The ***kinesthetic*** representation of the filled up position is announced by clamping of the right hand. An empty position is represented by the clamping of the left hand. The kinesthetic representation for this picture is clamping the left hand three times and the right hand also three times.

MATHEMATICS FIND AN ANSWER, EXERCISE 8

In Exercise 8, in a ***visual*** display of multiplication we see two pictures of boxes. At the bottom are three pictures with probable answers. In the first picture, the first three positions are empty, while position four is filled with a symbol. In the second picture, the first two positions are empty and the third position is filled with a symbol. The multiplication sign is between the first and second picures. Below the second picture is the equal line, which separates the problem from the answer. The ***audio*** representation for the pictures consists of a **double knock, double knock, double knock, knock, blick, double knock, double knock, knock, and cling.**

From the probable answers for the multiplication we need to find the right one. The audio signals are: 1)**double knock, knock, double knock, double knock, double knock and double knock**, 2) **double knock, double knock, knock, double knock, double knock and double knock**, 3) **double knock, double knock, double knock, double knock, double knock, and knock**.

Exercise 8

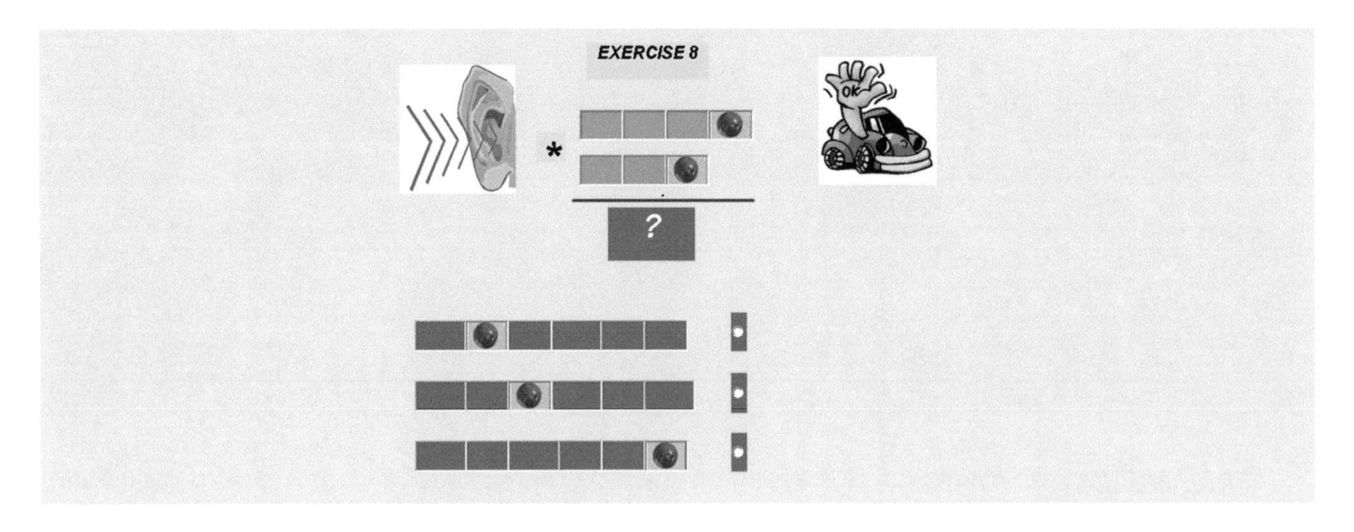

The ***kinesthetic*** representation of a filled position is represented by the clenching of the right hand. An empty position is represented by the clenching of the left hand. The multiplication sound, blick, is represented by both hands extended in front. The equal sound, cling, is represented by crossing both hands. The kinesthetic representation for both pictures and the answer is the clenching of the left hand three times, right hand once, both hands extended in front, the clenching of the left hand twice, right once, crossing of both hands, the clenching of the left hand five times and the right once.

MULTIPLICATION, EXERCISE 9

In exercise 9 in ***visual*** display of multiplication we see three pictures of boxes, where in the first picture, which has value of eight; we only have the fourth position filled up with a red ball. In the second picture, we have position one and three filled up with a red ball. We want to specify, that between the first and second picture we see the multiplication sign. Bellow the second picture we see the equal line, which separates the problem from the answer. In the third picture, which is the answer for the multiplication and has a value of fourthy, we see that position one, two, three, and five are empty but position four and six are filled up with a symbol.

The ***audio*** representation for three pictures consumes of a **double knock, double knock, double knock, knock, blick, knock, double knock, knock, cling, double knock, double knock, double knock, knock, double knock and knock.**

Figure 9

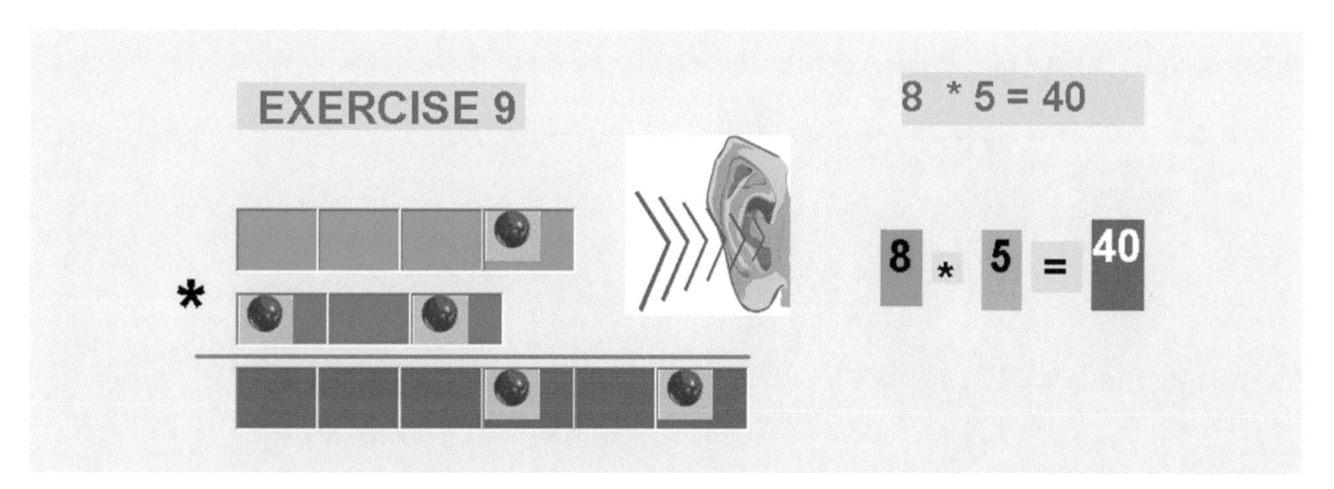

The ***kinesthetic*** representation of the filled up position is announced by clamping of the right hand. An empty position is represented by the clamping of the left hand. The multiplication sign; blick, is represented by both hands extended in front. The equal sign, cling, is represented by crossing both hands. The kinesthetic representation for both pictures are the clamping of the left hand three times, right once, extending both hands forward, right, left, right, crossing of both hands, left hand three times, right once, left once and the right hand once.

MATHEMATICS

MULTIPLICATION, EXERCISE 9, STEP 9.1

In exercise 9, step 9.1 in ***visual*** display of multiplication we see three pictures of boxes. We are moving the first picture on two positions to the right.

In the first picture, which has value of eight, where only the fourth position is filled up with a red ball. In the second picture, which has the value of one, we see position one is filled up with symbol. We want to specify, that between the first and second pictures we see the multiplication sign. Bellow the second picture we see the equal line, which separates the problem from the answer. In the third picture, which is the answer for step one of multiplication; we see that only position four is filled up with a red ball symbol, which has the value of eight.

The ***audio*** representation for three pictures consumes of a **double knock, double knock, double knock, knock, blick, knock, double knock, cling, double knock, double knock, double knock and knock**.

Step 9.1

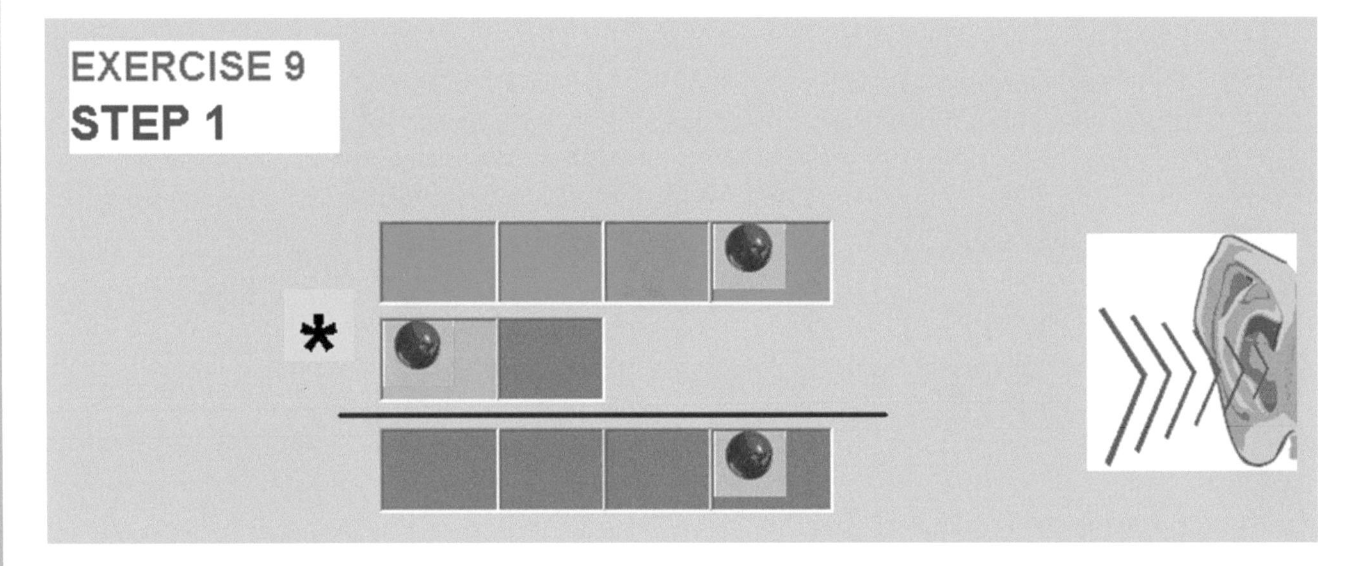

The ***kinesthetic*** representation of the filled up position is announced by clamping of the right hand. An empty position is represented by the clamping of the left hand. The multiplication sound; blick, is represented by both hands extended forward. The equal sound; cling, is represented by crossing hands. The kinesthetic representation for all three pictures is the clamping of the left hand three times, right one, extending both hands forward, right once, left once, crossing of both hands, left hand three times and the right hand once.

MULTIPLICATION, EXERCISE 9, STEP 9.2

In exercise 9 Step 9.2 in ***visual*** display for this step of multiplication we see three pictures of boxes. We are moving the first picture on two positions to the right. In the first picture, which has a value of eight, we see that only position four is filled up with a red ball. In the second picture, which has a value of four, we see that the third position is filled up with a symbol. We want to specify, that between the first and second pictures we see the multiplication sign. Bellow the second picture we see the equal line, which separates the problem from the answer. In the third picture, which is the answer for the multiplication and has a value of thirty two, we see that position six is filled up with the symbol.

The ***audio*** representation for the three pictures consumes of a **double knock, double knock, double knock, knock, blick, double knock, double knock, knock, cling, double knock, double knock, double knock, double knock, double knock and knock**.

Step 9.2

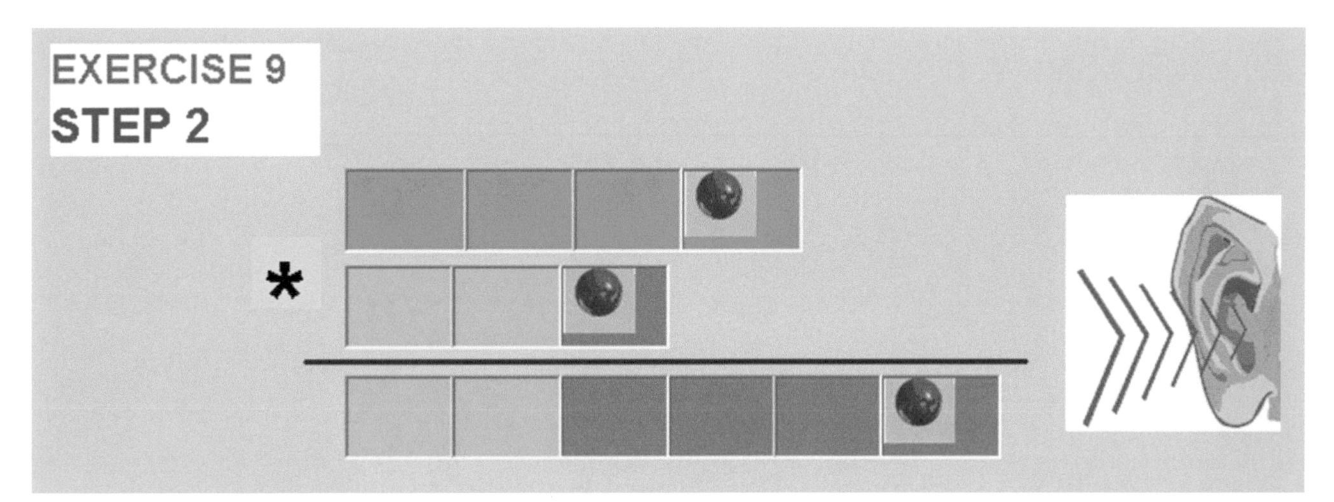

The ***kinesthetic*** representation of the filled up position is announced by clamping of the right hand. An empty position is represented by the clamping of the left hand. The multiplication sign, blick, is represented by both hands extended in front. The equal sign, cling, is represented by crossing both hands. The kinesthetic representation for three pictures is the clamping of the left hand three times, right once, both hands extended in front, left twice, right once, crossing of both hands, clamping the left hand five times and the right hand three times.

MATHEMATICS

MULTIPLICATION BY ADDITION, EXERCISE 9, STEP 9.3

In exercise 9, step 9.3 in ***visual*** display of the third step of multiplication we are adding together the result of the first and second steps. In the first picture, which has value of eight, we see that position four is filled up with a red ball. In the second picture, which has the value of thirty two, we see only the sixth position filled up with the red ball. We want to specify, that between the first and second pictures we see the add sign. Bellow the second picture we see the equal line, which separates the problem from the answer. In the third picture, which is an answer for the third step of multiplication, we see position four and six filled up with a symbol and the answer has a value of forty.

The ***audio*** representation for three pictures consumes of a **double knock, double knock, double knock, knock, tram, double knock, double knock, double knock, double knock, double knock, knock, cling, double knock, double knock, double knock, knock, double knock and knock**.

Step 9.3

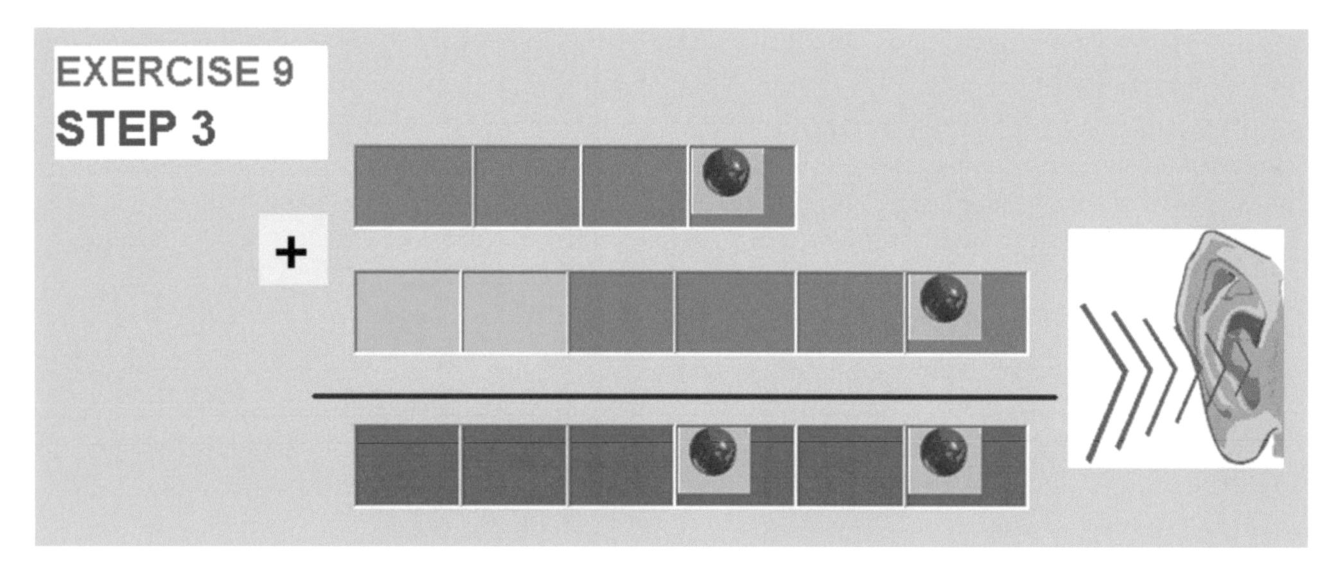

The ***kinesthetic*** representation of the filled up position is announced by clamping of the right hand. An empty position is represented by the clamping of the left hand. The addition sound, tram, is represented by extended left hand in front. The equal sound, cling, is represented by crossing both hands. The kinesthetic representation for the three pictures is the clamping of the left hand three times, right once, left hand extended in front, clamping the left hand five times, right once, crossing of both hands, left hand three times and the right hand once.

MULTIPLICATION, EXERCISE 9, STEP 9.4

In exercise 9, step 9.4 in ***visual*** display of the last step of multiplication, we see one picture of boxes, which is the answer of the problem. We see that the first, second, third and fifth positions are empty but position four and six are filled up with a red ball and the answer has a value of forty.

The ***audio*** representation for this picture consumes of a **double knock, double knock, double knock, knock, double knock and knock**. Step 9.4

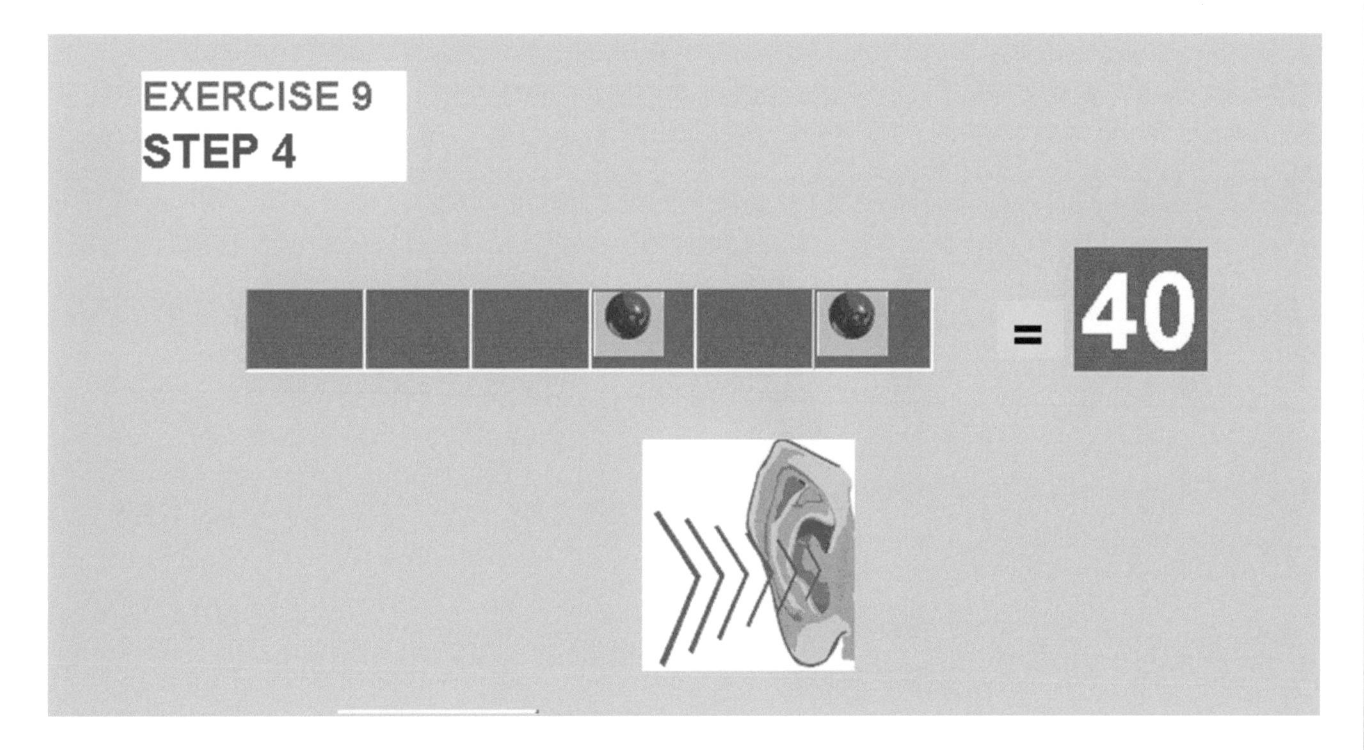

The ***kinesthetic*** representation of the filled up position is announced by clamping of the right hand. An empty position is represented by the clamping of the left hand. The kinesthetic representation for this picture is clamping of the left hand three times, right once, left once and the right hand once.

MATHEMATICS: FIND AN ANSWER, EXERCISE 9

In Exercise 9, in a ***visual*** display of multiplication we see two pictures of boxes. At the bottom are three pictures with probable answers. In the first picture, the first three positions are empty while position four is filled with a red ball. In the second picture, positions one and three filled with symbols. The multiplication sign is between the first and second pictures. Below the second picture is the equal line, which separates the problem from the answer. The ***audio*** representation for the pictures consists of a **double knock, double knock, double knock, knock, blick, knock, double knock, knock, and cling.**

From the probable answers for the multiplication we need to find the right one. The audio signals are: 1) **knock, double knock, double knock, double knock, double knock and knock**, 2) **double knock, double knock, knock, knock, double knock and double knock**, 3) **double knock, double knock, double knock, knock, double knock, and knock**.

Exercise 9

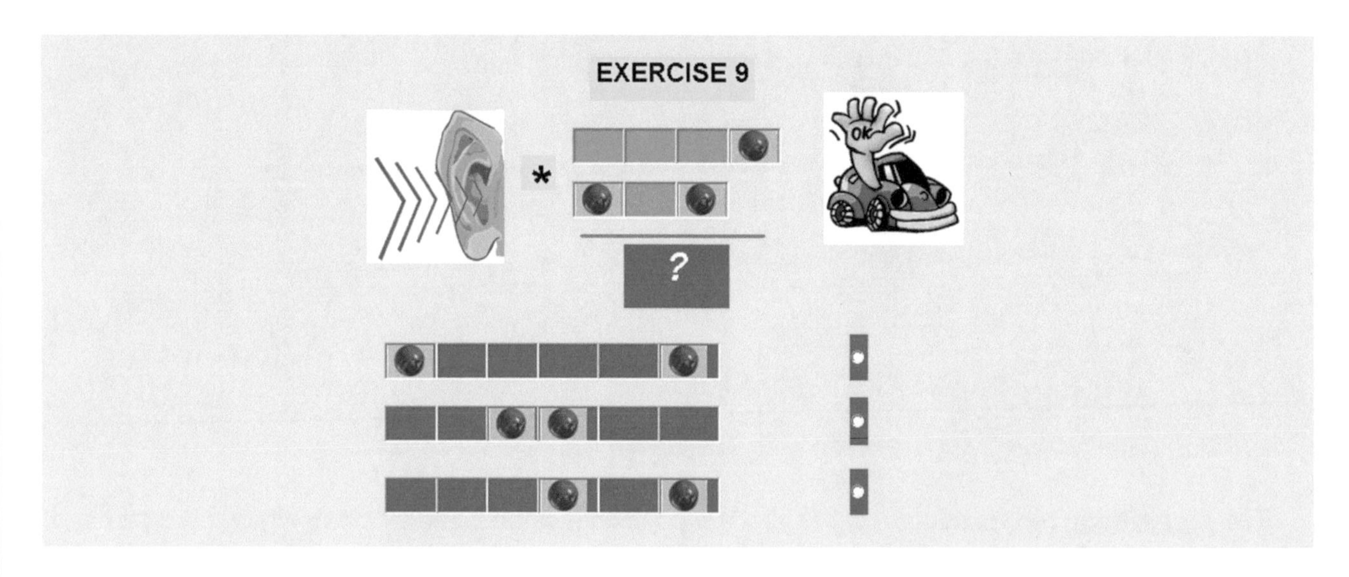

The ***kinesthetic*** representation of a filled position is represented by the clenching of the right hand. An empty position is represented by the clenching of the left hand. The multiplication sound, blick, is represented by both hands extended in front. The equal sound, cling, is represented by crossing both hands. The kinesthetic representation for both pictures and the answer is the clenching of the left hand three times, right once, both hands extended in front, the clenching of the right, left, right, crossing of both hands, the clenching of the left hand three times, right once, left once and the right once.

MULTIPLICATION, EXERCISE 10

In exercise 10 in ***visual*** display of multiplication we see three pictures of boxes, where in the first picture, which has value of nine; we have position one and four filled up with a red ball. In the second picture, we have both positions filled up with a red ball. We want to specify, that between the first and second picture we see the multiplication sign. Bellow the second picture we see the equal line, which separates the problem from the answer. In the third picture, which is the answer for the multiplication and has a value of twenty seven, we see that position three is empty but position one, two, four and five are filled up with a symbol.

The ***audio*** representation for the three pictures consumes of a **knock, double knock, double knock, knock, blick, knock, knock, cling, knock, knock, double knock, knock and knock.**

Figure 10

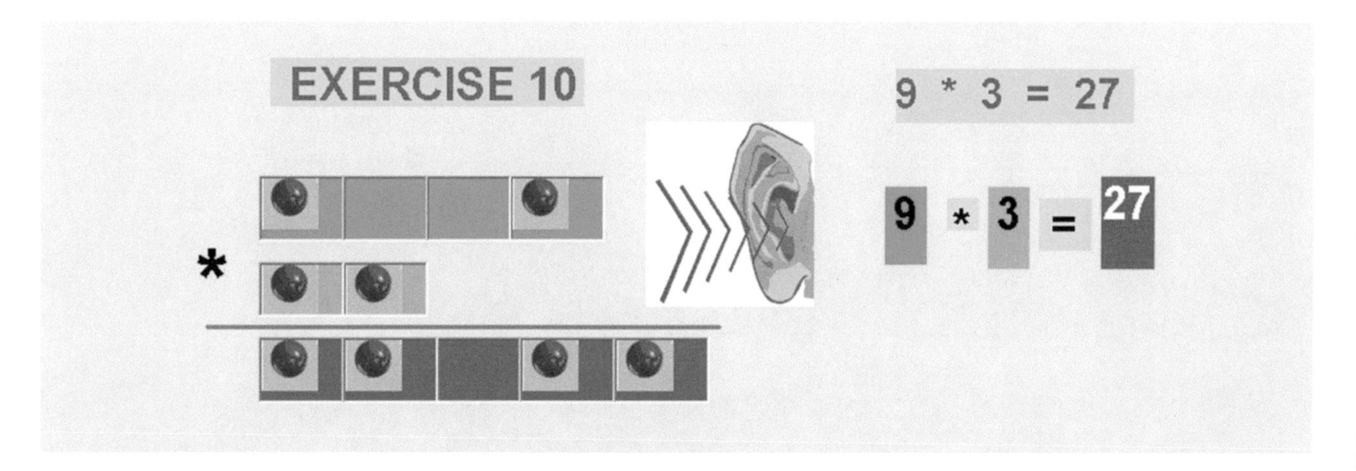

The ***kinesthetic*** representation of the filled up position is announced by clamping of the right hand. An empty position is represented by the clamping of the left hand. The multiplication sign; blick, is represented by both hands extended in front. The equal sign, cling, is represented by crossing both hands. The kinesthetic representation for all pictures is the clamping of the right, left twice, right once, extending both hands forward, right twice, crossing of both hands, right twice, left once and the right hand twice.

MULTIPLICATION, EXERCISE 10, STEP 10.1

In exercise 10, step 10.1 in ***visual*** display of multiplication we see three pictures of boxes. We are moving the first picture on two positions to the right.

In the first picture, which has value of eight, where only the fourth position is filled up with a red ball. In the second picture, which has the value of one, we see position one is filled up with symbol. We want to specify, that between the first and second pictures we see the multiplication sign. Bellow the second picture we see the equal line, which separates the problem from the answer. In the third picture, which is the answer for this step of multiplication; we see that only position four is filled up with a red ball symbol, which has the value of eight.

The ***audio*** representation for three pictures consumes of a **double knock, double knock, double knock, knock, blick, knock, double knock, cling, double knock, double knock, double knock and knock**.

Step 10.1

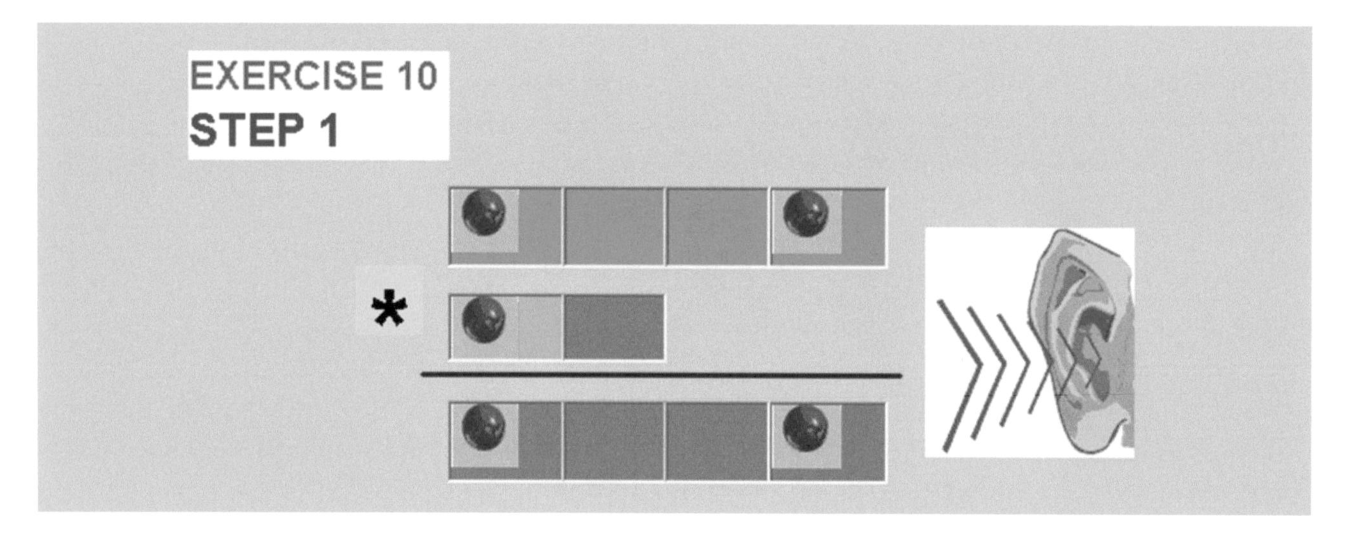

The ***kinesthetic*** representation of the filled up position is announced by clamping of the right hand. An empty position is represented by the clamping of the left hand. The multiplication sound; blick, is represented by both hands extended forward. The equal sound; cling, is represented by crossing hands. The kinesthetic representation for all three pictures is the clamping of the right, left twice, right one, extending both hands forward, right once, left once, crossing of both hands, right, left twice and the right hand once.

MULTIPLICATION, EXERCISE 10, STEP 10.2

In exercise 10 Step 10.2 in ***visual*** display for this step of multiplication we see three pictures of boxes. We are moving the second picture on one position to the right. In the first picture, which has a value of nine, we see that position one and four are filled up with a red ball. In the second picture, which has a value of two, we see that only the second position is filled up with a symbol. We want to specify, that between the first and second pictures we see the multiplication sign. Bellow the second picture we see the equal line, which separates the problem from the answer. In the third picture, which is the answer for the multiplication and has a value of eighteen, we see that position two and five are filled up with the symbol.

The ***audio*** representation for the three pictures consumes of a **knock, double knock, double knock, knock, blick, double knock, knock, cling, double knock, knock, double knock, double knock and knock**.

Step 10.2

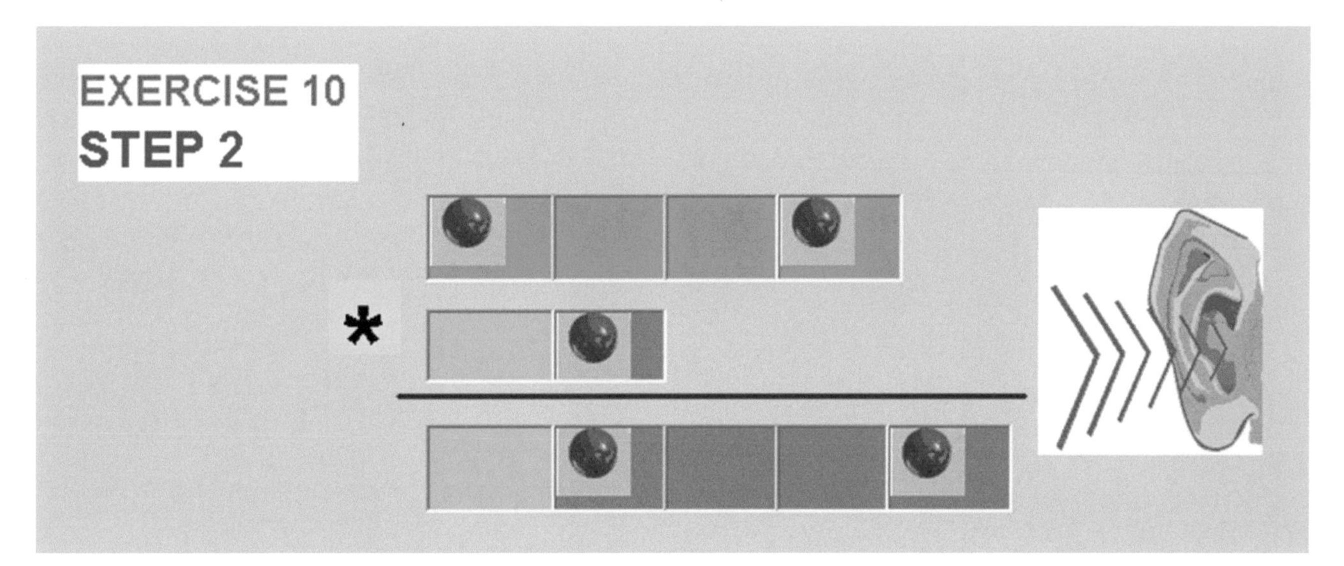

The ***kinesthetic*** representation of the filled up position is announced by clamping of the right hand. An empty position is represented by the clamping of the left hand. The multiplication sign, blick, is represented by both hands extended in front. The equal sign, cling, is represented by crossing both hands. The kinesthetic representation for the three pictures is the clamping of the right, left twice, right once, both hands extended in front, left, right, crossing of both hands, clamping the left, right, left twice and the right hand once.

MULTIPLICATION BY ADDITION, EXERCISE 10, STEP 10.3

In exercise 10, step 10.3 in ***visual*** display of the third step of multiplication we are adding together the result of the first and second steps. In the first picture, which has value of nine, we see that position one and four are filled up with a red ball. In the second picture, which has the value of eighteen, we see position two and five filled up with the red ball. We want to specify, that between the first and second pictures we see the add sign. Bellow the second picture we see the equal line, which separates the problem from the answer. In the third picture, which is an answer for the third step of multiplication, we see positions one, two, four and five are filled up with a symbol and the answer has a value of twenty seven.

The ***audio*** representation for three pictures consumes of a **knock, double knock, double knock, knock, tram, double knock, knock, double knock, double knock, knock, cling, knock, knock, double knock, knock and knock**.

Step 10.3

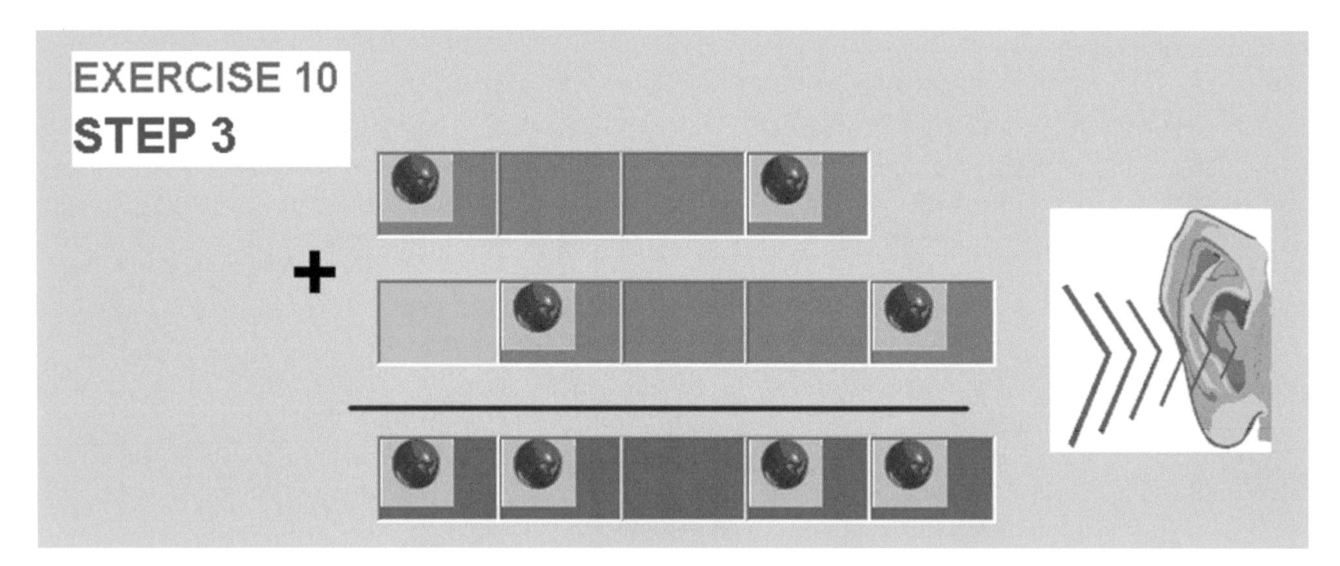

The ***kinesthetic*** representation of the filled up position is announced by clamping of the right hand. An empty position is represented by the clamping of the left hand. The addition sound, tram, is represented by extended left hand in front. The equal sound, cling, is represented by crossing both hands. The kinesthetic representation for the three pictures is the clamping of the right once, left twice, right once, left hand extended in front, clamping the left, right, left hand two times, right once, crossing of both hands, right twice, left once and the right hand twice.

MULTIPLICATION, EXERCISE 10, STEP 10.4

In exercise 10, step 10.4 in ***visual*** display of the last step of multiplication, we see one picture of boxes, which is the answer of the problem. We see that position two is empty but position one, two, four and five are filled up with a red ball and the answer has a value of twenty seven.

The ***audio*** representation for this picture consumes of a **knock, knock, double knock, knock and knock**.

Step 10.4

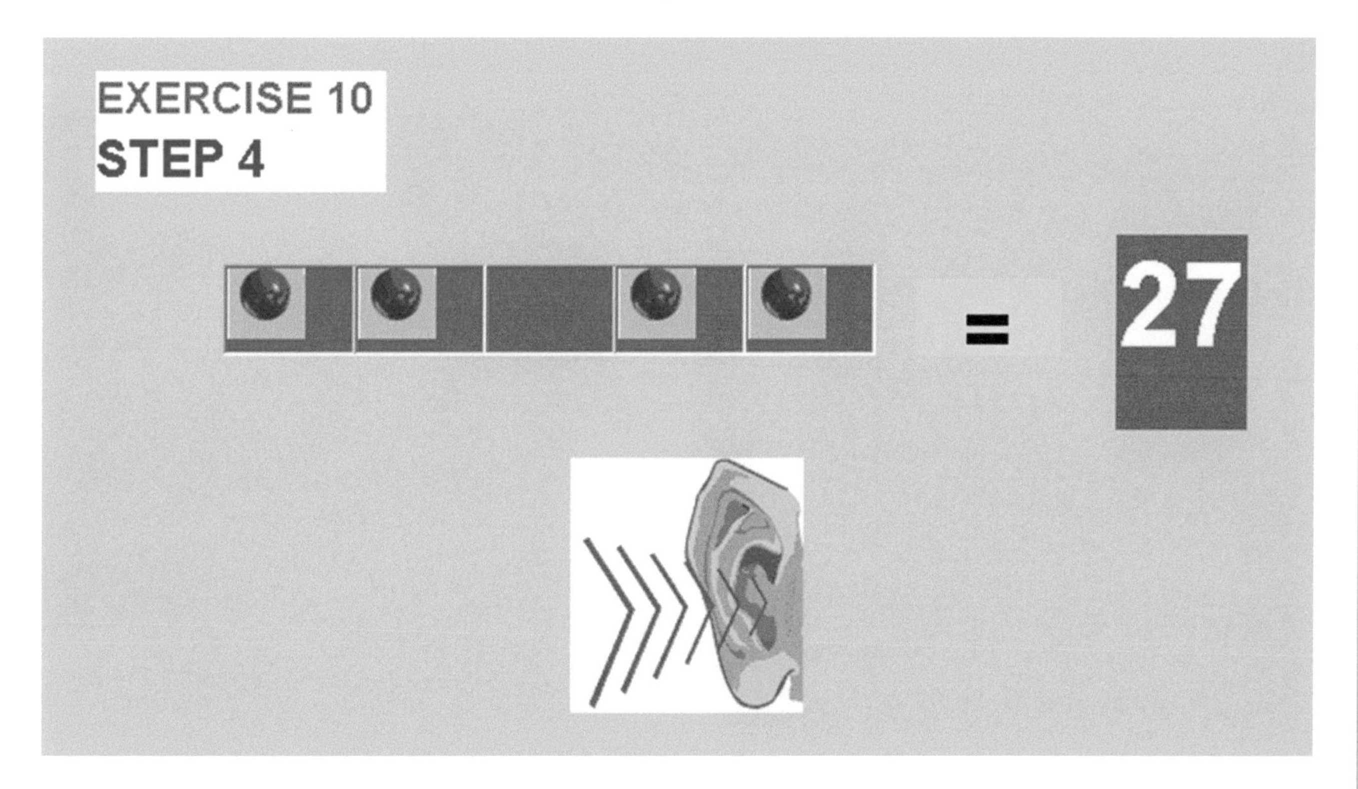

The ***kinesthetic*** representation of the filled up position is announced by clamping of the right hand. An empty position is represented by the clamping of the left hand. The kinesthetic representation for this picture is clamping of the right twice, left once and the right hand twice.

MATHEMATICS

FIND AN ANSWER, EXERCISE 10

In Exercise 10, in a ***visual*** display of multiplication we see two pictures of boxes. At the bottom are three probable answers. In the first picture, positions one and four are filled with red balls. In the second picture, positions one and two are filled with symbols. The multiplication sign is between the first and second pictures. Below the second picture is the equal line, which separates the problem from the answer. The ***audio*** representation for the pictures consists of a **knock, double knock, double knock, knock, blick, knock, knock, and cling.**

From the probable answers for the multiplication we need to find the right one. The audio signals are: 1) **knock, knock, double knock, knock and knock**, 2) **double knock, knock, knock, knock and knock**, 3) **knock, double knock, knock, knock, and knock**.

Exercise 10

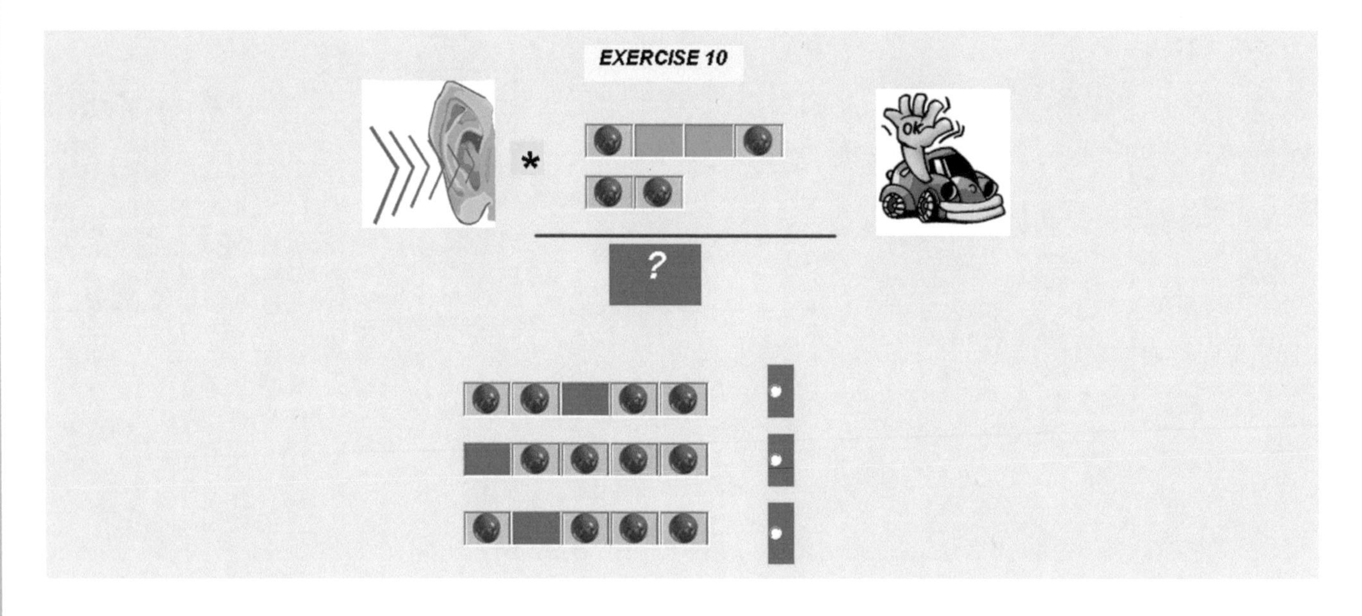

The ***kinesthetic*** representation of a filled position is represented by the clenching of the right hand. An empty position is represented by the clenching of the left hand. The multiplication sound, blick, is represented by both hands extended in front. The equal sound, cling, is represented by crossing both hands. The kinesthetic representation for both pictures and the answer is *the clenching of the right hand, left twice, right once, both hands extended in front, the clenching of the right twice, crossing of both hands, the clenching of the right twice, left once and the right twice.*

MULTIPLICATION, EXERCISE 11

In exercise 11 in ***visual*** display of multiplication we see three pictures of boxes, where in the first picture, which has value of nine; we have position one and four filled up with a red ball. In the second picture, we have position one and three filled up with a red ball. We want to specify, that between the first and second picture we see the multiplication sign. Bellow the second picture we see the equal line, which separates the problem from the answer. In the third picture, which is the answer for the multiplication and has a value of forty five, we see that position two and five are empty but position one, three, four and six are filled up with a symbol.

The ***audio*** representation for the three pictures consumes of a **knock, double knock, double knock, knock, blick, knock, double knock, knock, cling, knock, double knock, knock, knock, double knock, and knock.**

Figure 11

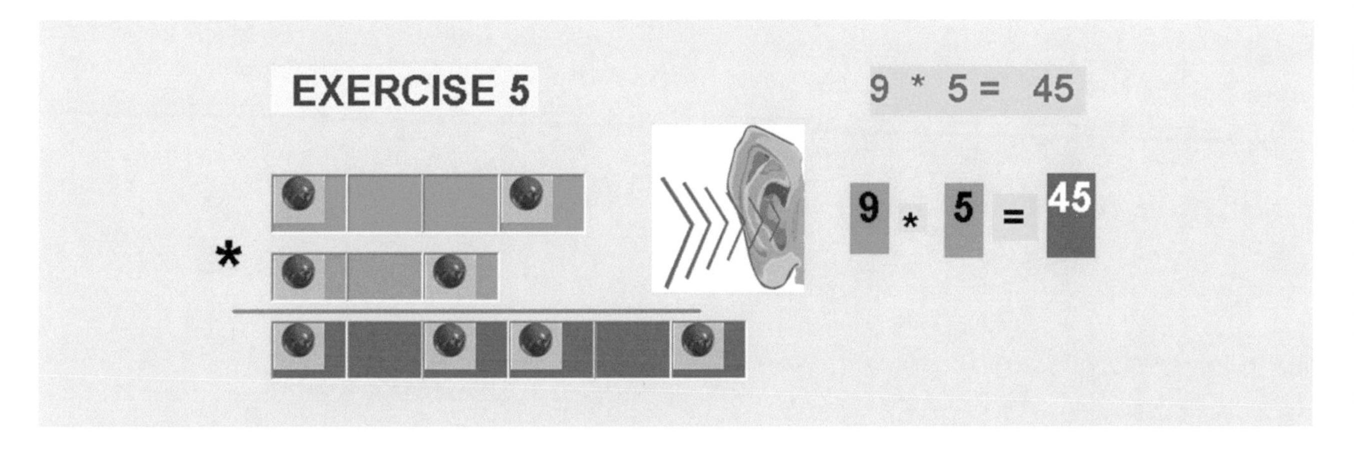

The ***kinesthetic*** representation of the filled up position is announced by clamping of the right hand. An empty position is represented by the clamping of the left hand. The multiplication sign; blick, is represented by both hands extended in front. The equal sign, cling, is represented by crossing both hands. The kinesthetic representation for all pictures is the clamping of the right, left twice, right once, extending both hands forward, right, left, right, crossing of both hands, right, left, right hand twice, left once and the right hand once.

MATHEMATICS

MULTIPLICATION, EXERCISE 11, STEP 11.1

In exercise 11, step 11.1 in ***visual*** display of multiplication we see three pictures of boxes. We are moving the first picture on two positions to the right. In the first picture, which has value of eight, where only the fourth position is filled up with a red ball. In the second picture, which has the value of one, we see position one is filled up with symbol. We want to specify, that between the first and second pictures we see the multiplication sign. Bellow the second picture we see the equal line, which separates the problem from the answer. In the third picture, which is the answer for this step of multiplication; we see that only position four is filled up with a red ball symbol, which has the value of eight.

The ***audio*** representation for three pictures consumes of a **double knock, double knock, double knock, knock, blick, knock, double knock, cling, double knock, double knock, double knock and knock**.

Step 11.1

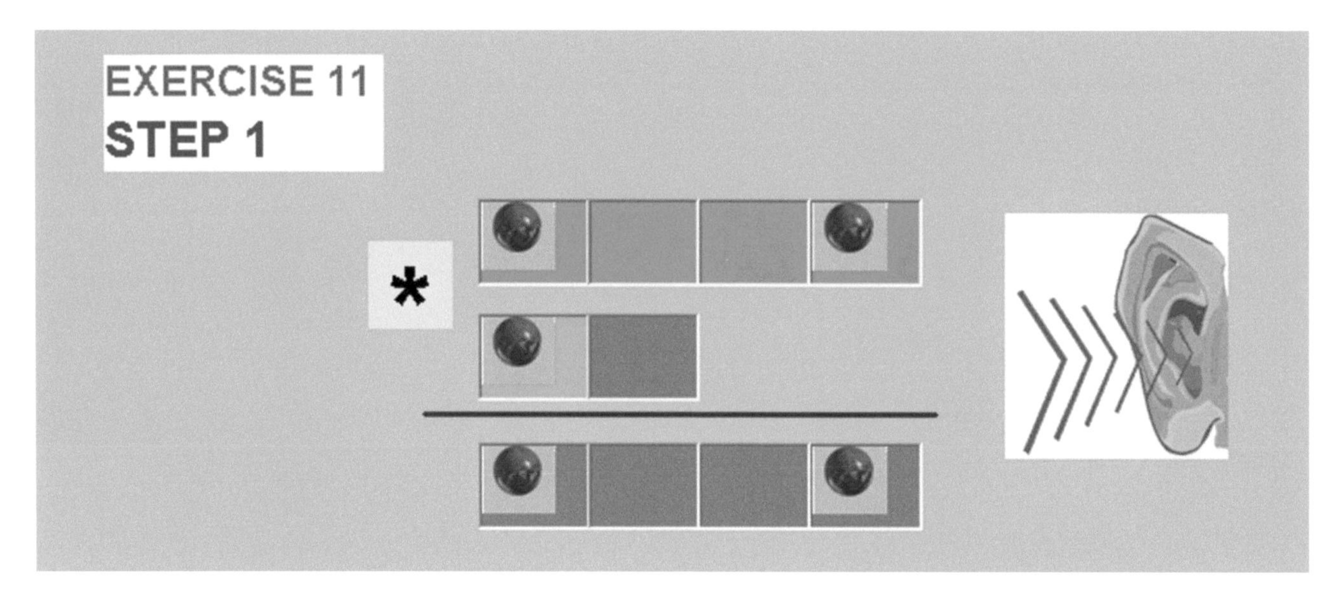

The ***kinesthetic*** representation of the filled up position is announced by clamping of the right hand. An empty position is represented by the clamping of the left hand. The multiplication sound; blick, is represented by both hands extended forward. The equal sound; cling, is represented by crossing hands. The kinesthetic representation for all three pictures is the clamping of the right, left twice, right one, extending both hands forward, right once, left once, crossing of both hands, right, left twice and the right hand once.

MULTIPLICATION, EXERCISE 11, STEP 11.2

In exercise 11 Step 11.2 in ***visual*** display for this step of multiplication we see three pictures of boxes. We are moving the second picture on two positions to the right. In the first picture, which has a value of nine, we see that position one and four are filled up with a red ball. In the second picture, which has a value of four, we see that only the third position is filled up with a symbol. We want to specify, that between the first and second pictures we see the multiplication sign. Bellow the second picture we see the equal line, which separates the problem from the answer. In the third picture, which is the answer for the multiplication and has a value of thirty six, we see that position three and six are filled up with the symbol.

The ***audio*** representation for the three pictures consumes of a **knock, double knock, double knock, knock, blick, double knock, double knock, knock, cling, double knock, double knock, knock, double knock, double knock and knock.**

Step 11.2

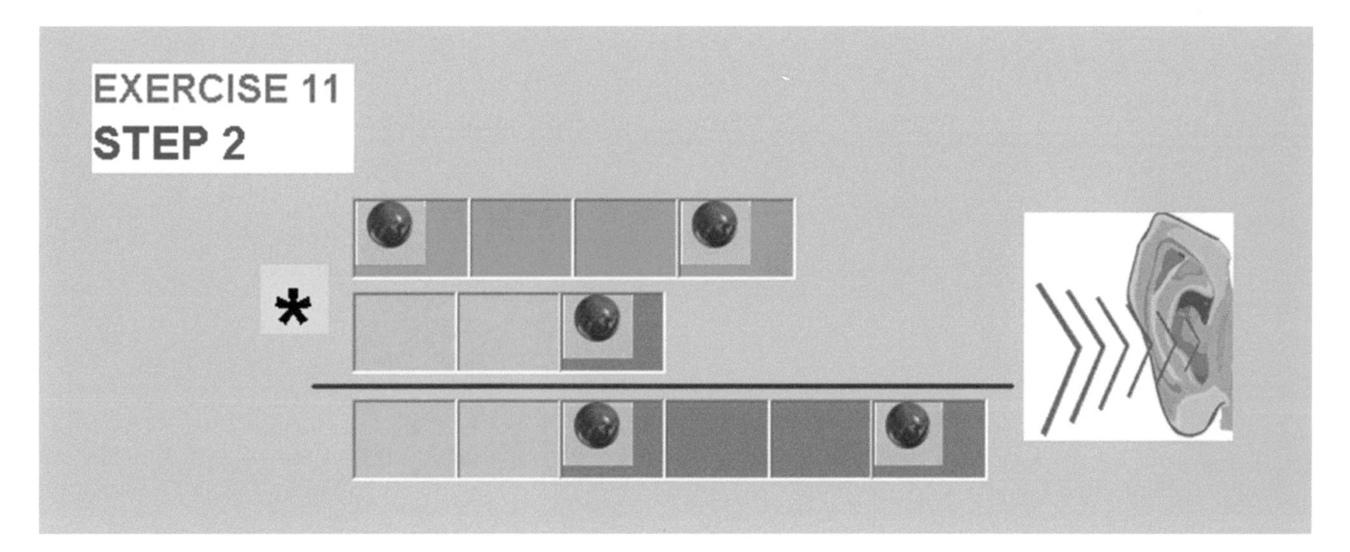

The ***kinesthetic*** representation of the filled up position is announced by clamping of the right hand. An empty position is represented by the clamping of the left hand. The multiplication sign, blick, is represented by both hands extended in front. The equal sign, cling, is represented by crossing both hands. The kinesthetic representation for the three pictures is the clamping of the right, left twice, right once, both hands extended in front, left twice, right once, crossing of both hands, clamping the left twice, right once, left twice and the right hand once.

MATHEMATICS **MULTIPLICATION BY ADDITION, EXERCISE 11, STEP 11.3**

In exercise 11, step 11.3 in ***visual*** display of the third step of multiplication we are adding together the result of the first and second steps. In the first picture, which has value of nine, we see that position one and four are filled up with a red ball. In the second picture, which has the value of thirty six, we see position three and six filled up with the red ball. We want to specify, that between the first and second pictures we see the add sign. Bellow the second picture we see the equal line, which separates the problem from the answer. In the third picture, which is an answer for the third step of multiplication, we see positions one, three, four and six are filled up with a symbol and the answer has a value of forty five.

The ***audio*** representation for three pictures consumes of a **knock, double knock, double knock, knock, tram, double knock, double knock, knock, double knock, double knock, knock, cling, knock, double knock, knock, knock, double knock and knock**.

Step 11.3

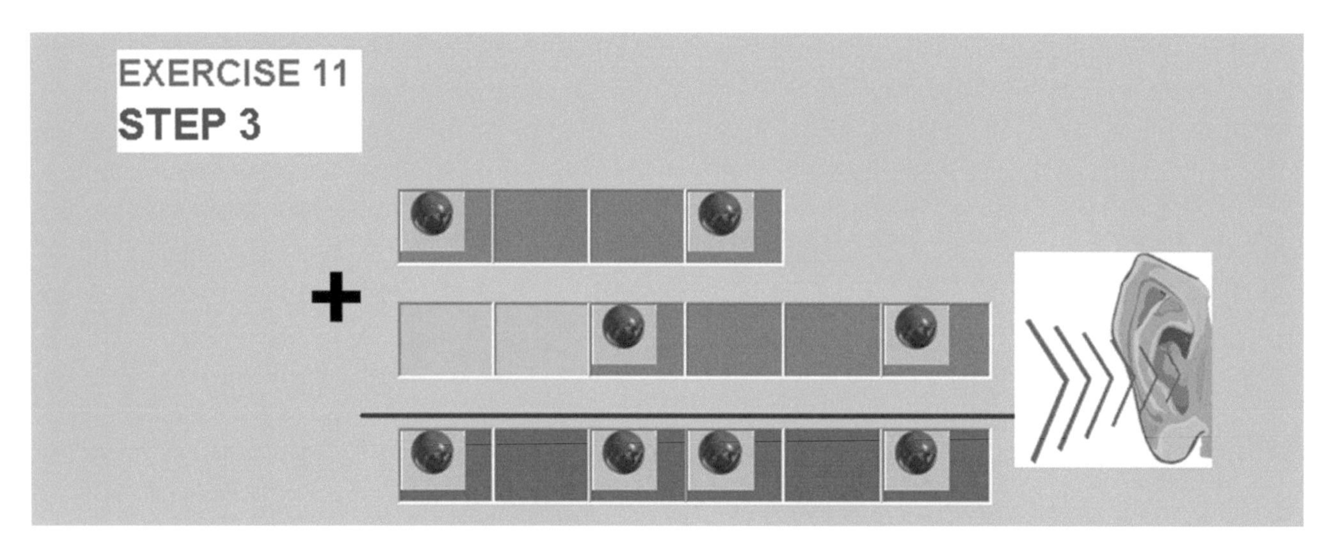

The ***kinesthetic*** representation of the filled up position is announced by clamping of the right hand. An empty position is represented by the clamping of the left hand. The addition sound, tram, is represented by extended left hand in front. The equal sound, cling, is represented by crossing both hands. The kinesthetic representation for the three pictures is the clamping of the right once, left twice, right once, left hand extended in front, clamping the left twice, right, left hand two times, right once, crossing of both hands, right once, left once, right twice, left once and the right hand once.

MULTIPLICATION, EXERCISE 11, STEP 11.4

In Exercise 11 Step 11.4, in a ***visual*** display of the last step of the multiplication, we see one picture of boxes, which is the answer of the problem. Positions two and five are empty, while positions one, two, three, four and six are filled with red balls. The answer has a value of forty five.

The ***audio*** representation for this picture consists of a **knock, double knock, knock, knock, double knock, and knock**.

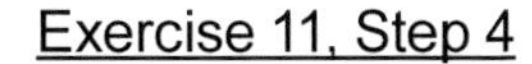

Exercise 11, Step 4

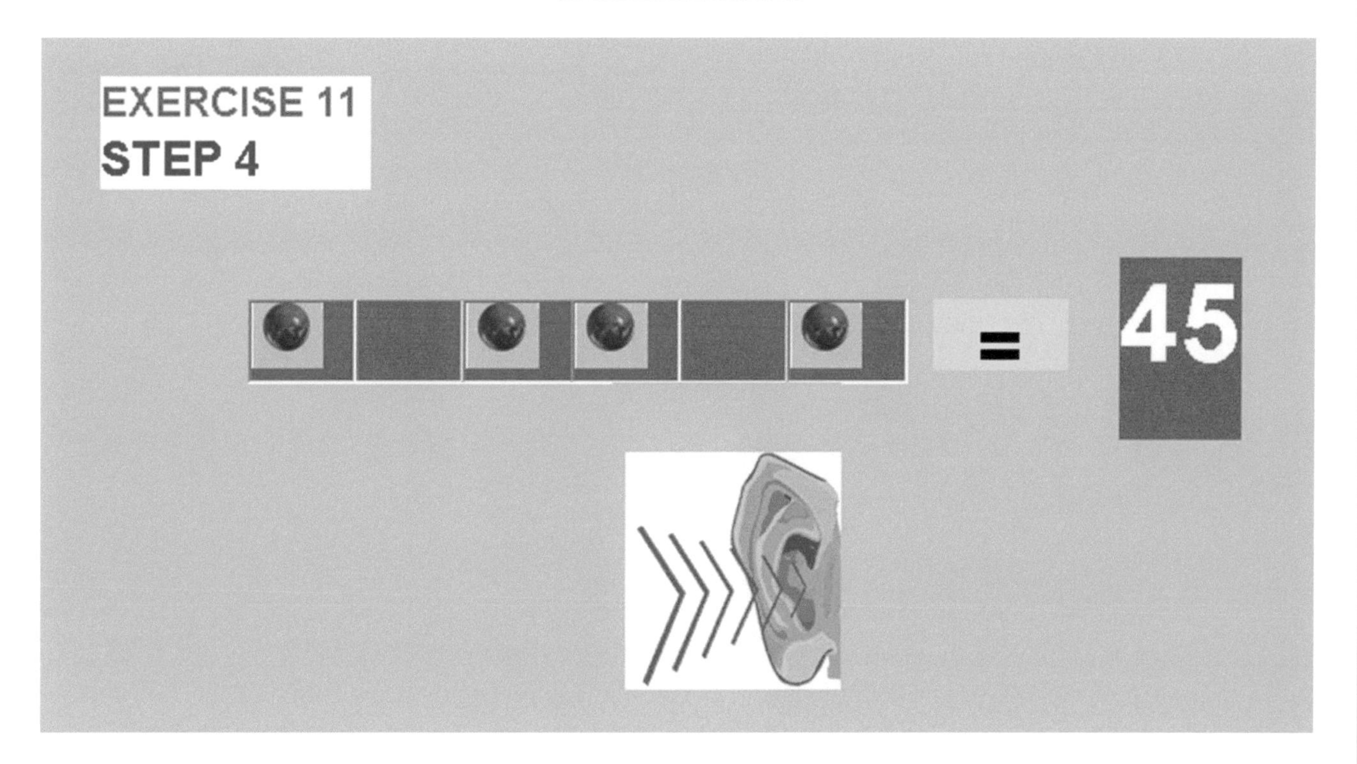

The ***kinesthetic*** representation of a filled position is represented by the clenching of the right hand. An empty position is represented by the clenching of the left hand. The kinesthetic representation for this picture is *the clenching of the right hand, left hand, right hand twice, left hand, and right hand.*

MATHEMATICS MULTIPLICATION, EXERCISE 12

In exercise 12 in ***visual*** display of multiplication we see three pictures of boxes, where in the first picture, which has value of ten; we have the second and fourth position filled up with a red ball. In the second picture, we have position three filled up with a red ball. We want to specify, that between the first and second picture we see the multiplication sign. Bellow the second picture we see the equal line, which separates the problem from the answer. In the third picture, which is the answer for the multiplication and has a value of forty, we see that position one, two, three and five are empty but positions four and six are filled up with a symbol.

The ***audio*** representation for three pictures consumes of a **double knock, knock, double knock, knock, blick, double knock, double knock, knock, cling, double knock, double knock, double knock, knock, double knock and knock.**

Figure 12

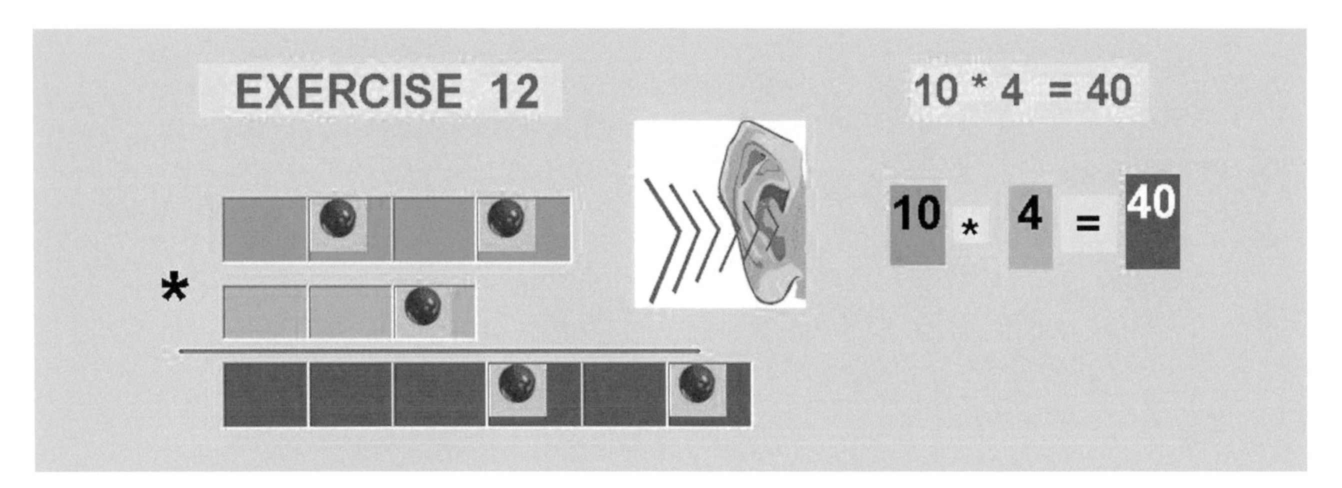

The ***kinesthetic*** representation of the filled up position is announced by clamping of the right hand. An empty position is represented by the clamping of the left hand. The multiplication sign; blick, is represented by both hands extended in front. The equal sign, cling, is represented by crossing both hands. The kinesthetic representation for both pictures are the clamping of the left, right, left, right, extending both hands forward, left hand twice, right once, crossing of both hands, left hand three times, right once, left once and the right hand once.

MULTIPLICATION, EXERCISE 12, STEP 12.1

In exercise 12, step 12.1 in ***visual*** display of multiplication we see three pictures of boxes. We are moving the second picture on two positions to the right. In the first picture, which has value of ten, where the second and fourth positions are filled up with a red ball. In the second picture, which has the value of four, we see that only the third position is filled up with symbol. We want to specify, that between the first and second pictures we see the multiplication sign. Bellow the second picture we see the equal line, which separates the problem from the answer. In the third picture, which is the answer for this step of multiplication; we see that position four and six are filled up with a red ball symbol, which has the value of forty.

The ***audio*** representation for three pictures consumes of a **double knock, knock, double knock, knock, blick, double knock, double knock, knock, cling, double knock, double knock, double knock, knock, double knock and knock**.

Step 12.1

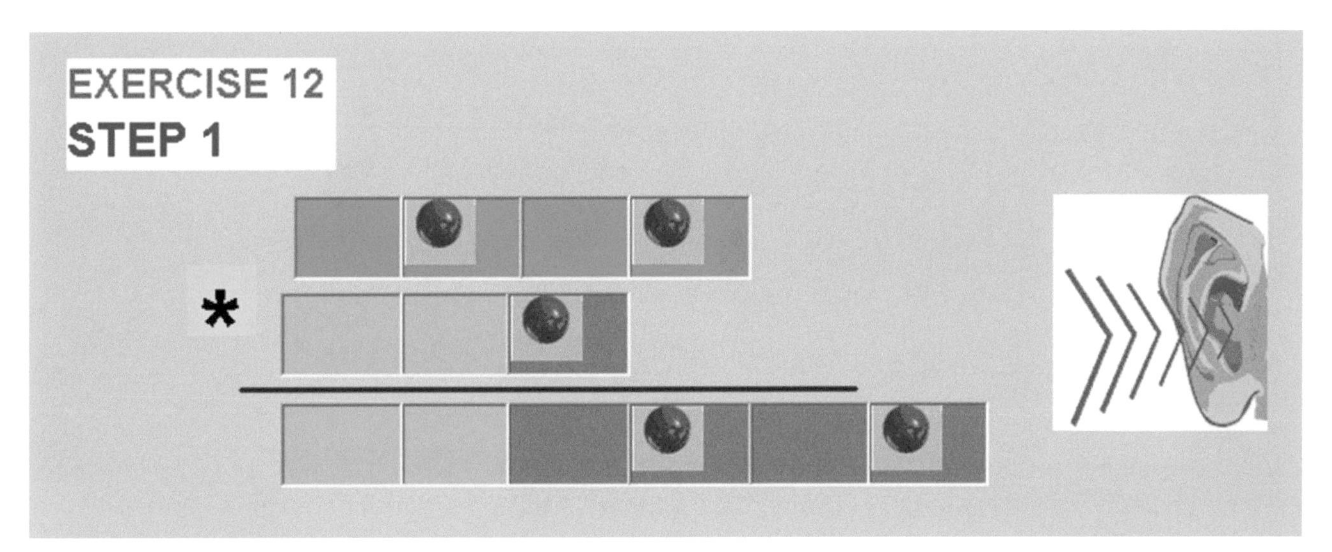

The ***kinesthetic*** representation of the filled up position is announced by clamping of the right hand. An empty position is represented by the clamping of the left hand. The multiplication sound; blick, is represented by both hands extended forward. The equal sound; cling, is represented by crossing hands. The kinesthetic representation for both pictures are the clamping of the left, right, left, right, extending both hands forward, left hand twice, right hand once, crossing of both hands, left hand three times, right, left and the right hand once.

MULTIPLICATION, EXERCISE 12, STEP 12.2

In exercise 12, step 12.2 in ***visual*** display of the last step of multiplication, we see one picture of boxes, which is the answer of the problem. We see that the first, second, third and fifth positions are empty but position four and six are filled up with a red ball and the answer has a value of forty.

The ***audio*** representation for this picture consumes of a **double knock, double knock, double knock, knock, double knock and knock**.

Figure 12.2

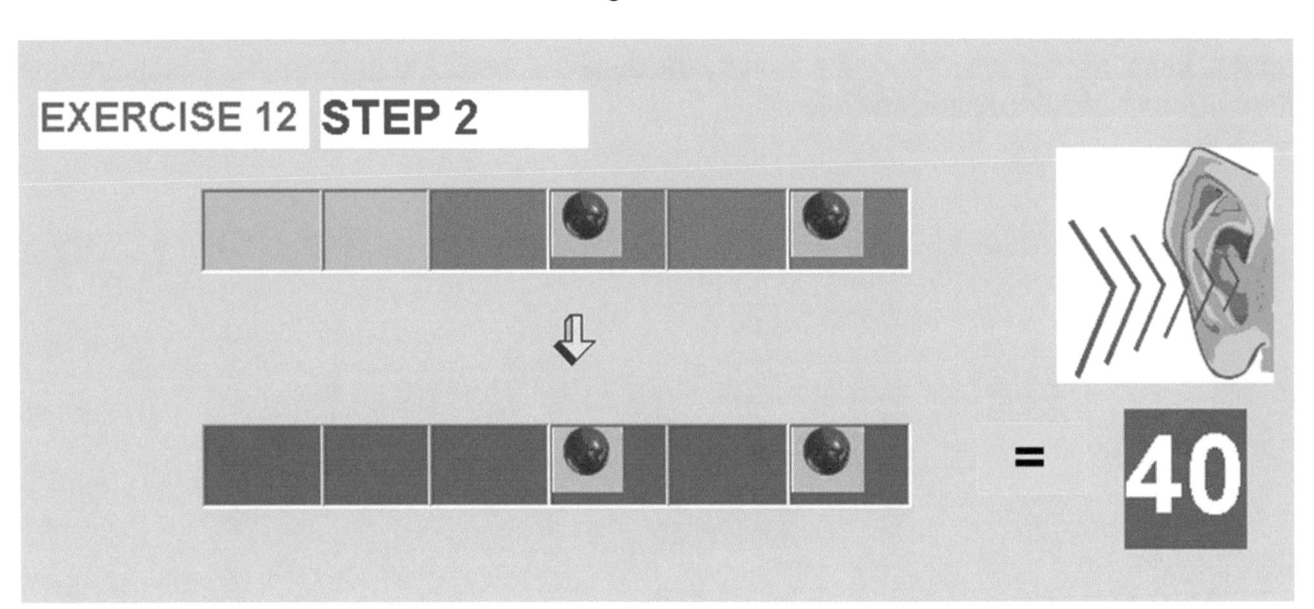

The ***kinesthetic*** representation of the filled up position is announced by clamping of the right hand. An empty position is represented by the clamping of the left hand. The kinesthetic representation for this picture is clamping the left hand three times, right once, left once and the right hand once.

FIND AN ANSWER, EXERCISE 12

In Exercise 12, in a ***visual*** display of multiplication we see two pictures of boxes. At the bottom are the probable answers. In the first picture, positions two and four are filled with red balls. In the second picture, position three is filled with a symbol. The multiplication sign is between the first and second pictures. Below the second picture is the equal line, which separates the problem from the answer. The ***audio*** representation for the pictures consists of a **double knock, knock, double knock, knock, blick, double knock, double knock, knock, and cling.**

From the probable answers for multiplication we need to find the right one. The audio signals are: 1) **double knock, double knock, double knock, knock, double knock and knock**, 2) **double knock, double knock, knock, double knock, double knock and knock**, 3) **double knock, double knock, double knock, double knock, knock, and knock**.

Exercise 12

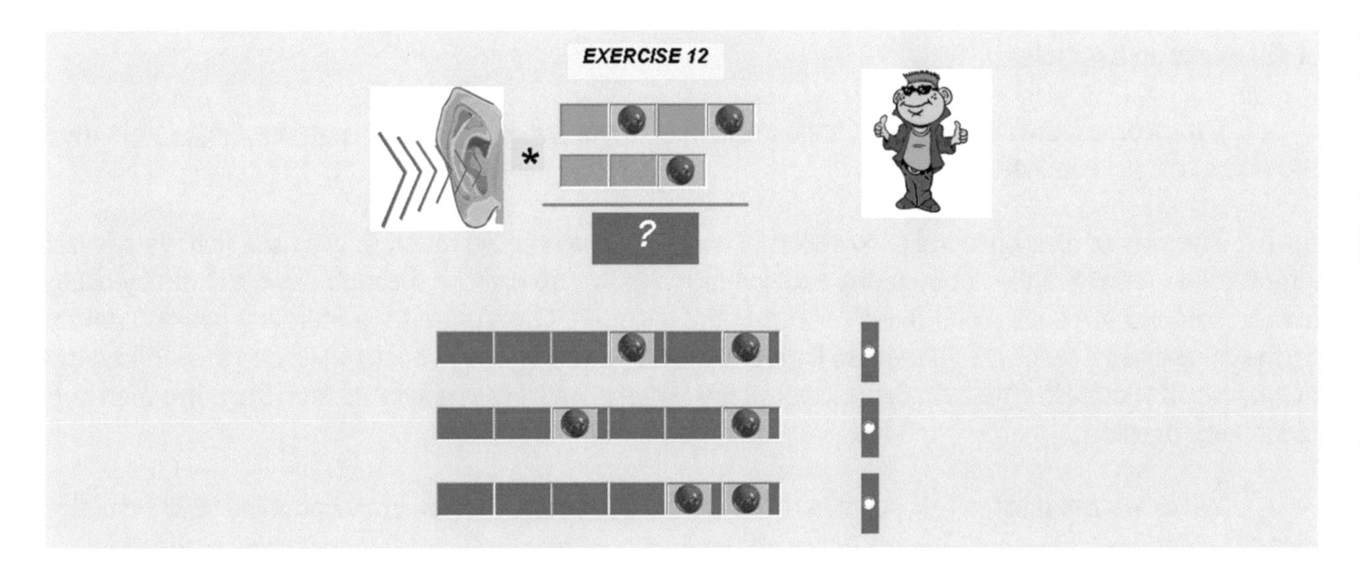

The ***kinesthetic*** representation of a filled position is represented by the clenching of the right hand. An empty position is represented by the clenching of the left hand. The multiplication sound, blick, is represented by both hands extended in front. The equal sound, cling, is represented by crossing both hands. The kinesthetic representation for both pictures and the answer is the clenching of the left hand, right, left, right, both hands extended in front, the clenching of the left twice, right once, crossing of both hands, the clenching of the left hand four times and the right hand twice.

DIVISION

Part I

- Litvin's Code is different from division in the standard decimal numbers, where we are trying to find the answer by multiplying the divisor by an approximate number to get closer to the dividend. In the decimal division after we are multiplying the approximate number by the divisor, we are arriving to an approximate answer, which should be less than dividend. We are subtracting the approximate number from the dividend. We continue in the next step the process of multiplication until we get the results of subtraction which yield to zero.

- In Litvin's code we don't need to multiply approximate number on divisor to get approximate answer. In division by Litvin's Code we are right away subtracting the divisor from the dividend until the result yields to zero. The division by Litvin's Code is a process of subtracting two numbers, which are represented by symbols in different positions. The division of the two numbers required, that the first number was bigger or equal to the second number, because we are not using fractions in present format in the follow up examples.

- If the corresponding positions in the beginning of both dividend and divisor are empty then they are automatically eliminated.

- The way of division is the subtraction of two numbers until we reach 0. We usually have several intermediate results of the subtractions until we reach 0. We do division through Litvin's Code by using subtraction and are subtracting the divisor from the dividend. Our goal is to get the first position empty on the answer as a result of subtraction. It means that subtraction was successful. When the subtraction in the step is successful, then on the answer of the division we fill with a symbol (dot) the corresponding to the step position.

- When we see that the first position of the dividend is empty and the corresponding position of the divisor is filled up with symbol, we are moving the corresponding empty position to answer of the division, and in the mean time eliminating it on the dividend.

- When after the subtraction we have empty first position on the result of the subtraction, we are eliminating the first empty position from the result of subtraction, and filling up with symbol (dot) the corresponding position on the answer of the division. In another example, if after subtraction we have several positions empty on the result of subtraction then we have a different situation. First of all, we fill up with a symbol (dot) the first empty corresponding position on the answer of the division and eliminate the first position on the result of subtraction. When more positions in front on the result of the subtraction are empty, we move the corresponding position without symbol (dot) to corresponding position on the answer of the division and eliminating those positions from the result of the subtraction.

- When subtraction is not successful, and the intermediate result has the first position still filled with a symbol (dot), we are keeping the first position on the result of subtraction and continuing subtract divisor in the next step. We are also keeping empty the corresponding position on the answer of the division. To be more precise, when after the subtraction the first position of the result of intermediate step is filled up with a symbol, then we keep the corresponding position on the answer of the division empty. It could be that the divisor is too big and we need to continue subtracting this number in the next steps until the first position on the result of subtraction become empty. Until the first position result of subtraction is empty, we continuously keep empty corresponding positions of the answer until the result of the subtraction has the first position filled up with a symbol.

Rules of the Division:

- When in the beginning both numbers have equal number of empty positions, then we eliminate them from both numbers.

- When during the division, the subtraction is producing the empty position on the beginning of the result of the subtraction, we eliminate the empty position from the result of subtraction for this step. We place the symbol to the corresponding positions on the answer of this division.

- When the subtraction does not produce empty space on the beginning of the result for the step, then we are leaving empty the corresponding position on the answer for this division, and continuing to subtract divisor.

- When the beginning of the first number is an empty position and the second number does not have empty positions, we are automatically transferring the empty position from the first number to the corresponding position to the answer of the division.

Example 5.1

First number is 12 “0 0 x x”

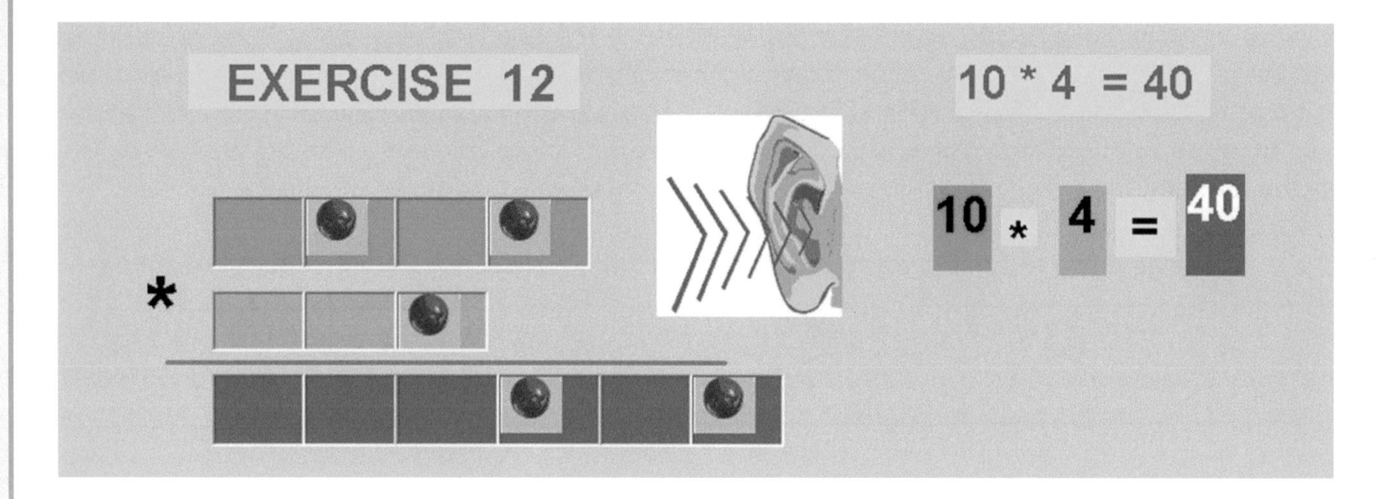

Second number is 2 “0 x”

Answer is 6 “0 x x”

```
      |  6
    2 |12  ---
  -----
    0
```

In example 5.1 we divide two numbers, where dividend has four positions and divider has two positions. The positions three and four on the dividend are filled up with symbols. The divider has second position filled up with a symbol. The dividend has two empty positions in front, and the divisor has one empty position. Before the subtraction we are eliminating the equally empty corresponding position in the beginning of both numbers. The equally empty positions are in the first positions of both numbers and we are eliminating them. The dividend had four boxes and divisor had two boxes. After elimination of the corresponding empty positions, we have three boxes in dividend and one box in divisor. After the first step dividend has an empty position in front but divisor does not, and we are moving the empty position to the answer.

Explanation

Both numbers have the empty space on the first position. We are comparing this to the digit "0" of the regular decimal division. In the regular division, when we are eliminating 0 at the end of the number, the number becomes ten times smaller. When we are eliminating the first position from both numbers and then both numbers become twice lower.

New first number is 6 "0 x x"

New second number is 1 " x "

We know, by dividing any number into 1, we will have the same number we started with. In this example we just want to demonstrate below how the Litvin's Code applies to this division. We also are eliminating the empty position from the first number, and bringing the first empty position to the answer. When the dividend has the first empty position and the first position of the divisor is filled up with a symbol (dot) we are moving the first empty position from dividend to the answer of the division and in mean time are eliminating the empty position on the dividend.

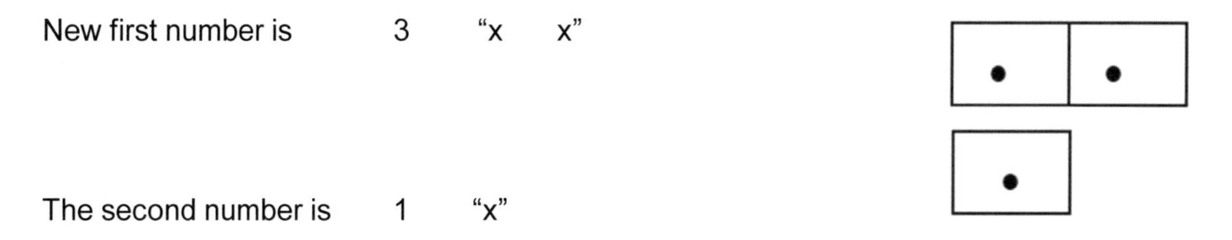

New first number is 3 "x x"

The second number is 1 "x"

Now we have two steps for subtraction. ***In the first step,*** after erasing the corresponding empty positions, we are subtracting the new divisor from the new first number and as a result of subtraction there is an empty space. When we get an empty space as a result of subtraction this indicates that first number was bigger than divisor, and the subtraction was successful. When subtraction is successful, we are filling up with a symbol (dot) the corresponding position on the answer of the division. ***In step two*** we are also subtracting the divisor by the new first number and in the result of subtraction is an empty space. As we mentioned before, when we have an empty space as a result of the subtraction, we fill up with a symbol (dot) the corresponding position on the answer of the division.

In Summary:

- When we are dividing 12 by 2 by using Litvin's Code, we have on the beginning of both numbers one empty position and we are eliminating this position from both numbers and the numbers became 6 and 1.

- Since number 6 has an empty first position and the divisor has no empty spaces, then we move the empty space in the dividend to the answer. When we are doing division using using Litvin's Code and the first number equal 6, we are moving the first empty position to the answer of the division and the numbers become 3 and 1.

- When in division by using Litvin's Code, we are subtracting 1 from 3, and we have an empty position of the result of subtraction. We are filling up with a symbol the next corresponding position on the answer of the division. The numbers became 1 and 1.

- When by using Litvin's Code we are subtracting 1 from 1 and we are getting 0 or empty position on the result of the subtraction, we are filling up with a symbol the next position on the answer of the division.

End.

Example 5.2

First number is 24 "0 0 0 x x"

Second number is 6 "0 x x"

The answer is 4 "0 0 x"

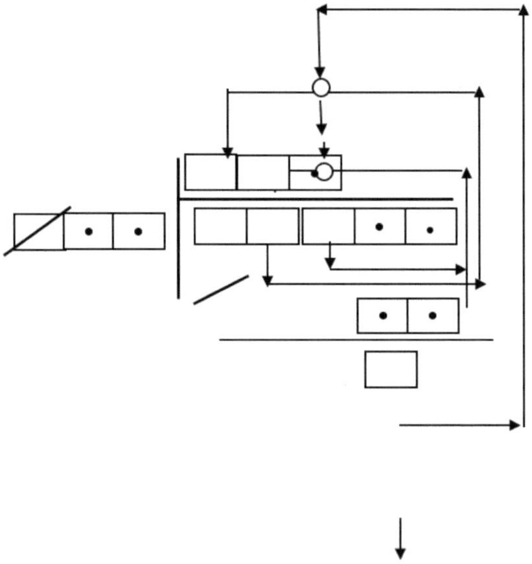

In example 5.2 there is the division of two numbers. The first number has five boxes and the second number has three boxes. The dividend has positions four and five filled up with symbols. The divisor has positions two and three filled up with symbols. The dividend has three empty spaces, and divisor has one. In division by using Litvin"s Code before subtraction we are eliminating or transferring the empty positions.

Explanation

The both numbers have the empty positions in the front. We are eliminating the corresponding empty positions in the front of both numbers. After the eliminating of the first corresponding empty positions, the first number left four boxes and the second number two boxes.

The second number does not have any empty position in the front, but the first number still has in front the two empty positions, which are automatically moved into the answer of the division. We are eliminating the empty position from the first number. Now we have in the first number only two boxes remind. The remained two boxes on first number are equal to the decimal number 3. The second number is equal to 3. Now, if we subtract 3 out 3, we will have zero or an empty position as a result of the subtraction. As we have discussed about the successful subtraction, the next position is filled up with a symbol on the answer of the division.

In Summary:

- When we divide 24 by 6, we eliminate one empty position from both numbers and the numbers become 12 and 3.

- When we divide 12 by 3, we are transferring two empty positions of the number 12 to the answer of the division the numbers will be 3 and 3.

- When we subtract 3 from 3, the result for this step becomes an empty position. We are filling the next position on the answer of the division with a symbol.

End.

Example 5.3

First number is 65 “x 0 0 0 0 0 x”

Second number is 5 "x 0 x"

The answer is 13 "x 0 x x"

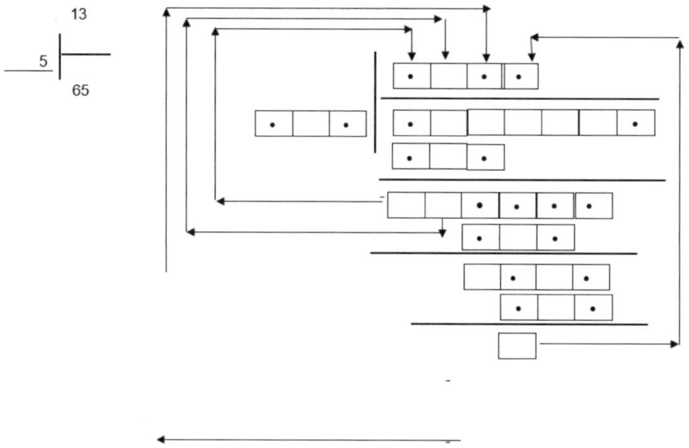

13

5

65

In the Example 5.3 we are dividing two numbers with seven positions on the first number and three positions on the second number. On dividend the positions one and seven are filled up with symbols. On divisor positions one and three filled up with symbols. The five positions of the first number in the middle are empty, and the divisor has an empty position in the middle.

The way to divide numbers in the Litvin's Code is to do subtraction step by step until 0. The division could be done as subtractions. In step one after subtraction we have six boxes as a result of subtraction. The first two boxes on the result of the subtraction are empty. If, after subtraction, the first position on the result of the subtraction is empty, we erase this position and fill with a symbol in the corresponding position on the answer of the division. Now we left with five boxes on the result of subtraction. We still have an empty box in front of the result of the subtraction. We are transferring the next empty position from the result of the subtraction to the answer of the division as an empty position. We left with four positions on the result of the subtraction. In other words, if after the first subtraction the first position of the result of subtraction is empty, we erase the empty position and fill with a symbol the corresponding position of the answer of the division.

Explanation

In the example 5.3, in the first step we subtract the second number from the first. We borrow the last symbol from the last position of the first number to the prior positions. The symbol from the last position does not appear anymore in the calculation because the last number is twice as big than prior, it is the same as having two symbols in the prior position. We leave one symbol in this position, but the other symbol we are borrowing again to the prior position. Again, we have the same situation that this symbol is twice as big as the regular one and is appears as two symbols on the prior position. We are leaving one symbol (dot) in this position and are moving one symbol to the descending position. We are continuously doing this processing of the next symbol to the prior position. In other words the position seven has one symbol, which we borrowed, and then positions four, five and six have one symbol, but position three has two symbols. In our example the third position on the divisor has a symbol two. In the meantime, on the third position of the dividend we have two borrowed symbols from the next position. We are subtracting one symbol from the first number and then are moving the second symbol to the result of the subtraction. In step two the new first number, which is the result of the subtraction, becomes sixty. We see that the two first positions on the result of the subtraction are empty. If the first position of the results of subtraction is empty, it means that the subtraction was successful. We are eliminating the one empty position from the result of the subtraction and are moving the symbol to the first position of the answer. After we are eliminating the second empty position from the result of the subtraction, and placing the empty space on a corresponding position on the answer of the division. The new first number became four times smaller. We are seeing that number 60 has two empty positions in the front.

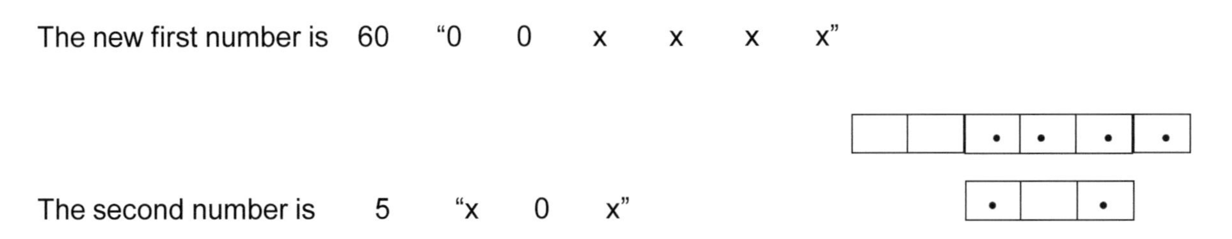

In step three we see that after eliminating the first two positions the number becomes 15. We are subtracting from the new first number the second number. We are eliminating the empty space on the result of the subtraction. We are moving the symbol (dot) to the corresponding position on the answer of the division. When we are subtracting 5 from 15 the result is 10.

The new first number is 15 “x x x x”

The second number is 5 “x 0 x”

The result for step is 10 “0 x 0 x”

In step four we are repeating the same sequence. We have one empty position in front of the results of the subtraction. We are transferring this position to the answer of the division, which is filled up with a symbol (dot). This is showing that the result of the subtraction was successful.

The new first number is 10 “0 x 0 x”

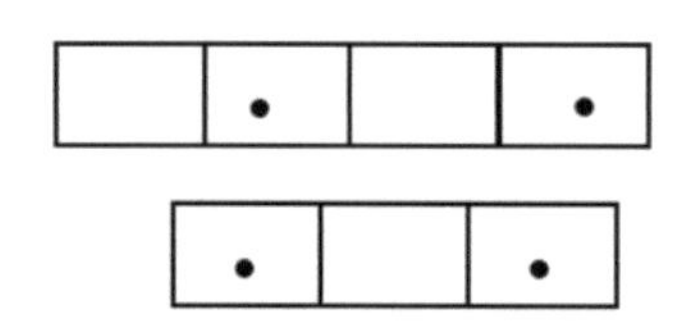

The second number is 5 “x 0 x”

As a result of eliminating the first position of number ten, we have five as our first number.

In another words, during the subtraction we are eliminating the empty position from the result of the subtraction. We are moving the symbol to the corresponding position on the answer of the division.

The new first number is 5 “x 0 x”

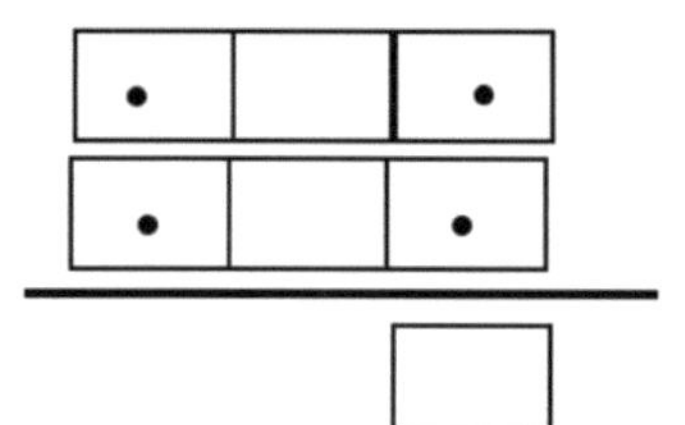

The second number is 5 “x 0 x”

The result for this step is 0 “0”

Once again we are filling with a symbol (dot) the corresponding position on the answer of the division.

The final answer is 13 “x 0 x x”

•		•	•

In Summary:

- When we do division, we are subtracting two numbers.

- When we do division and the beginning positions (first, second, ext.) of the first number are empty, we are transferring the positions to the corresponding positions on the answer of the division. After the subtraction, the first position on the result of subtraction is empty, then we are erasing this position and filling up with the symbol the corresponding position on the answer of the division, and the second position we are just transferring as empty.

- When we are dividing 65 into 5, in the first step we are subtracting 5 from 65 and the answer is 60. The result of the subtraction for the first step is the number sixty, and is represented by the Litvin’s Code. The first position of the first number is empty. We are eliminating it from the result of subtraction and are filling with a symbol (dot), the corresponding position on the answer of the division. The next position of the result of subtraction has the empty space too. We are eliminating the next empty position from the result of the subtraction. We are transferring the second empty position to the answer of the division as empty. After the transfer the number 60 became 15.

- When we are subtracting 5 from 15, the result of subtraction for this step is 10 and has the empty space in the first position. We are eliminating the first position from the result of subtraction for this step. We are simultaneously filling up with a symbol (dot) the corresponding position on the answer of the division. In the meantime, after the transfer the 10 became 5.

- When we are subtracting 5 from 5 the result of the subtraction is 0 or empty space. We are filling up with a symbol the corresponding position on the answer of the division. The answer of the division is 13.

End.

Example 5.4

Dividend is 21 “x 0 x 0 x“

Divisor is 3 “x x”

-

The final answer is 7 “x x x”

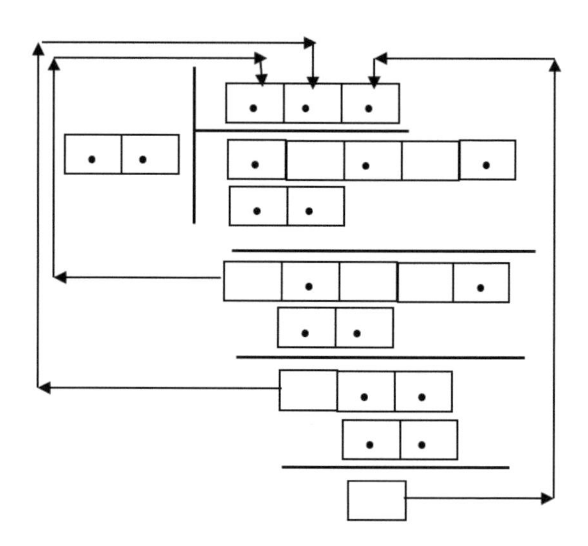

In example 5.4 we are dividing the two numbers with five positions on the first number and two positions on the second number. The dividend has positions one three and five filled with the symbols. The divisor has positions one and two filled with symbols. The positions two and four on dividend are empty. This division we do step by step. It requires three steps to complete this calculation and get the final answer:

Explanation

First step for Example 5.4

Subtract from dividend 21 “x 0 x 0 x”

-

divisor 3 ”x x 0”

first result 18 “0 x 0 0 x”

Because the first position of the result of subtraction is empty, we are filling with a symbol the first position of the answer of the division. In other words, we are erasing the first position from the result of subtraction and are filling with a symbol for the answer of the division.

Result of subtraction

After the erasing of the first position, the result is

9 “x 0 0 x”

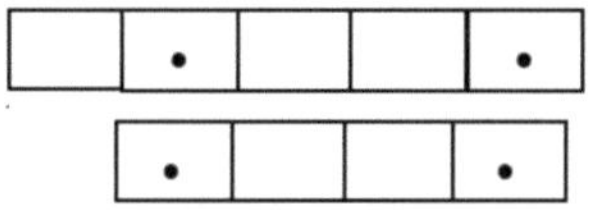

When we are erasing the empty position of the result of subtraction, we are filling with a symbol the corresponding position on the answer of the division.

Answer for the results of division after the subtraction for step one is

1 “x 0“

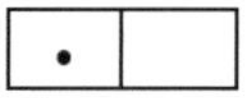

Second step

Subtract from the new first number

9 “x 0 0 x”

divisor - 3 “x x”

second result 6 “0 x x”

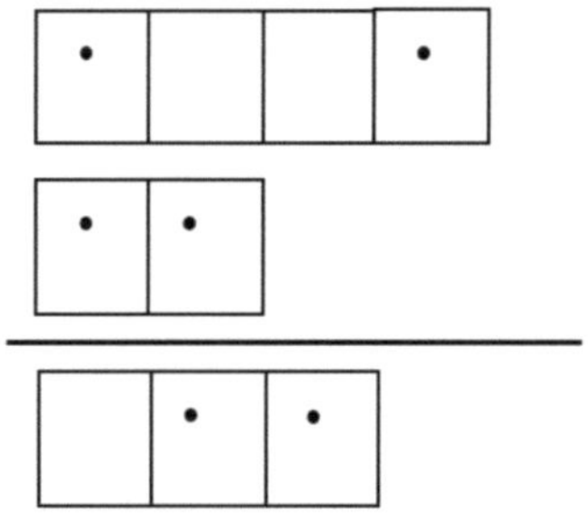

After the subtraction 3 from 9 the answer is 6. The first position on the answer of the subtraction is empty. We are eliminating the first position of the result of the subtraction and filling the answer of the division with a symbol (dot) in corresponding position. After eliminating the empty position the number 6 become number 3. If the first position is empty on the result of the subtraction, then the second position of the answer of the division is filled up with a symbol. The first position of the second result of subtraction is erased.

Erase the first position and the result is

3 "x x"

When we are erasing the empty position of the result of subtraction, we are filling the results of the division with a symbol.

The answer for the results of division after the subtraction for step two is

3 "x x"

Third step

Subtract from the new first number

3 "x x"

-

divisor 3 "x x"

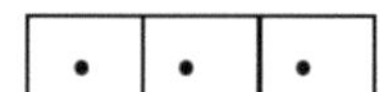

The result of third subtraction is

0 "–"

When the result of the subtraction is zero, we are filling up with a symbol (dot) corresponding to the position of the answer of the division.

answer for results of the division is

7 "x x x"

When we look at the answer of the division, we could see that three subtractions were successful, and we have three positions filled up with a symbol (dot) of the result of the division.

In Summary:

- Whenever we get the first empty position in the result of the subtraction, then the corresponding position on the answer of the division is filled up with a symbol. The first position of the result of the step is erased.

- When there are no more steps for subtraction and the result of subtraction is 0, then it is the last step and we have arrived at the final answer.

- When we are dividing 21 by 3, we are first subtracting 3 from 21 and result is 18. In the number 18 expressed in Litvin's Code, the first position is empty. We are eliminating this position from the result of the subtraction and filling up with a symbol, (dot) the corresponding position of the answer. After eliminating the first position of the result of subtraction, which is number 18, it becomes 9.

- When we are subtracting 3 from 9, the result is 6, and the first position of the number 6, which is expressed in Litvis Code, is empty. We are eliminating this position from the result of the subtraction and we are filling up with a symbol on the second position of the answer of the division. The number 6 became 3. We are subtracting 3 from 3. The result of subtraction is an empty space or 0. We are filling up with a symbol the third position of the answer of the division. The answer of the division is 7.

End.

Example 5.5

Dividend is 65 "x 0 0 0 0 0 x"

Divisor is 13 "x 0 x x"

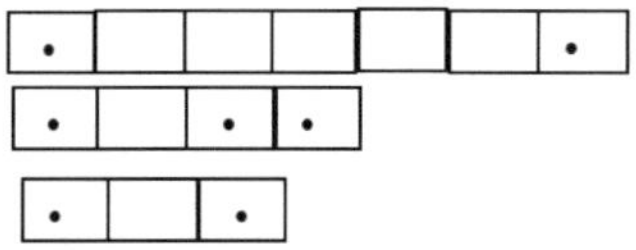

•						•

•		•	•

•		•

The final answer is 5 "x 0 x"

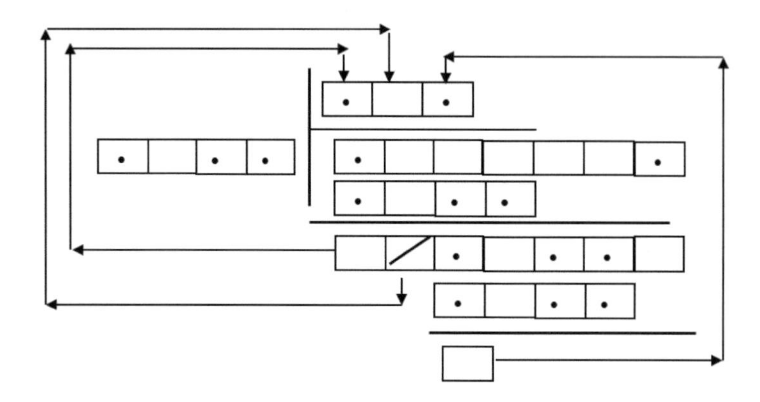

In example 5.5 we are dividing two numbers. Dividend is equal to 65. It means that the first and last position in this number is occupied with symbols. The last position is the 7th position. Between first and last position we have five empty positions. To do division it means to subtract from dividend the divisor until we get the result of zero. When we have number 65 as a dividend and 13 as a divisor we are borrowing a symbol from the last position to the prior positions. This division requires two steps of subtraction to complete this calculation and to get the final answer.

Explanation

First step for example 5.5

Subtract from dividend 65 "x 0 0 0 0 0 x"

Divisor 13 "x 0 x x" -

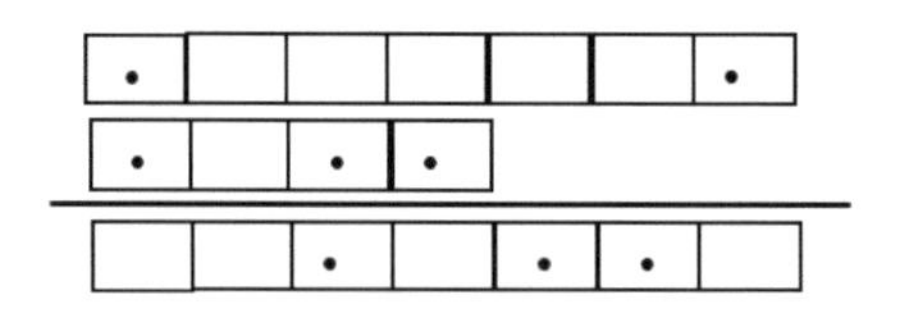

first result of subtraction is

52 "0 x 0 0 x x"

When we subtract 13 from 65, the result of the subtraction is 52. We see that the first position on the subtraction is number 52 and its empty. It means the subtraction was successful. We are eliminating the first position of number 52 and number 52 become number 26. We also see that number 26 has an empty position in the front. We are moving this empty position to the corresponding position of the answer of the division.

After the move of the first position the number 52

Becomes half and the result is

26 "0 x 0 x x"

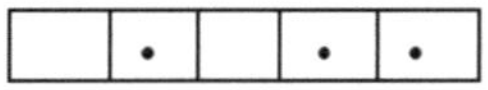

After the step the answer of division has one filled up position and one empty.

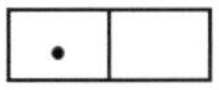

In our first subtraction in this division the result is 52. The number 52 has two empty positions in front. The first empty position indicates successful subtraction, which transferred the first position of the results of the subtraction as filled up with a symbol (dot) to the corresponding position on the answer of the division. When we erase the first position in number 52 it yields to 26. The number 26 also has the first position empty. This empty position is transferred to the corresponding position of the answer of the division as empty. Now, the answer of the division has one filled up position and one empty. After erasing the first position from 26 the result is 13.

Second step for Example 5.5

subtract from third step result

13 "x 0 x x"

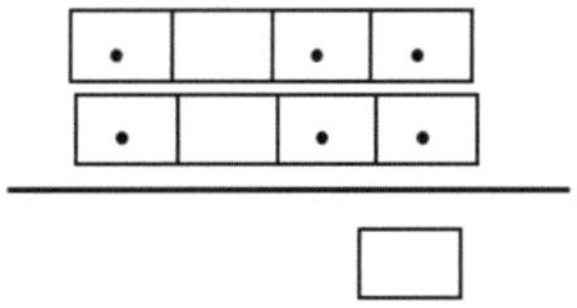

divisor 13 "x 0 x x"

second step result 0 "0"

Now we subtract the divisor (13) from the result (13). The result of the second step is an empty space after the subtraction. When we have an empty space as a result of the subtraction for a final result, we are filling with a symbol (dot) corresponding position on the answer of the division.

Final answer is 5 "x 0 x"

The final answer shows that two steps had successful subtraction and two corresponding positions on the answer of the division are filled up with a symbol (dot). We see that in the middle of the final answer of the division is an empty position. It means that after the first step, as a result of subtraction we had two empty positions, one of which was transferred as a filled corresponding position in the answer of the division with a symbol (dot). The other, which was left, will transfer as an empty position to the answer of the division.

In Summary:

- When we are dividing 65 by 13, we are subtracting from the first number equal to 65 from the second number equal to13.

- When we have the result of subtraction equal to 52, expressed in Litvin's Code, the two front positions are empty.

- When we got the result of 52, we erased the first empty position and the number became 26. This fills up the first position of the answer of the division with a symbol (dot).

- When we see the first position on the number 26 is empty, which is not the result of subtraction, we transfer it as empty to the corresponding position on the answer of the division. The result after the transfer becomes 13.

- When we are subtracting 13 from 13 the answer is 0. When we have the result of 0 for the subtraction, then it was successful, and then we are filling up with a symbol (dot) on the third position of the answer.

Example 5.6

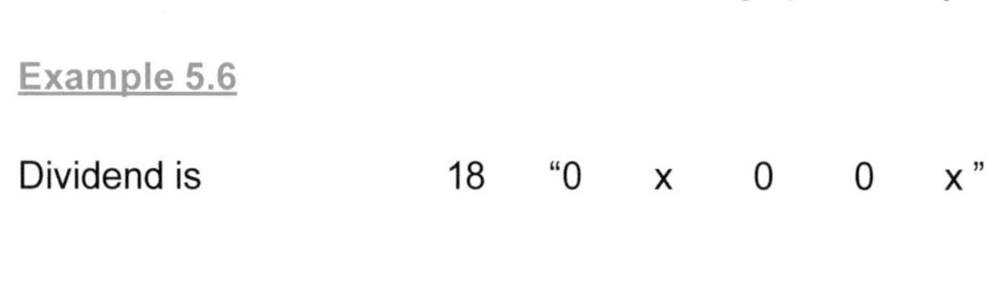

Dividend is 18 "0 x 0 0 x"

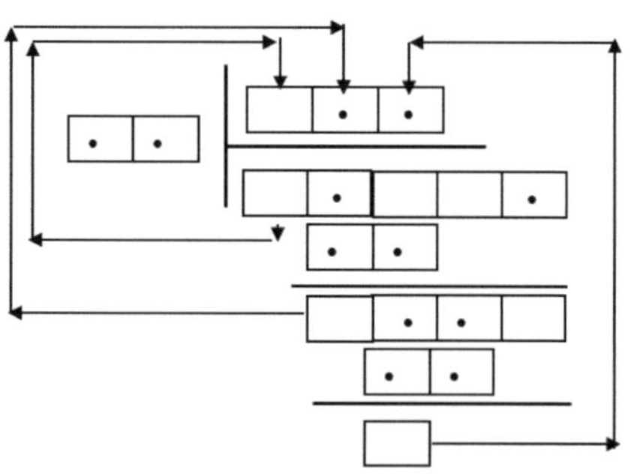

Divisor is 3 "x x"

The final answer for it is 6 “0 x x“

In example 5.6 we are dividing two numbers. Dividend is equal to 18. The first position on the number 18 is empty. The second position is filled up with a symbol (dot) and the last position is the 5th position that is also filled up with a symbol (dot). Between the second and last position we have two empty positions. To do division it means to subtract the divisor from the dividend until we get the result of zero. The first position of the number 18 is empty and it is not the result of subtraction, then we transfer this position as an empty position to the answer of the division. The number 18 after the transfer of the first position become 9. When we have number 9 as a dividend and 3 as a divisor we are borrowing a symbol from the last position to the prior positions. This division required three steps to achieve this calculation or final answer:

Explanation

First step for Example 5.6

Because the first position on the dividend is empty and the first position of the divisor is filled up with a symbol, we do not do any subtractions. The first position of the dividend is empty then the first position of the dividend is erased from the dividend.

Erase first position of dividend

18 “0 x 0 0 x”

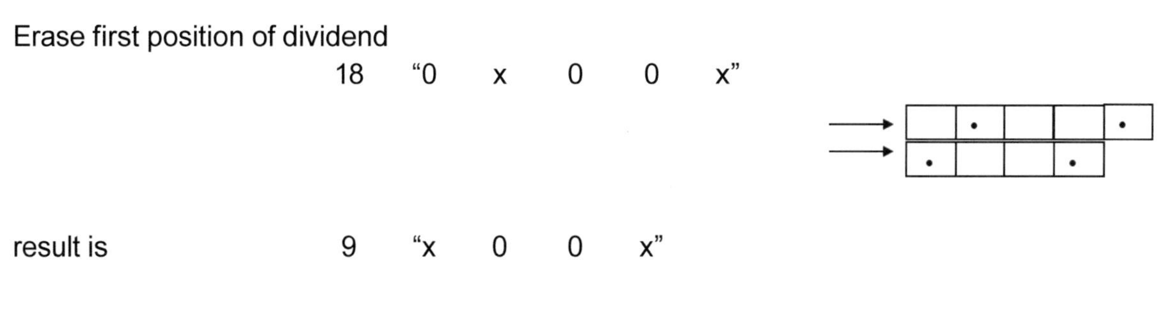

result is 9 “x 0 0 x”

For the first step we are erasing the first empty position of the dividend and then

result for step one is 9 “x 0 0 x”

answer for step one is 0 “0”

Second step for Example 5.6

When we are subtracting 9 from 3 the answer for the second step is 6. When we see number 6 using Litvin's Code, we understand that the first position of this number is empty. It means that the subtractions were successful, it means that the answer for the division will have the second position filled up with a symbol (dot). This is a reverse operation. We are erasing the first position of the result of subtraction and filling up with a symbol (dot) to the corresponding position on the answer to the division.

Subtract from the new first result

9 “x 0 0 x”

\- 3 “x x”

result is 6 “0 x x”

erase the first position and the result for the second step is

3 “x x”

Answer for step two 1 “x”

Third step for Example 5.6

When we subtract from the second step

3 “x x”

Divisor 3 “x x” -

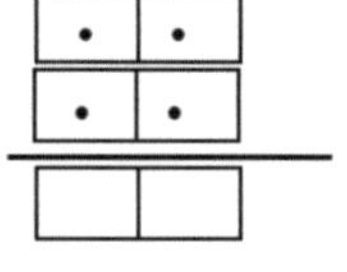

third result 0 “0”

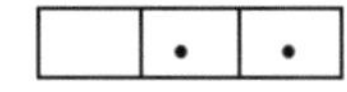

answer for the step three 6 “0 x x”

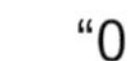

In Summary:

- When we are dividing 18 by 3, we see that number 18 has the first position empty but the number 3 has the first position filled up with a symbol. We then eliminate the first position from the number 18 and we move the first empty position to the answer of the division. The number 18 became number 9. We are subtracting 3 from 9 and the answer is 6. The number 6 has the first position – empty. It means that subtraction was successful and we need to eliminate this position.

- When we are eliminating the empty position from the result of the subtraction and are filling up the second position on the answer with a symbol, the number six became three. We are subtracting 3 from 3. The result of subtraction is an empty space or 0. We are filling up with the third position of the answer with a symbol. The result of division is 6.

End.

Example 5.7

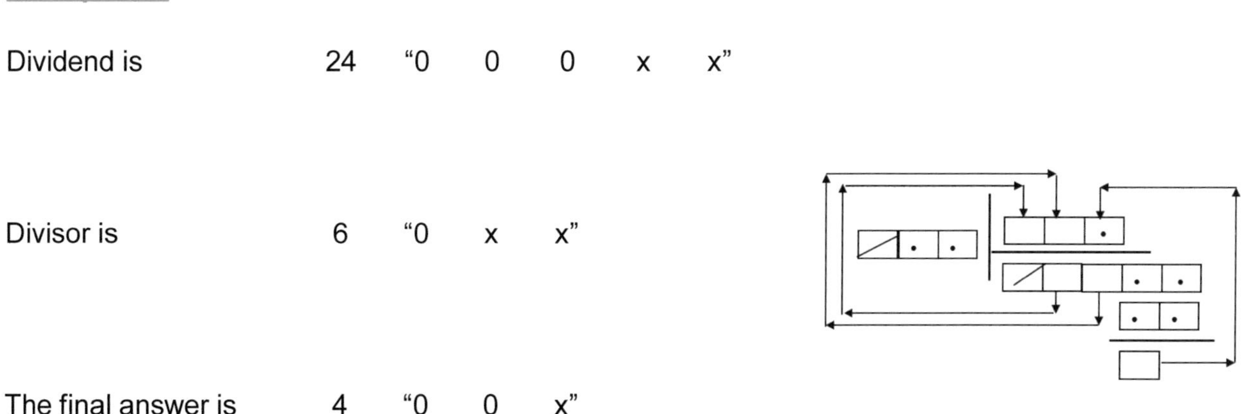

In example 5.7 we are dividing two numbers. The dividend is equal to 24. The first three positions on the number 24 are empty. The last two positions are filled up with a symbol (dot) and the last positions are the 4th & 5th position. The divisor is equal to 6 and has one empty position in front. Positions two and three are filled up with a symbol (dot). When the first position on the dividend and divisor is empty, we are eliminating them without adding any positions to the answer of the division. Now the divisor has only two positions filled with a symbol (dot) and this positions are one and two which is equivalent to number three. After eliminating the first position of the dividend, on the dividend we still have two empty positions in the front. Now we must move these two empty positions from the dividend to the answer of the division. After the elimination of the first position in the dividend, the number is equivalent to 12. When we are eliminating the second position, the number becomes six. Then we eliminate the third position by transferring to the answer of the division as an empty position as well. After eliminating these three positions on the dividend we are left with two filled up positions which is equal to the number 3. To do division it means to subtract the divisor from the dividend until we are getting the result of zero. This division requires two steps to achieve this calculation or final answer.

Explanation

First step for Example 5.7

Erase first, second and third positions on the dividend and first position of the divisor.

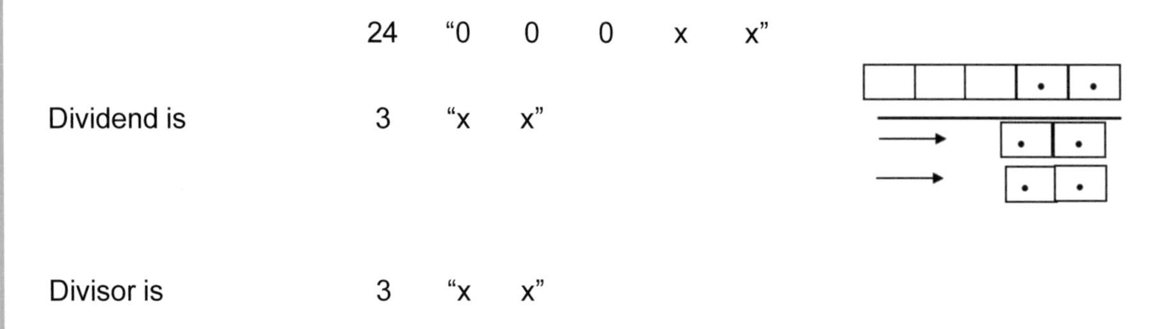

Because the first, second and third positions on the dividend are empty and first position of the divisor is empty too, we do not do any subtractions. We eliminate the empty positions on the dividend and divisor. We are transferring the second and third empty positions of the dividend to the answer of division as empty. The first, second, third positions of the dividend are erased from the dividend.

Second step for Example 5.7

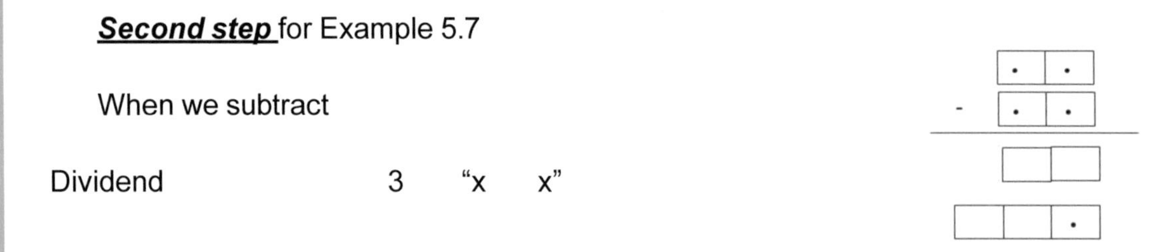

Divisor 3 “x x”

The result of subtraction is
0 “0 0”

answer for the division 4 “0 0 x”

end

Because the first position of the result of the subtraction in step two is empty then the third position in the answer of the division will be filled up with a symbol (dot). Because after the subtraction the last step’s result is empty, then the answer of the division in the third position is filled up with a symbol (dot).

In Summary:

- When we are dividing 24 by 6, we see that the first position on the number 24 and the number 6 is empty. In Litvin’s Code we automatically eliminate the first empty positions of the dividend and divisor, and we are left with two empty positions on the dividend and the divisor has no empty positions left to eliminate.. The number 24 and 6 became 12 and 3. We see that number 12 has the two empty positions in front but number 3 has not.

- When we are eliminating those empty positions from the dividend and are transferring two empty positions to the answer, the number 12 becomes 3. We are subtracting 3 from 3, which is equal to zero. Division was successful and result zero is transferred to the answer of the division as a symbol (dot). The result of the division is equal to 4.

End.

MATHEMATICS

DIVISION, EXERCISE 1

In Exercise 1, in a ***visual*** display of division we see three pictures of boxes. All three pictures have the first position filled with a red ball while the second position is empty. The division sign is between the first and second pictures. Over the second picture is the equal line, which separates the problem from the answer. In the third picture, which is the answer for the division and has a value of one, the first position is filled with a symbol and second is empty.

The ***audio*** representation for the pictures consists of a **knock, double knock, double click, knock, double knock, cling, knock, and double knock.**

Exercise 1

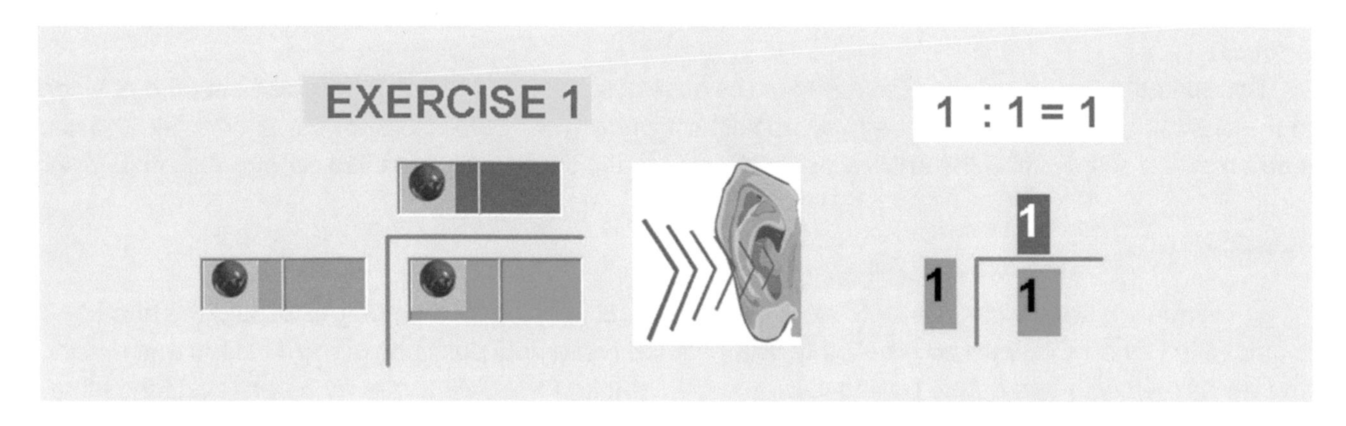

The ***kinesthetic*** representation of a filled up position is represented by the clenching of the right hand. The clenching of the left hand represents an empty position. Both hands extended up represent the division sign, double click. The equal sign, cling, is represented by crossing both hands. The kinesthetic representation for both pictures is the clenching of the right hand, left hand, extending both hands up, right hand, left hand, crossing of hands, right hand, and the left hand.

DIVISION, EXERCISE 1, STEP 1.1

In Exercise 1 Steps 1.1, in a ***visual*** display of division we see three pictures of boxes - the dividend, the divisor and the answer. Between the first and second pictures is the division sign. Over the dividend is the equal line, which separates the problem from the answer. On top of the dividend is the third picture, which is the answer of the division. The ***audio*** representation for the pictures consists of a **knock, double click, knock, double knock, cling, knock.**

In the next part of Step 1.1, we have a **visual** display of how we arrive at the answer. We subtract picture four from picture two, which is the dividend. The subtraction sign is between the second and fourth pictures. Below picture four is the equal sign which separates the problem from the answer. Picture five is the result of the subtraction. When the result of the subtraction is an empty position, then a filled position is transferred to the answer of the division. The ***audio*** representation for the subtraction consists of a **knock, double tram, knock, cling, double knock.**

Exercise 1

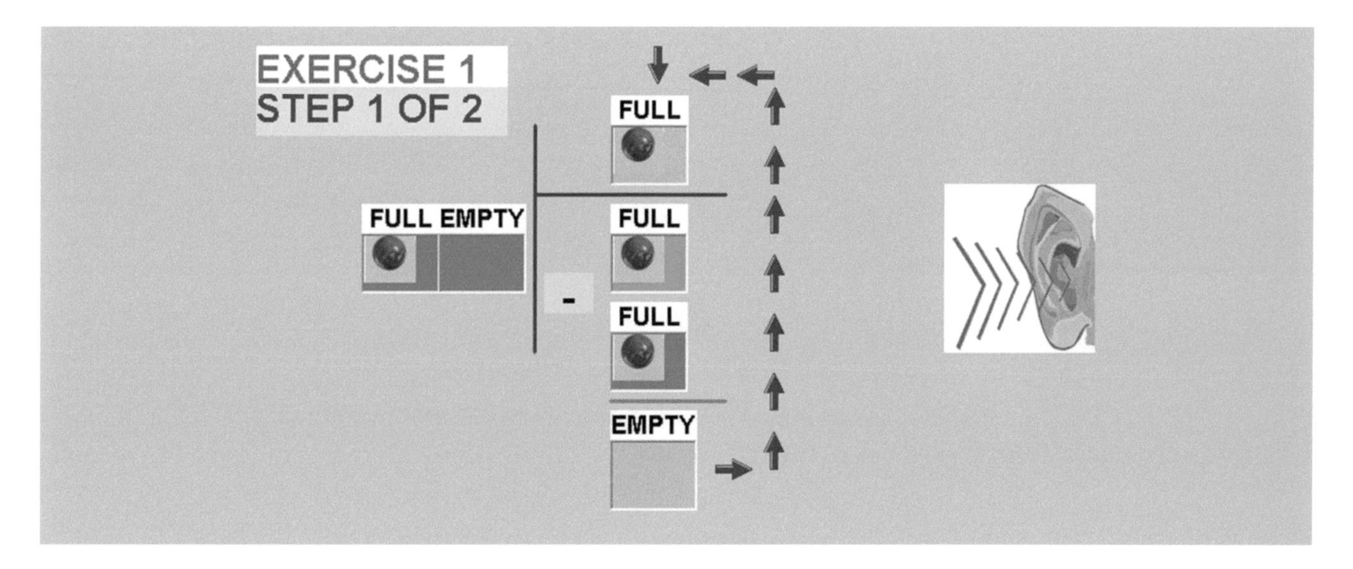

The ***kinesthetic*** representation of a filled up position is represented by the clenching of the right hand. The clenching of the left hand represents an empty position. The division sound, double click, is represented by both hands extended up. The equal sound, cling, is represented by crossing both hands. The kinesthetic representation for the pictures is the clenching of the right hand, both hands extended up, right hand, left hand, crossing both hands, and the right hand.

DIVISION, EXERCISE 1, STEP 1.2

In Exercise 1 Step 1.2, in a ***visual*** display of the last step of the division there are two pictures of boxes. In both pictures the first position is filled with a red ball while the second position is empty. The colors and the placement of the boxes and the balls are different in the pictures. The form, shape, and color of the symbols do no change the mathematical value of the pictures, which is one.

The ***audio*** representation for the two pictures consists of a **knock, double knock, and cling.**

Exercise 2

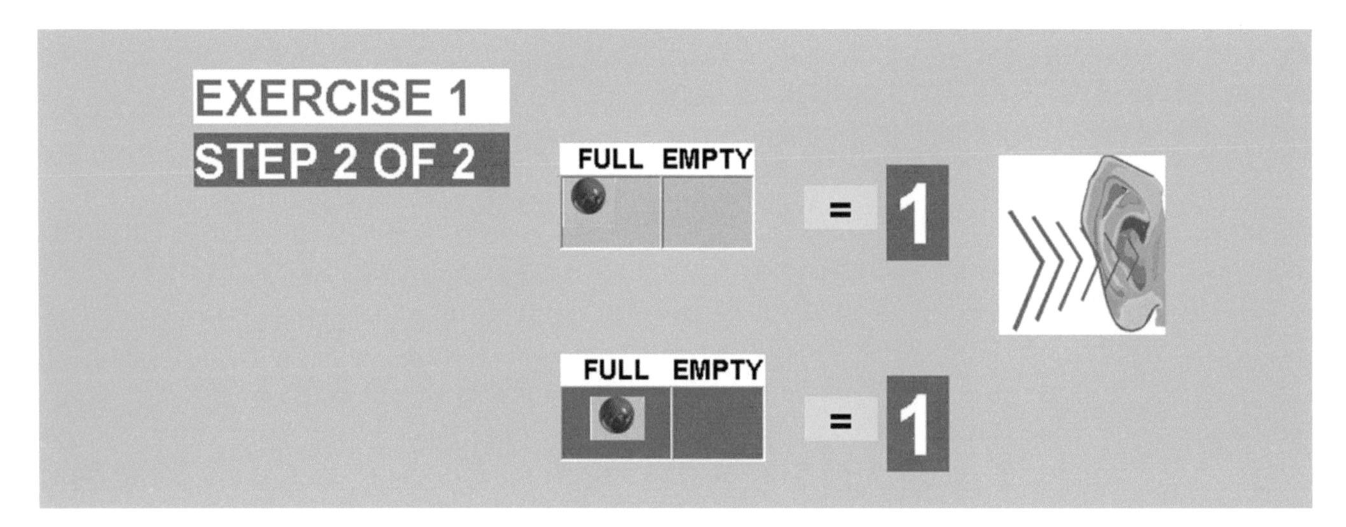

The ***kinesthetic*** representation of a filled up position is represented by the clenching of the right hand. The clenching of the left hand represents an empty position. The equal sound, cling, is represented by crossing the hands. The kinesthetic representation for the two pictures is the clenching of the right hand, left hand, crossing of hands, right hand, and left hand.

DIVISION, EXERCISE 1, ANSWER 1

In Exercise 1, in a ***visual*** display of division we see two pictures of boxes and the division sign in-between. In the first picture, the first position is filled with a red ball and the second is empty. In the second picture, the first position is also filled with a ball and the second is empty. The division sign is between the first and second pictures. On top of the second picture is the equal line, which separates the problem from the answer. The ***audio*** representation for the division and equal signs consists of a **knock, double knock, double click, knock, double knock, and cling.**

At the bottom we see three pictures with probable answers. From the probable answers for the division we need to find the right one. The audio signals are: 1) **knock, and double knock,** 2) **double knock and knock**, 3) **knock and knock**.

Exercise 1

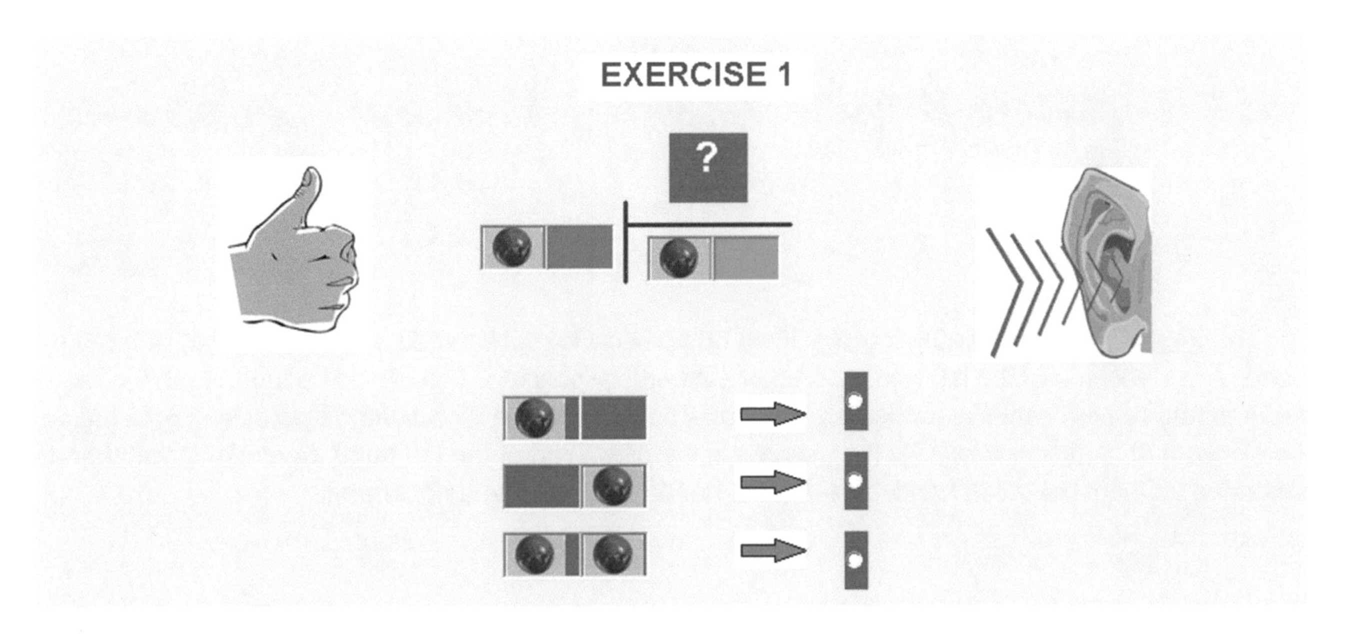

The ***kinesthetic*** representation of a filled up position is represented by the clenching of the right hand. An empty position is represented by the clenching of the left hand. The division sound, double click, is represented by both hands extended up. The equal sound, cling, is represented by crossing both hands. The kinesthetic representation for both pictures and the answer is the clenching of the right, left, both hands extended up, the clenching of the right, left, crossing the hands, right hand and the left hand.

MATHEMATICS DIVISION, EXERCISE 2

In Exercise 2, in a ***visual*** display of division we see three pictures of boxes - the dividend, the divisor and the answer. Between the first and second pictures is the division sign. Over the dividend is the equal line, which separates the problem from the answer. On top of the dividend is the third picture, which is the answer of the division. The ***audio*** representation for the pictures consists of a **double knock, knock, double click, knock, double knock, cling, double knock, and knock.**

Exercise 2

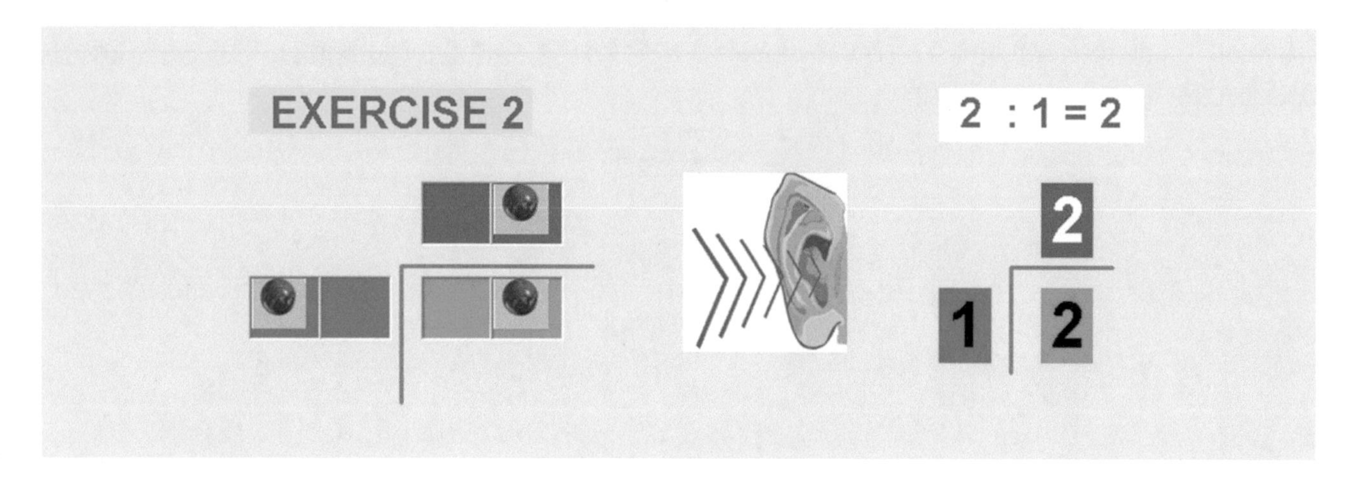

The ***kinesthetic*** representation of a filled up position is represented by the clenching of the right hand. The clenching of the left hand represents an empty position. The division sound, double click, is represented by both hands extended up. The equal sound, cling, is represented by crossing both hands. The kinesthetic representation for the pictures is the clenching of the left hand, right hand, both hands extended up, right hand, left hand, crossing the hands, left hand, and right hand.

DIVISION, EXERCISE 2, STEP 2.1

In Exercise 2 Step 2.1, in a ***visual*** display of division we see three pictures of boxes - the dividend, the divisor and the answer. Between the first and second pictures is the division sign. Over the dividend is the equal line, which separates the problem from the answer. On top of the dividend is the third picture, which is the answer of the division problem. The ***audio*** representation for the pictures consists of a **double knock, knock, double click, knock, cling, and double knock.**

In the next part of Step 2.1, we have a **visual** display of how we arrive at the answer. In the divisor, the first empty box is automatically taken to the top.

Exercise 2

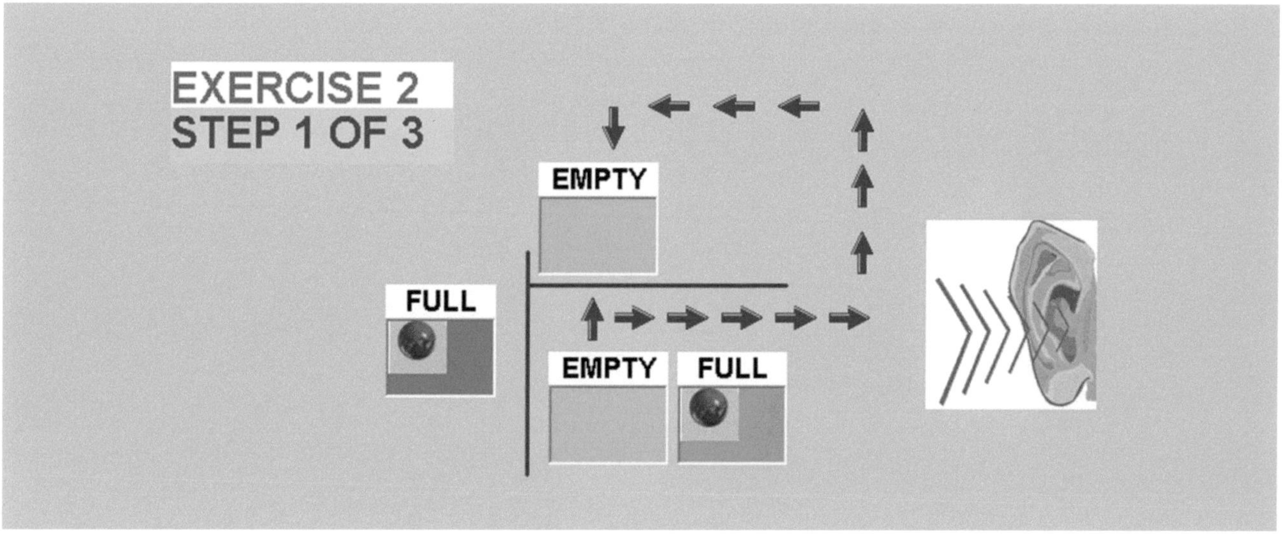

EXERCISE 2
STEP 1 OF 3

EMPTY

FULL

EMPTY FULL

The ***kinesthetic*** representation of a filled up position is represented by the clenching of the right hand. The clenching of the left hand represents an empty position. The division sound, double click, is represented by both hands extended up. Both hands extended in front represent the subtraction sound, double tram. The equal sound, cling, is represented by crossing both hands. The kinesthetic representation for the pictures is the clenching of the left hand, right hand, both hands extended up, right hand, crossing of the hands, and left hand.

MATHEMATICS

DIVISION, EXERCISE 2, STEP 2.2

In Exercise 2 Step 2.2, in a ***visual*** display of division we see three pictures of boxes - the dividend, the divisor and the answer. Between the first and second pictures is the division sign. Over the dividend is the equal line, which separates the problem from the answer. On top of the dividend is the third picture, which is the answer of the division. The ***audio*** representation for the pictures consists of a **knock, double click, knock, cling, double knock, and knock.**

In the next part of Step 2.2, we have a **visual** display of how we arrive at the answer. We subtract picture four from picture two, which is the dividend. Between pictures two and four is the subtraction sign. Below picture four is the equal sign which separates the problem from the answer. Picture five is the result of the subtraction. When the result of the subtraction is an empty position, then a filled position is transferred to the answer of the division. The ***audio*** representation for the subtraction consists of a **knock, double tram, knock, cling, double knock.**

Exercise 2

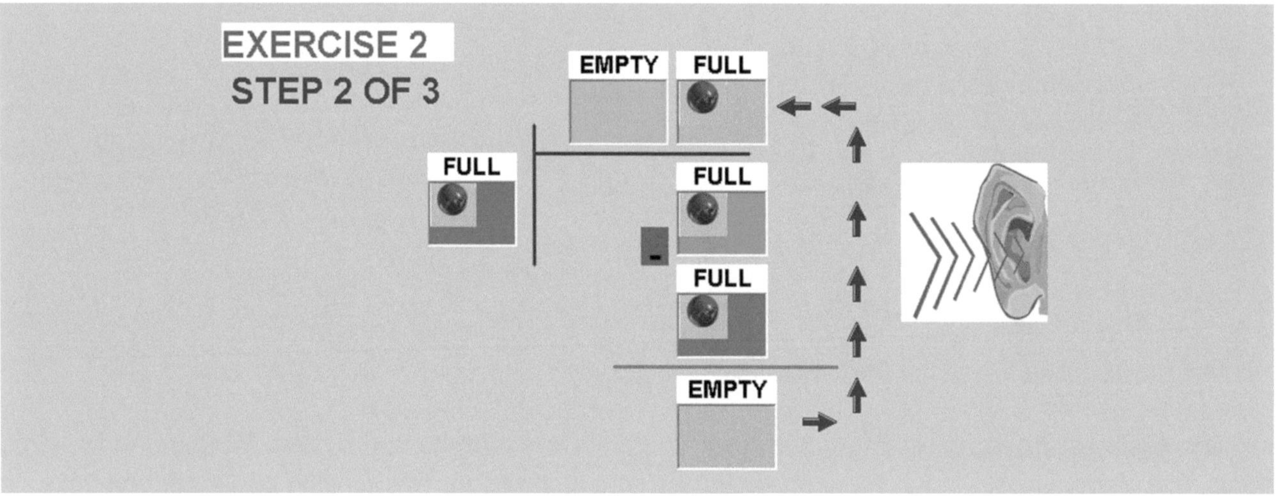

The ***kinesthetic*** representation of a filled up position is represented by the clenching of the right hand. The clenching of the left hand represents an empty position. The division sound, double click, is represented by both hands extended up. Both hands extended in front represents the subtraction sound, double tram. The equal sound, cling, is represented by crossing both hands. The kinesthetic representation for the pictures is the clenching of the right hand, both hands extended up, right hand, crossing of hands, left hand, and right hand.

DIVISION, EXERCISE 2, STEP 2.3

In Exercise 2 Step 2.3, in a ***visual*** display of the last step of the division we see two pictures of boxes. In both pictures the first position is empty while the second position is filled with a red ball. The colors and the placement of the boxes and the balls are different in the pictures. The form, shape, and color of the symbols do no change the mathematical values of the pictures, which is two.

The ***audio*** representation for the pictures consists of a **double knock, knock, and cling.**

Exercise 2

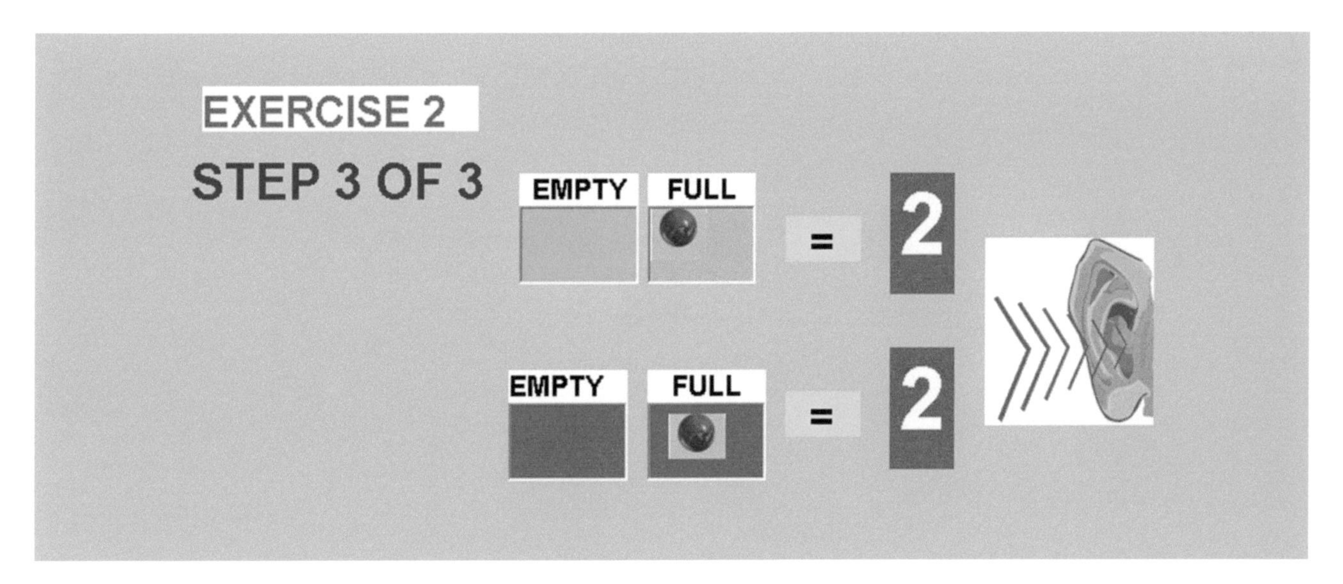

The ***kinesthetic*** representation of the filled up position is represented by the clenching of the right hand. The clenching of the left hand represents an empty position. The equal sound, cling, is represented by the crossing of hands. The kinesthetic representation for the pictures is the clenching of the left hand, right hand, crossing of hands, left hand, and right hand.

MATHEMATICS DIVISION, EXERCISE 2, ANSWER 2

In Exercise 2, in a ***visual*** display of division we see two pictures of boxes and the division sign in-between. In the first picture, position one is filled with a ball and position two is empty. In the second picture, position one is empty and position two is filled with a red ball. The division sign is between the first and second pictures. On top of the second picture is the equal line, which separates the problem from the answer. The ***audio*** representation for the division and consists of a **double knock, knock, double click, knock, double knock, and cling.**

At the bottom we see three pictures with probable answers. From the probable answers for the division we need to find the right one. The audio signals are: 1) **knock, and double knock,** 2) **double knock and knock**, 3) **knock and knock**.

Exercise 2

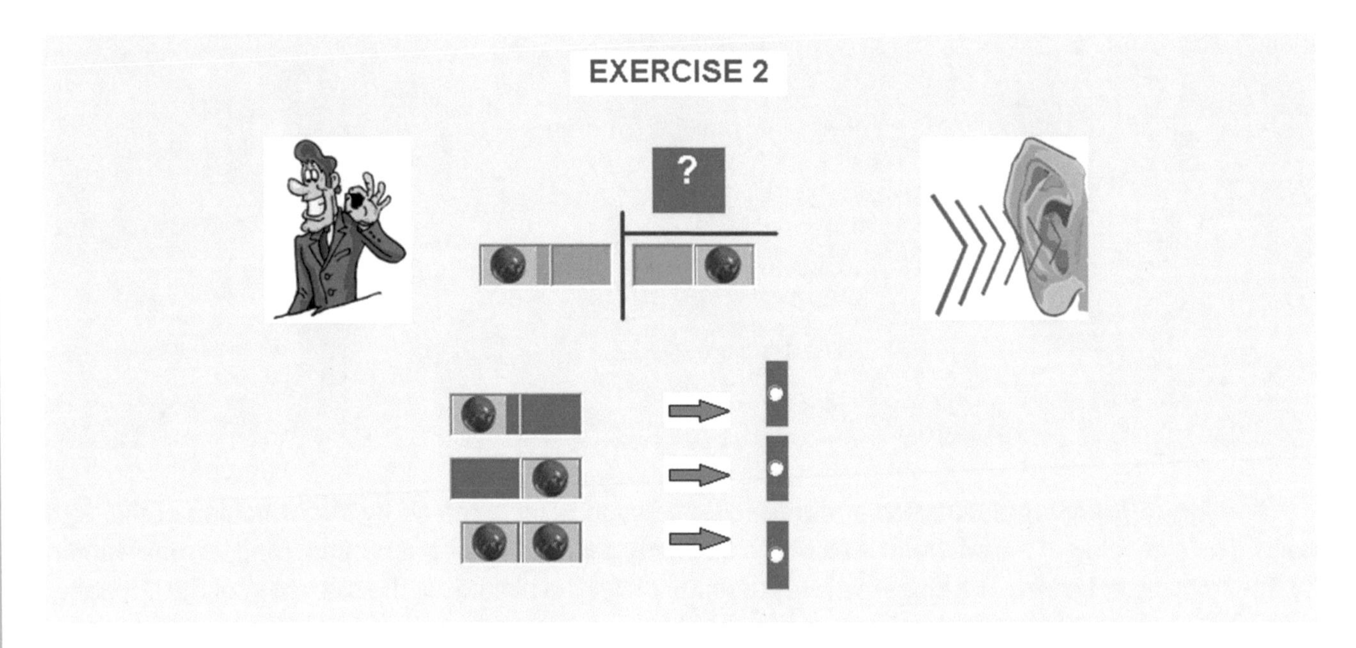

The ***kinesthetic*** representation of a filled up position is represented by the clenching of the right hand. An empty position is represented by the clenching of the left hand. The division sound, double click, is represented by both hands extended up. The equal sound, cling, is represented by crossing both hands. The kinesthetic representation for both pictures and the answer is the clenching of the left hand, then right, both hands extended up, the clenching of the right, left, crossing both hands, the clenching of the left hand and the right hand.

DIVISION, EXERCISE 3

In Exercise 3, in a ***visual*** display of division we see three pictures of boxes - the dividend, the divisor and the answer. Between the first and second pictures is the division sign. Over the dividend is the equal line, which separates the problem from the answer. On top of the dividend is the third picture, which is the answer of the division. The ***audio*** representation for the pictures consists of a **double knock, double knock, knock, double click, double knock, knock, cling, double knock, and knock.**

Exercise 3

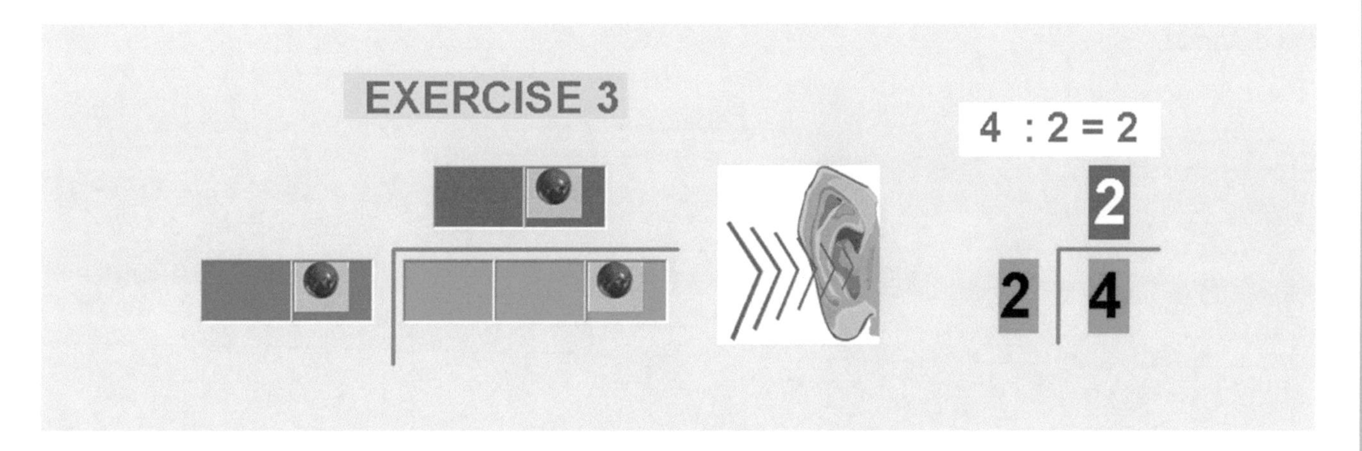

The ***kinesthetic*** representation of a filled up position is represented by the clenching of the right hand. The clenching of the left hand represents an empty position. The division sound, double click, is represented by both hands extended up. The equal sound, cling, is represented by crossing both hands. The kinesthetic representation for the pictures is the clenching of left hand twice, right hand, both hands extended up, left hand, right hand, crossing of the hands, left hand, and right hand.

DIVISION, EXERCISE 3, STEP 3.1

In Exercise 3 Step 3.1, in a ***visual*** display of division we see two pictures of boxes - the dividend and the divisor. Between the first and second pictures is the division sign. Over the dividend is the equal line, which separates the problem from the answer. On top of the dividend is the third picture, which is the answer of the division problem. The ***audio*** representation for the pictures consists of a **double knock, double knock, knock, double click, double knock, knock, and cling.**

In the next part of Step 3.1, we have a **visual** display of how we arrive at the answer. If the divisor has an empty position before the filled position, then that position cancels out the first empty position of the dividend.

Exercise 3

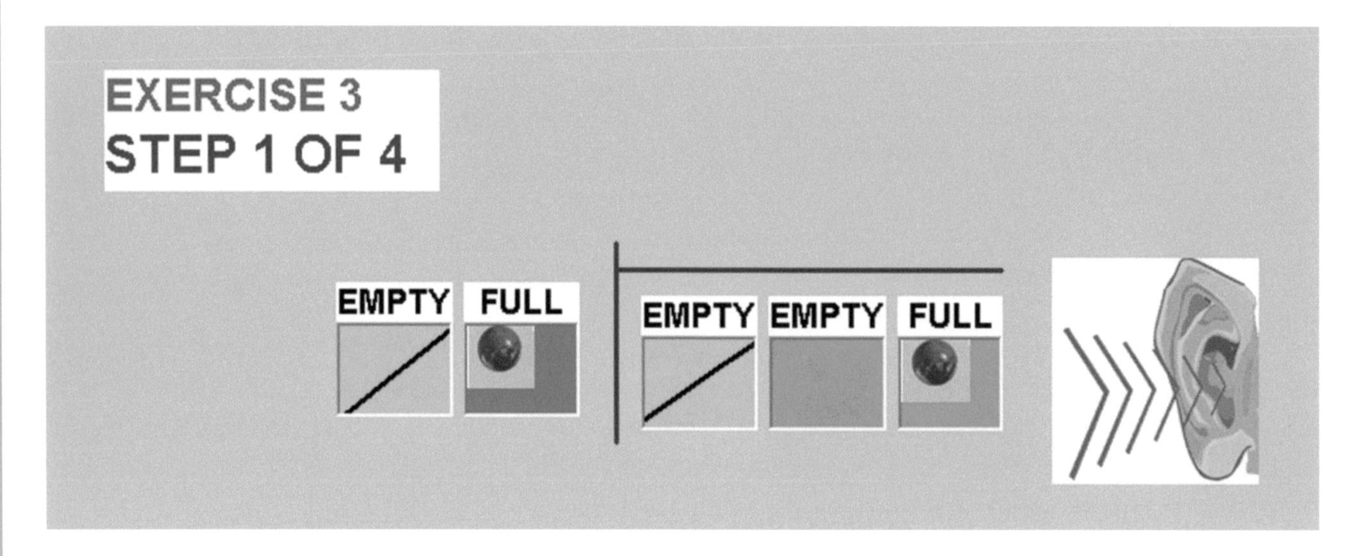

The ***kinesthetic*** representation of a filled up position is represented by the clenching of the right hand. The clenching of the left hand represents an empty position. The division sound, double click, is represented by both hands extended up. Both hands extended in front represent the subtraction sound, double tram. The equal sound, cling, is represented by crossing both hands. The kinesthetic representation for the pictures is the clenching of the left hand twice, right hand, both hands extended up, left hand, right hand, and crossing the hands.

DIVISION, EXERCISE 3, STEP 3.2

In Exercise 3 Step 3.2, in a ***visual*** display of division we see three pictures of boxes - the dividend, the divisor and the answer. Between the first and second pictures is the division sign. Over the dividend is the equal line, which separates the problem from the answer. On top of the dividend is the third picture, which is the answer of the division problem. The ***audio*** representation for the pictures consists of **double knock, knock, double click, knock, cling, and double knock.**

In the next part of Step 3.2, we have a **visual** display of how we arrive at the answer. In the dividend, the empty positions before the filled up position are automatically taken to the top to the answer.

Exercise 3

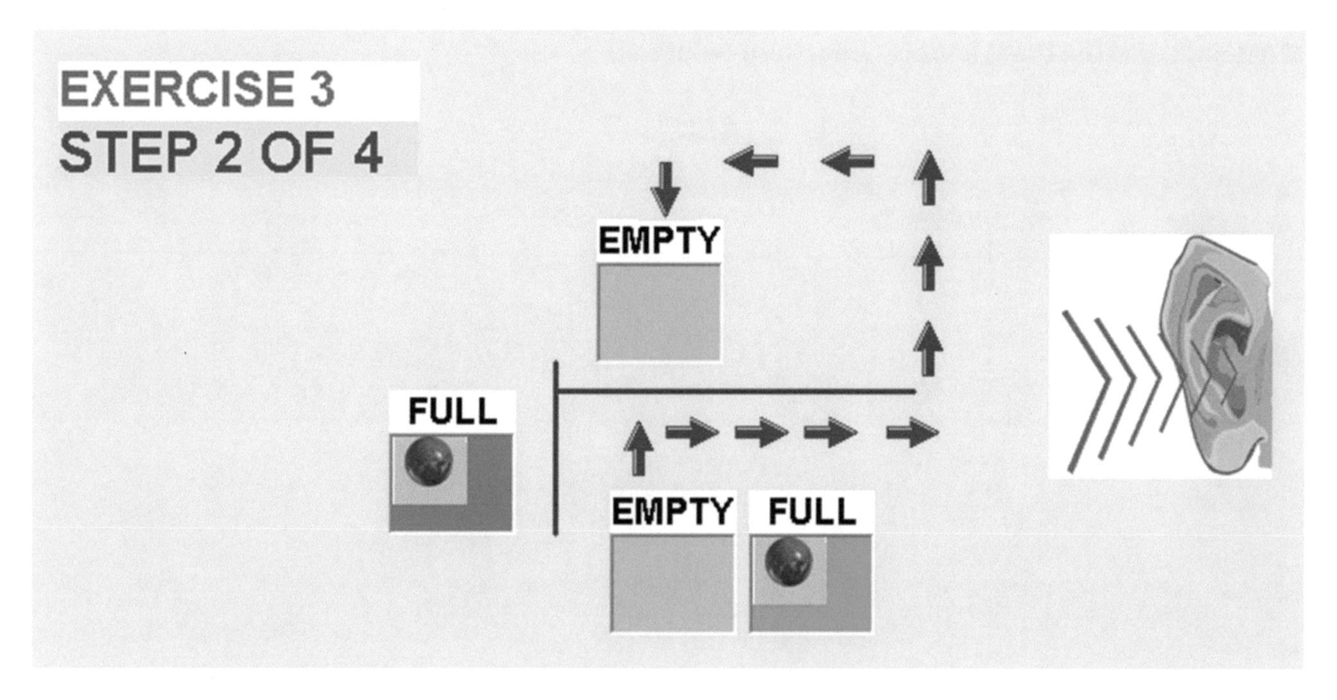

The ***kinesthetic*** representation of a filled up position is represented by the clenching of the right hand. The clenching of the left hand represents an empty position. The division sound, double click, is represented by both hands extended up. Both hands extended in front represent the subtraction sound, double tram. The equal sound, cling, is represented by crossing of hands. The kinesthetic representation for the pictures is the clenching of the left hand, right hand, both hands extended up, right hand, crossing of the hands, and left hand.

MATHEMATICS

DIVISION, EXERCISE 3, STEP 3.3

In Exercise 3 Step 3.3, in a ***visual*** display of division we see three pictures of boxes - the dividend, the divisor and the answer. Between the first and second pictures is the division sign. Over the dividend is the equal line, which separates the problem from the answer. On top of the dividend is the third picture, which is the answer of the division. The ***audio*** representation for the pictures consists of a **knock, double click, knock, cling, double knock, and knock.**

In the next part of Step 3.3, we have a **visual** display of how we arrive at the answer. We subtract picture four from picture two, which is the dividend. Between pictures two and four is the subtraction sign. Below picture four is the equal sign, which separates the problem from the answer. Picture five is the result of the subtraction. When the result of a subtraction is equal to an empty position, then a filled position is transferred to the answer of the division. The ***audio*** representation for the subtraction consists of a **knock, double tram, knock, cling, double knock.**

Exercise 3

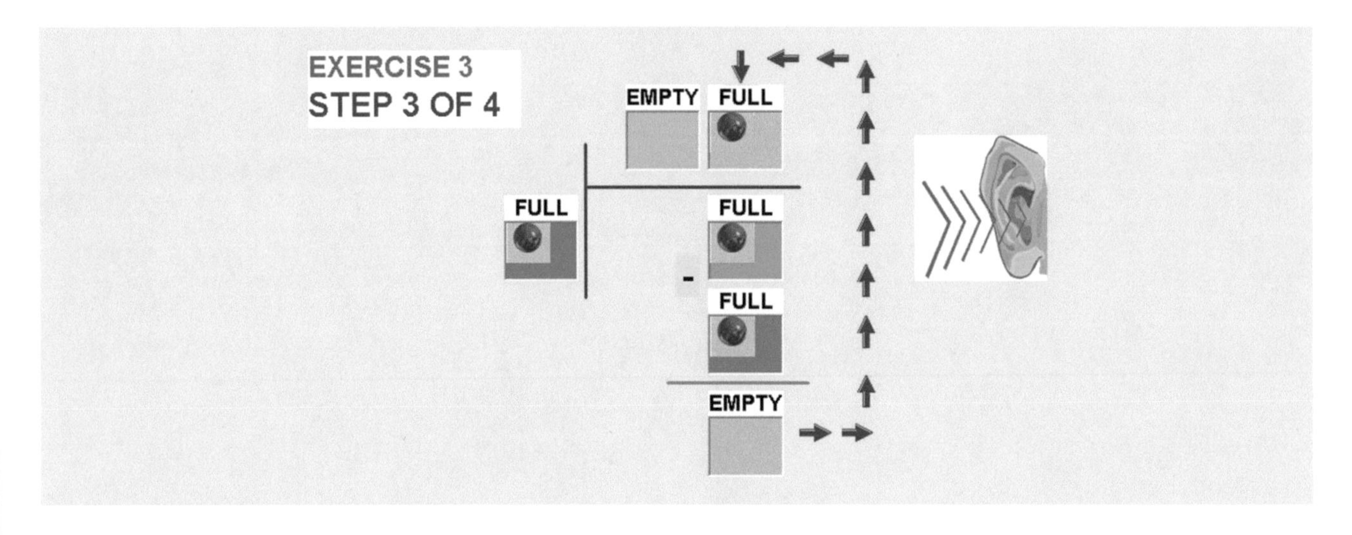

The ***kinesthetic*** representation of a filled up position is represented by the clenching of the right hand. The clenching of the left hand represents an empty position. The division sound, double click, is represented by both hands extended up. Both hands extended in front represent the subtraction sound, double tram. The equal sound, cling, is represented by crossing both hands. The kinesthetic representation for the pictures is the clenching of the right hand, both hands extended up, right hand, crossing the hands, left hand, and right hand.

DIVISION, EXERCISE 3, STEP 3.4

In Exercise 3 Step 3.4, in a ***visual*** display of the last step of the division we see two pictures of boxes. In both pictures, the first position is empty while the second position is filled with a red ball. The colors and the placement of the boxes and the balls are different in the pictures. The form, shape, and color of the symbols do no change the mathematical value of the pictures, which is two.

The ***audio*** representation for the pictures consists of a **double knock, knock, and cling.**

Exercise 3

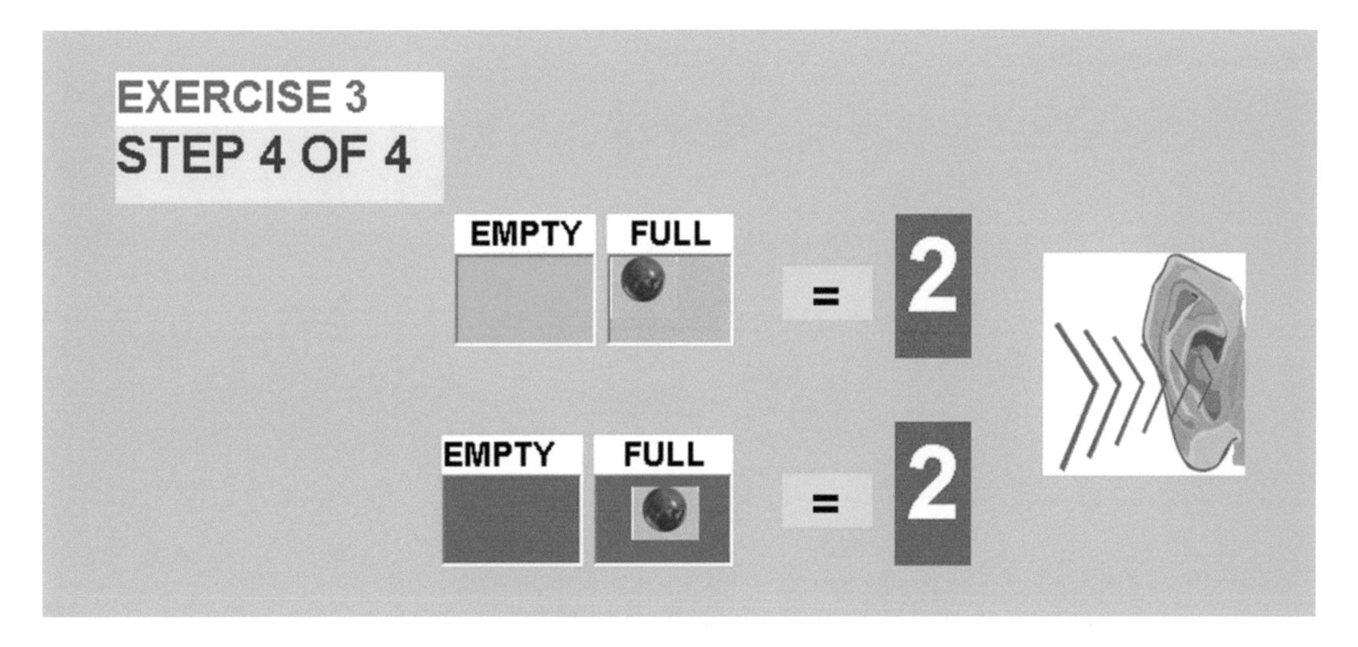

The ***kinesthetic*** representation of a filled up position is represented by the clenching of the right hand. The clenching of the left hand represents an empty position. The equal sound, cling, is represented by crossing of hands. The kinesthetic representation for the pictures is the clenching of the left hand, right hand, crossing the hands, left hand, and right hand.

MATHEMATICS

DIVISION, EXERCISE 3, ANSWER 3

In Exercise 3, in a ***visual*** display of division we see two pictures of boxes and the division sign in-between. In the first picture, the first position is empty and the second is filled with a ball. In the second picture, the first two positions are empty while the last position has a red ball. The division sign is between the first and second pictures. On top of the second picture is the equal line, which separates the problem from the answer. The ***audio*** representation for the division consists of a **double knock, double knock, knock, double click, double knock, knock, knock, and cling.**

At the bottom we see three pictures with probable answers. From the probable answers for the division we need to find the right one. The audio signals are: 1) **knock and double knock,** 2) **double knock and knock**, 3) **knock and knock**.

Exercise 3

The ***kinesthetic*** representation of a filled up position is represented by the clenching of the right hand. An empty position is represented by the clenching of the left hand. The division sound, double click, is represented by both hands extended up. The equal sound, cling, is represented by crossing both hands. The kinesthetic representation for both pictures and the answer is the clenching of the left hand twice, right hand, both hands extended up, the clenching of the left hand, right hand, crossing the hands, left hand and right hand.

DIVISION, EXERCISE 4, STEP 4.1

In Exercise 4, in a ***visual*** display of division we see three pictures of boxes - the dividend, the divisor and the answer. Between the first and second pictures is the division sign. Over the dividend is the equal line, which separates the problem from the answer. On top of the dividend is the third picture, which is the answer of the division. The ***audio*** representation for the pictures consists of a **double knock, knock, knock, double click, double knock, knock, cling, knock, and knock.**

Exercise 4

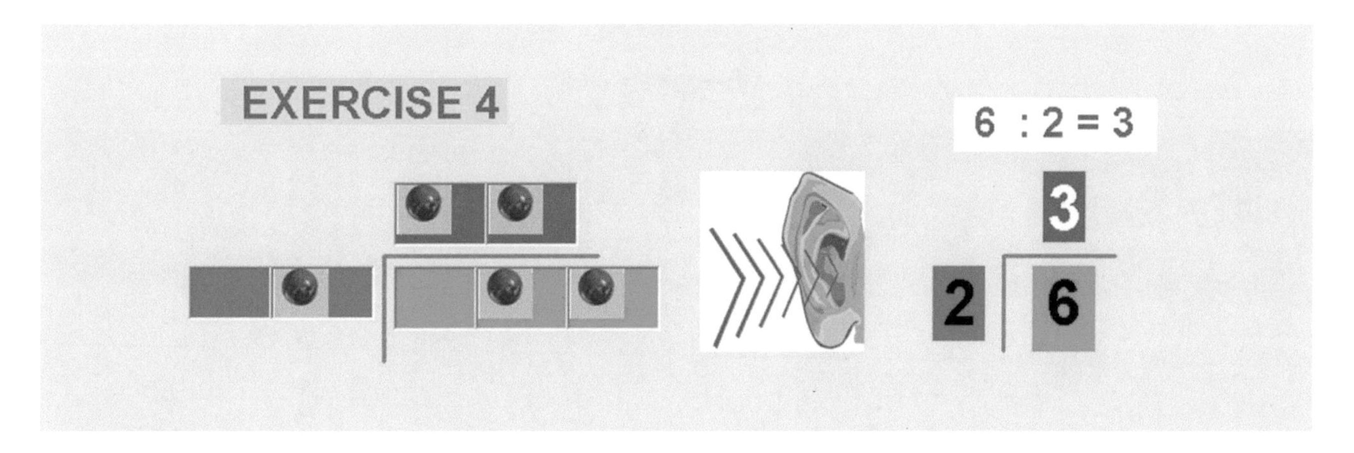

The ***kinesthetic*** representation of a filled up position is represented by the clenching of the right hand. The clenching of the left hand represents an empty position. The division sound, double click, is represented by both hands extended up. The equal sound, cling, is represented by crossing both hands. The kinesthetic representation for the pictures is the clenching of the left hand, right hand twice, both hands extended up, left hand, right hand, crossing both hands, and right hand twice.

DIVISION, EXERCISE 4, STEP 4.1

In Exercise 4 Step 4.1, in a ***visual*** display of division we see two pictures of boxes - the dividend and the divisor. Between the first and second pictures is the division sign. Over the dividend is the equal line, which separates the problem from the answer. On top of the dividend is the third picture, which is the answer of the division. The ***audio*** representation for the pictures consists of a **double knock, knock, knock, double click, double knock, knock, and cling.**

In the next part of Step 4.1, we have a **visual** display of how we arrive at the answer. If the divisor has an empty position before the filled position, then that position cancels out the first empty position of the dividend.

Exercise 4

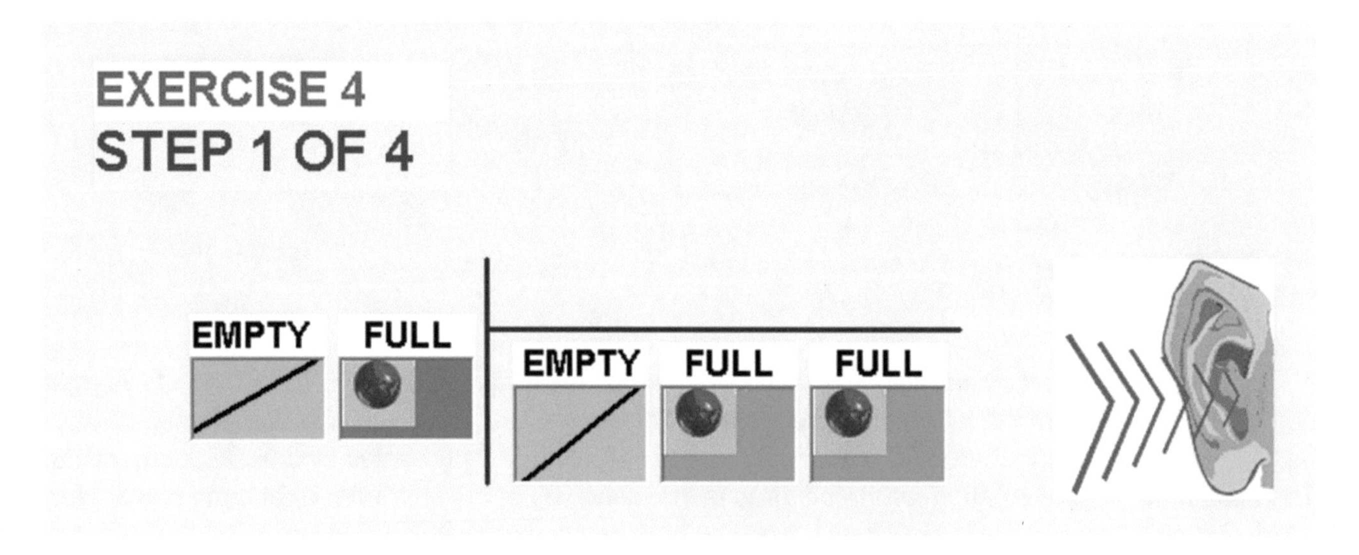

The ***kinesthetic*** representation of a filled up position is represented by the clenching of the right hand. The clenching of the left hand represents an empty position. The division sound, double click, is represented by both hands extended up. Both hands extended in front represent the subtraction sound, double tram. The equal sound, cling, is represented by crossing both hands. The kinesthetic representation for the pictures is the clenching of the left hand, right hand twice, both hands extended up, left hand, right hand, and crossing both hands.

DIVISION, EXERCISE 4, STEP 4.2

In Exercise 4 Step 4.2, in a ***visual*** display of division we see three pictures of boxes - the dividend, the divisor and the answer. Between the first and second pictures is the division sign. Over the dividend is the equal line, which separates the problem from the answer. On top of the dividend is the third picture, which is the answer of the division problem. The ***audio*** representation for the pictures consists of a **knock, knock, double click, knock, cling, and knock.**

In the next part of Step 4.2, we have a **visual** display of how we arrive at the answer. We subtract picture four from picture two, which is the dividend. Between pictures two and four is the subtraction sign. Below picture four is the equal sign which separates the problem from the answer. Picture five is the result of the subtraction. When the result of the subtraction is an empty position then a filled position is transferred to the answer of the division. The ***audio*** representation for the subtraction consists of a **knock, knock, double tram, knock, cling, double knock, and knock.**

Exercise 4

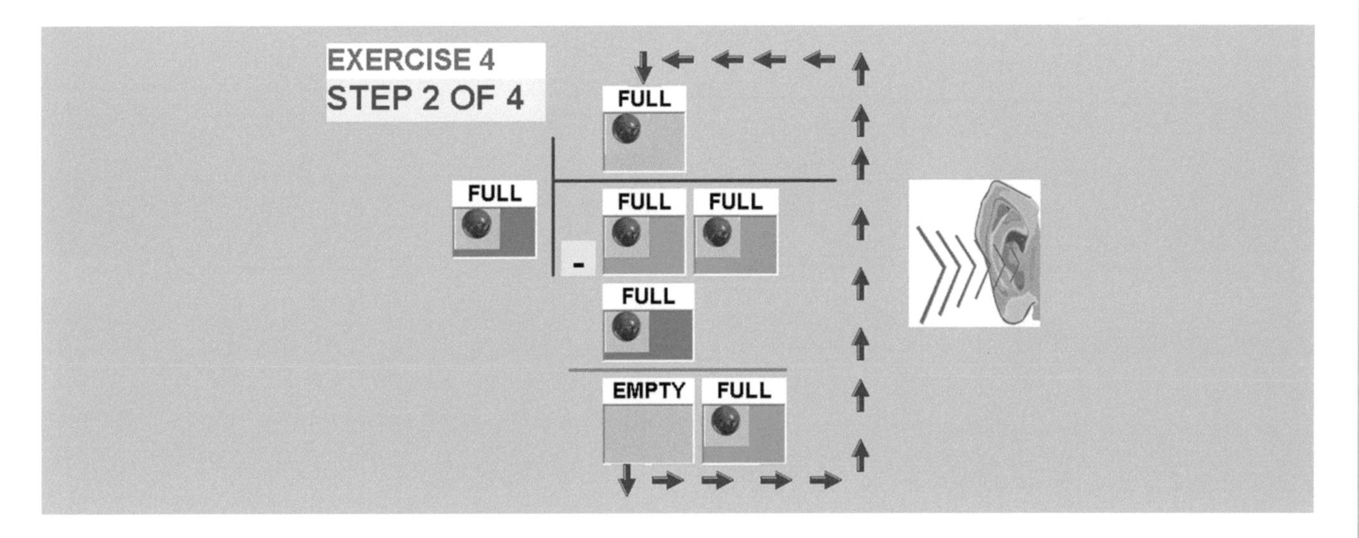

The ***kinesthetic*** representation of a filled up position is represented by the clenching of the right hand. The clenching of the left hand represents an empty position. The division sound, double click, is represented by both hands extended up. Both hands extended in front represent the subtraction sound, double tram. The equal sound, cling, is represented by crossing both hands. The kinesthetic representation for the pictures is the clenching of the right hand twice, both hands extended up, right hand, crossing both hands, and right hand.

MATHEMATICS DIVISION, EXERCISE 4, STEP 4.3

In Exercise 4 Step 4.3, in a ***visual*** display of division we see three pictures of boxes - the dividend, the divisor and the answer. Between the first and second pictures is the division sign. Over the dividend is the equal line, which separates the problem from the answer. On top of the dividend is the third picture, which is the answer of the division. The ***audio*** representation for the pictures consists of a **knock, double click, knock, cling, knock, and knock.**

In the next part of Step 4.3, we have a **visual** display of how we arrive at the answer. We subtract picture four from picture two, which is the dividend. Between pictures two and four is the subtraction sign. Below picture four is the equal sign, which separates the problem from the answer. Picture five is the result of the subtraction. When the result of the subtraction is an empty position, then a filled position is transferred to the answer of the division. The ***audio*** representation for the subtraction consists of a **knock, double tram, knock, cling, double knock.**

Exercise 4

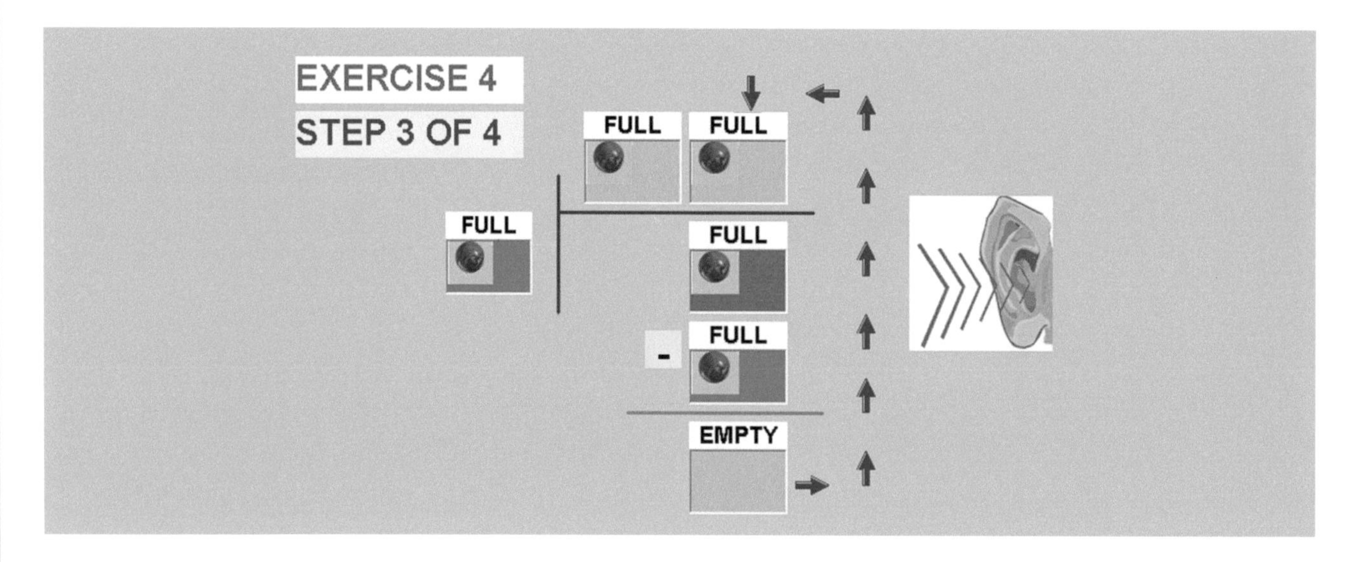

The ***kinesthetic*** representation of a filled up position is represented by the clenching of the right hand. The clenching of the left hand represents an empty position. The division sound, double click, is represented by both hands extended up. Both hands extended in front represent the subtraction sound, double tram. The equal sound, cling, is represented by crossing both hands. The kinesthetic representation for the pictures is the clenching of the right hand, both hands extended up, right hand, crossing both hands, right hand twice.

DIVISION, EXERCISE 4, STEP 4.4

In Exercise 4 Step 4.4, in a ***visual*** display of the last step of the division we see two pictures of boxes. In both pictures, both positions are filled with red balls. The colors and the placement of the boxes and the balls are different in the pictures. The form, shape, and color of the symbols do no change the mathematical value of the pictures, which is equal to three.

The ***audio*** representation for the pictures consists of a **knock, knock, and cling.**

Exercise 4

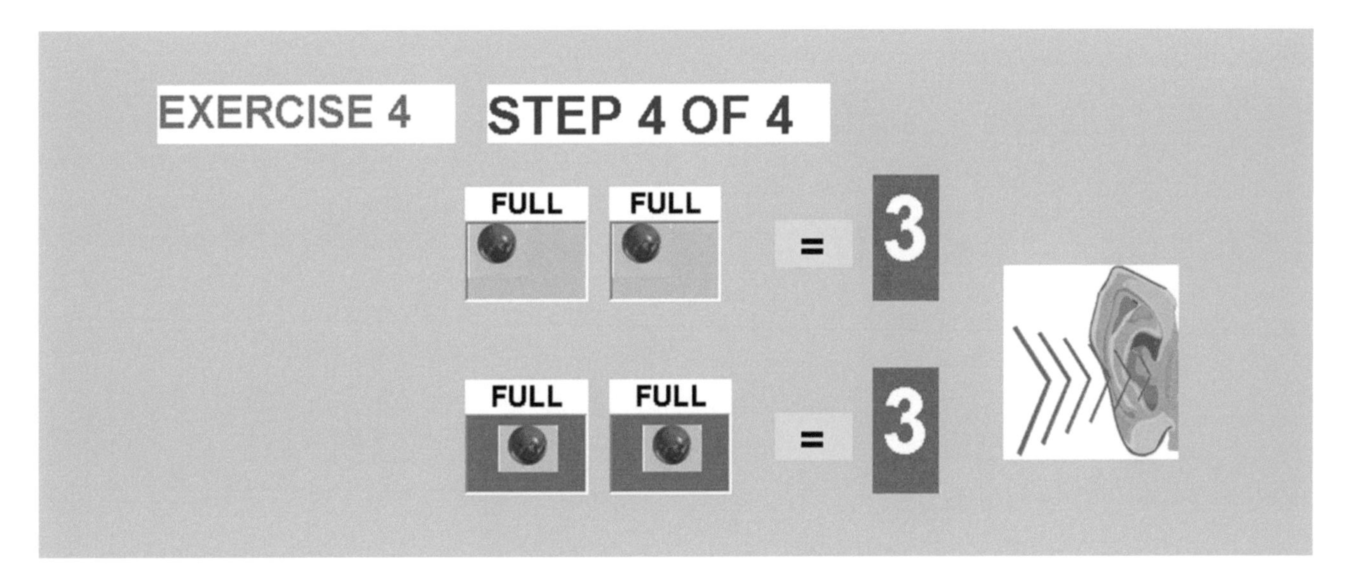

The ***kinesthetic*** representation of a filled up position is represented by the clenching of the right hand. The clenching of the left hand represents an empty position. The equal sound, cling, is represented by crossing both hands. The kinesthetic representation for the pictures is the clenching of the right hand twice, crossing both hands, and right hand twice.

MATHEMATICS

DIVISION, EXERCISE 4, ANSWER 4

In Exercise 4, in a ***visual*** display of division we see two pictures of boxes and the division sign in-between. In the first picture, the first position is empty while the second position has a red ball. In the second picture, the first position is empty while the last two positions have red balls. The division sign is between the first and second pictures. On top of the second picture is the equal line, which separates the problem from the answer. The ***audio*** representation for the division consists of a **double knock, knock, knock, double click, double knock, knock, and cling.**

At the bottom we see three pictures with probable answers. From the probable answers for the division we need to find the right one. The audio signals are: 1) **double knock and knock,** 2) **knock and double knock**, 3) **knock and knock**.

Exercise 4

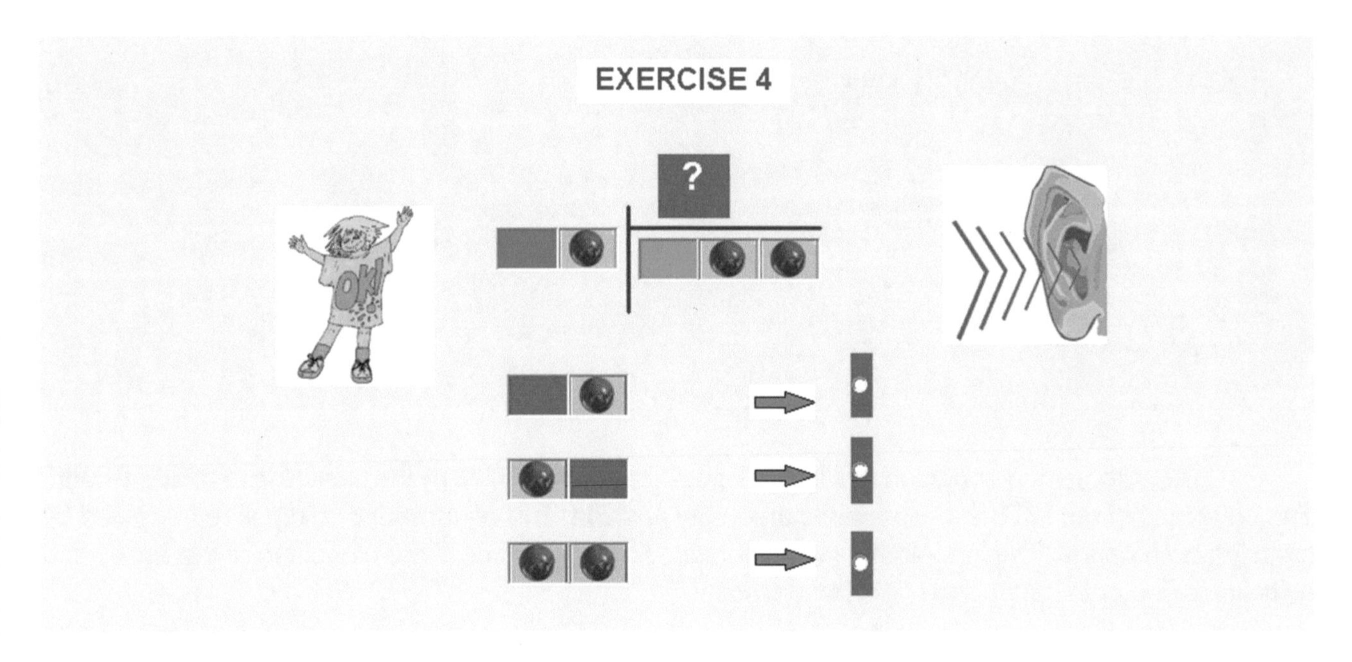

The ***kinesthetic*** representation of a filled up position is represented by the clenching of the right hand. The clenching of the left hand represents an empty position. Both hands extended up represent the division sound, double click. The equal sound, cling, is represented by crossing both hands. The kinesthetic representation for both pictures and the answer is the clenching of the left hand, right hand twice, both hands extended up, the clenching of the left hand, right hand, crossing of hands, and right hand twice.

DIVISION, EXERCISE 5

In Exercise 5, in a ***visual*** display of division we see three pictures of boxes. In the first picture, which has a value of three, both positions are filled with red balls. In the second picture, which has a value of six, position one is empty and positions two and three are filled with balls. The division sign is between the first and second pictures. Over the second picture is the equal line, which separates the problem from the answer. In the third picture, which is the answer for the division and has a value of two, the first position is empty while the second has a symbol.

The ***audio*** representation for the pictures consists of a **double knock, knock, knock, double click, knock, knock, cling, double knock, and knock.**

Exercise 5

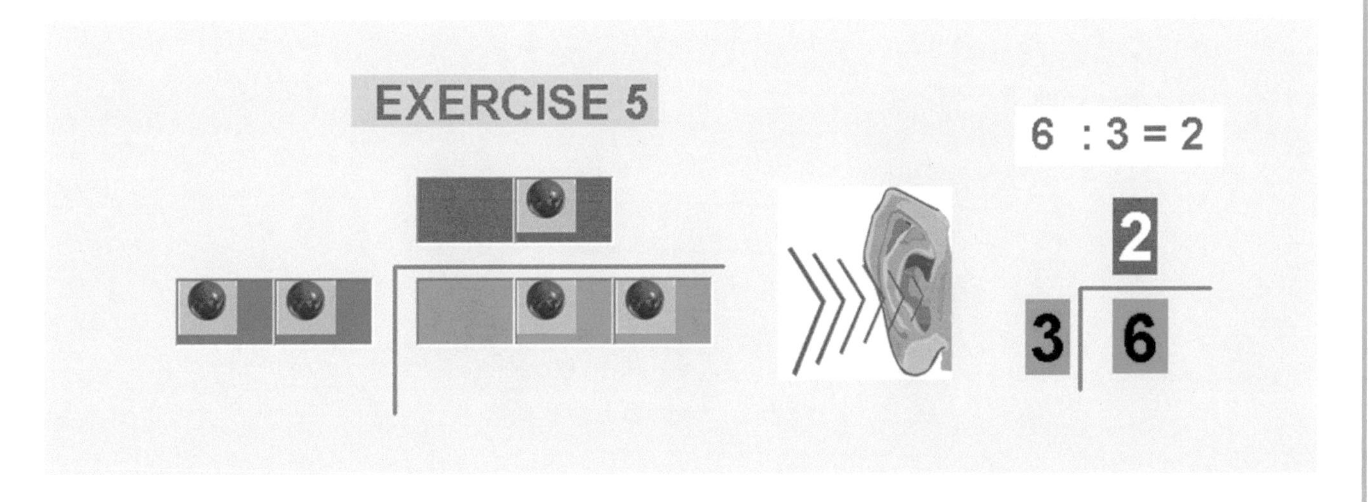

The ***kinesthetic*** representation of a filled up position is represented by the clenching of the right hand. The clenching of the left hand represents an empty position. Both hands extended up represent the division sign, double click. The equal sign, cling, is represented by crossing both hands. The kinesthetic representation for both pictures is the clenching of the left hand, right hand twice, extending both hands up, right hand twice, crossing both hands, left hand and right hand.

MATHEMATICS DIVISION, EXERCISE 5, STEP 5.1

In Exercise 5 Step 5.1, in a ***visual*** display of division we see three pictures of boxes - the dividend, the divisor and the answer. Between the first and second pictures is the division sign. Over the dividend is the equal line, which separates the problem from the answer. On top of the dividend is the third picture, which is the answer of the division. The ***audio*** representation for the pictures consists of a **double knock, knock, knock, double click, knock, knock, cling, and double knock.**

In the next part of Step 5.1, we have a **visual** display of how we arrive at the answer. In the dividend, the empty positions before the filled up position are automatically taken to the top to the answer.

Exercise 5

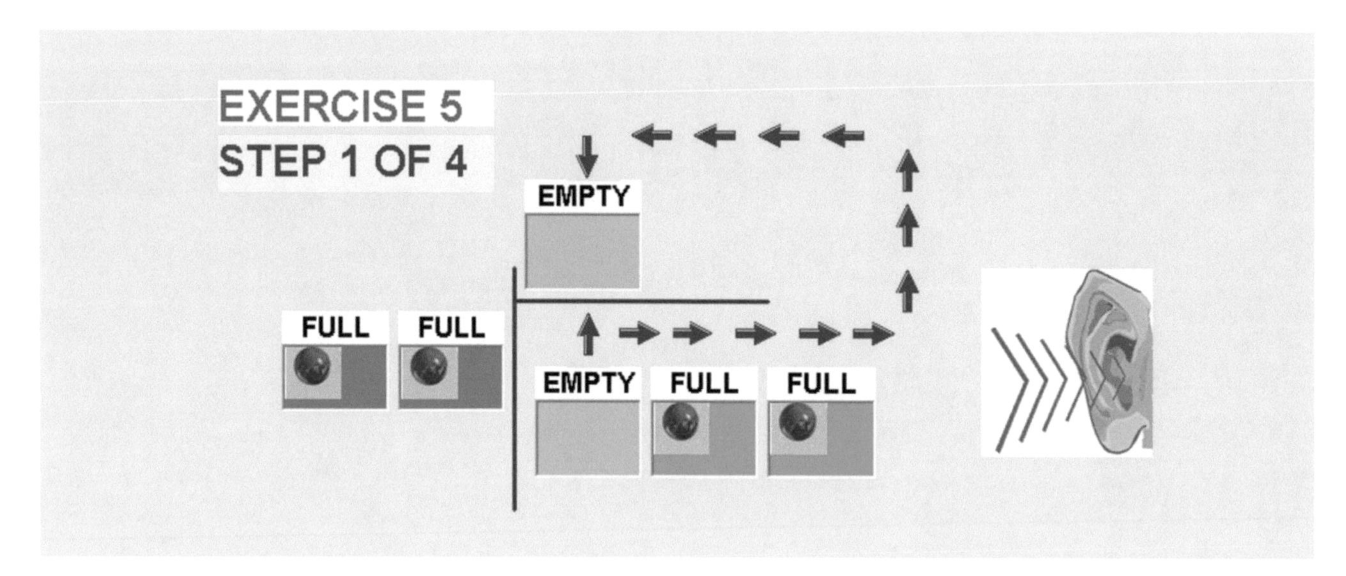

The ***kinesthetic*** representation of a filled up position is represented by the clenching of the right hand. The clenching of the left hand represents an empty position. The division sound, double click, is represented by both hands extended up. Both hands extended in front represent the subtraction sound, double tram. The equal sound, cling, is represented by crossing both hands. The kinesthetic representation for the pictures is the clenching of the left hand, right hand twice, both hands extended up, right hand twice, crossing of both hands, and left hand.

DIVISION, EXERCISE 5, STEP 5.2

In Exercise 5 Step 5.2, in a ***visual*** display of division we see three pictures of boxes - the dividend, the divisor, and the answer. Between the first and second pictures is the division sign. Over the dividend is the equal line, which separates the problem from the answer. On top of the dividend is the third picture, which is the answer of the division problem. The ***audio*** representation for the pictures consists of a **knock, knock, double click, knock, knock, cling, and double knock.**

In the next part of Step 5.2, we have a **visual** display of how we arrive at the answer. After having eliminated the empty positions, the next step can be applied to the problem.

Exercise 5

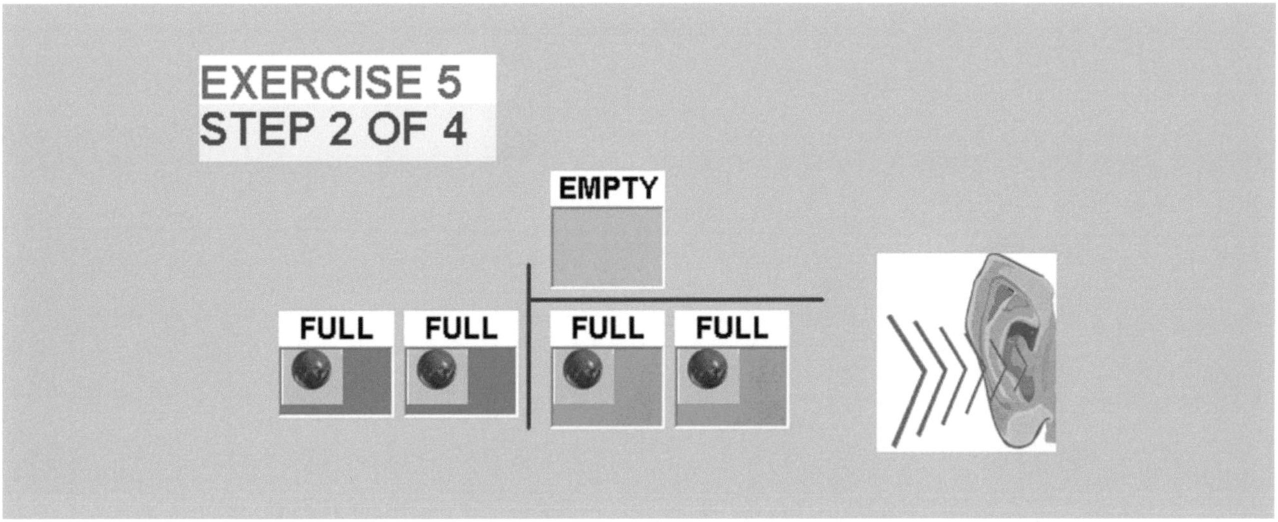

EXERCISE 5
STEP 2 OF 4

EMPTY

FULL FULL FULL FULL

The ***kinesthetic*** representation of a filled up position is represented by the clenching of the right hand. The clenching of the left hand represents an empty position. The division sound, double click, is represented by both hands extended up. Both hands extended in front represent the subtraction sound, double tram. The equal sound, cling, is represented by crossing both hands. The kinesthetic representation for the pictures is the clenching of the right hand twice, hands extended up, right hand twice, crossing both hands, and left hand.

MATHEMATICS

DIVISION, EXERCISE 5, STEP 5.3

In Exercise 5 Step 5.3, in a ***visual*** display of division we see three pictures of boxes - the dividend, the divisor and the answer. Between the first and second pictures is the division sign. Over the dividend is the equal line, which separates the problem from the answer. On top of the dividend is the third picture, which is the answer of the division problem. The ***audio*** representation for the pictures consists of a **knock, knock, double click, knock, knock, cling, double knock, and knock.**

In the next part of Step 5.3, we have a **visual** display of how we arrive at the answer. We subtract picture four from picture two, which is the dividend. Between pictures two and four is the subtraction sign. Below picture four is the equal sign which separates the problem from the answer. Picture five is the result of the subtraction. When the result of the subtraction is an empty position, then a filled position is transferred to the answer of the division. The ***audio*** representation for the subtraction consists of a **knock, knock, double tram, knock, knock, cling, double knock, and double knock.**

Exercise 5

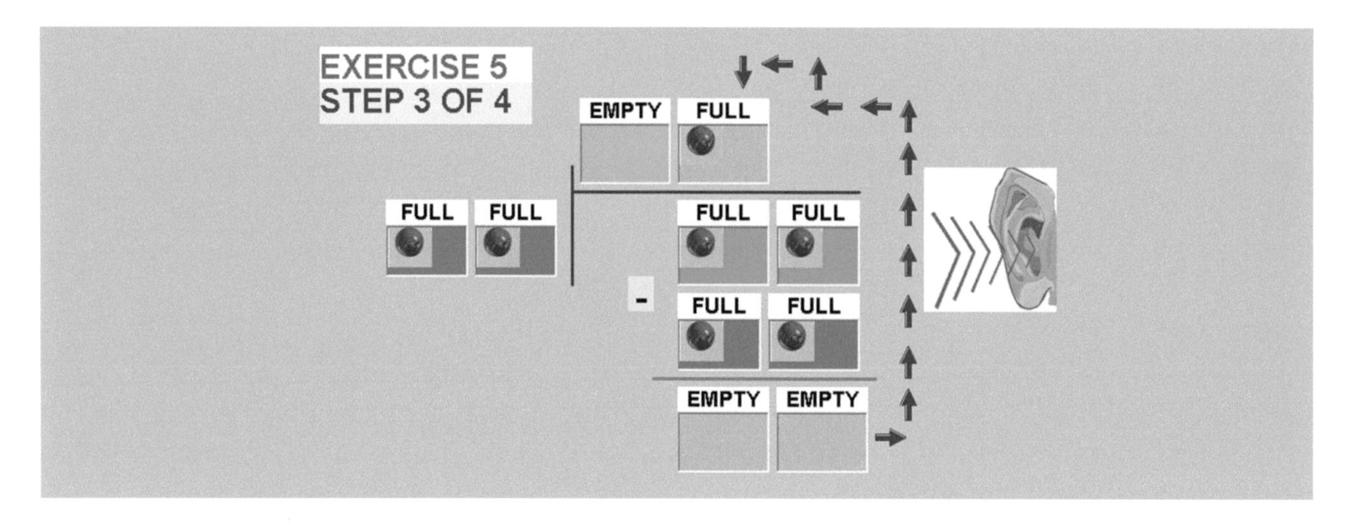

The ***kinesthetic*** representation of a filled up position is represented by the clenching of the right hand. The clenching of the left hand represents an empty position. The division sound, double click, is represented by both hands extended up. Both hands extended in front represent the subtraction sound, double tram. The equal sound, cling, is represented by crossing both hands. The kinesthetic representation for the pictures is the clenching of the right hand twice, hands extended up, right hand twice, crossing both hands, left hand, and right hand.

DIVISION, EXERCISE 5, STEP 5.4

In Exercise 5 Step 5.4, in a ***visual*** display of the last step of the division we see two pictures of boxes. In both pictures, the first position is empty while the second position is filled with a red ball. The colors and the placement of the boxes and the balls are different in the pictures. The form, shape, and color of the symbols do no change the mathematical value of the pictures, which is two.

The ***audio*** representation for the picture consists of a **double knock, knock, and cling.**

Exercise 5

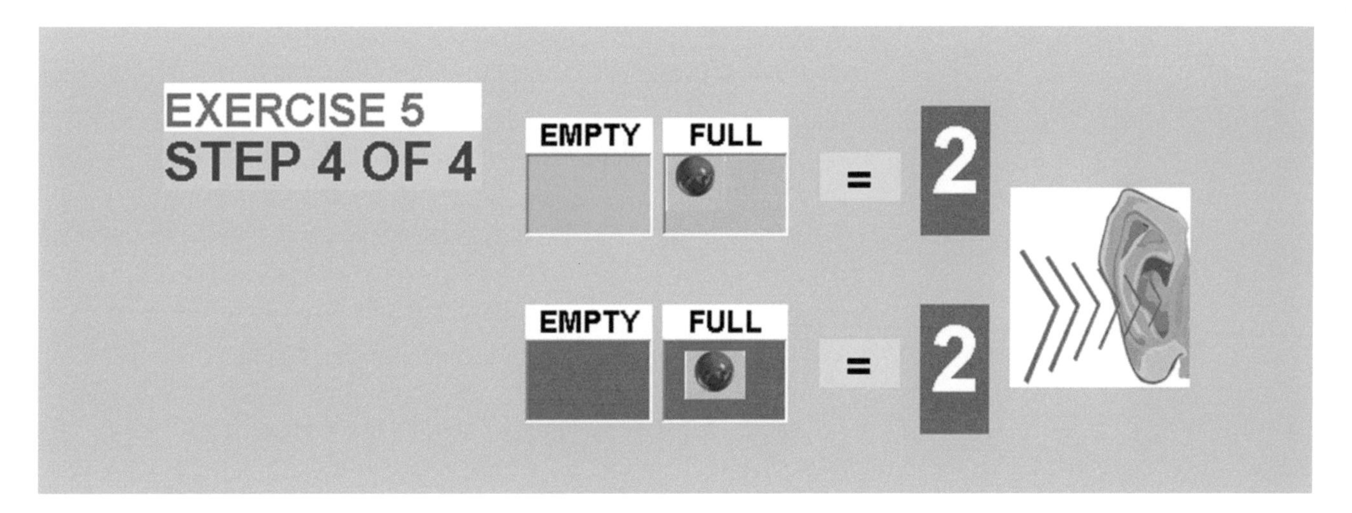

The ***kinesthetic*** representation of a filled position is represented by the clenching of the right hand. The clenching of the left hand represents an empty position. The equal sound, cling, is represented by crossing hands. The kinesthetic representation for the pictures is the clenching of the left hand, right hand, crossing the hands, left hand, and right hand.

DIVISION, EXERCISE 5, ANSWER 5

In Exercise 5, in a ***visual*** display of division we see two pictures of boxes and the division sign in-between. In the first picture, both positions have a red ball. In the second picture, the first position is empty while the last two positions have red balls. The division sign is between the first and second pictures. On top of the second picture is the equal line, which separates the problem from the answer. The ***audio*** representation for the division consists of a **double knock, knock, knock, double click, knock, knock, and cling.**

At the bottom we see three pictures with probable answers. From the probable answers for the division we need to find the right one. The audio signals are: 1) **double knock and knock,** 2) **knock and double knock**, 3) **knock and knock**.

Exercise 5

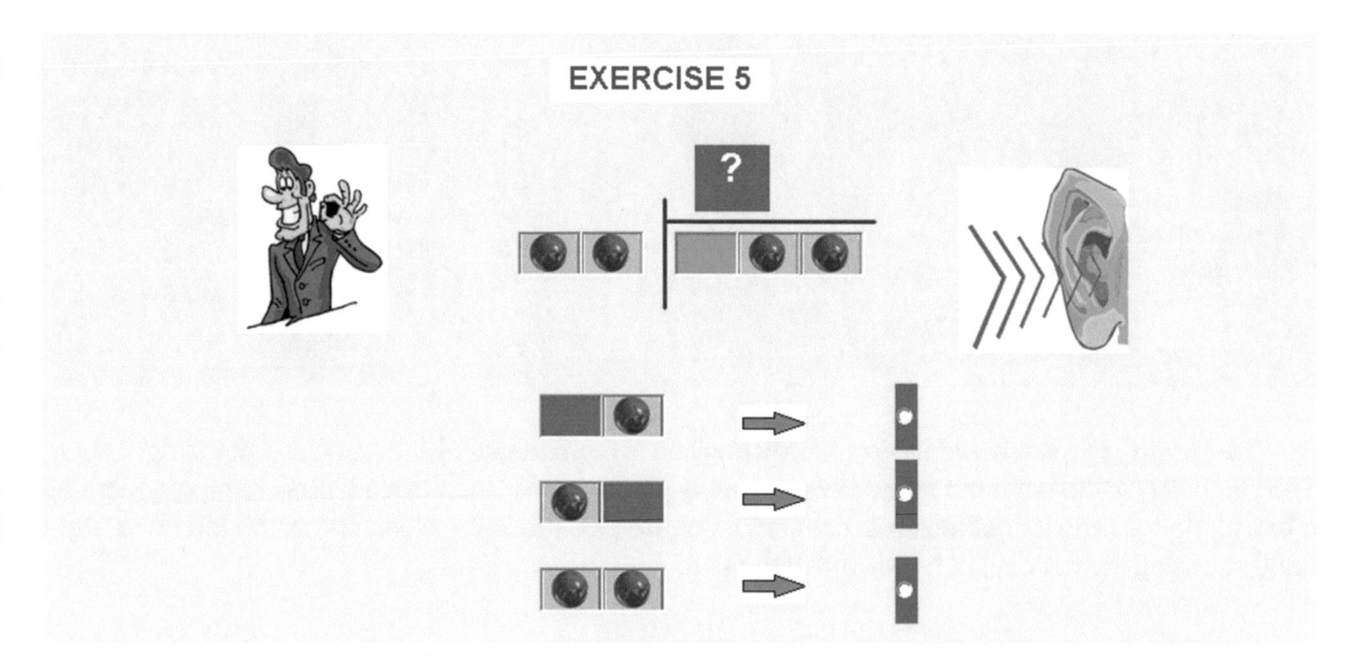

The ***kinesthetic*** representation of a filled up position is represented by the clenching of the right hand. The clenching of the left hand represents an empty position. Both hands extended up represent the division sound, double click. The equal sound, cling, is represented by crossing both hands. The kinesthetic representation for both pictures and the answer is the clenching of the left hand, right hand twice, both hands extended up, the clenching of the right hand twice, crossing the hands, left hand, and right hand.

DIVISION, EXERCISE 6

In Exercise 6, in a ***visual*** display of division we see three pictures of boxes. In the first picture, which has a value of five, the first position is filled with a red ball, the second position empty, and the last position has a red ball. In the second picture, which has a value of ten, position one is empty, position two is filled with red a ball, position three is empty, and position four is filled with a red ball. The division sign is between the first and second pictures. Over the second picture is the equal line, which separates the problem from the answer. In the third picture, which is the answer for the division and has a value of two, the first position is empty while the second has a symbol.

The ***audio*** representation for the pictures consists of a **double knock, knock, double knock, knock, double click, knock, double knock, knock, cling, double knock, and knock.**

Exercise 6

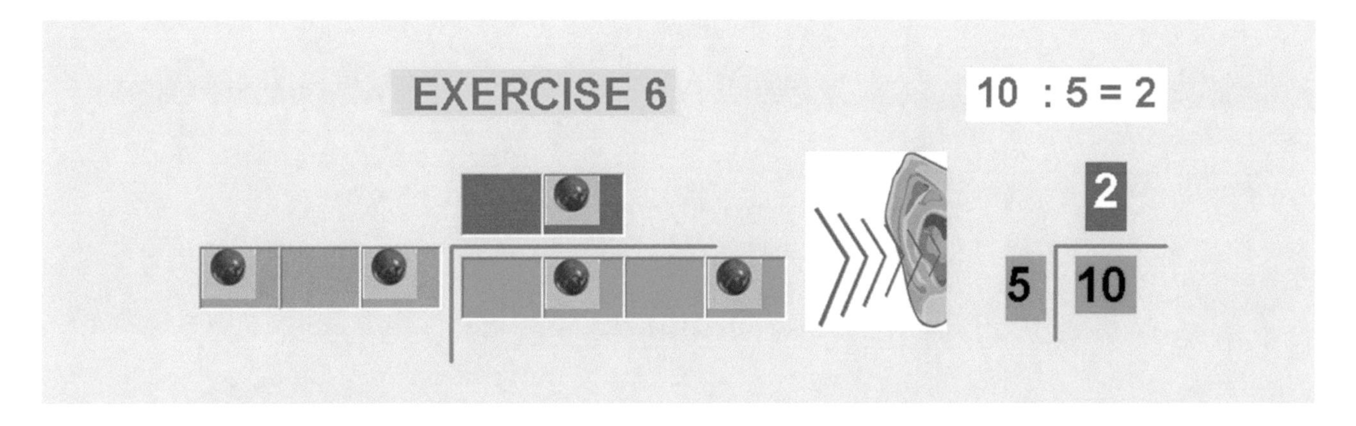

The ***kinesthetic*** representation of a filled up position is represented by the clenching of the right hand. The clenching of the left hand represents an empty position. Both hands extended up represent the division sign, double click. The equal sign, cling, is represented by crossing both hands. The kinesthetic representation for both pictures is the clenching of the left hand, right hand, left hand, right hand, extending both hands up, right hand, left hand, right hand, crossing both hands, left hand and right hand.

MATHEMATICS

DIVISION, EXERCISE 6, STEP 6.1

In Exercise 6 Step 6.1, in a ***visual*** display of division we see three pictures of boxes - the dividend, the divisor and the answer. Between the first and second pictures is the division sign. Over the dividend is the equal line, which separates the problem from the answer. On top of the dividend is the third picture, which is the answer of the division problem. The ***audio*** representation for the pictures consists of a **double knock, knock, double knock, knock, double click, knock, double knock, knock, cling, and double knock.**

In the next part of Step 6.1, we have a **visual** display of how we arrive at the answer. In the dividend, the empty positions before the filled up position are automatically taken to the top to the answer.

Exercise 6

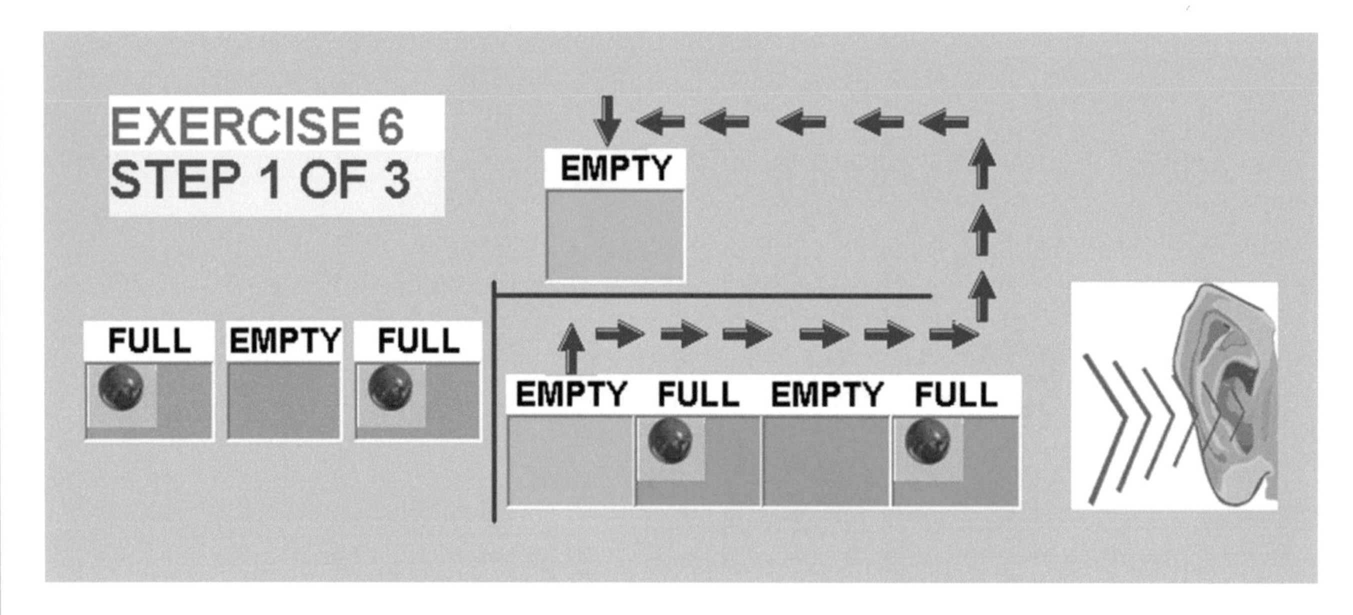

The ***kinesthetic*** representation of a filled up position is represented by the clenching of the right hand. The clenching of the left hand represents an empty position. The division sound, double click, is represented by both hands extended up. Both hands extended in front represent the subtraction sound, double tram. The equal sound, cling, is represented by crossing both hands. The kinesthetic representation for the pictures is the clenching of the left hand, right hand, left hand, right hand, both hands extended up, right hand, left hand, right hand, crossing the hands, and left hand.

DIVISION, EXERCISE 6, STEP 5.2

In Exercise 6 Step 6.2, in a ***visual*** display of division we see three pictures of boxes - the dividend, the divisor and the answer. Between the first and second pictures is the division sign. Over the dividend is the equal line, which separates the problem from the answer. On top of the dividend is the third picture, which is the answer of the division problem. The ***audio*** representation for the pictures consists of a **knock, double knock, knock, double click, knock, double knock, knock, cling, double knock, and knock.**

In the next part of Step 5.3, we have a **visual** display of how we arrive at the answer. We subtract picture four from picture two, which is the dividend. Between pictures two and four is the subtraction sign. Below picture four is the equal sign which separates problem from the answer. Picture five is the result of the subtraction. When the result of subtraction is an empty position, then a filled position is transferred to the answer of the division. The ***audio*** representation for the subtraction consists of a **knock, double knock, knock, double tram, knock, double knock, knock, cling, double knock, double knock, and double knock.**

Exercise 6

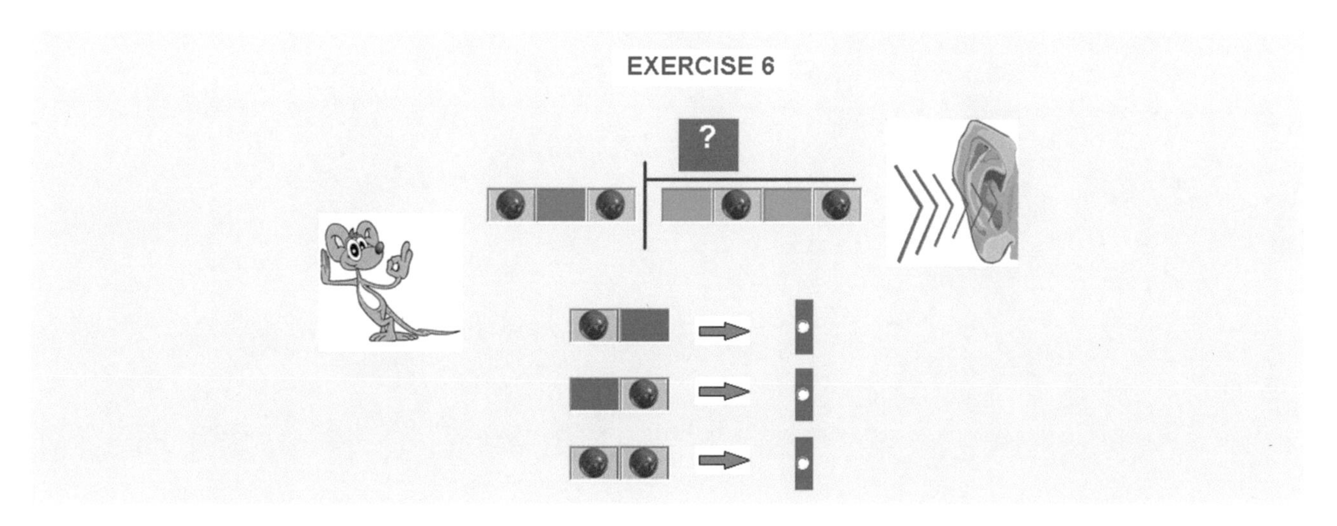

The ***kinesthetic*** representation of a filled up position is represented by the clenching of the right hand. The clenching of the left hand represents an empty position. The division sound, double click, is represented by both hands extended up. Both hands extended in front represent the subtraction sound, double tram. The equal sound, cling, is represented by crossing both hands. The kinesthetic representation for the pictures is the clenching of the right hand, left hand, right hand, both hands extended up, right hand, left hand, right hand, crossing the hands, left hand, and right hand.

MATHEMATICS

DIVISION, EXERCISE 6, ANSWER 6

In Exercise 6, in a ***visual*** display of division we see two pictures of boxes and the division sign in-between. In the first picture, the first position has a red ball, the second box is empty, and the last position is filled with a red ball. In the second picture, the first position is empty, the next position has a red ball, the third position is empty, and the last position has a red ball. The division sign is between the first and second pictures. On top of the second picture is the equal line, which separates the problem from the answer. The ***audio*** representation for the division consists of a **double knock, knock, double knock, knock, double click, knock, double knock, knock, and cling.**

At the bottom we see three pictures with probable answers. From the probable answers for the division we need to find the right one. The audio signals are: 1) **double knock and knock,** 2) **knock and double knock**, 3) **knock and knock**.

Exercise 6

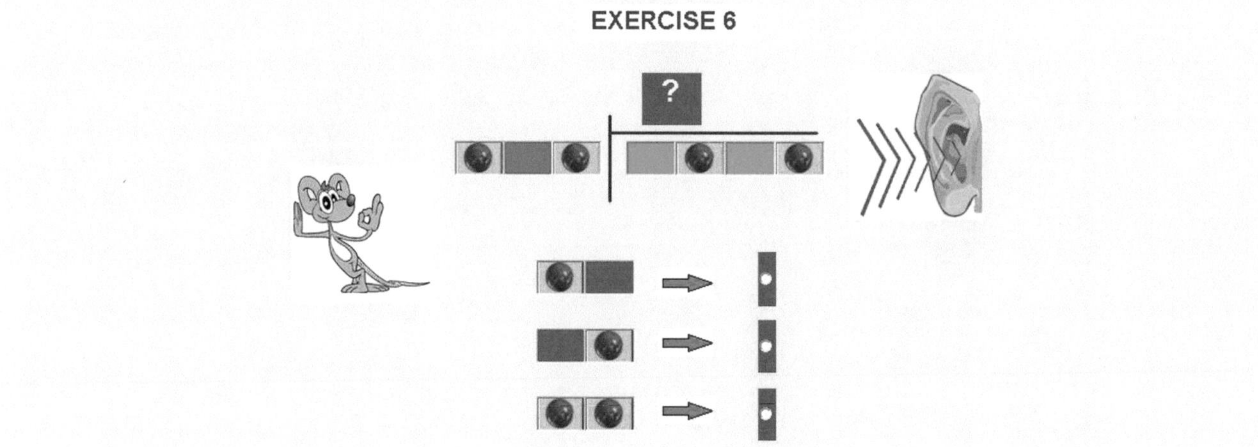

EXERCISE 6

The ***kinesthetic*** representation of a filled up position is represented by the clenching of the right hand. The clenching of the left hand represents an empty position. Both hands extended up represent the division sound, double click. The equal sound, cling, is represented by crossing both hands. The kinesthetic representation for both pictures and the answer is the clenching of the left hand, right hand, left hand, right hand, hands extended up, the clenching of the right hand, left hand, right hand, crossing of hands, left hand, and right hand.

DIVISION, EXERCISE 7

In Exercise 7, in a ***visual*** display of division we see three pictures of boxes and the division sign in-between. In the first picture there is one empty position and one position filled with a red ball. In the second picture, the first three positions are empty while the last position is filled with a red ball. In the third picture, the first two positions are empty while the last position is filled with a red ball. The division sign is between the first and second pictures. On top of the second picture is the equal line, which separates the problem from the answer. The ***audio*** representation for the division consists of a **double knock, double knock, double knock, knock, double click, double knock, knock, cling, double knock, double knock, and knock.**

Exercise 7

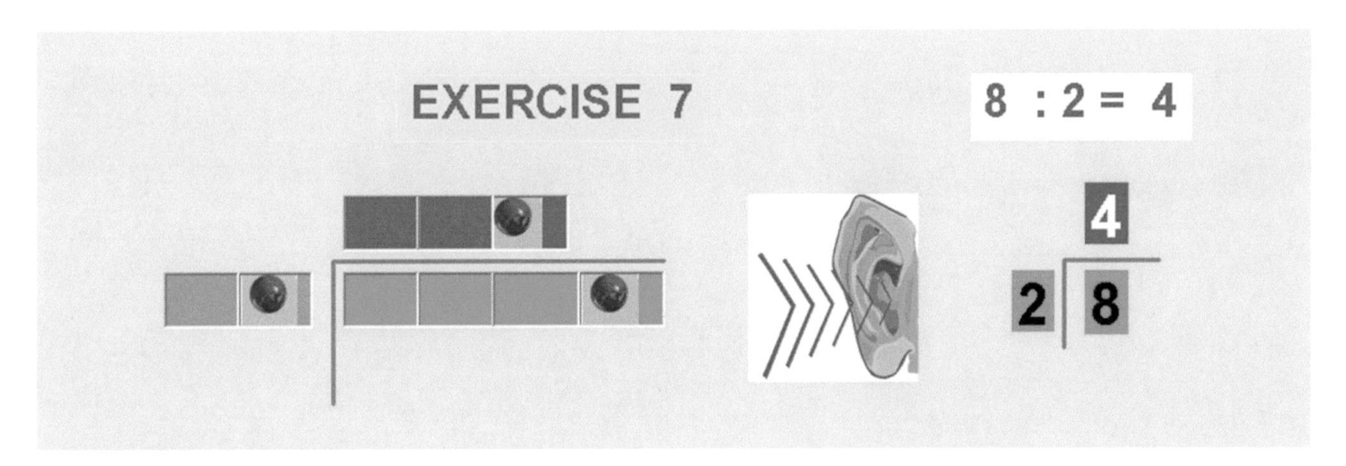

The ***kinesthetic*** representation of a filled up position is represented by the clenching of the right hand. The clenching of the left hand represents an empty position. Both hands extended up represent the division sound, double click. The equal sound, cling, is represented by crossing both hands. The kinesthetic representation for both pictures and the answer is the clenching of the left hand three times, right, both hands extended up, left hand, right hand, crossing of hands, left hand twice, and right hand.

MATHEMATICS

DIVISION, EXERCISE 7, STEP 7.1

In Exercise 7 Step 7.1, in a ***visual*** display of division we see two pictures of boxes - the dividend and the divisor. Between the first and second pictures is the division sign. Over the dividend is the equal line, which separates the problem from the answer. On top of the dividend is the third picture, which is the answer of the division. The ***audio*** representation for the pictures consists of a **double knock, double knock, double knock, knock, double click, double knock, knock, and cling.**

In the next part of Step 7.1, we have a **visual** display of how we arrive at the answer. If the divisor has an empty position before the filled up position, then that position cancels out the first empty position of the dividend.

Exercise 7

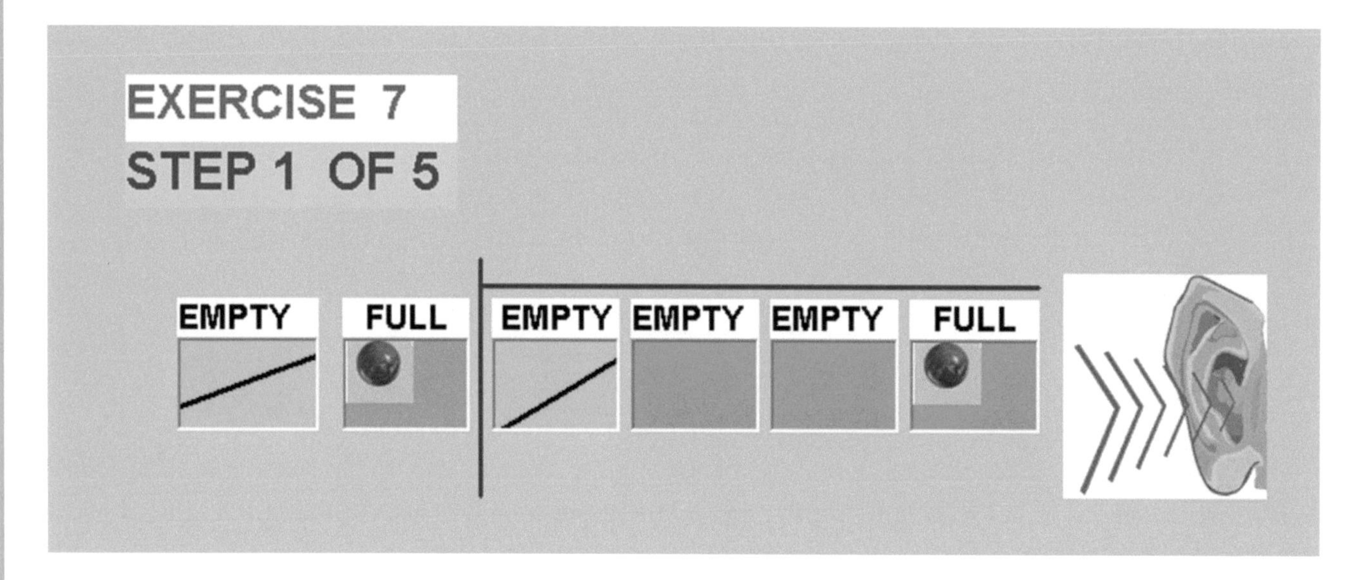

The ***kinesthetic*** representation of a filled up position is represented by the clenching of the right hand. The clenching of the left hand represents an empty position. The division sound, double click, is represented by both hands extended up. Both hands extended in front represent the subtraction sound, double tram. The equal sound, cling, is represented by crossing both hands. The kinesthetic representation for the pictures is the clenching of the left hand three times, right hand, both hands extended up, left hand, right hand, and the crossing of both hands.

DIVISION, EXERCISE 7, STEP 7.2

In Exercise 7 Step 7.2, in a ***visual*** display of division we see three pictures of boxes - the dividend, the divisor and the answer. Between the first and second pictures is the division sign. Over the dividend is the equal line, which separates the problem from the answer. On top of the dividend is the third picture, which is the answer of the division. The ***audio*** representation for the pictures consists of a **double knock, double knock, knock, double click, knock, cling, and double knock.**

In the next part of Step 7.2, we have a **visual** display of how we arrive at the answer. In the dividend, the empty positions before the filled up position are automatically taken to the top to the answer.

Exercise 7

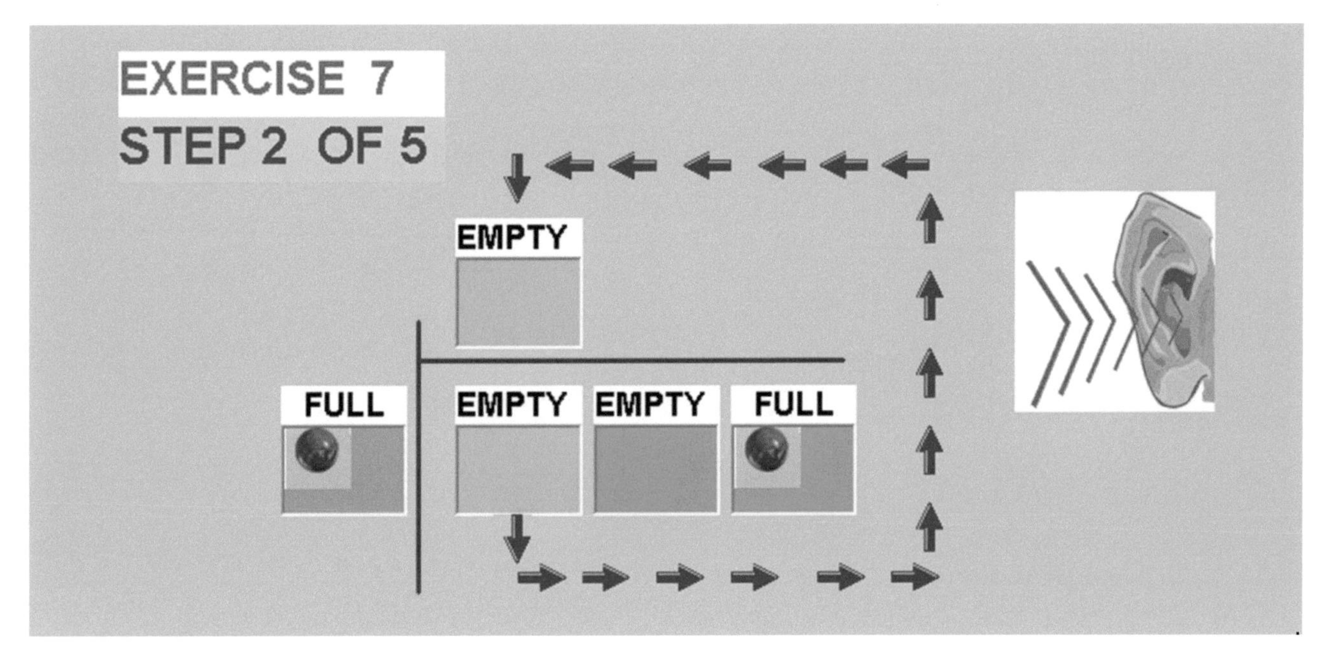

The ***kinesthetic*** representation of a filled up position is represented by the clenching of the right hand. The clenching of the left hand represents an empty position. The division sound, double click, is represented by both hands extended up. Both hands extended in front represent the subtraction sound, double tram. The equal sound, cling, is represented by crossing both hands. The kinesthetic representation for the pictures is the clenching of the left hand twice, right hand, both hands extended up, right hand, crossing both hands, and left hand.

MATHEMATICS

DIVISION, EXERCISE 7, STEP 7.3

In Exercise 7 Step 7.3, in a ***visual*** display of division we see three pictures of boxes - the dividend, the divisor and the answer. Between the first and second pictures is the division sign. Over the dividend is the equal line, which separates the problem from the answer. On top of the dividend is the third picture, which is the answer of the division problem. The ***audio*** representation for the pictures consists of a **double knock, knock, double click, knock cling, double knock, and double knock.**

In the next part of Step 8.3, we have a **visual** display of how we arrive at the answer. In the dividend, the empty positions before the filled up position are automatically taken to the top to the answer.

Exercise 7

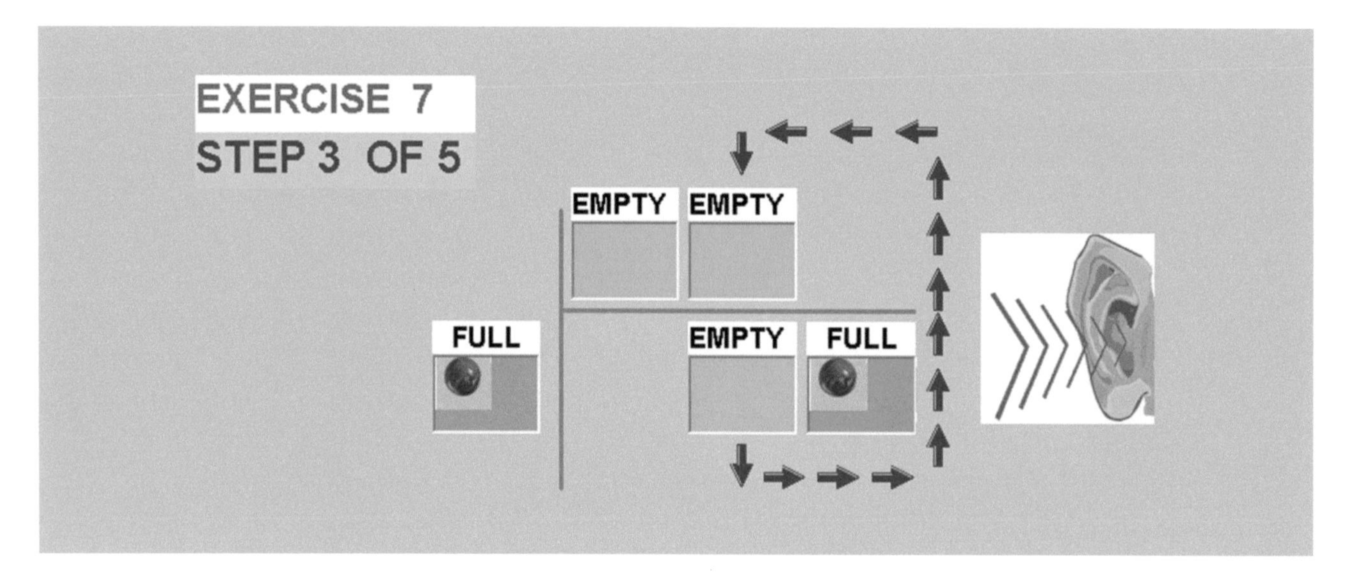

The ***kinesthetic*** representation of a filled up position is represented by the clenching of the right hand. The clenching of the left hand represents an empty position. The division sound, double click, is represented by both hands extended in up. Both hands extended in front represent the subtraction sound, double tram. The equal sound, cling, is represented by crossing both hands. The kinesthetic representation for the pictures is the clenching of the left hand, right hand, both hands extended up, right hand, crossing of both hands, and left hand twice.

DIVISION, EXERCISE 7, STEP 7.4

In Exercise 7 Step 7.4, in a ***visual*** display of division we see three pictures of boxes - the dividend, the divisor and the answer. Between the first and second pictures is the division sign. Over the dividend is the equal line, which separates the problem from the answer. On top of the dividend is the third picture, which is the answer of the division problem. The ***audio*** representation for the pictures consists of a **knock, double click, knock, cling, double knock, double knock, and knock.**

In the next part of Step 7.4, we have a **visual** display of how we arrive at the answer. We subtract picture four from picture two, which is the dividend. Between pictures two and four is the subtraction sign. Below picture four is the equal sign which separates the problem from the answer. Picture five is the result of the subtraction. When the result of subtraction is an empty position, then a filled position is transferred to the answer of the division. The ***audio*** representation for the subtraction consists of a **knock, double tram, knock, cling, and double knock.**

Exercise 7

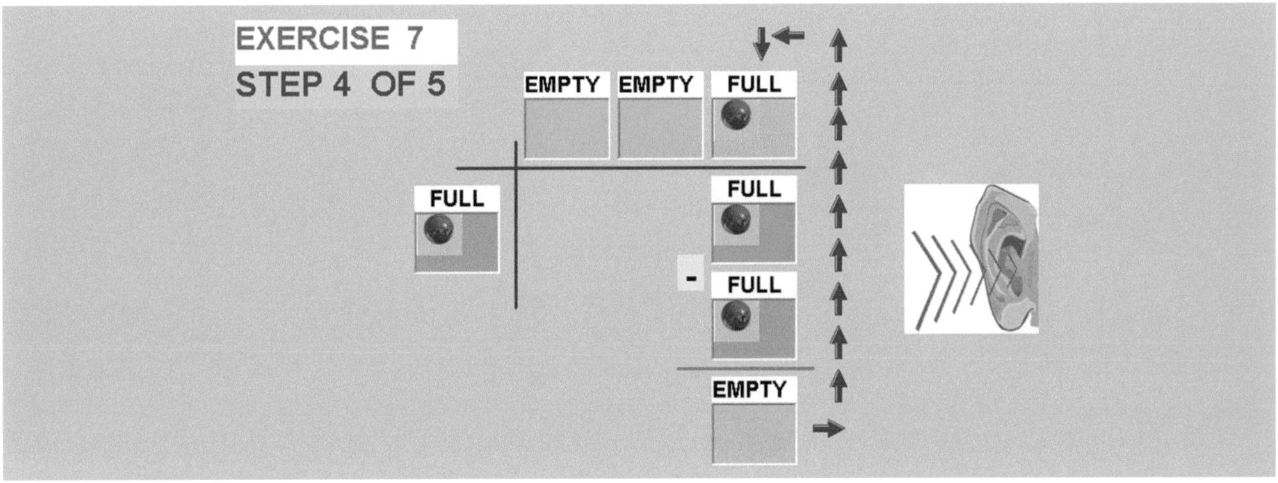

The ***kinesthetic*** representation of a filled up position is represented by the clenching of the right hand. The clenching of the left hand represents an empty position. The division sound, double click, is represented by both hands extended up. Both hands extended in front represent the subtraction sound, double tram. The equal sound, cling, is represented by crossing both hands. The kinesthetic representation for the pictures is the clenching of the right hand, both hands extended up, right hand, crossing of hands, left hand twice, and right hand.

MATHEMATICS DIVISION, EXERCISE 7, STEP 7.5

In Exercise 7 Step 7.5, in a ***visual*** display of the last step of division we see two pictures of boxes. In both pictures the first two positions are empty, and the last position is filled with a red ball. The colors and placement of the boxes and the balls are different in the pictures. The form, shape, and color of the symbols do no change the mathematical value of the pictures, which is five.

The ***audio*** representation for the pictures consists of a **double knock, double knock, knock, and cling.**

Exercise 7

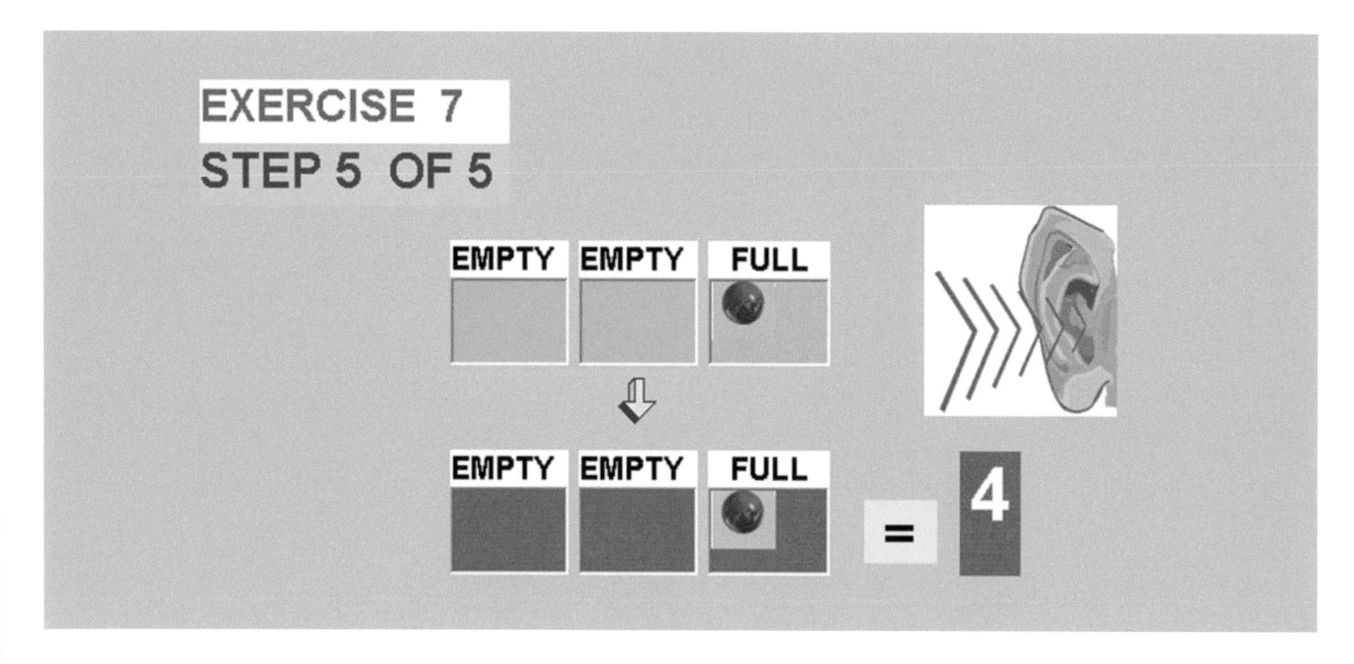

The ***kinesthetic*** representation of a filled up position is represented by the clenching of the right hand. The clenching of the left hand represents an empty position. The equal sound, cling, is represented by crossing the hands. The kinesthetic representation for the pictures is the clenching of the left hand twice, right hand, crossing of hands, left hand twice, and right hand.

DIVISION, EXERCISE 7

In Exercise 7, in a ***visual*** display of division we see two pictures of boxes and the division sign in-between. In the first picture, the first position is empty and the second is filled with a red ball. In the second picture, the first three positions are empty while the last position is filled with a red ball. The division sign is between the first and second pictures. On top of the second picture is the equal line, which separates the problem from the answer. The ***audio*** representation for the division consists of a **double knock, double knock, double knock, knock, double click, double knock, knock, and cling.**

At the bottom we see three pictures with probable answers. From the probable answers for the division we need to find the right one. The audio signals are: 1) **knock, double knock, and double knock,** 2) **double knock, double knock, and knock**, 3) **double knock, knock, and double knock.**

Exercise 7

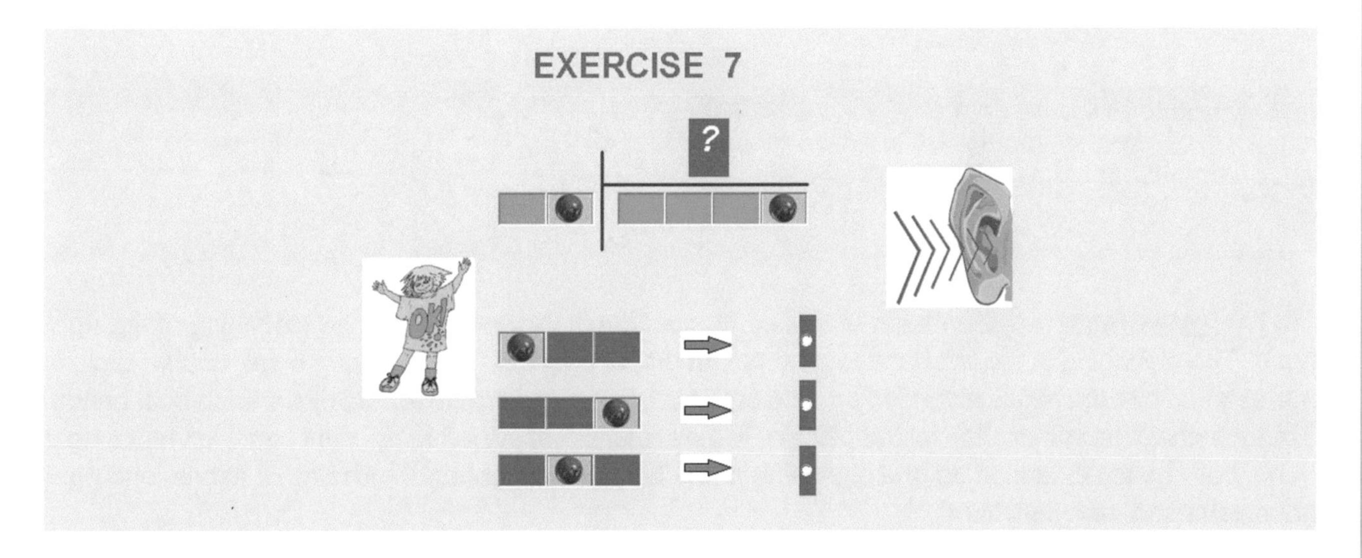

The ***kinesthetic*** representation of a filled up position is represented by the clenching of the right hand. The clenching of the left hand represents an empty position. The sound of double click is represented in kinesthetic by both hands extended up. The equal sound, cling, is represented by crossing both hands. The kinesthetic representation for the pictures and the answer is the clenching of the left hand three times, right hand, both hands extended up, the clenching of the left hand, right hand, crossing the hands, left hand twice, and right hand.

DIVISION, EXERCISE 8,

In Exercise 8, in a ***visual*** display of division we see three pictures of boxes - the dividend, the divisor and the answer. Between the first and second pictures is the division sign. Over the dividend is the equal line, which separates the problem from the answer. On top of the dividend is the third picture, which is the answer of the division. The ***audio*** representation for the pictures consists of a **double knock, knock double knock, knock, double click, double knock, knock, cling, knock, double knock, and knock.**

Exercise 8

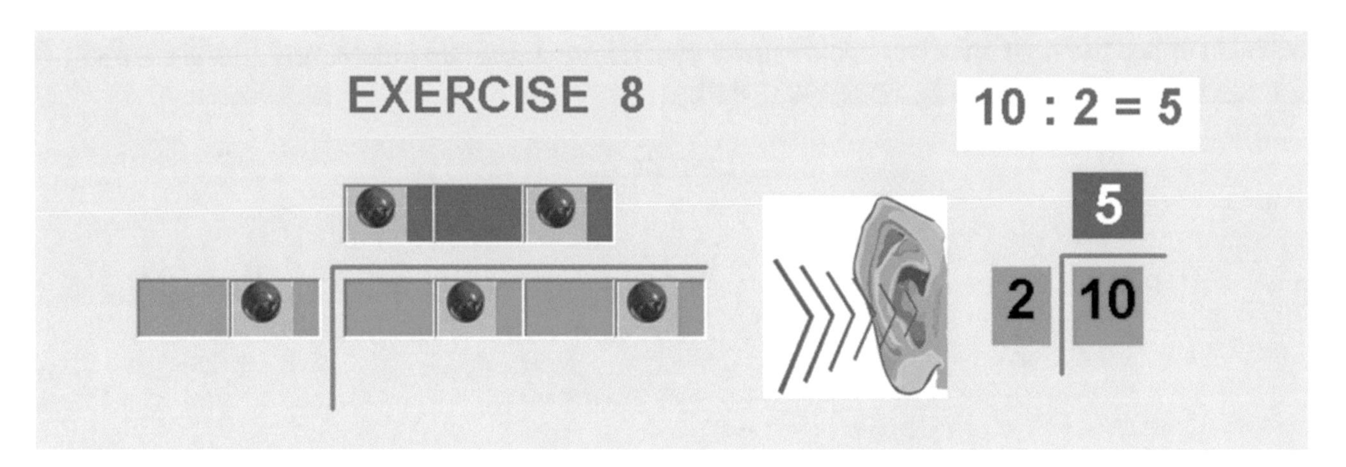

The ***kinesthetic*** representation of a filled up position is represented by the clenching of the right hand. The clenching of the left hand represents an empty position. The division sound, double click, is represented by both hands extended up. The equal sound, cling, is represented by crossing both hands. The kinesthetic representation for the pictures is the clenching of the left hand, right hand, left hand, right hand, both hands extended up, the clenching of the left hand, right hand, crossing of hands, and right hand, left hand, and right hand.

DIVISION, EXERCISE 8, STEP 8.1

In Exercise 8 Step 8.1, in a ***visual*** display of division we see two pictures of boxes - the dividend and the divisor. Between the first and second pictures is the division sign. Over the dividend is the equal line, which separates the problem from the answer. On top of the dividend is the third picture, which is the answer of the division. The ***audio*** representation for the pictures consists of a **double knock, knock, double knock, knock, double click, double knock, knock, and cling.**

In the next part of Step 8.1, we have a **visual** display of how we arrive at the answer. If the divisor has an empty position before the filled up position, then that position cancels out the first empty position of the dividend.

Exercise 8

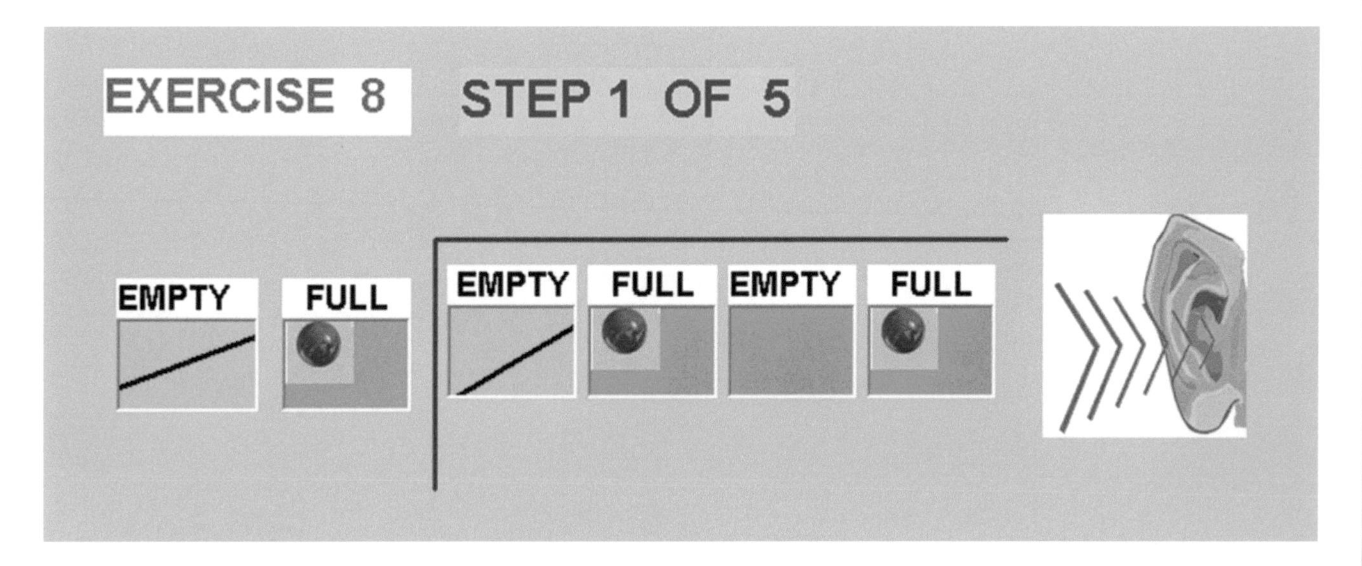

The ***kinesthetic*** representation of a filled position is represented by the clenching of the right hand. The clenching of the left hand represents an empty position. The division sound, double click, is represented by both hands extended up. Both hands extended in front represent the subtraction sound, double tram. The equal sound, cling, is represented by crossing both hands. The kinesthetic representation for the pictures is the clenching of the left hand, right hand, left hand, right hand, both hands extended up, the clenching of the left hand, right, and then crossing of hands.

MATHEMATICS

DIVISION, EXERCISE 8, STEP 8.2

In Exercise 8 Step 8.2, in a ***visual*** display of division we see three pictures of boxes - the dividend, the divisor and the answer. Between the first and second pictures is the division sign. Over the dividend is the equal line, which separates the problem from the answer. On top of the dividend is the third picture, which is the answer of the division. The ***audio*** representation for the pictures consists of a **knock, double knock, knock, double click, knock, cling, and knock.**

In the next part of Step 8.2, we have a **visual** display of how we arrive at the answer. We subtract picture four from picture two, which is the dividend. Between pictures two and four is the subtraction sign. Below picture four is the equal sign, which separates the problem from the answer. Picture five is the result of the subtraction. When the result of the subtraction is an empty position, then a filled position is transferred to the answer of the division. The ***audio*** representation for the subtraction consists of a **knock, double knock, knock, double tram, knock, cling, double knock, double knock, knock.**

Exercise 8

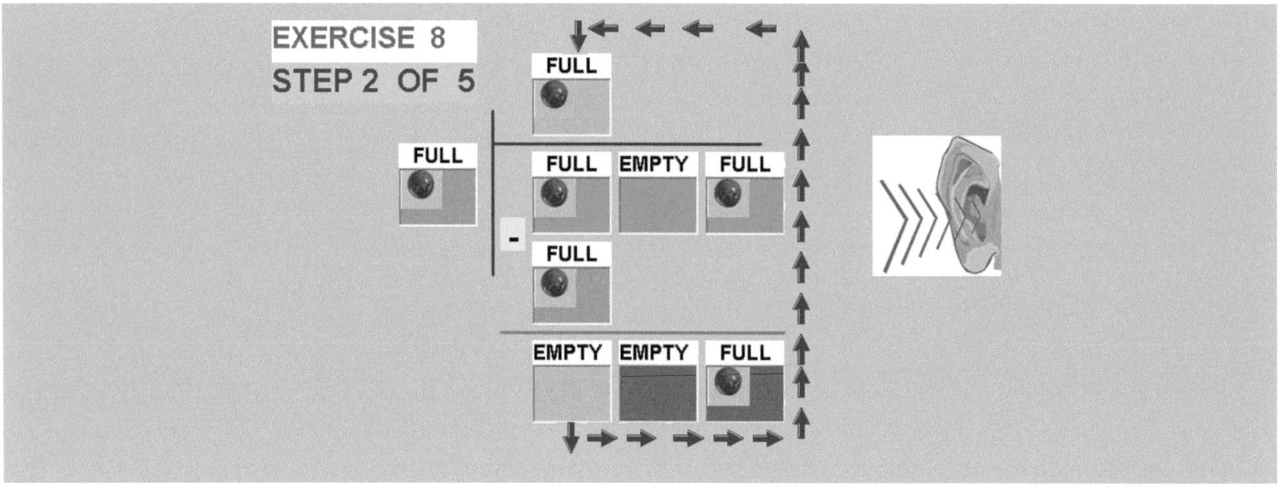

The ***kinesthetic*** representation of a filled up position is represented by the clenching of the right hand. The clenching of the left hand represents an empty position. The division sound, double click, is represented by both hands extended up. Both hands extended in front represent the subtraction sound, double tram. The equal sound, cling, is represented by crossing both hands. The kinesthetic representation for the pictures is the clenching of the right hand, left hand, right hand, both hands extended up, right hand, crossing of hands, and right hand.

DIVISION, EXERCISE 8, STEP 8.3

In Exercise 8 Step 8.3, in a ***visual*** display of division we see three pictures of boxes - the dividend, the divisor and the answer. Between the first and second pictures is the division sign. Over the dividend is the equal line, which separates the problem from the answer. On top of the dividend is the third picture, which is the answer of the division. The ***audio*** representation for the pictures consists of a **double knock, knock, double click, knock, cling, knock, and double knock.**

In the next part of Step 8.3, we have a **visual** display of how we arrive at the answer. In the dividend, the empty positions before the filled up position are automatically taken to the top to the answer.

Exercise 8

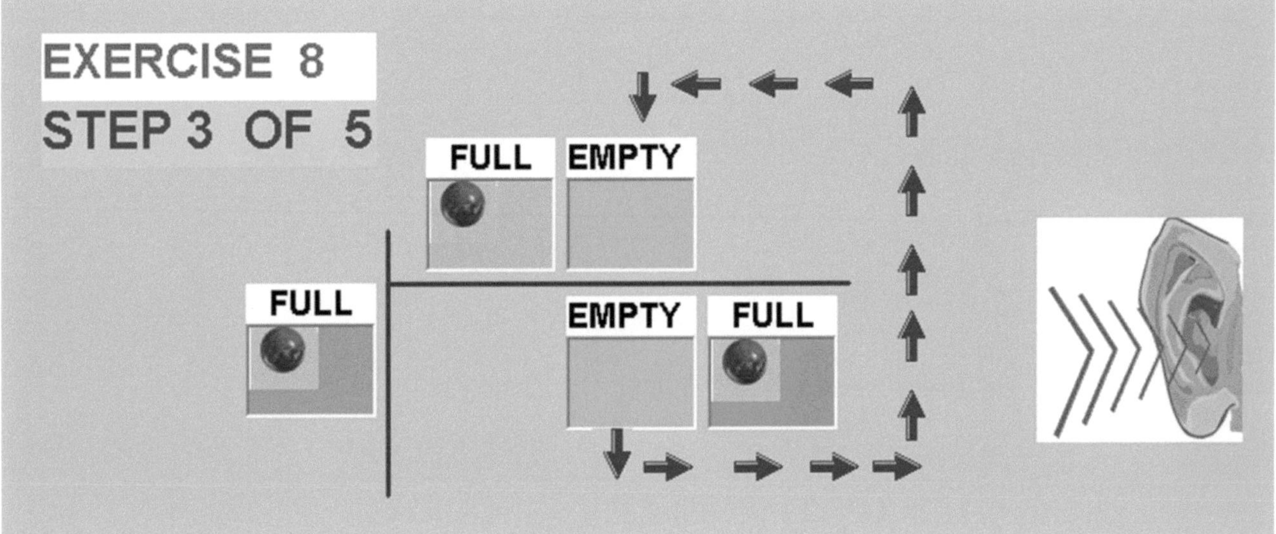

The ***kinesthetic*** representation of a filled up position is represented by the clenching of the right hand. The clenching of the left hand represents an empty position. The division sound, double click, is represented by both hands raised up. Both hands extended in front represent the subtraction sound, double tram. The equal sound, cling, is represented by crossing both hands. The kinesthetic representation for the pictures is the clenching of the left hand, right hand, both hands raised up, right hand, crossing of hands, right hand, and left hand.

MATHEMATICS

DIVISION, EXERCISE 8, STEP 8.4

In Exercise 8 Step 8.4, in a ***visual*** display of division we see three pictures of boxes - the dividend, the divisor and the answer. Between the first and second pictures is the division sign. Over the dividend is the equal line, which separates the problem from the answer. On top of the dividend is the third picture, which is the answer of the division. The ***audio*** representation for the pictures consists of a **knock, double click, knock, cling, knock, double knock, and knock.**

In the next part of Step 8.4, we have a **visual** display of how we arrive at the answer. We subtract picture four from picture two, which is the dividend. Between pictures two and four is the subtraction sign. Below picture four is the equal sign which separates the problem from the answer. Picture five is the result of the subtraction. When the result of the subtraction is an empty position, then a filled position is transferred to the answer of the division. The ***audio*** representation for the subtraction consists of a **knock, double tram, knock, cling, and double knock.**

Exercise 8

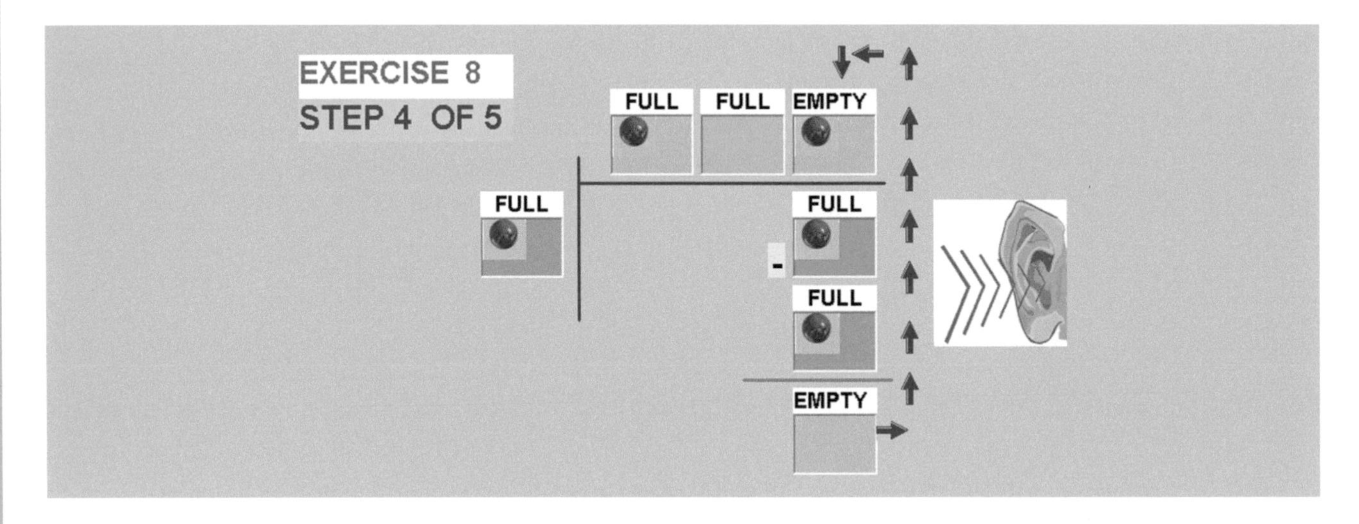

The ***kinesthetic*** representation of a filled up position is represented by the clenching of the right hand. The clenching of the left hand represents an empty position. The division sound, double click, is represented by both hands raised up. Both hands extended in front represent the subtraction sound, double tram. The equal sound, cling, is represented by crossing both hands. The kinesthetic representation for the pictures is the clenching of the right hand, both hands extended up, right hand, crossing of hands, right hand, left hand, and right hand.

DIVISION, EXERCISE 8, STEP 8.5

In Exercise 8 Step 8.5, in a ***visual*** display of the last step of the division we see two pictures of boxes. In both pictures, the first position is filled, the second position is empty, and the last position is filled with a red ball. The colors and placement of the boxes and the balls are different in the pictures. The form, shape, and color of the symbols do no change the mathematical value of the pictures, which is five.

The ***audio*** representation for the pictures consists of a **knock, double knock, knock, and cling.**

Exercise 8

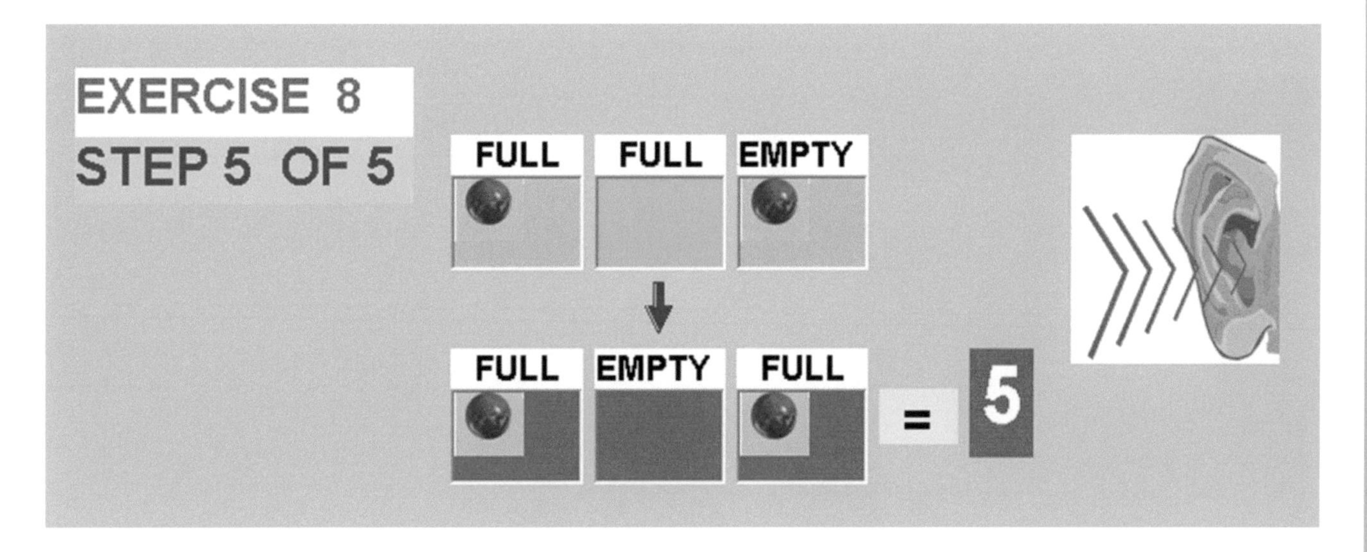

The ***kinesthetic*** representation of a filled up position is represented by the clenching of the right hand. The clenching of the left hand represents an empty position. The equal sound, cling, is represented by crossing of hands. The kinesthetic representation for the pictures is the clenching of the right hand, left hand, right hand, crossing of hands, right hand, left hand, and right hand.

MATHEMATICS

DIVISION, EXERCISE 8

In Exercise 8, in a ***visual*** display of division we see two pictures of boxes and the division sign in-between. In the first picture, the first position is empty while the second position is filled with a red ball. In the second picture, the first position is empty, the second position is filled with a red ball, the third position is empty, and the last position is filled with a red ball. The division sign is between the first and second pictures. On top of the second picture is the equal line, which separates the problem from the answer. The ***audio*** representation for the division consists of a **double knock, knock, double knock, knock double click, double knock, knock, and cling.**

At the bottom we see three pictures with probable answers. From the probable answers for the division we need to find the right one. The audio signals are: 1) **knock, knock, and double knock**, 2) **double knock, knock, and knock**, 3) **knock, double knock, and knock**.

Exercise 8

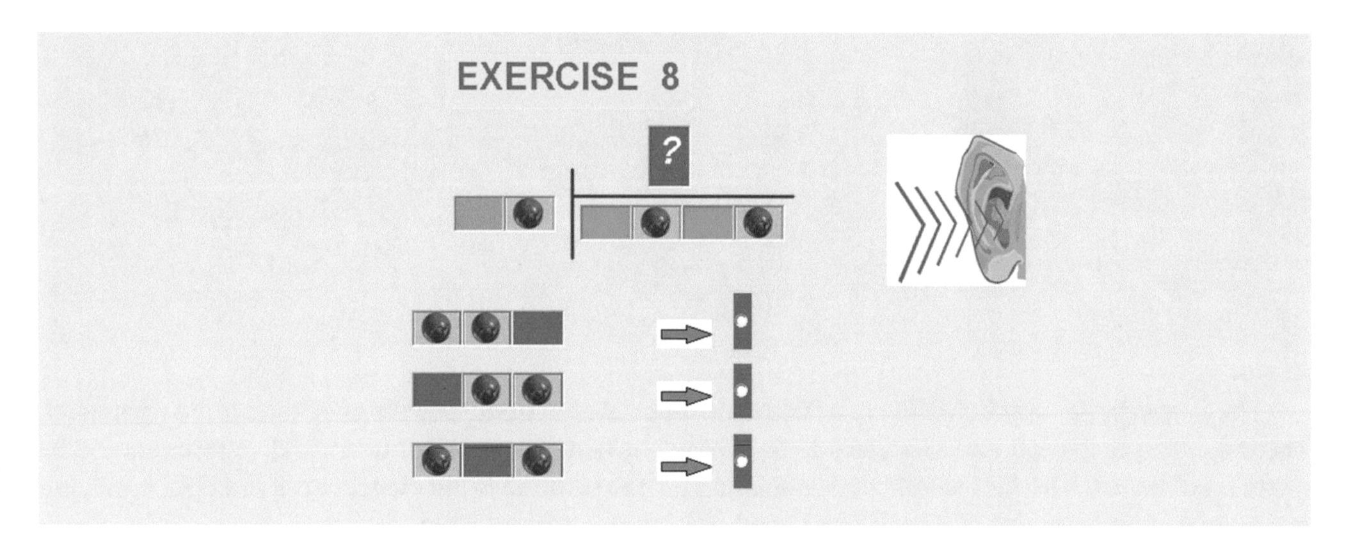

The ***kinesthetic*** representation of a filled up position is represented by the clenching of the right hand. The clenching of the left hand represents an empty position. Both hands extended up represent the division sound, double click. The equal sound, cling, is represented by crossing both hands. The kinesthetic representation for both pictures and the answer is the clenching the left hand, right hand, the clenching the left hand, right hand, both hands extended up, the clenching of the left hand, right hand, crossing of hands, right hand, left hand, and right hand.

DIVISION, EXERCISE 9,

In Exercise 9, in a ***visual*** display of division we see three pictures of boxes - the dividend, the divisor and the answer. Between the first and second pictures is the division sign. Over the dividend is the equal line, which separates the problem from the answer. On top of the dividend is the third picture, which is the answer of the division problem. The ***audio*** representation for the pictures consists of a **double knock, double knock, knock, knock, double click, double knock, knock, cling, double knock, knock, and knock.**

Exercise 9

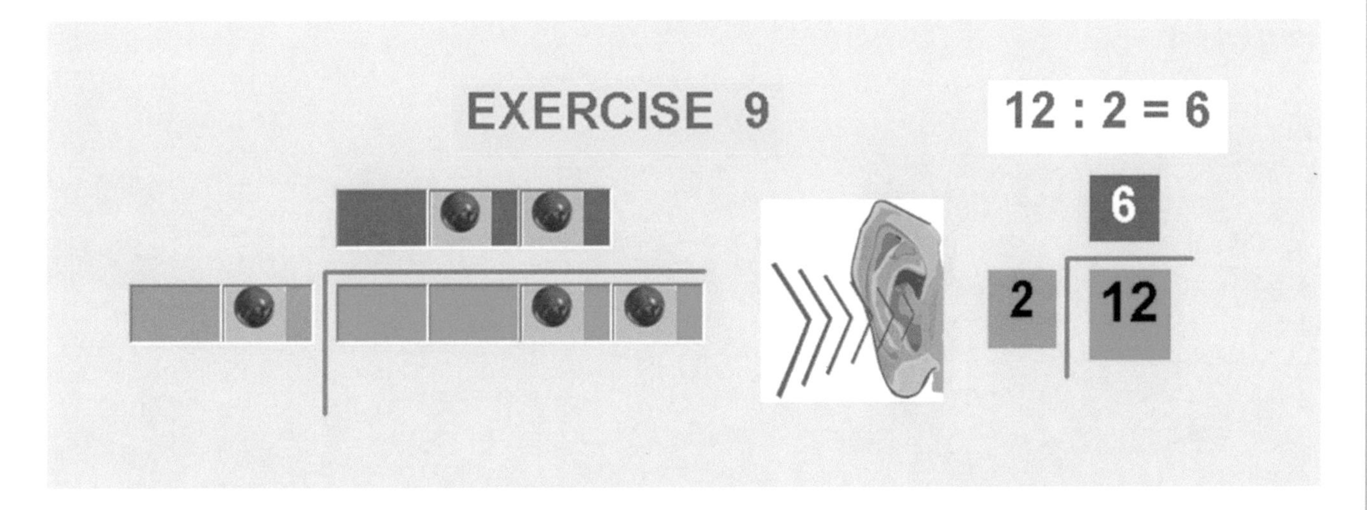

The ***kinesthetic*** representation of a filled up position is represented by the clenching of the right hand. The clenching of the left hand represents an empty position. The division sound, double click, is represented by both hands extended up. The equal sound, cling, is represented by crossing both hands. The kinesthetic representation for the pictures is the clenching of the left hand twice, right hand twice, both hands extended up, left hand, right hand, crossing of hands, left hand, and right hand twice.

MATHEMATICS DIVISION, EXERCISE 9, STEP 9.1

In Exercise 9 Step 9.1, in a ***visual*** display of division we see two pictures of boxes - the dividend and the divisor. Between the first and second pictures is the division sign. Over the dividend is the equal line, which separates the problem from the answer. On top of the dividend is the third picture, which is the answer of the division. The ***audio*** representation for the pictures consists of a **double knock, double knock, knock, knock, double click, double knock, knock, cling.**

In the next part of Step 9.1, we have a **visual** display of how we arrive at the answer. If the divisor has an empty position before the filled up position, then that position cancels out the first empty position of the dividend.

Exercise 9

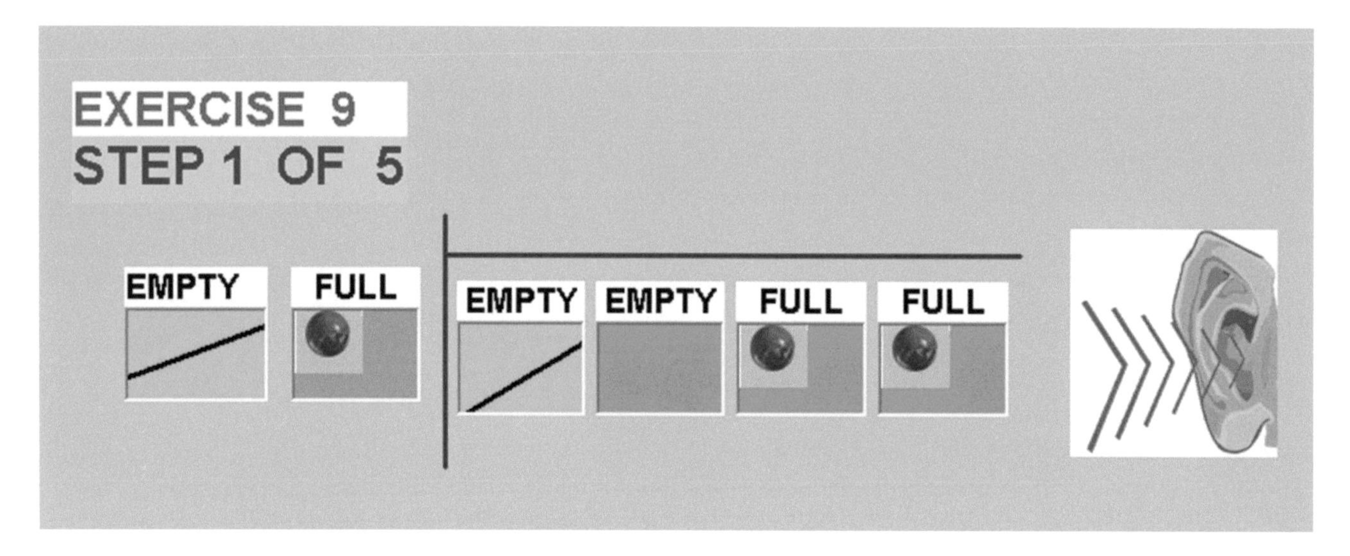

The ***kinesthetic*** representation of a filled up position is represented by the clenching of the right hand. The clenching of the left hand represents an empty position. The division sound, double click, is represented by both hands extended up. Both hands extended in front represent the subtraction sound, double tram. The equal sound, cling, is represented by crossing both hands. The kinesthetic representation for the pictures is the clenching of the left hand twice, right hand twice, both hands extended up, left hand, right hand, and crossing of hands.

DIVISION, EXERCISE 9, STEP 9.2

In Exercise 9 Step 9.2, in a ***visual*** display of division we see three pictures of boxes - the dividend, the divisor and the answer. Between the first and second pictures is the division sign. Over the dividend is the equal line, which separates the problem from the answer. On top of the dividend is the third picture, which is the answer of the division problem. The ***audio*** representation for the pictures consists of a **double knock, knock, knock, double click, knock, cling, and double knock.**

In the next part of Step 9.2, we have a **visual** display of how we arrive at the answer. In the dividend, the empty positions before the filled up position are automatically taken to the top to the answer.

Exercise 9

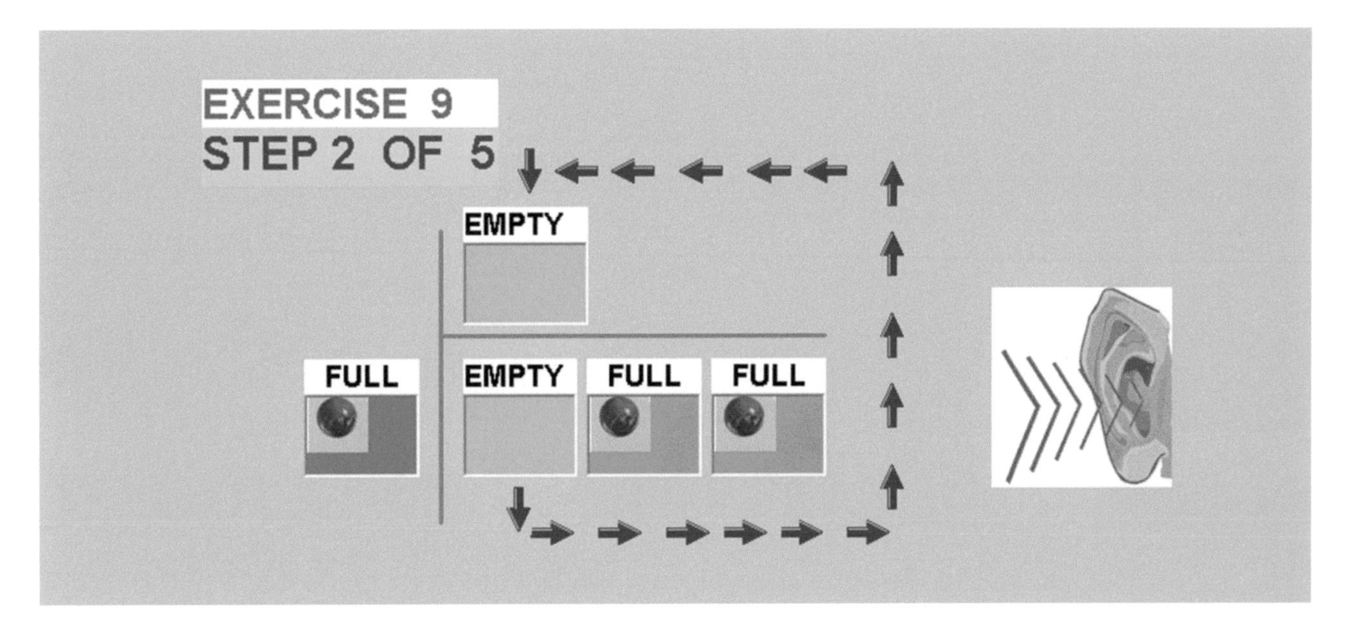

The ***kinesthetic*** representation of a filled up position is represented by the clenching of the right hand. The clenching of the left hand represents an empty position. The division sound, double click, is represented by both hands extended up. Both hands extended in front represent the subtraction sound, double tram. The equal sound, cling, is represented by crossing both hands. The kinesthetic representation for the pictures is the clenching of the left hand, right hand twice, hands extended up, right hand, crossing of both hands, and left hand.

DIVISION, EXERCISE 9, STEP 9.3

In Exercise 9 Step 9.3, in a ***visual*** display of division we see three pictures of boxes - the dividend, the divisor and the answer. Between the first and second pictures is the division sign. Over the dividend is the equal line, which separates the problem from the answer. On top of the dividend is the third picture, which is the answer of the division problem. The ***audio*** representation for the pictures consists of a **knock, knock, double click, knock, cling, double knock, and knock.**

In the next part of Step 9.3, we have a **visual** display of how we arrive at the answer. We subtract picture four from picture two, which is the dividend. Between pictures two and four is the subtraction sign. Below picture four is the equal sign, which separates the problem from the answer. Picture five is the result of the subtraction. When the result of the subtraction is an empty position, then a filled position is transferred to the answer of the division. The ***audio*** representation for the subtraction consists of a **knock, knock, double tram, knock, cling, double knock, and knock.**

Exercise 9

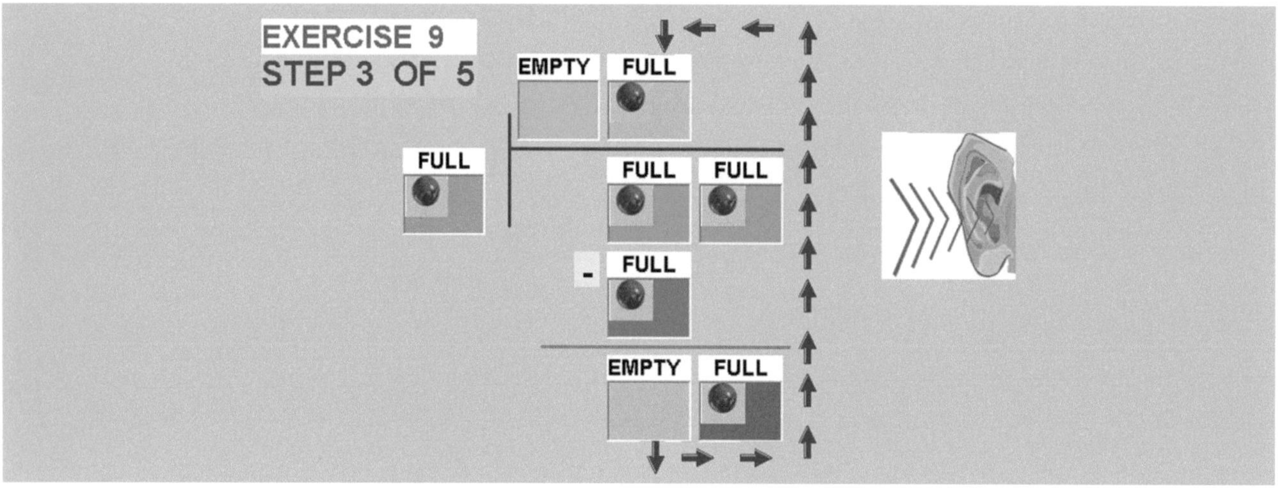

The ***kinesthetic*** representation of a filled up position is represented by the clenching of the right hand. The clenching of the left hand represents an empty position. The division sound, double click, is represented by both hands extended up. Both hands extended in front represent the subtraction sound, double tram. The equal sound, cling, is represented by crossing both hands. The kinesthetic representation for the pictures is the clenching of the right hand twice, both hands extended up, right hand, crossing of hands, left hand, and right hand.

DIVISION, EXERCISE 9, STEP 9.4

In Exercise 9 Step 9.4, in a ***visual*** display of division we see three pictures of boxes - the dividend, the divisor and the answer. Between the first and second pictures is the division sign. Over the dividend is the equal line, which separates the problem from the answer. On top of the dividend is the third picture, which is the answer of the division problem. The ***audio*** representation for the pictures consists of a **knock, double click, knock, cling, double knock, knock, and knock.**

In the next part of Step 9.4, we have a **visual** display of how we arrive at the answer. We subtract picture four from picture two, which is the dividend. Between pictures two and four is the subtraction sign. Below picture four is the equal sign, which separates the problem from the answer. Picture five is the result of the subtraction. When the result of the subtraction is an empty position, then a filled position is transferred to the answer of the division. The ***audio*** representation for the subtraction consists of a **knock, double tram, knock, cling, and double knock.**

Exercise 9

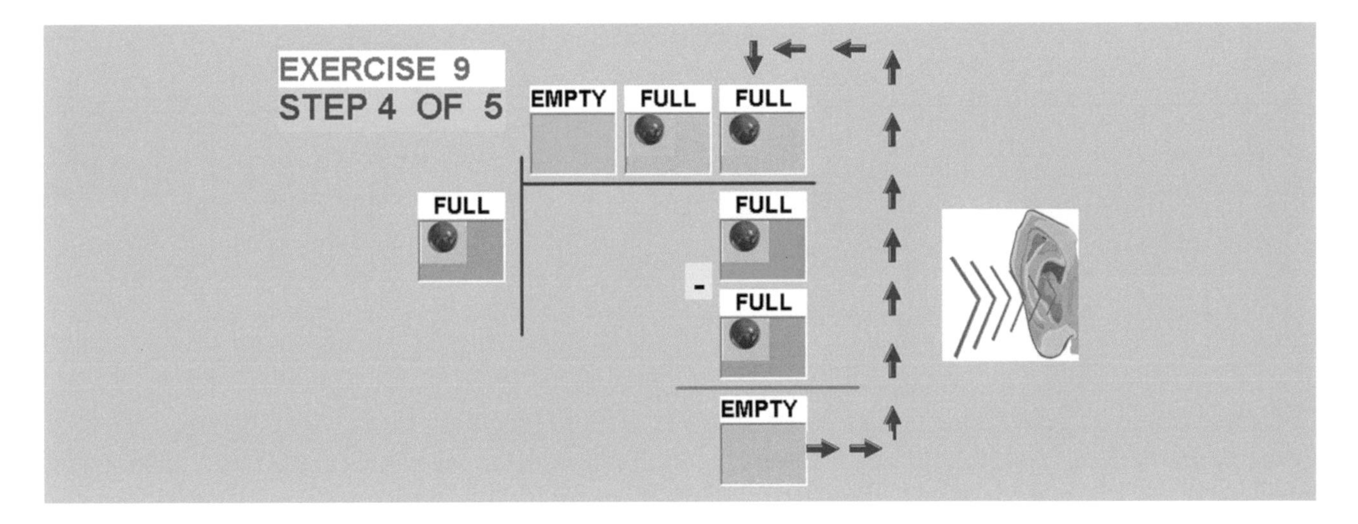

The ***kinesthetic*** representation of a filled up position is represented by the clenching of the right hand. The clenching of the left hand represents an empty position. The division sound, double click, is represented by both hands extended up. Both hands extended in front represent the subtraction sound, double tram. The equal sound, cling, is represented by crossing both hands. The kinesthetic representation for the pictures is the clenching of the right hand, both hands extended up, right hand, crossing of the hands, left hand, and right hand twice.

MATHEMATICS

DIVISION, EXERCISE 9, STEP 9.5

In Exercise 9 Step 9.5, in a ***visual*** display of the last step of the division we see two pictures of boxes. In both pictures, the first position is empty, while the next two positions are filled with red balls. The colors and placement of the boxes and the balls are different in the pictures. The form, shape, and color of the symbols do no change the mathematical value of the pictures, which is six.

The ***audio*** representation for the pictures consists of a **double knock, knock, knock, and cling.**

Exercise 9

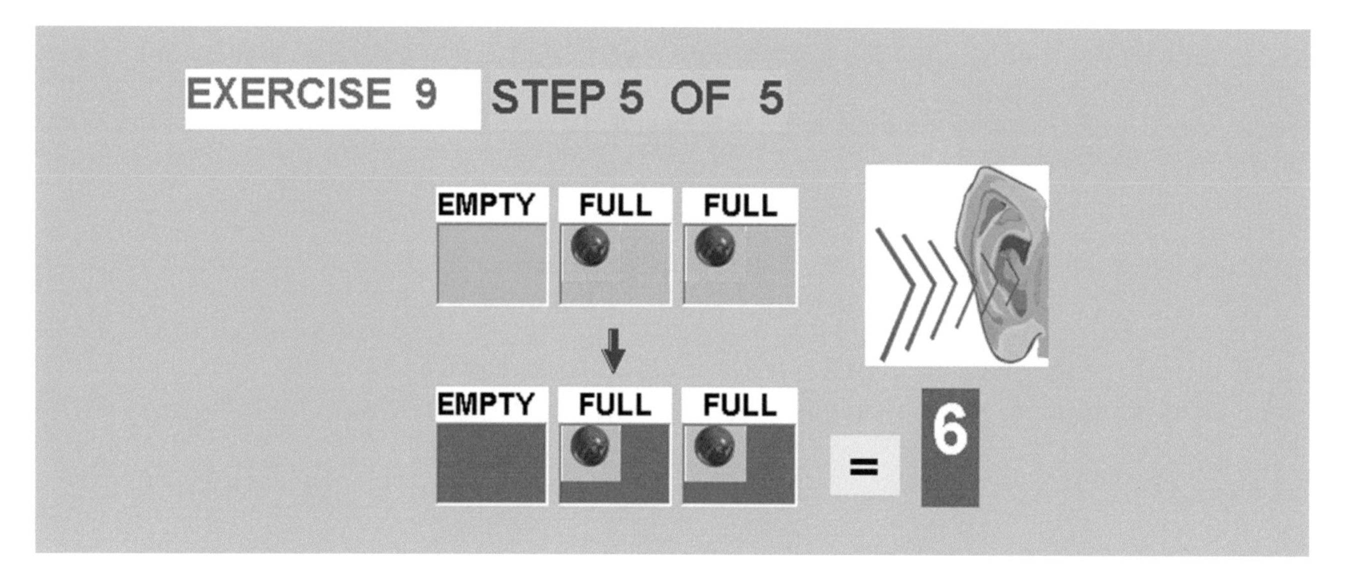

The ***kinesthetic*** representation of a filled up position is represented by the clenching of the right hand. The clenching of the left hand represents an empty position. The equal sound, cling, is represented by crossing the hands. The kinesthetic representation for the pictures is the clenching of the left hand, right hand twice, crossing of hands, left hand, and right hand twice.

DIVISION, EXERCISE 9

In Exercise 9, in a ***visual*** display of division we see two pictures of boxes and the division sign in-between. In the first picture, we see one empty position and one position filled with a red ball. In the second picture, the first two positions are empty and the second two are filled with red balls. The division sign is between the first and second pictures. On top of the second picture is the equal line, which separates the problem from the answer. The ***audio*** representation for the division consists of a **double knock, double knock, knock, knock, double click, double knock, knock, and a cling.**

At the bottom are three pictures with probable answers. From the probable answers for the division we need to find the right one. The audio signals are: 1) **double knock, knock, and knock,** 2) **knock, knock, and double knock**, 3) **knock, double knock, and knock.**

Exercise 9

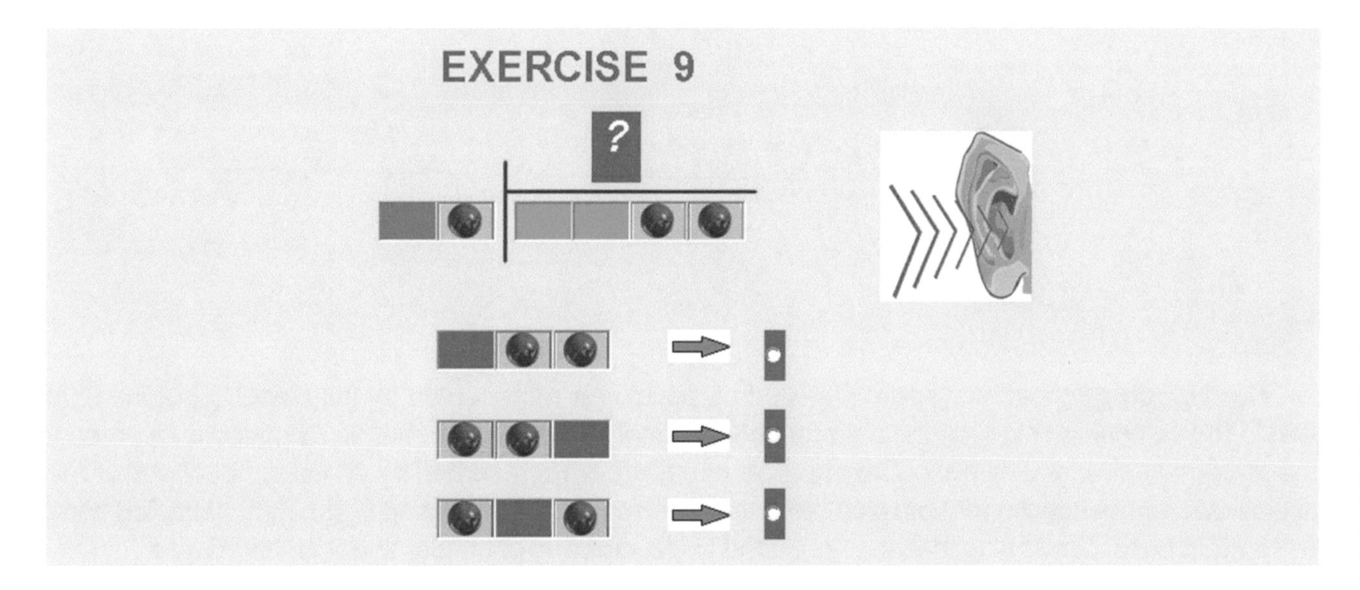

The ***kinesthetic*** representation of a filled up position is represented by the clenching of the right hand. The clenching of the left hand represents an empty position. Both hands extended up represent the division sound, double click. The equal sound, cling, is represented by crossing both hands. The kinesthetic representation for both pictures and the answer is the clenching of the left hand twice, right hand twice, both hands extended up, left hand, right hand, crossing the both hands, left hand, and right hand twice.

MATHEMATICS

DIVISION, EXERCISE 10

In Exercise 10, in a ***visual*** display of division we see three pictures of boxes and the division sign in-between. In the first picture, both positions filled with red balls. In the second picture, the first position is filled with a red ball, the next two positions are empty, and the last position is filled with a red ball. The third picture has two positions filled with red balls. The division sign is between the first and second pictures. On top of the second picture is the equal line, which separates the problem from the answer. The ***audio*** representation for the division consists of a **knock, double knock, double knock, knock, double click, knock, knock, cling, knock, and knock.**

Exercise 10

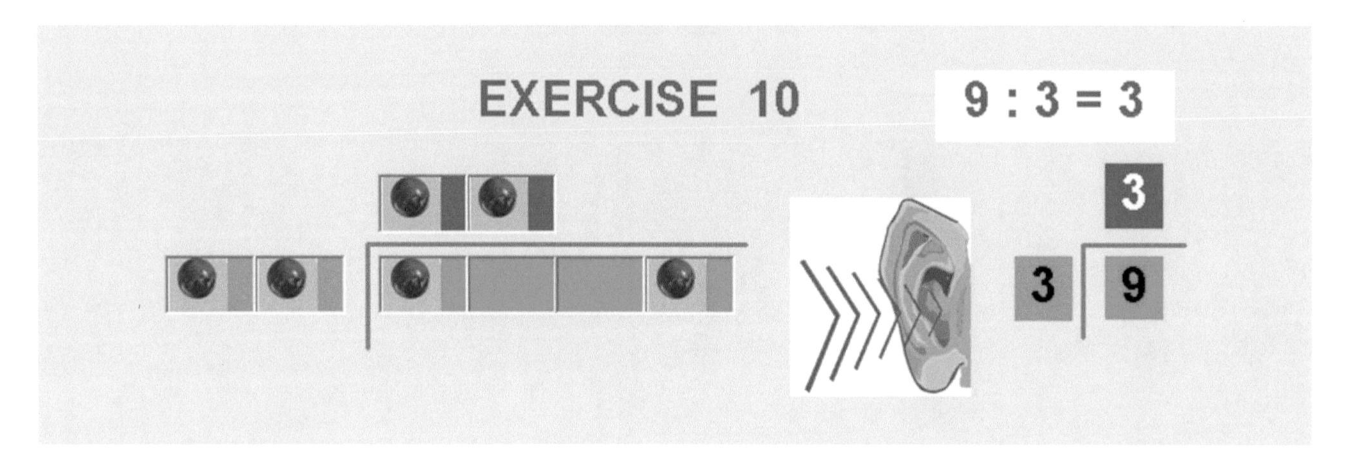

The ***kinesthetic*** representation of a filled up position is represented by the clenching of the right hand. The clenching of the left hand represents an empty position. Both hands extended up represent the division sound, double click. The equal sound, cling, is represented by crossing both hands. The kinesthetic representation for both pictures and the answer is the clenching of the right hand, left hand twice, right hand, hands extended up, right hand twice, crossing of hands, and right hand twice.

DIVISION, EXERCISE 10

In Exercise 10, in a ***visual*** display of division we see two pictures of boxes and the division sign in-between. In the first picture, both positions are filled with red balls. In the second picture, the first position filled with a red ball, the next two positions are empty, and the last position is filled with a red ball. The division sign is between the first and second pictures. On top of the second picture is the equal line, which separates the problem from the answer. The ***audio*** representation for the division consists of a **knock, double knock, double knock, knock, double click, knock, knock, and cling.**

In the next part of Exercise 10, we have a **visual** display of how we arrive at the answer. We subtract picture three from picture two, which is the dividend. Between pictures two and three is the subtraction sign. Below picture three is the equal sign, which separates the problem from the answer. When the result of the subtraction is an empty position, then a filled position is transferred to the answer of the division. The ***audio*** representation for the subtraction consists of a **knock, double knock, double knock, knock, double tram, knock, knock, and cling.**

Exercise 10

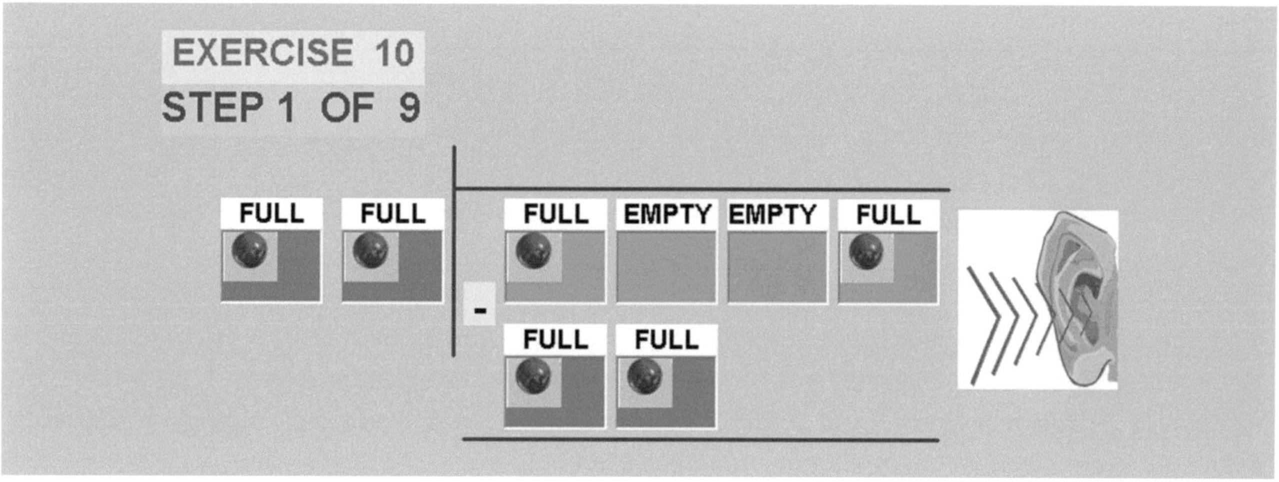

The ***kinesthetic*** representation of a filled up position is represented by the clenching of the right hand. The clenching of the left hand represents an empty position. Both hands extended up represent the division sound, double click. The equal sound, cling, is represented by crossing both hands. The kinesthetic representation for the pictures is the clenching of the right hand, left hand twice, right hand, both hands extended up, right hand twice,

DIVISION, EXERCISE 10, STEP 10.2

In Exercise 10 Step 10.2, in a ***visual*** display of the last step of the division we see two pictures of boxes. In both pictures, the first position is filled with a red ball, the second and third positions are empty, and the last position is filled with a red ball. The colors and placement of the boxes and the balls are different in the pictures. The form, shape, and color of the symbols do no change the mathematical value of the pictures, which is equal to five.

The ***audio*** representation for the pictures consists of a **knock, double knock, double knock, and knock.**

Exercise 10

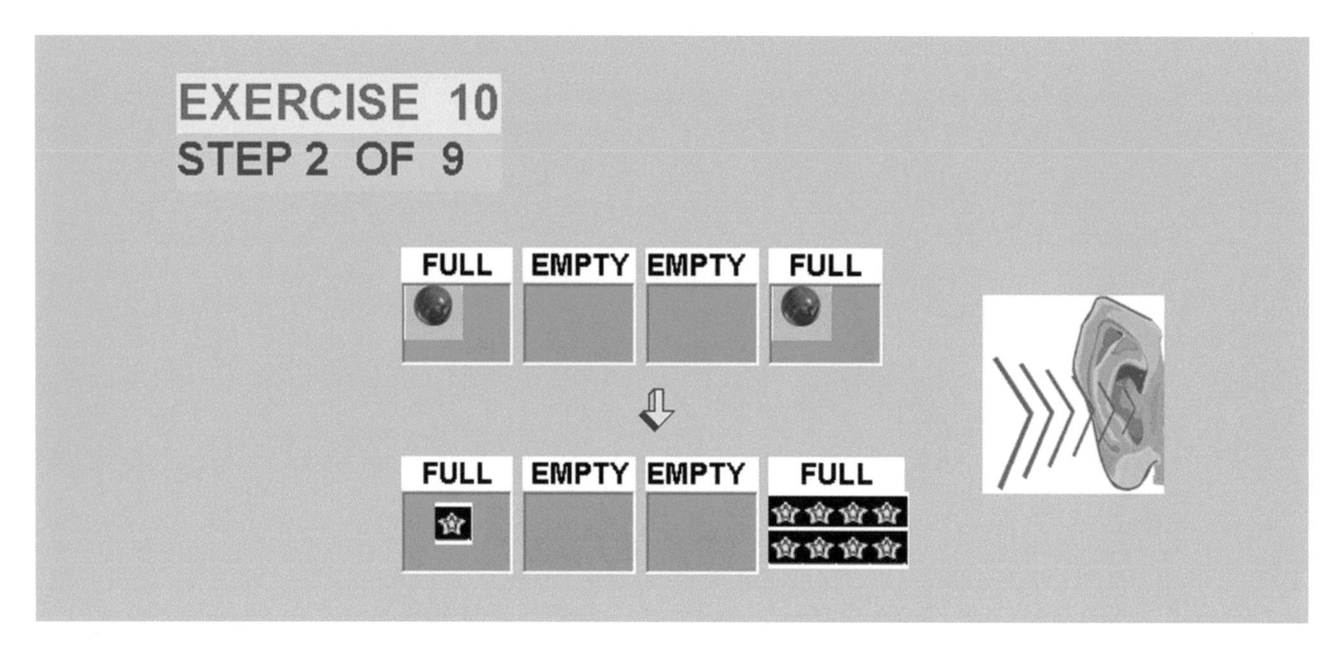

The ***kinesthetic*** representation of a filled up position is represented by the clenching of the right hand. The clenching of the left hand represents an empty position. The equal sound, cling, is represented by crossing the hands. The kinesthetic representation for the pictures is the clenching of the right hand, left hand twice, right hand, and crossing of hands.

DIVISION, EXERCISE 10, STEP 10.3

In Exercise 10 Step 10.3, in a ***visual*** display of division we see three pictures of boxes. In the first two pictures, the first position is filled with a symbol, the second and third positions are empty, and the last position is filled with a symbol. For the purpose of the division, we introduce the concept of borrowing. We borrow from the fourth to the third position. The audio representation for the pictures does not change, because the numerical value does not change. The ***audio*** representation for the pictures with the equal sign consists of a **knock, double knock, double knock, knock, and cling.**

Exercise 10

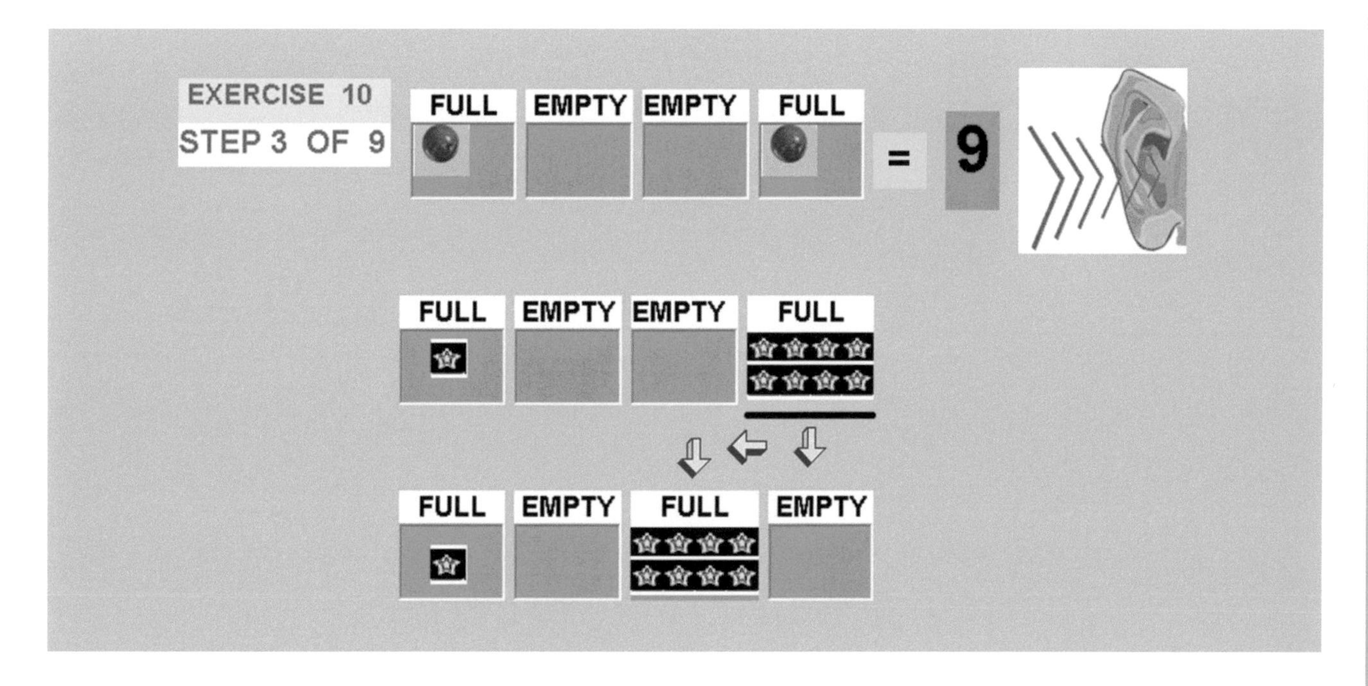

The ***kinesthetic*** representation of a filled up position is represented by the clenching of the right hand. The clenching of the left hand represents an empty position. The equal sound, cling, is represented by crossing both hands. The kinesthetic representation for both pictures and the answer is the clenching of the right hand, left hand twice, right hand, and crossing of hands.

MATHEMATICS

DIVISION, EXERCISE 10, STEP 10.4

In Exercise 10 Step 10.4, in a ***visual*** display of division we see three pictures of boxes. In first picture, the first position is filled with a symbol, the second and third positions are empty, and the last position is filled with a symbol. In the second picture, the first position is filled, the second position is empty, the third position is filled, and the last position is empty. The third picture has the first three positions filled and the last position empty. According to the concept of borrowing, we borrow from the third to the second and third positions of the next picture. The audio representation for the pictures does not change, because the numerical value does not change. The ***audio*** representation for the three pictures with the equal sign consists of a **knock, double knock, double knock, knock, and cling.**

Exercise 10

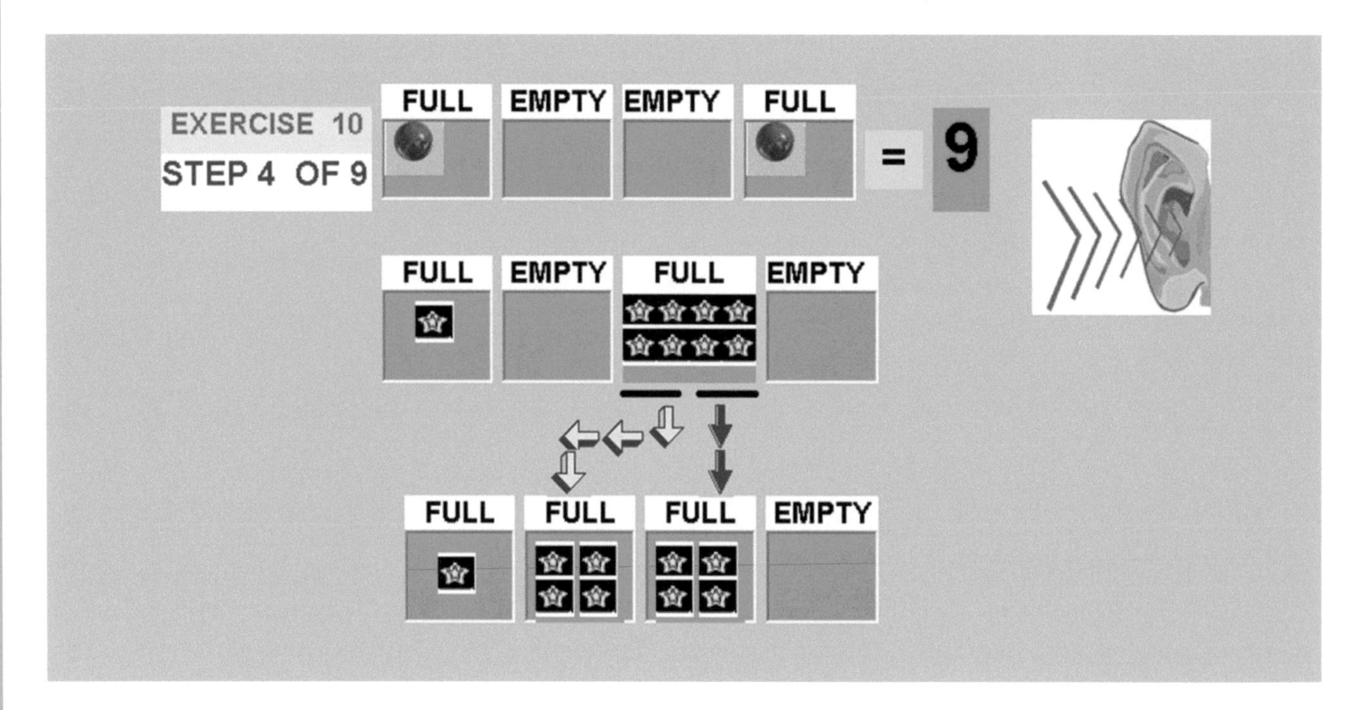

The ***kinesthetic*** representation of a filled up position is represented by the clenching of the right hand. The clenching of the left hand represents an empty position. The equal sound, cling, is represented by crossing both hands. The kinesthetic representation for both pictures and the answer is the clenching of the right hand, left hand twice, right hand, and crossing of the hands.

DIVISION, EXERCISE 10, STEP 10.5

In Exercise 10 Step 10.5, in a ***visual*** display of division we see three pictures of boxes - the dividend, the divisor and the answer. Between the first and second pictures is the division sign. Over the dividend is the equal line, which separates the problem from the answer. On top of the dividend is the third picture, which is the answer of the division problem. The ***audio*** representation for the pictures consists of a **knock, knock, knock, double click, knock, knock, cling.**

In the next part of Step 10.5, we have a **visual** display of how we arrive at the answer. We subtract picture four from picture two, which is the dividend. Between pictures two and four is the subtraction sign. Below picture four is the equal sign which separates the problem from the answer. Picture five is the result of the subtraction. The ***audio*** representation for the subtraction consists of a **knock, knock, knock, double tram, knock, knock, cling, double knock, knock, and knock.**

Exercise 10

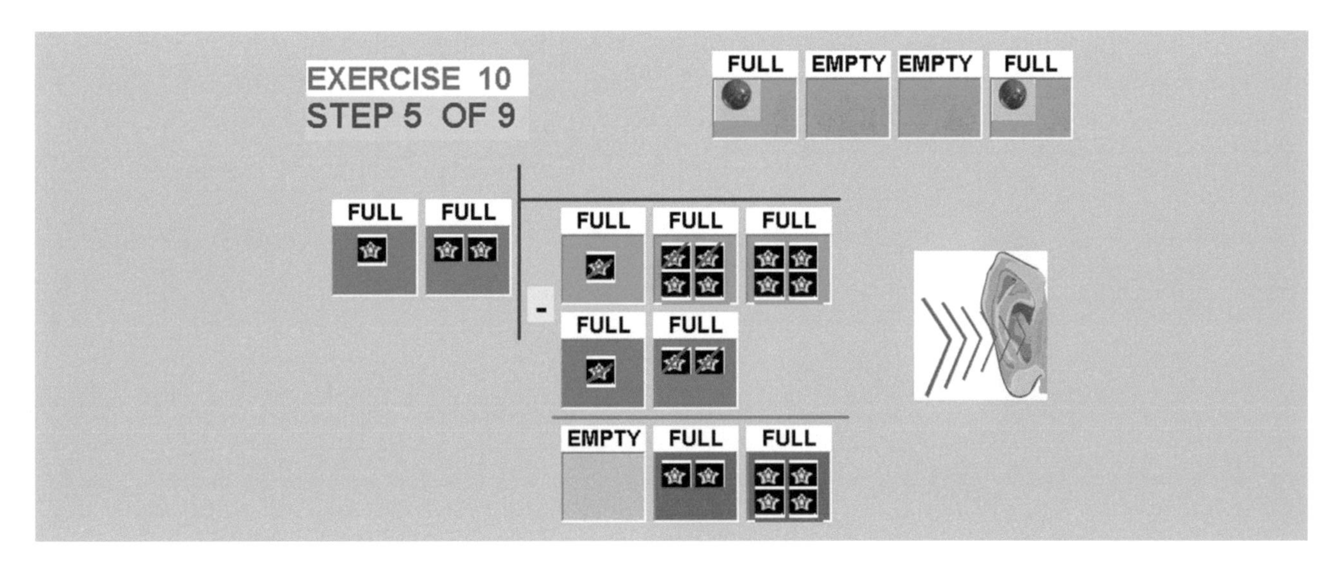

The ***kinesthetic*** representation of a filled up position is represented by the clenching of the right hand. The clenching of the left hand represents an empty position. The division sound, double click, is represented by both hands extended up in the air. Both hands extended in front represent the subtraction sound, double tram. The equal sound, cling, is represented by crossing both hands. The kinesthetic representation for the pictures is the clenching of the right hand three times, both hands extended up, right hand twice, crossing the hands.

MATHEMATICS

DIVISION, EXERCISE 10, STEP 10.6

In Exercise 10 Step 10.6, in a ***visual*** display of division we see three pictures of boxes - the dividend, the divisor and the answer. Between the first and second pictures is the division sign. Over the dividend is the equal line, which separates the problem from the answer. On top of the dividend is the third picture, which is the answer of the division problem. The ***audio*** representation for the pictures consists of a **double knock, knock, knock, double click, knock, knock, cling, and knock.**

In the next part of Step 10.6, we have a **visual** display of how we arrive at the answer. In the dividend, the empty positions before the filled up position are automatically taken to the top to the answer as a filled position.

Exercise 10

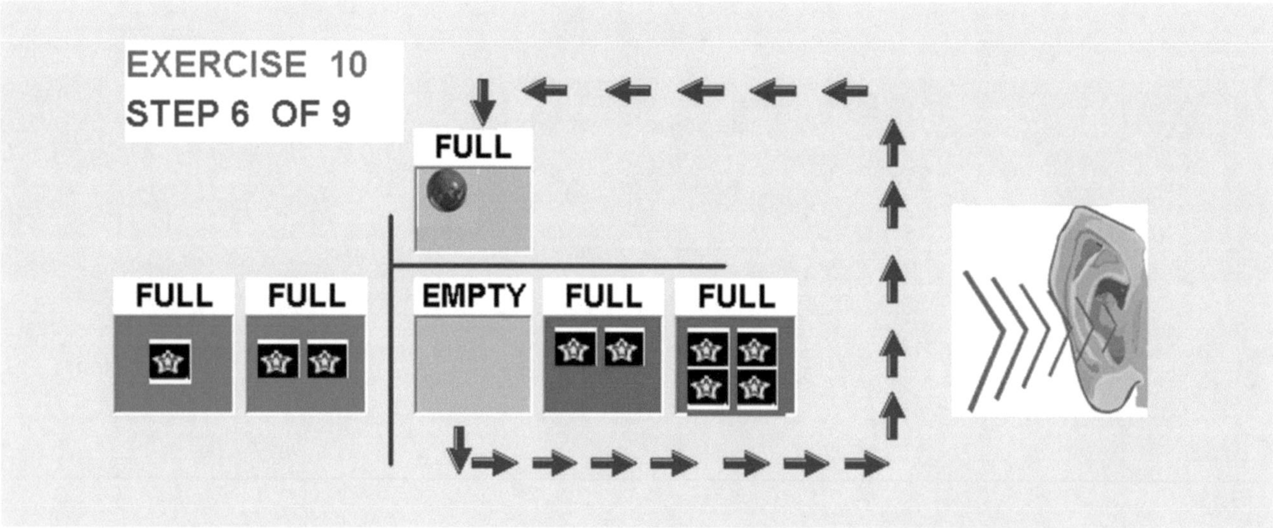

EXERCISE 10
STEP 6 OF 9
FULL
FULL
FULL
EMPTY
FULL
FULL

The ***kinesthetic*** representation of a filled up position is represented by the clenching of the right hand. The clenching of the left hand represents an empty position. The division sound, double click, is represented by both hands extended up in the air. Both hands extended in front represent the subtraction sound, double tram. The equal sound, cling, is represented by crossing both hands. The kinesthetic representation for the pictures is the clenching of the left hand, right hand twice, both hands extended up, right hand twice, crossing both hands, and right hand.

DIVISION, EXERCISE 10, STEP 10.8

In Exercise 10 Step 10.8, in a ***visual*** display of division we see three pictures of boxes - the dividend, the divisor and the answer. Between the first and second pictures is the division sign. Over the dividend is the equal line, which separates the problem from the answer. On top of the dividend is the third picture, which is the answer of the division problem. The ***audio*** representation for the pictures consists of a **knock, knock, double click, knock, knock, cling, knock, and knock.**

In the next part of Step 10.8, we have a **visual** display of how we arrive at the answer. We subtract picture four from picture two, which is the dividend. Between pictures two and four is the subtraction sign. Below picture four is the equal sign, which separates problem from the answer. Picture five is the result of the subtraction. When the result of the subtraction is an empty position, then a filled position is transferred to the answer of the division. The ***audio*** representation for the subtraction consists of a **knock, knock, double tram, knock, knock, cling, and double knock.**

Exercise 10

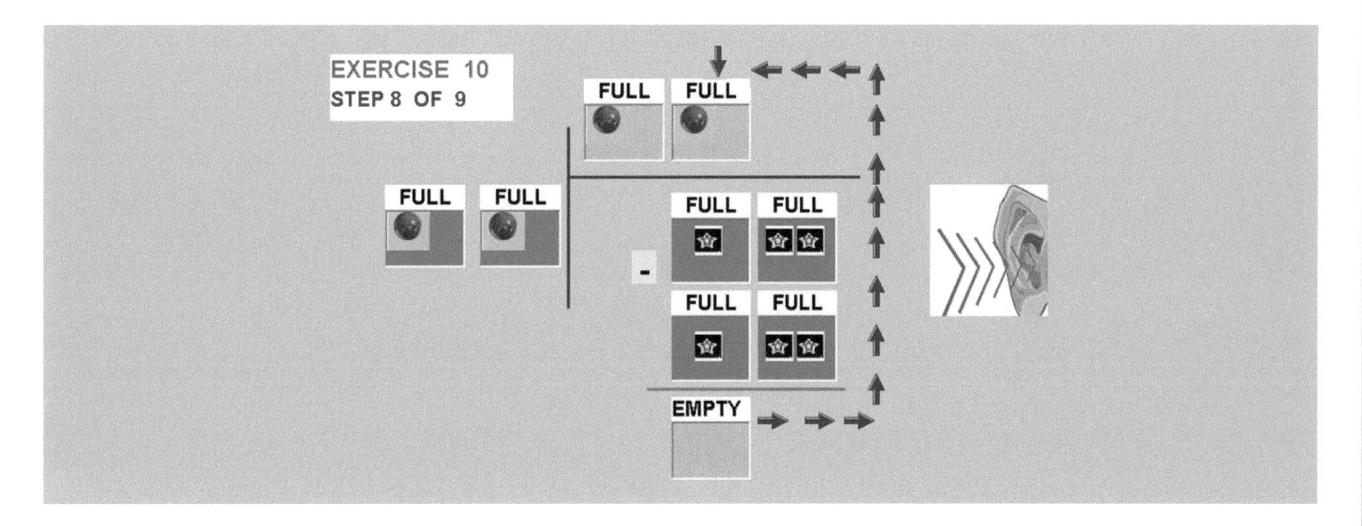

The ***kinesthetic*** representation of a filled up position is represented by the clenching of the right hand. The clenching of the left hand represents an empty position. The division sound, double click, is represented by both hands extended up. Both hands extended in front represent the subtraction sound, double tram. The equal sound, cling, is represented by crossing both hands. The kinesthetic representation for the pictures is the clenching of the right hand twice, hands extended up, right hand twice, crossing of hands, and right hand twice.

DIVISION, EXERCISE 10, STEP 10.9

In Exercise 10 Step 10.9, in a ***visual*** display of the last step of the division we see two pictures of boxes. In both pictures, both positions are filled with red balls. The colors of the boxes are different in the pictures. The form, shape, and color of the symbols do no change the mathematical value of the pictures, which is three.

The ***audio*** representation for the pictures consists of a **knock, knock, and cling.**

Exercise 10

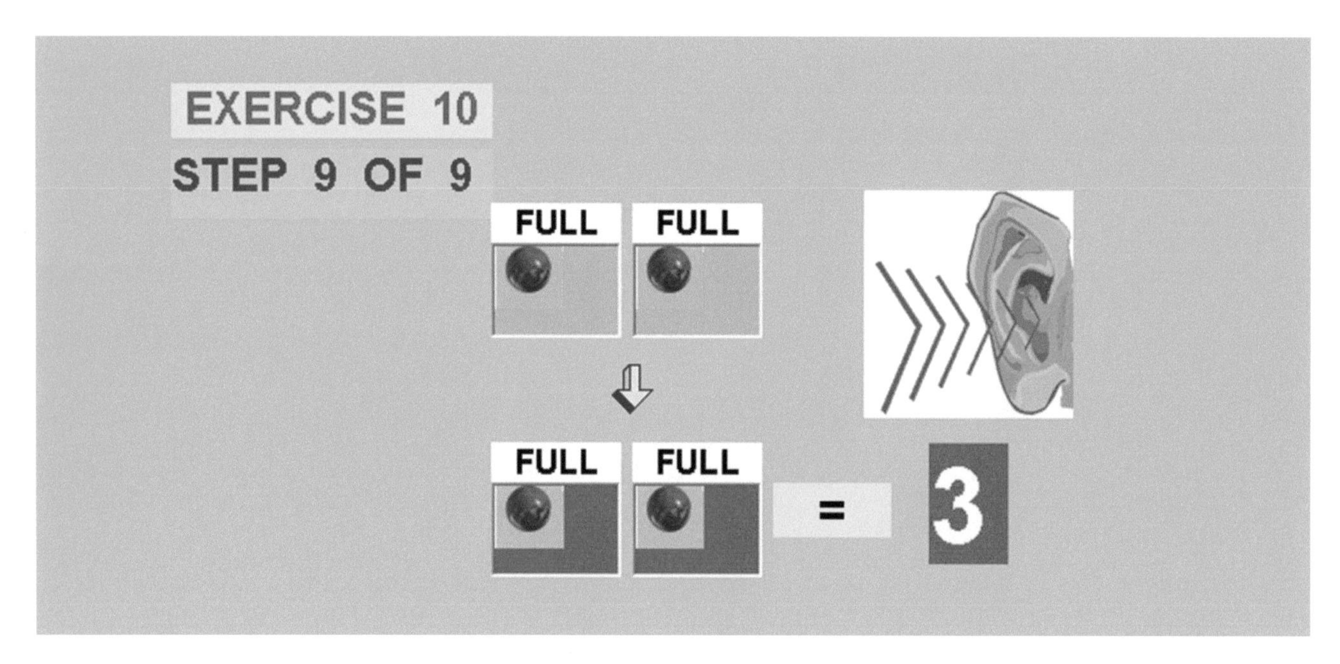

The ***kinesthetic*** representation of a filled up position is represented by the clenching of the right hand. The clenching of the left hand represents an empty position. The equal sound, cling, is represented by crossing the hands. The kinesthetic representation for the pictures is the clenching of the right hand twice, crossing the hands, and the clenching of the right hand twice again.

DIVISION, EXERCISE 10

In Exercise 10, in a ***visual*** display of division we see two pictures of boxes and the division sign in-between. In the first picture, both positions are filled with a red ball. In the second picture, the first position is filled with a red ball, the next two positions are empty, and the last position is filled with a red ball. The division sign is between the first and second pictures. On top of the second picture is the equal line, which separates the problem from the answer. The ***audio*** representation for the division consists of a **knock, double knock, double knock, knock, double click, knock, knock, cling, knock, and knock.**

Exercise 10

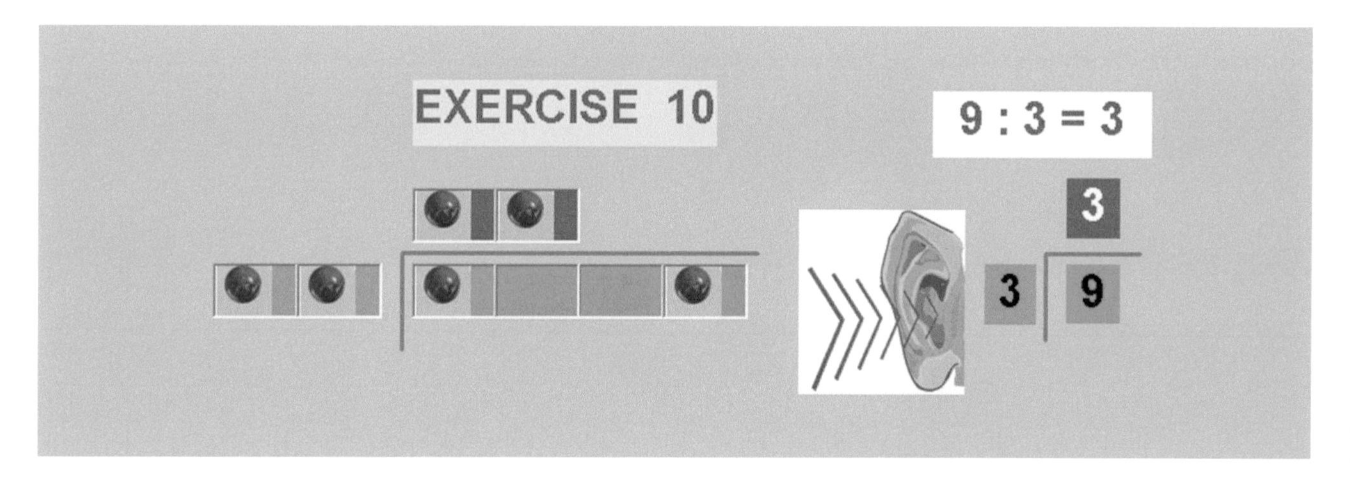

The ***kinesthetic*** representation of a filled up position is represented by the clenching of the right hand. The clenching of the left hand represents an empty position. Both hands extended up represent the division sound, double click. The equal sound, cling, is represented by crossing both hands. The kinesthetic representation for both pictures and the answer is the clenching of the right hand, left hand twice, right hand, both hands extended up, right hand twice, crossing both hands, and right hand twice.

MATHEMATICS

DIVISION, EXERCISE 12

In Exercise 12, in a ***visual*** display of division we see two pictures of boxes and the division sign in-between. In the first picture, the first position is empty while the next two are filled with red balls. In the second picture, the first position is empty, the second position is filled with red ball, the third and fourth positions are empty, and the last position is filled with a red ball. The division sign is between the first and second pictures. On top of the second picture is the equal line, which separates the problem from the answer. The ***audio*** representation for the division consists of a **double knock, knock, double knock, double knock, knock, double click, double knock, knock, knock, cling, knock, and knock.**

Exercise 12

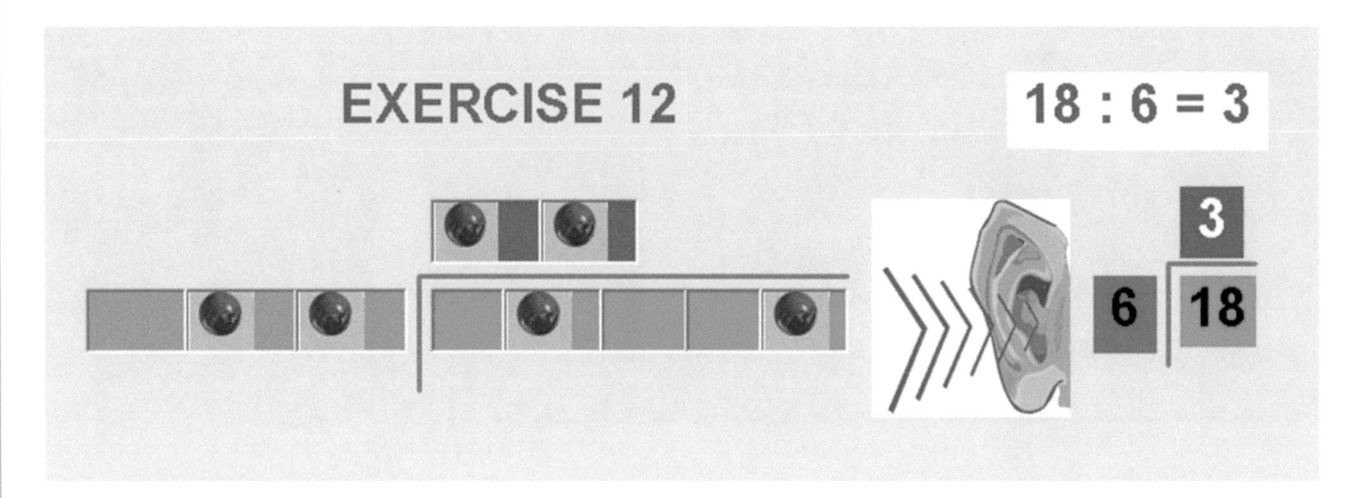

The ***kinesthetic*** representation of a filled up position is represented by the clenching of the right hand. The clenching of the left hand represents an empty position. Both hands extended up represent the division sound, double click. The equal sound, cling, is represented by crossing both hands. The kinesthetic representation for both pictures and the answer is the clenching of the left hand, right hand, left hand twice, right hand, both hands extended up, left hand, right hand twice, crossing both hands, and right hand twice.

DIVISION, EXERCISE 12, STEP 12.1

In Exercise 12 Step 12.1, in a ***visual*** display of division we see two pictures of boxes - the dividend and the divisor. Between the first and second pictures is the division sign. Over the dividend is the equal line, which separates the problem from the answer. The ***audio*** representation for the pictures consists of a **double knock, knock, double knock, double knock, knock double click, double knock, knock, knock, cling.**

In the next part of Step 8.1, we have a **visual** display of how we arrive at the answer. If the divisor has an empty position before the filled up position, then that position cancels out the first empty position of the dividend.

Exercise 12

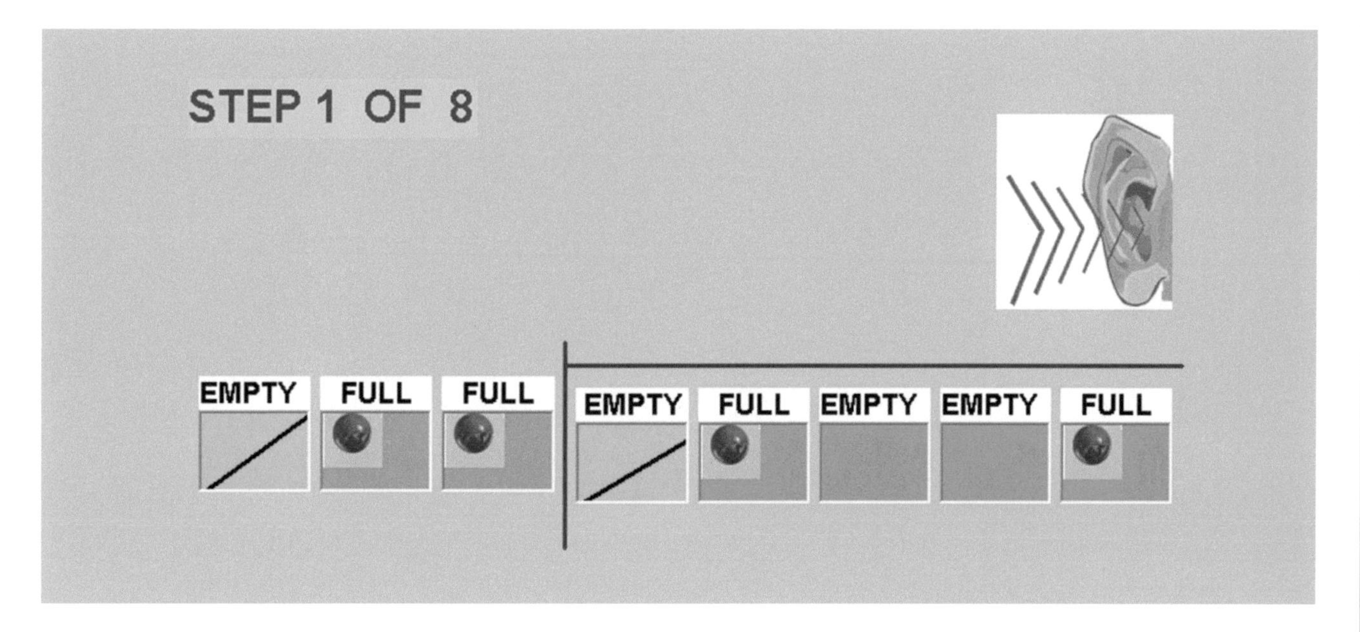

The ***kinesthetic*** representation of a filled up position is represented by the clenching of the right hand. The clenching of the left hand represents an empty position. The division sound, double click, is represented by both hands extended up. Both hands extended in front represent the subtraction sound, double tram. The equal sound, cling, is represented by crossing both hands. The kinesthetic representation for the pictures is the clenching of the left hand, right hand, left hand twice, right hand, both hands extended up, left hand, right hand twice, crossing the hands.

DIVISION, EXERCISE 12, STEP 12.2

In Exercise 12 Step 12.2, in a ***visual*** display of the next step of the division we see two pictures of boxes. In both pictures, the first position is filled with a symbol, the second and third positions are empty, and the last position is filled with a symbol. The colors and placement of the boxes and balls are different in the pictures. The form, shape, and color of the symbols do no change the mathematical value of the pictures, which is equal to five.

The ***audio*** representation for the pictures consists of a **knock, double knock, double knock, and knock.**

Exercise 12

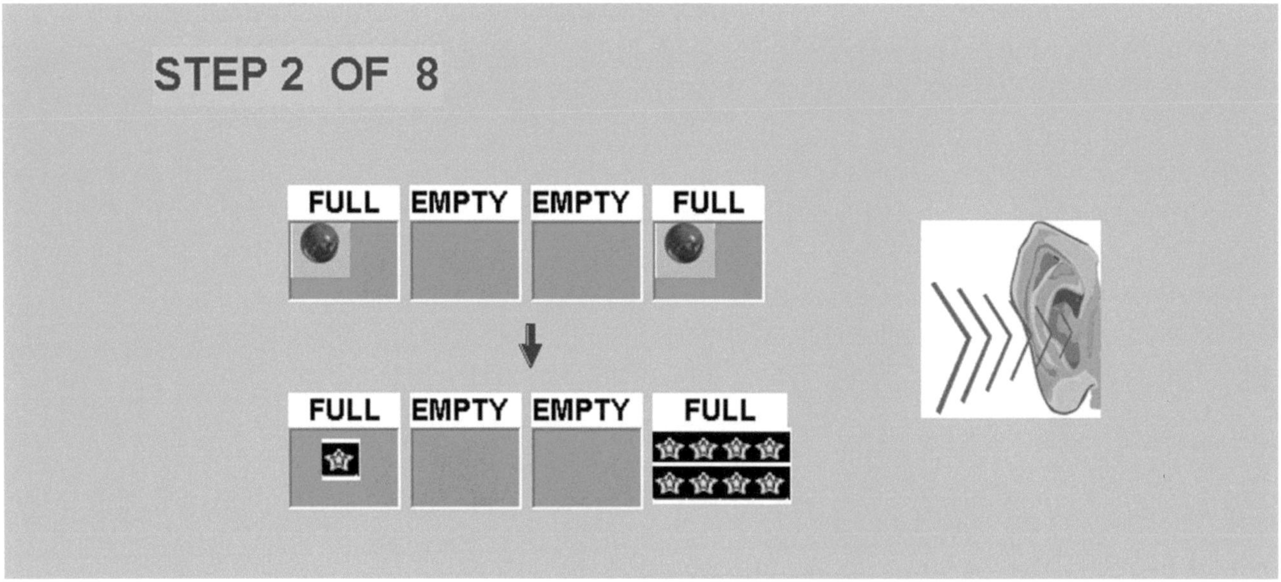

The ***kinesthetic*** representation of a filled up position is represented by the clenching of the right hand. The clenching of the left hand represents an empty position. The equal sound, cling, is represented by crossing the hands. The kinesthetic representation for the pictures is the clenching of the right hand, left hand twice, right hand, and crossing of the hands.

DIVISION, EXERCISE 12, STEP 12.3

In Exercise 12 Step 12.3, in a ***visual*** display of division we see three pictures of boxes. In the first two pictures, the first position is filled with a symbol, the second and third positions are empty, and the last position is filled a symbol. For the purpose of division, we introduce the concept of borrowing. We borrow from the fourth to the third position. The audio representation for the pictures does not change, because the numerical value does not change. The ***audio*** representation for the pictures with the equal sign consists of a **knock, double knock, double knock, knock, and cling.**

Exercise 12

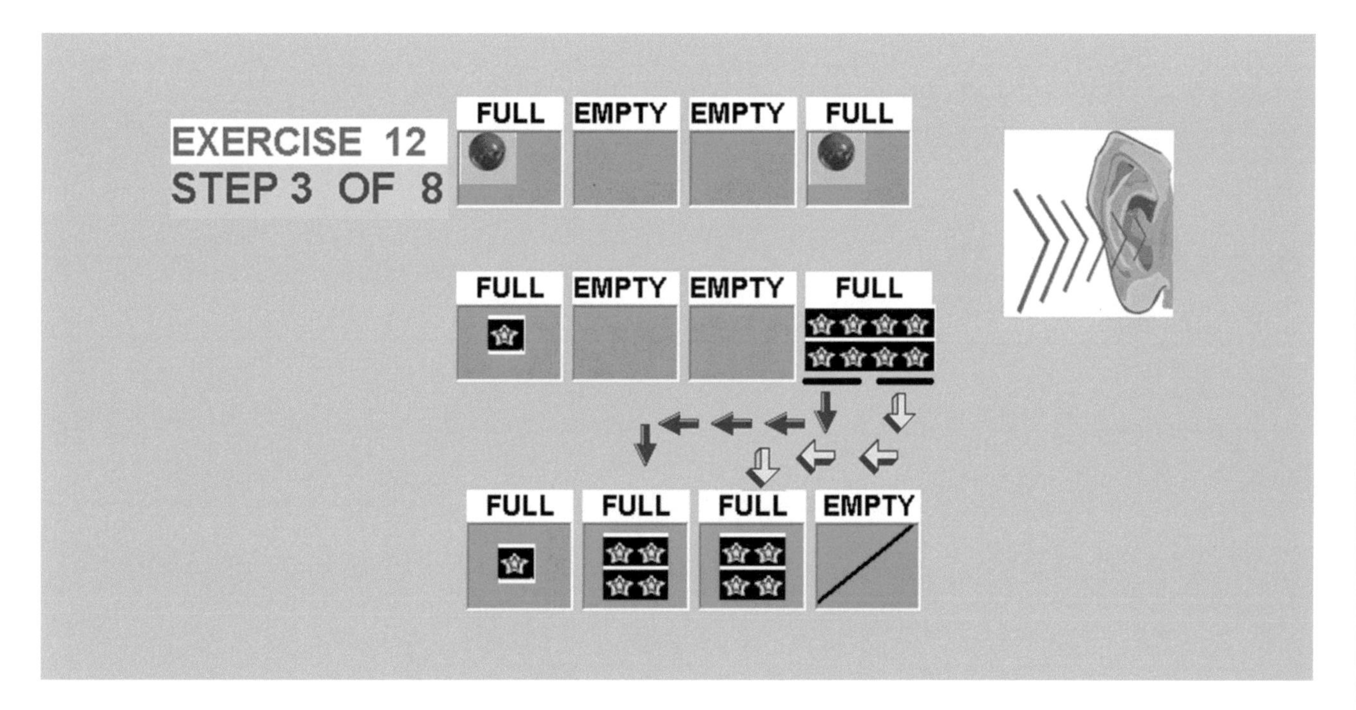

The ***kinesthetic*** representation of a filled up position is represented by the clenching of the right hand. The clenching of the left hand represents an empty position. The equal sound, cling, is represented by crossing both hands. The kinesthetic representation for the pictures and the answer is the clenching of the right hand, left hand twice, right hand, and crossing the hands.

DIVISION, EXERCISE 12, STEP 12.4

In Exercise 12 Step 12.4, in a ***visual*** display of division we see three pictures of boxes - the dividend, the divisor and the answer. Between the first and second pictures is the division sign. Over the dividend is the equal line, which separates the problem from the answer. On top of the dividend is the third picture, which is the answer of the division problem. The ***audio*** representation for the pictures consists of a **knock, knock, knock double click, knock, knock, cling, and knock.**

In the next part of Step 12.4, we have a **visual** display of how we arrive at the answer. We subtract picture four from picture two, which is the dividend. Between pictures two and four is the subtraction sign. Below picture four is the equal sign, which separates the problem from the answer. Picture five is the result of the subtraction. The ***audio*** representation for the subtraction consists of a **knock, knock, knock, double tram, knock, knock, cling, double knock, knock, and knock.**

Exercise 12

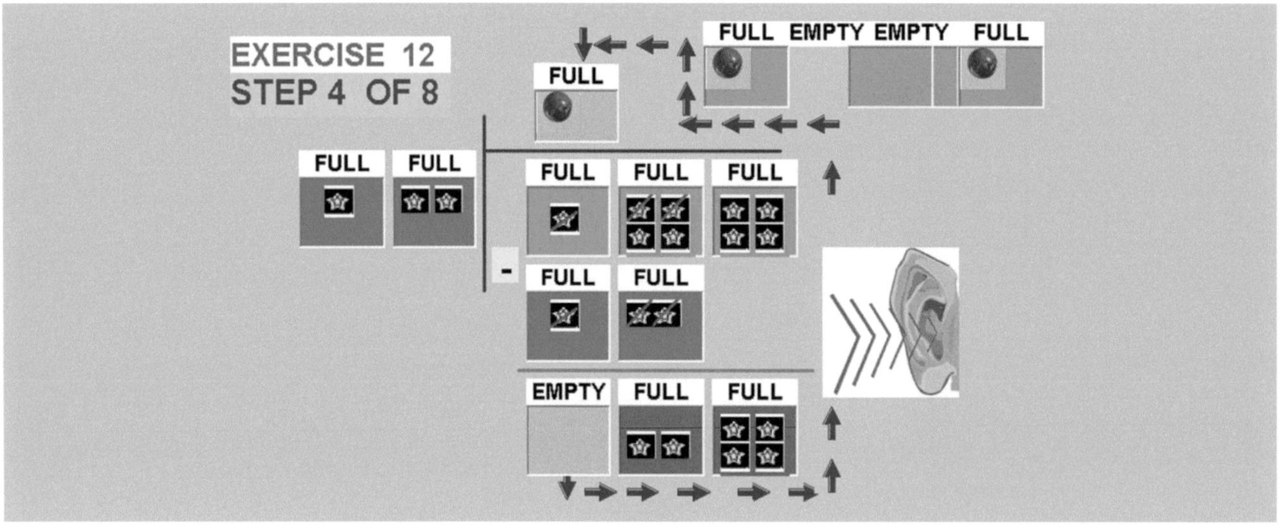

The ***kinesthetic*** representation of a filled up position is represented by the clenching of the right hand. The clenching of the left hand represents an empty position. The division sound, double click, is represented by both hands extended up. Both hands extended in front represent the subtraction sound, double tram. The equal sound, cling, is represented by crossing both hands. The kinesthetic representation for the pictures is the clenching of the right hand three times, hands extended up, right hand twice, crossing of the hands, and right hand.

DIVISION, EXERCISE 12, STEP 12.5

In Exercise 12 Step 12.5, in a ***visual*** display of division we see three pictures of boxes - the dividend, the divisor and the answer. Between the first and second pictures is the division sign. Over the dividend is the equal line, which separates the problem from the answer. On top of the dividend is the third picture, which is the answer of the division. The ***audio*** representation for the pictures consists of a **double knock, knock, knock, double click, knock, knock, cling, and knock.**

In the next part of Step 12.1, we have a **visual** display of how we arrive at the answer. In the dividend, the empty positions before the filled up position are automatically taken to the top to the answer as a filled up position.

Exercise 12

EXERCISE 12
STEP 5 OF 8
FULL
FULL FULL
EMPTY FULL FULL

The ***kinesthetic*** representation of a filled up position is represented by the clenching of the right hand. The clenching of the left hand represents an empty position. The division sound, double click, is represented by both hands extended up. Both hands extended in front represent the subtraction sound, double tram. The equal sound, cling, is represented by crossing both hands. The kinesthetic representation for the pictures is the clenching of the left hand, right hand twice, hands extended up, right hand twice, crossing of the hands, and right hand.

MATHEMATICS

DIVISION, EXERCISE 12, STEP 12.6

In Exercise 12 Step 12.6, in a ***visual*** display of the next step of the division we see two pictures of boxes. In both pictures, both positions are filled with red balls. The colors and placement of the boxes and balls are different in the pictures. The form, shape, and color of the symbols do no change the mathematical value of the pictures, which is equal to five.

The ***audio*** representation for the pictures consists of a **knock, knock, cling, knock, and knock.**

Exercise 12

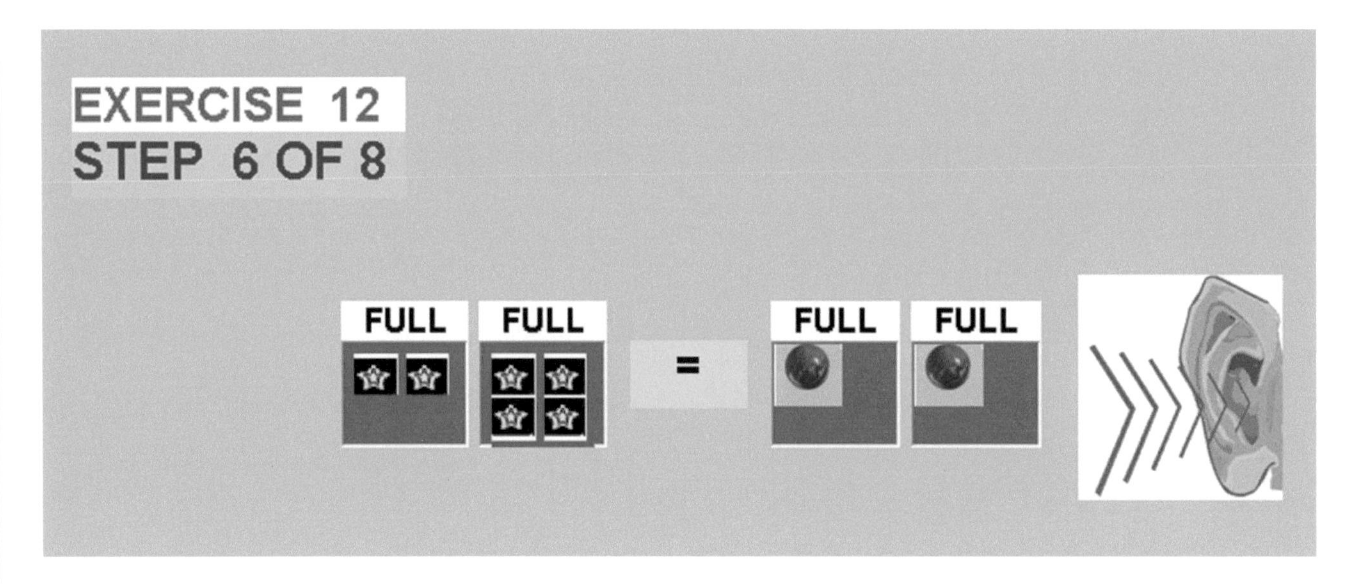

The ***kinesthetic*** representation of a filled up position is represented by the clenching of the right hand. The clenching of the left hand represents an empty position. The equal sound, cling, is represented by crossing the hands. The kinesthetic representation for the two pictures is the clenching of the right hand twice, crossing the hands, and right hand twice.

DIVISION, EXERCISE 12, STEP 12.7

In Exercise 12 Step 12.7, in a ***visual*** display of division we see three pictures of boxes - the dividend, the divisor and the answer. Between the first and second pictures is the division sign. Over the dividend is the equal line, which separates the problem from the answer. On top of the dividend is the third picture, which is the answer of the division. The ***audio*** representation for the pictures consists of a **knock, knock, double click, knock, knock, cling, knock, and knock.**

In the next part of Step 12.7, we have a **visual** display of how we arrive at the answer. We subtract picture four from picture two, which is the dividend. Between pictures two and four is the subtraction sign. Below picture four is the equal sign, which separates the problem from the answer. Picture five is the result of the subtraction. When the result of the subtraction is an empty position, then a filled position is transferred to the answer of the division. The ***audio*** representation for the subtraction consists of a **knock, knock, double tram, knock, knock, cling, and double knock.**

Exercise 12

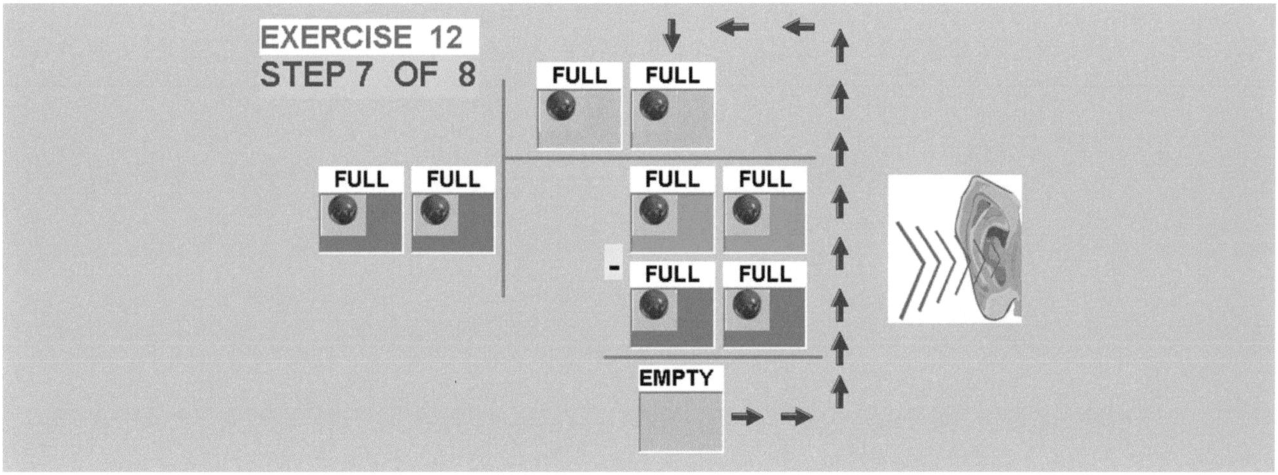

EXERCISE 12
STEP 7 OF 8

FULL FULL

FULL FULL

FULL FULL

\- FULL FULL

EMPTY

The ***kinesthetic*** representation of a filled up position is represented by the clenching of the right hand. The clenching of the left hand represents an empty position. The division sound, double click, is represented by both hands extended up. Both hands extended in front represent the subtraction sound, double tram. The equal sound, cling, is represented by crossing both hands. The kinesthetic representation for the pictures is the clenching of the right hand twice, hands extended up, right hand twice, crossing of the hands, and right hand twice.

DIVISION, EXERCISE 12, STEP 12.8

In Exercise 12 Step 12.9, in a ***visual*** display of the last step of the division we see two pictures of boxes. In both pictures, both positions are filled with red balls. The colors and placement of the boxes and balls are different in the pictures. The form, shape, and color of the symbols do no change the mathematical value of the pictures, which is three.

The ***audio*** representation for the pictures consists of a **knock, knock, and cling.**

Exercise 12

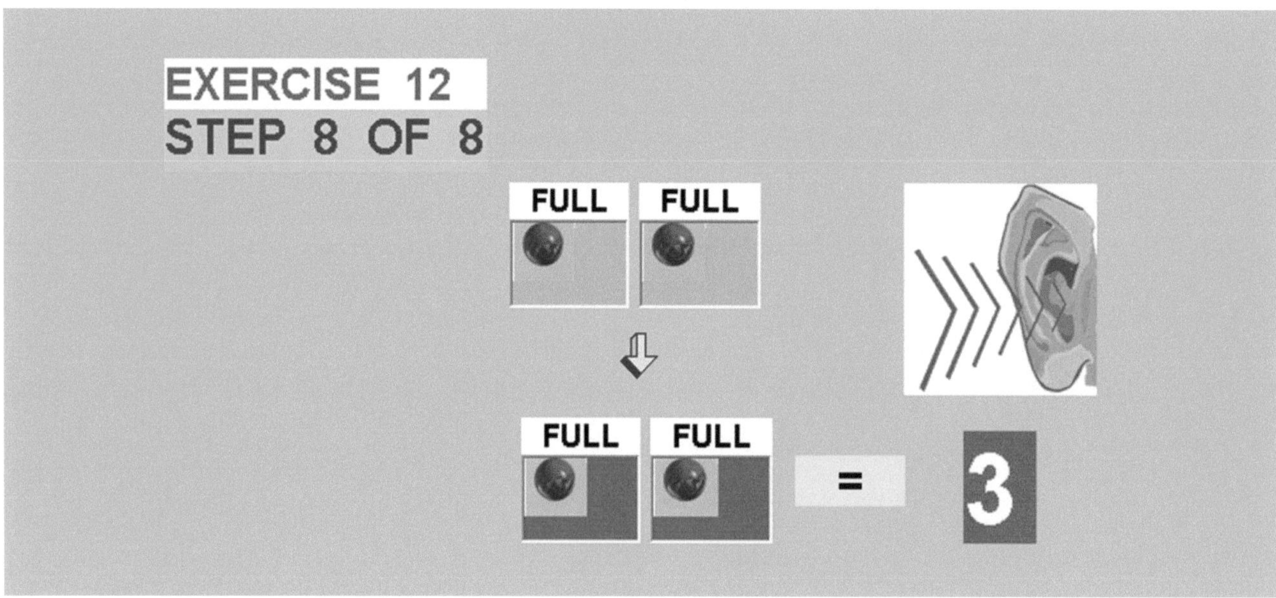

The ***kinesthetic*** representation of a filled up position is represented by the clenching of the right hand. The clenching of the left hand represents an empty position. The equal sound, cling, is represented by the crossing of hands. The kinesthetic representation for the pictures is the clenching of the right hand twice, crossing of hands and right hand twice.

DIVISION, EXERCISE 12

In Exercise 12, in a ***visual*** display of division we see two pictures of boxes and the division sign in-between. In the first picture, we see one empty position and the next two positions are filled with a red ball. In the second picture, the first position is empty, the second position is filled with a red ball, the third and fourth positions are empty, and the last position is filled with a red ball. The division sign is between pictures one and two. On top of the second picture is the equal line, which separates the problem from the answer. The ***audio*** representation for the division consists of a **double knock, knock, double knock, double knock, knock, double click, double knock, knock, knock, and cling.**

At the bottom are three pictures with probable answers. From the probable answers for the division we need to find the right one. The audio signals are: 1) **knock, double knock, and knock,** 2) **knock, knock, and double knock**, 3) **double knock, knock, and knock.**

Exercise 12

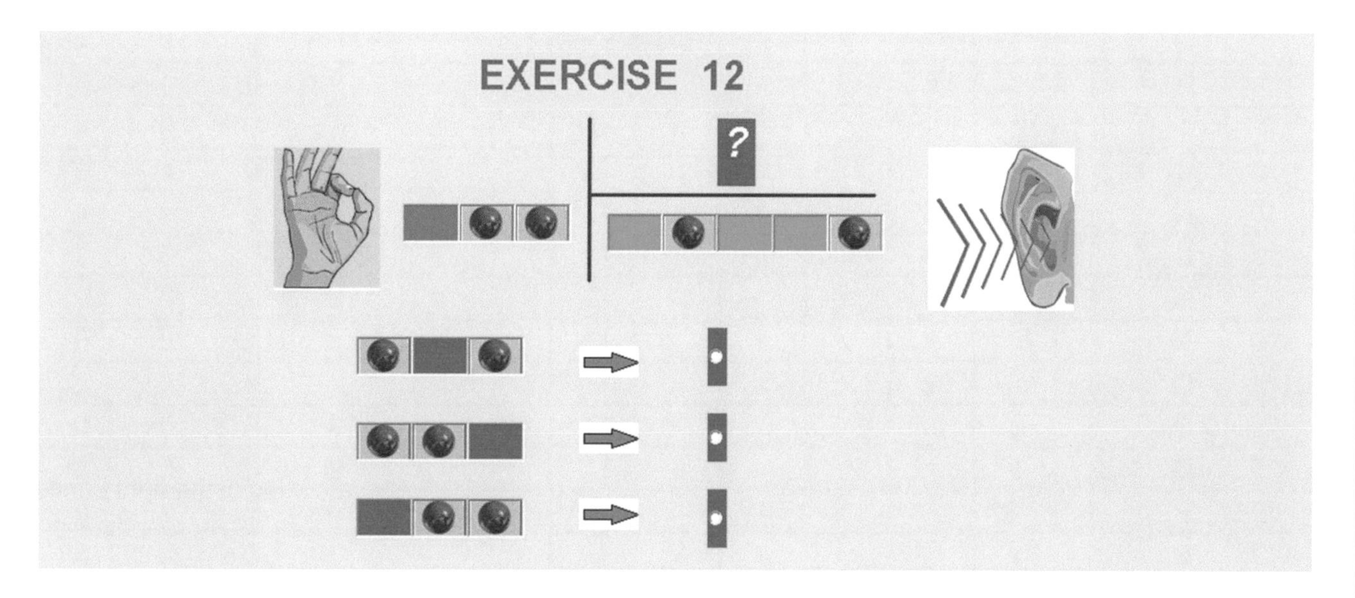

The ***kinesthetic*** representation of a filled up position is represented by the clenching of the right hand. The clenching of the left hand represents an empty position. Both hands extended up represent the division sound, double click. The equal sound, cling, is represented by crossing both hands. The kinesthetic representation for both pictures and the answer is the clenching of the left hand, right hand, left hand twice, right hand, hands extended up, left hand, right hand twice, crossing of hands, right hand twice, and left hand.

Reversed Patterns Used in Litvin's Code

In examples below, we count positions from right to left

Alphabets		***LITVIN'S CODE*** **Using only crosses**							***LITVIN'S CODE*** **Using "x" and "0"**					
		32	16	8	4	2	1		32	16	8	4	2	1
1	A						x	Or					0	x
2	B					x		Or					x	0
3	C					x	x	Or					x	x
4	D				x			Or				x	0	0
5	E				x		x	Or				x	0	x
6	F				x	x		Or				x	x	0
7	G				x	x	x	Or				x	x	x
8	H			x				Or			x	0	0	0
9	I			x			x	Or			x	0	0	x
10	J			x		x		Or			x	0	x	0
11	K			x		x	x	Or			x	0	x	x
12	L			x	x			Or			x	x	0	0
13	M			x	x		x	Or			x	x	0	x
14	N			x	x	x		Or			x	x	x	0
15	O			x	x	x	x	Or			x	x	x	x
16	P		x					Or		x	0	0	0	0
17	Q		x				x	Or		x	0	0	0	x
18	R		x			x		Or		x	0	0	x	0
19	S		x			x	x	Or		x	0	0	x	x
20	T		x		x			Or		x	0	x	0	0
21	U		x		x		x	Or		x	0	x	0	x
22	V		x		x	x		Or		x	0	x	x	0
23	W		x		x	x	x	Or		x	0	x	x	x

		32	16	8	4	2	1		32	16	8	4	2	1
24	X		x	x				Or		x	x	0	0	0
25	Y		x	x			x	Or		x	x	0	0	x
26	Z		x	x		x		Or		x	x	0	x	0
27			x	x		x	x	Or		x	x	0	x	x
28			x	x	x			Or		x	x	x	0	0
29			x	x	x		x	Or		x	x	x	0	x
30			x	x	x	x		Or		x	x	x	x	0
31			x	x	x	x	x	Or		x	x	x	x	x
32		x						Or	x	0	0	0	0	0
33		x					x	Or	x	0	0	0	0	x
34		x				x		Or	x	0	0	0	x	0
35		x				x	x	Or	x	0	0	0	x	x
36		x			x			Or	x	0	0	x	0	0
37		x			x		x	Or	x	0	0	x	0	x
38		x			x	x		Or	x	0	0	x	x	0
39		x			x	x	x	Or	x	0	0	x	x	x
40		x		x				Or	x	0	x	0	0	0

The presence of a symbol (dot) is a representation of a binary number. One can keep track of all positions with this symbol. In Litvin's Code, to get the desired number or letter, one needs to add the assigned numbers to the positions where the symbol (dot) is present.

In the examples below, the presence of symbols and empty positions is used to establish a connection between filled and not filled boxes and to show the difference between them. Litvin's Code uses a symbol to inform that an assigned binary number is present and an empty space to acknowledge that a position is equal to 0.

- In *position one*, the binary number 2^0 is always expected to equal to 1. Number one is represented by the presence of a dot.

•

- In the first position, there is an "x." The symbol (dot) in the first position is a representation of number 1 as well as letter A. The absence of this symbol represents the number 0.

•	

- In the *second position,* 2^1 is equal to 2. In Litvin's Code, this digit is represented by the presence of a symbol in the first position and empty space in the second. The first position is filled with a symbol and the second is empty. This is the reverse representation of the number 2 and letter B .

- **The combination of the *first and second positions* filled with symbols (dots) is equal to $2^1 + 2^0 = 3$. In Litvin's Code, it is represented by symbols "x x " that stand for number 3 or letter C .**

•	•

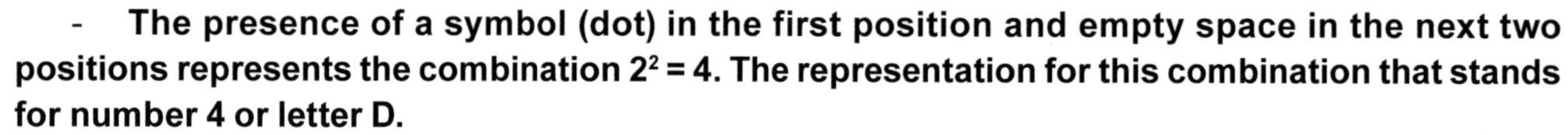

- **The presence of a symbol (dot) in the first position and empty space in the next two positions represents the combination $2^2 = 4$. The representation for this combination that stands for number 4 or letter D.**

•		

Example 1.0

When a symbol (dot) is in the first and the third positions, and the second position is empty, we add numbers in the first and the third positions: $2^2 + 2^0 = 4 + 1 = 5$. In Litvin's Code, it looks like "x 0 x".

•		•

This is the representation of number 5 and letter E.

End.

Example 1.1

When a symbol (dot) is in the first and second positions and the third position does not have a symbol (dot), we only add the assigned binary numbers in the first and the second positions: $2^2 + 2^1 = 4 + 2 = 6$. In Litvin's Code, it looks like "x x 0 " . This is the reverse representation of number 6 and letter F.

•	•	

End.

Example 1.2

When a symbol (dot) is in the first, second and third positions, we add the corresponding numbers in the first, second and third positions: $2^2 + 2^1 + 2^0 = 4 + 2 + 1 = 7$. The representation is "x x x "

•	•	•

This is the reverse representation for number 7 and letter G.

End.

Example 5.1 Reverse

First number is 12 “x x 0 0”

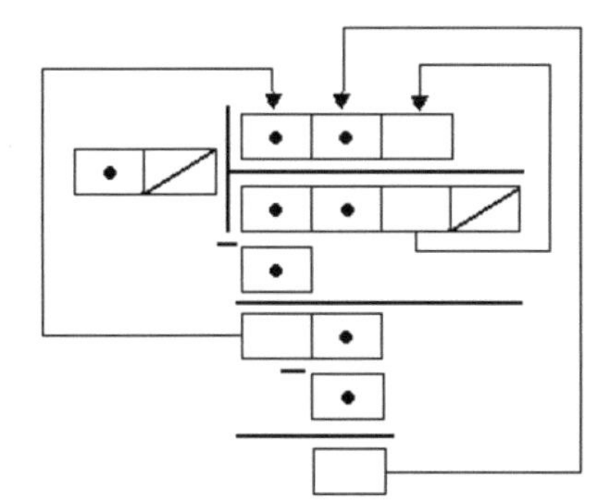

Second number is 2 “x 0”

Answer is 6 “x x 0”

$$\begin{array}{r|l} & 6 \\ \hline 2 & 12 \\ \hline 0 & \end{array}$$

Explanation

In the example above, the dividend is the first picture and the divisor is the second. In the dividend, the first two positions are filled with symbols and the next two are empty. In the divisor, position one is filled with a symbol and position two is empty. Since the configurations are in reverse, their mathematical values are 12 and 2. Both numbers have an empty space in the last position. We compare this to digit “0” in a regular division. In a regular division, when we eliminate 0 at the end of the number, the number becomes ten times smaller. In binary arithmetic used in Litvin’s Code, when we eliminate the last position from both numbers, then the numbers become twice as small.

New first number 6 “x x o”

New second number is 1 “x”

By dividing any number by 1, we expect to get the same number. In this example, the goal is to demonstrate how Litvin's Code applies to this division.

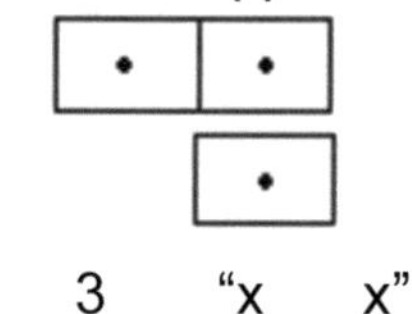

New first number is 3 "x x"

The second number is 1 "x"

To continue the division, one needs to complete two subtractions. In the first step, we subtract the divisor from the new first number and get an empty space in the first position of the result. When the result of a subtraction has an empty space, we fill the corresponding position in the answer of the division with a symbol (dot). In step two, we go on to subtract the divisor by the new first number and in the result of subtraction is an empty space. Similarly, there is an empty space as a result of the subtraction and we fill the corresponding position of the answer with a symbol, and eliminate the empty position. The division was successful.

In Summary:

- When dividing 12 by 2 using Litvin's Code, we have one empty position in the end of each number, the dividend and the divisor. We eliminate this position from both numbers and the numbers become 6 and 1.

- Number six has another empty ending position, so we move the empty space from the dividend to the answer, and the new numbers becomes 3 and 1.

- Using Litvin's Code, we subtract 1 from 3, and have the first position empty in the result of subtraction. We fill with a symbol the next corresponding position in the answer of the division. The numbers to subtract become 1 and 1.

- We subtract 1 from 1 and we get 0 or an empty position as the result of the subtraction. We fill the corresponding position in the answer of the division with a symbol.

End.

Example 5.2 Reverse

First number is 24 "x x o o o"

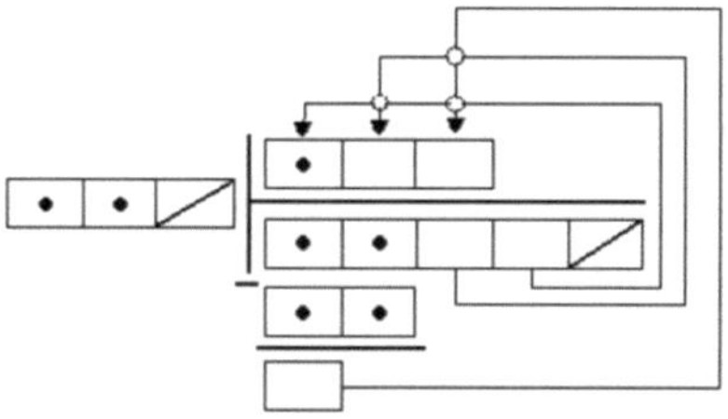

Second number is 6 "x x o"

Answer is 4 "x o o"

4
6|24
0

In example 5.2, in a division of two numbers, the first number has five boxes and the second number has three boxes. The dividend has positions one and two filled up with symbols and the next three positions empty. The divisor has positions one and two filled with symbols and position three empty. Using Litvin's Code, we eliminate the empty positions that correspond in both the dividend and the divisor in their positions.

Explanation

Both numbers have empty positions at the end of the configurations. We eliminate the corresponding empty positions at the end of both numbers. After eliminating the last corresponding empty positions, the first number is left with four boxes and the second number with two.

The second number does not have any empty position at the end, but the first number still has two empty positions, which are automatically moved into the end of the answer of the division. After moving the two empty positions, we eliminate the empty positions from the dividend. Now we have only two boxes remaining in the dividend. Those two boxes are equal to the decimal number 3. The second number is equal to 3 also. If we subtract 3 from 3, we will have zero or an empty position as the result of the subtraction. Since this is considered to be a successful subtraction, the remaining last empty box is moved to the answer of the division.

In Summary:

- When we divide 24 by 6, we eliminate one empty position from the end of both numbers and the numbers become 12 and 3.

- When we divide 12 by 3, we transfer two empty positions of the number 12 to the answer of the division. The dividend and the divisor become 3 and 3.

- When we subtract 3 from 3, the result for this step becomes an empty position. We fill the next position in the answer of the division with a symbol.

End.

Example 5.3 Reverse

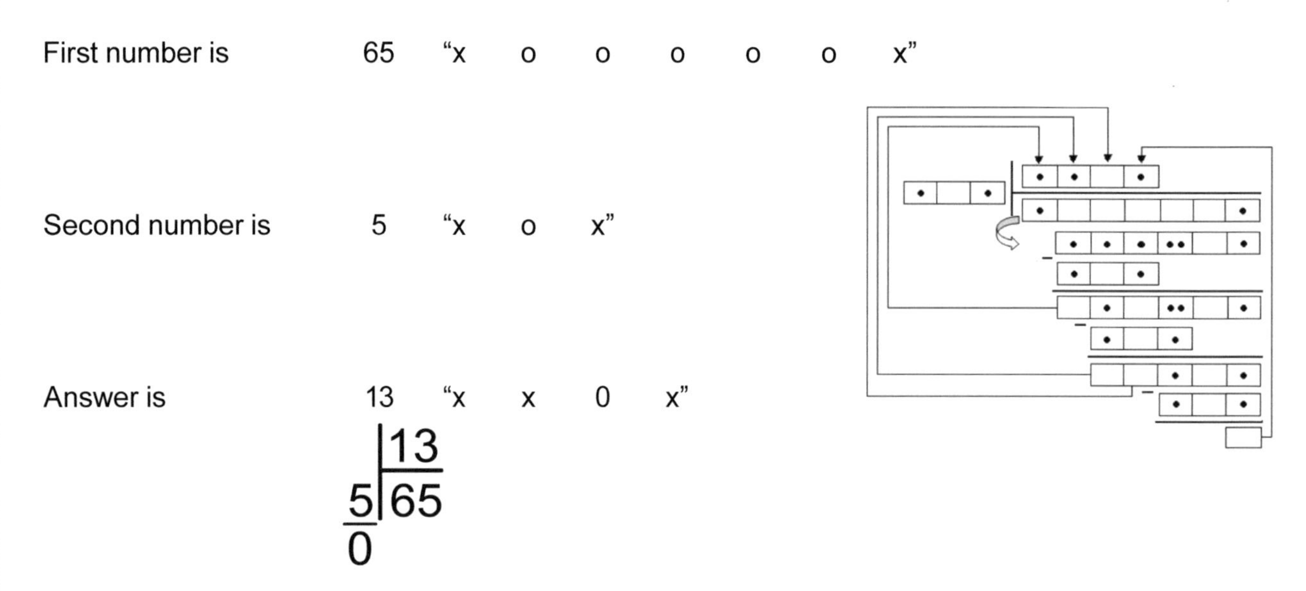

In Example 5.3, we are dividing two numbers with seven positions in the first number and three positions on the second. In the dividend, positions one and seven are filled with symbols, while positions two through six are empty. In the divisor, positions one and three filled with symbols and position two is empty.

The way to divide these numbers using Litvin's Code is to subtract step by step until the result is equal to 0.

Explanation

In the first step of Example 5.3, we subtract the second number from the first. Before the borrowing process, the dividend is transformed into a more compact configuration, by moving the first position over to the next repeatedly until the new configuration has its first three positions filled with single symbols, the fourth position filled with two symbols, position four empty, and position five filled with a symbol. The symbol from the first position does not appear anymore in the calculation. When borrowing the symbol from the prior position to the next empty one, the next position is equal to the prior, because it includes two symbols.

After the new configuration is formed, we subtract the divisor from the dividend twice, until the result is equal to zero.

The final answer is 13 "x 0 x x" 13 "x 0 x x"

•		•	•

In Summary:

- When completing the division above, we do two subtractions.
- During the subtraction, when the first position in the result of the subtraction is empty, then we eras this position and fill with a symbol the corresponding position in the answer of the division. The second position is transferred over to the answer of the division as an empty position.
- The answer of the division is 13.

End.

Example 5.4 Reverse

First number is 21 "x o x o x"

Second number is 3 "x x"

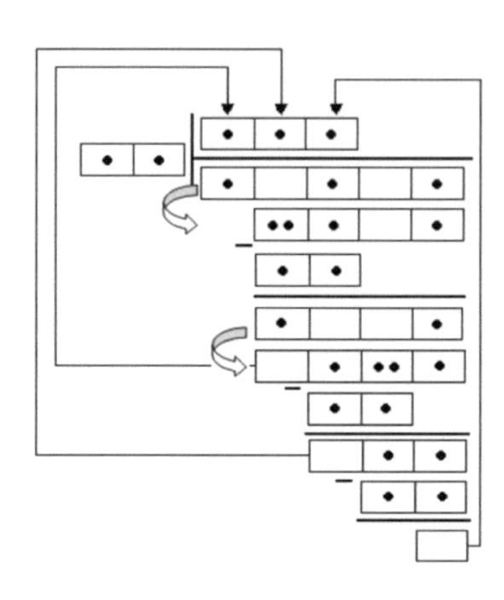

Answer is 7 "x x x"

$$\begin{array}{r|l} & 7 \\ \hline 3 & 21 \\ \hline 0 & \end{array}$$

Explanation

In Example 5.4, we divide two numbers, where picture one is the dividend and picture two is the divisor. The dividend has its first position filled with a symbol, the second position empty, the third position filled with a symbol, the fourth position empty, and the fifth position filled with a symbol. In the divisor, two positions are filled with symbols. This division is completed by subtracting the divisor from the dividend repeatedly until we reach 0 or an empty position. Before the subtraction, it is useful to compress the dividend into a smaller configuration by moving the first symbol to the next position, until the new configuration consists of two symbols in the first position, one symbol in the second position, an empty position in the third place and a symbol in the fourth position. After the subtraction from this new configuration, one can reconfigure the result of this subtraction also. We move the symbol from the first position to the next, just like before, after which we complete yet another subtraction, resulting in a zero.

In Summary:

- After reconfiguring the dividend, its first position is empty and is moved to fill a position in the answer, and this is done one more time.

- When there are no more steps for the subtraction and the result of subtraction is 0, then it is the last step and we have arrived at the final answer.

- The answer of the division is 7.

End.

Example 5.5 Reverse

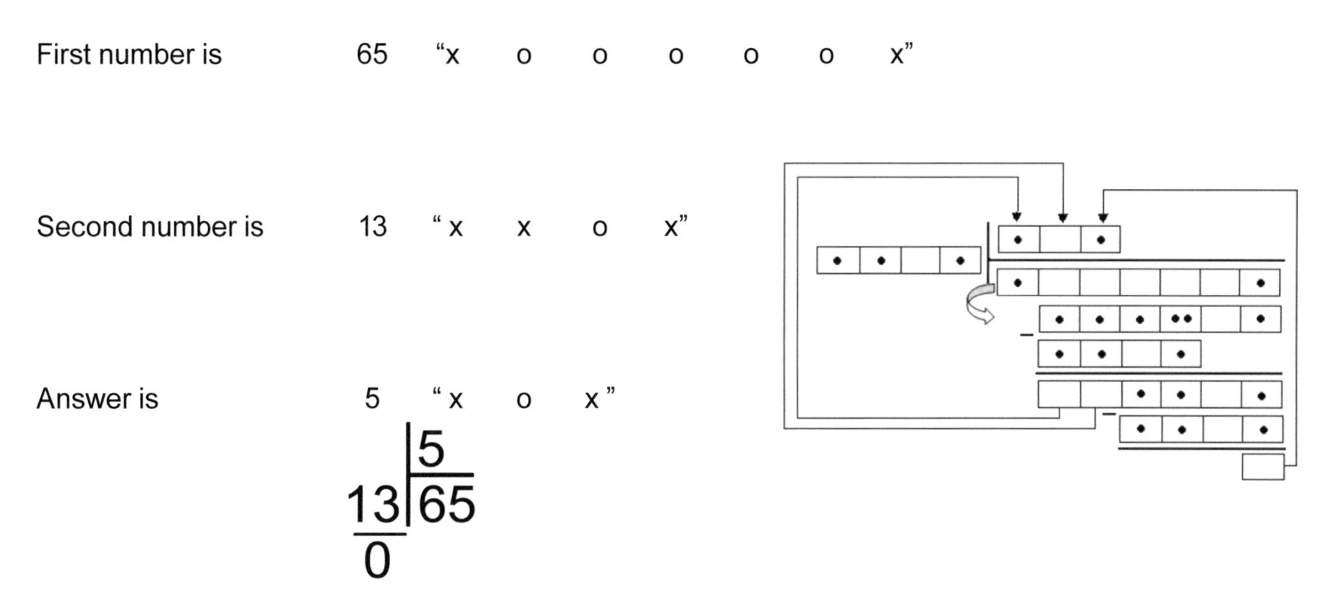

In Example 5.5R, we are dividing two numbers, where picture one is the dividend and picture two is the divisor. In the dividend, position one is filled with a symbol, positions two through six are empty, and position seven is filled with a symbol. In the divisor, positions one and two are filled with symbols, position three is empty, and position four is filled with a symbol. Before beginning the subtraction process of this division, we convert the dividend to a more compressed unit that is easier to subtract the divisor from. We move the first filled position of the dividend to the next position, filling it with two symbols instead of one. We repeat this step until all possible positions are filled. The new configuration has its first three positions filled with single symbols, position four filled with two symbols, position five empty, and position six filled with a symbol. Now we go on to subtract the divisor from the new configuration, ending up with a result that has its first two positions empty. We move the first empty position to the answer of the division as the first position filled with a symbol. After eliminating the first empty position, the second empty position is moved to the answer of the division. However, since this empty position is not the first position of the original result of the subtraction, it moves to the answer of the division as an empty unit. Then we subtract the divisor from the result of the subtraction once more, and end up with a zero or an empty position. The subtraction was successful, and the empty position is moved to the answer as a filled position.

In Summary:

- First we reconfigure the dividend, after which we subtract the divisor from the new configuration.

- Since the result of the subtraction has its first two positions empty, we move them over to the answer of the division as one filled and one empty position.

- We subtract from the new result the divisor one more time, and end up with a zero or an empty position, which is transferred to fill the last position of the answer of the division.

Example 5.6 Reverse

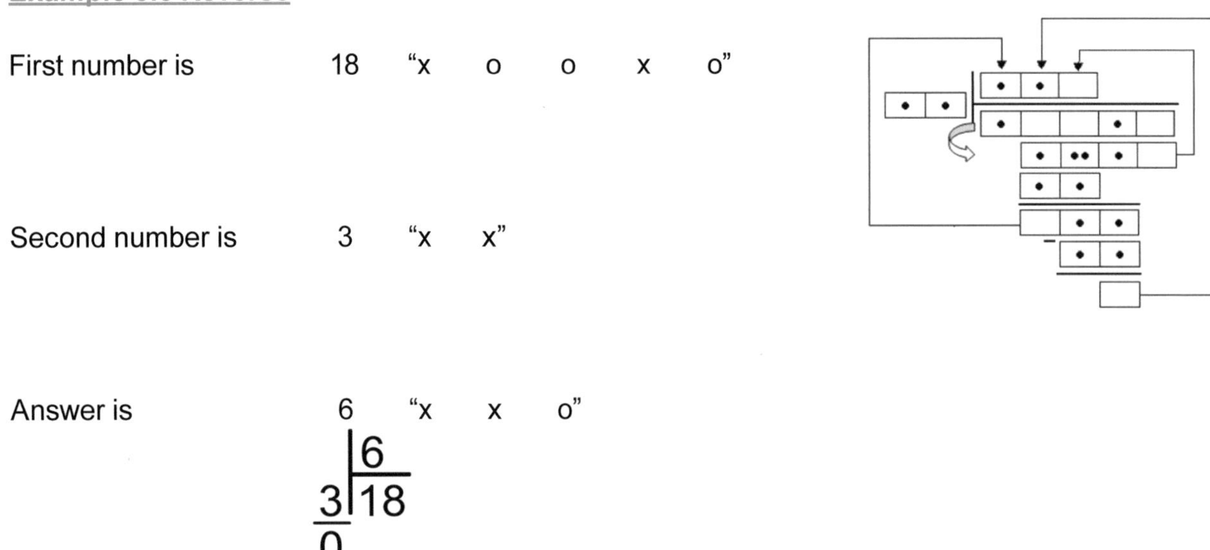

In Example 6.5R, we are dividing two numbers, where picture one is the dividend and picture two is the divisor. In the dividend, position one is filled with a symbol, positions two and three are empty, position four is filled with a symbol, and position five is empty. The divisor has two positions, both of which are filled with symbols. Before beginning the subtraction process, we change the configuration of the dividend. To do this, we move the first filled position to the next position, filling it with two symbols. The new compressed configuration is necessary for the subtraction process. Since we have one empty position in the end of the new dividend, we move this to the answer as an empty position. Since this division is a reverse process, we add the empty position from the right, instead of the left. The result of the subtraction has one empty position in the beginning and we move this to fill the first position of the answer with a symbol. We go on to complete the second subtraction from the result, and end up with an empty position. We move this last empty position to fill the second position of the answer of the division with a symbol, and declare our division to be successful.

First step for Example 5.6

Because the last position in the dividend is empty and the last position of the divisor is filled with a symbol, we do not yet do any subtractions. We move the last position of the dividend to the last position of the answer of the division, which is empty.

Second Step for Example 5.6

We subtract the divisor from the dividend first and then the result of the subtraction, until the final result is zero or an empty position.

We eventually subtract from the new result

	3	"x	x"		[• •] − [• •] = [ ]
The Divisor	3	"x	x"		
result	0	"0"			
Answer for the division	6	"x	x	0"	[• •]

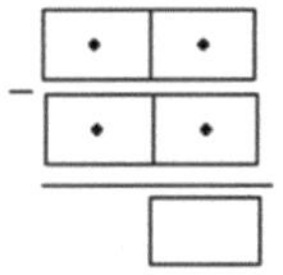

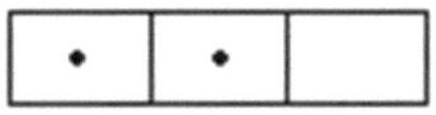

In Summary:

- We begin the division by reconfiguring the dividend to make it compatible with the upcoming subtraction. Since the last position of the new dividend is empty, we move it to the last position of the answer of the division.

- We subtract the divisor from the new dividend and end up with a result that has its first position empty. We move this empty position to fill the first position of the answer of the division with a symbol, and eliminate it from the result.

- We subtract the divisor from the new result and end up with an empty position, which is moved to fill the second position of the answer of the division.

Example 5.7 Reverse

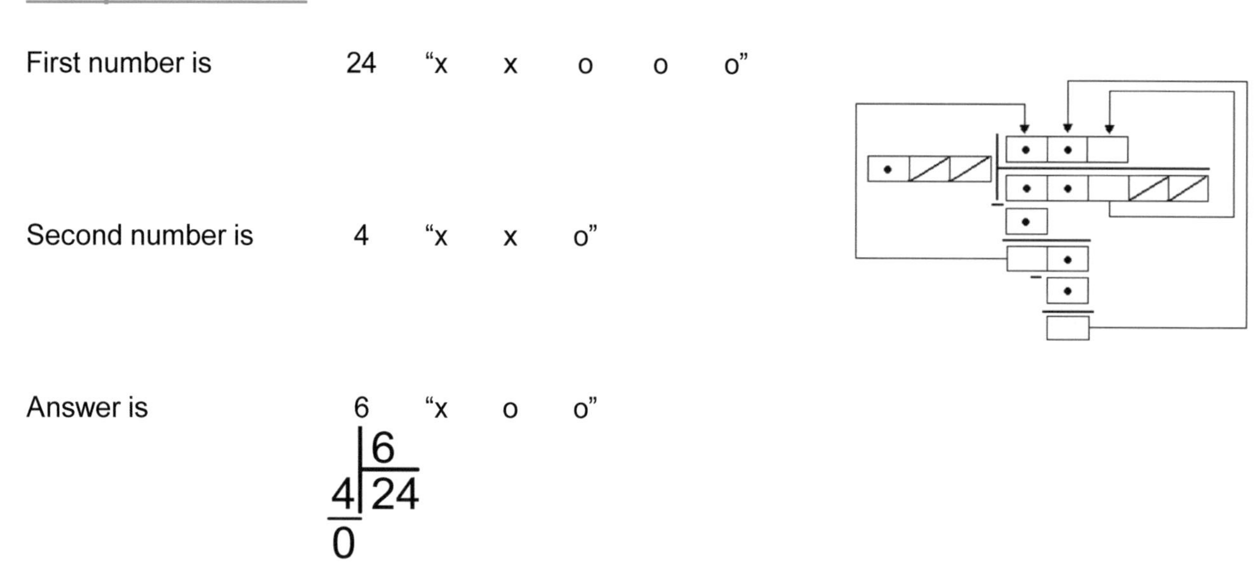

In Example 5.7, we are dividing two numbers. The dividend is equal to 24, and the divisor to 4. In the dividend, positions one and two are filled with symbols, while positions three, four and five are empty. In the divisor, position one is filled with a symbol, while positions two and three are empty. When the last positions in the dividend and the divisor are empty, we eliminate them without adding any positions to the answer of the division. Now the divisor has only one position filled with a symbol and this position is the first position and is equivalent to number one. After eliminating the two last positions of the dividend, we still have one empty position at the end. We must move this empty position from the dividend to the answer of the division.

After the elimination of the fifth position in the dividend, the number is equivalent to 12. When we eliminate the fourth position, the number becomes six. Then we eliminate the third empty position by transferring it to the answer of the division. After eliminating the three positions in the dividend, we are left with two filled positions which are equal to number 3. To complete the division, we need to subtract the divisor from the dividend until the result of is zero or an empty position. This division requires two steps to reach the final answer.

First step for Example 5.7

Erase third, fourth and fifth positions on the dividend and third position of the divisor.

Old dividend 24 "0 0 0 x x"

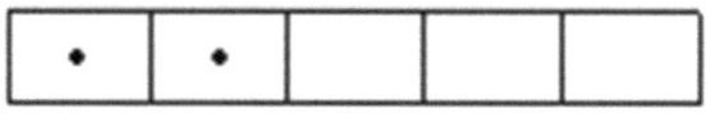

New Dividend is 3 "x x"

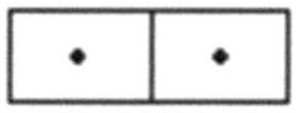

New Divisor is 1 "x 0"

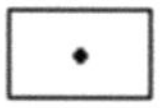

Because the third, fourth, and fifth positions on the dividend are empty and second and third positions of the divisor are empty too, we do not do any subtractions until we deal with all empty positions. We eliminate the empty positions on the dividend and divisor. We are transferring the third empty position of the dividend to the answer of the division as an empty position. The third position of the dividend is erased and we are subtracting the devisor from the new result.

In Summary:

- When we are dividing 24 by 4, we see that the positions, three, four and fifth on the number 24 and position two and three on the number 6 are empty. In Litvin's Code we automatically eliminate the last two empty positions of the dividend and divisor, and we are left with one empty position on the dividend and the divisor has no empty positions left to eliminate. The number 24 and 4 became 6 and 1. We see that number six has the one ending empty position but number 1 has not.

- When we are eliminating this empty position from the dividend and are transferring two empty positions to the answer, the number 12 becomes 3. We are subtracting 1 from 3, and then 1 from 1, which is equal to zero. Division was successful and twice the zeros from the subtractions were transferred to the answer of the division as symbols. The result of the division is equal to 6.

End.

MATHEMATICS

REVERSE DIVISION, EXERCISE 1R

In Exercise 1R, in a ***visual*** display of division we see three pictures of boxes. All three pictures have the first position empty, while the second positions are filled with a red ball. The division sign is between the first and second sets of boxes. Over the second picture is the equal line, which separates the problem from the answer. In the third picture, which is an answer for the division and has a value of one, the first position is filled with a symbol and the second is empty.

The ***audio*** representation for the division consists of a **double knock, knock, double click, double knock, knock, cling, double knock, and knock.**

Exercise 1

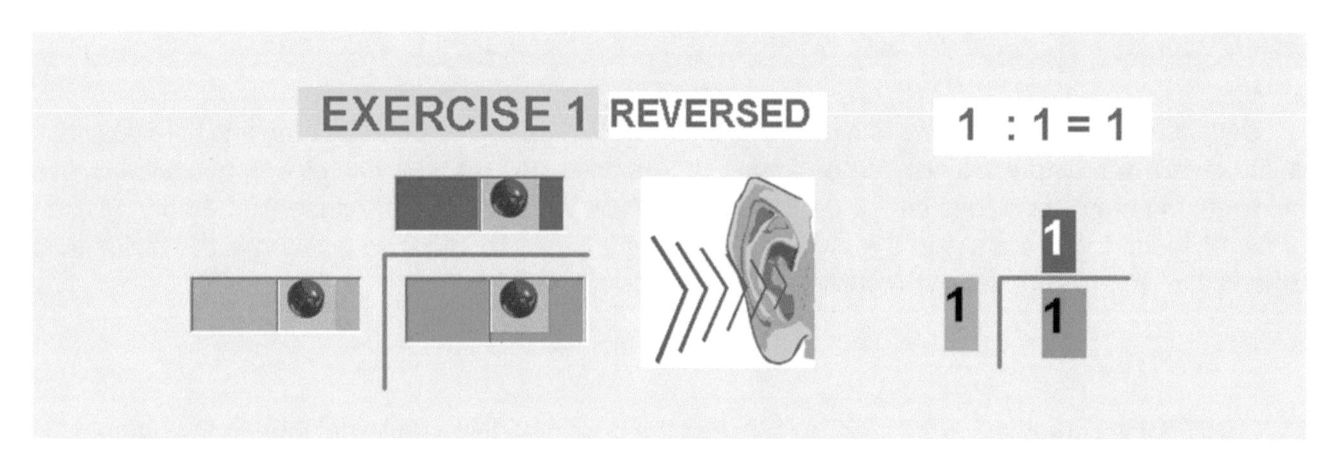

The ***kinesthetic*** representation of the filled up position is represented by the clenching of the right hand. The clenching of the left hand represents an empty position. Both hands extended up represent the division sign, double click. The equal sign, cling, is represented by crossing both hands. The kinesthetic representation for both pictures is the clenching of the left hand, right hand, extending both hands up, left hand, right hand, crossing the hands, left hand, and the right hand.

REVERSE DIVISION, EXERCISE 1.1 R

In Exercise 1R Step 1.1R, in a ***visual*** display of division we see three pictures of boxes - the dividend, the divisor and the answer. Between the first and second pictures is the division sign. Over the dividend is the equal line, which separates the problem from the answer. On top of the dividend is the third picture, which is the answer to the division. The ***audio*** representation for the division consists of a **knock, double click, double knock, kock, clings, knock.**

In the next part of Step 1.1R we also have a **visual** display of how we arrive to the answer. From picture two, which is dividend, we subtract picture four. Between picture two and four is the subtraction sign. Below picture four is the equal sign which separates the problem from the answer. Picture five is the result of the subtraction. When the result of a subtraction is an empty position, the filled position is transferred to the answer of the division on the top. The ***audio*** representation for the subtraction consists of a **knock, double tram, knock, cling, and double knock.**

Step 1.1

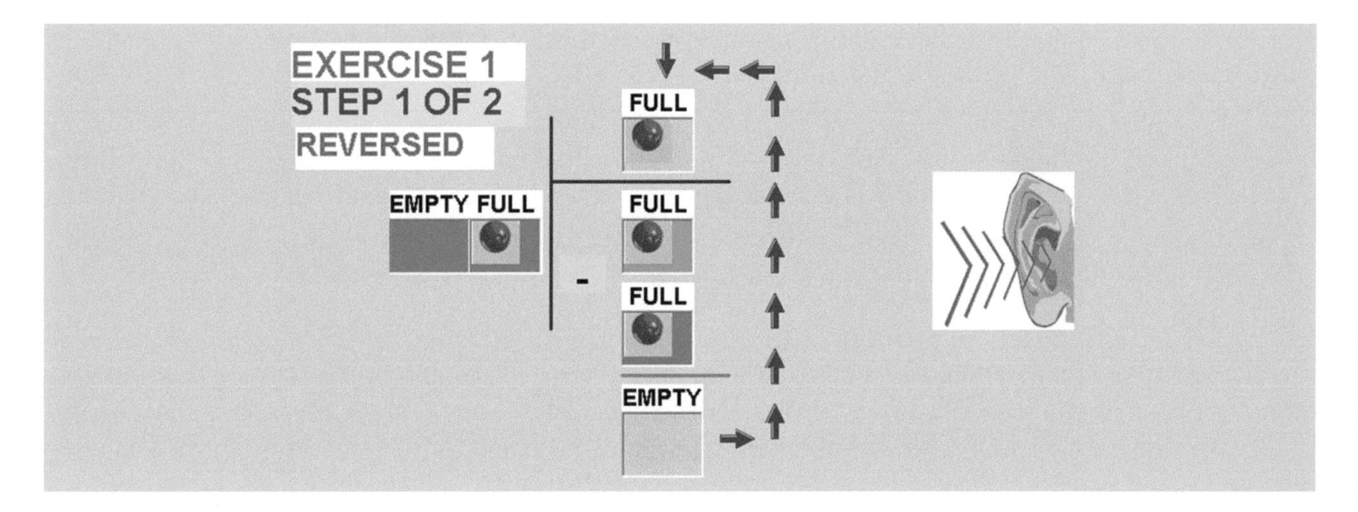

The ***kinesthetic*** representation of the filled up position is represented by the clenching of the right hand. The clenching of the left hand represents an empty position. The division sound, double click, is represented by both hands extended in up. The equal sound, cling, is represented by crossing both hands. The kinesthetic representation for the picture is the clenching of the right hand, both hands extended up, left hand, right hand, crossing both hands, and the right hand.

MATHEMATICS REVERSE DIVISION, EXERCISE R1.2

In Exercise 1 Step 1.2R, in a ***visual*** display of the last step of the division we see two pictures of boxes. In both pictures the first position is empty while the second position is filled with a red ball. The colors and the placement of the boxes and the balls are different in the pictures. The form, shape, and color of the symbols do no change their mathematical value, which is equal to one.

The ***audio*** representation for the pictures consists of a **double knock, knock, and cling.**

Step 1.2

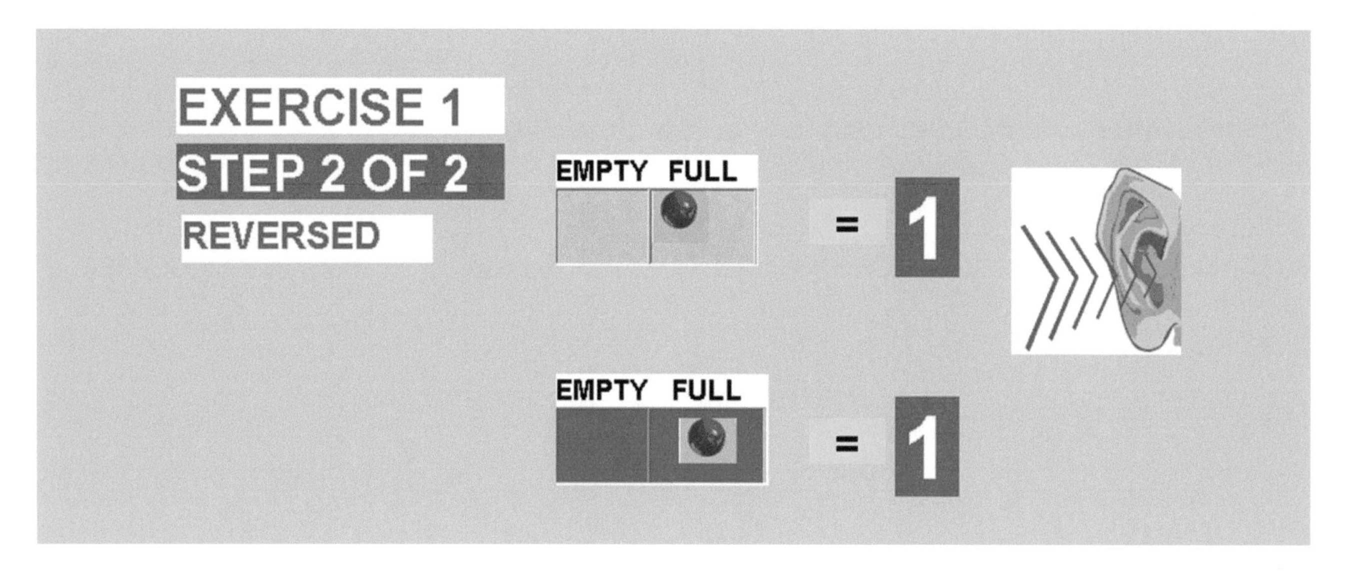

The ***kinesthetic*** representation of the filled up position is represented by the clenching of the right hand. The clenching of the left hand represents an empty position. The equal sound, cling, is represented by crossing the hands. The kinesthetic representation for the two pictures is the clenching of the left hand, right hand, crossing the hands, left hand, and right hand.

REVERSE DIVISION, EXERCISE 1R, ANSWER 1R

In Exercise 1R, in a ***visual*** display of division we see two pictures of boxes with the division sign in-between. In the first picture there is one empty position, and one position filled with a red ball. There is also one empty position in the second picture and one position filled with a red ball. On the top of the second picture is the equal line, which separates the problem from the answer. The ***audio*** representation for the below consists of a **double knock, knock, double click, double knock, knock, and cling.**

At the bottom are three probable answers. From the probable answers for the division we need to find the right one. The audio signals are: 1) **knock, and double knock,** 2) **double knock and knock**, 3) **knock and knock**.

Exercise 1

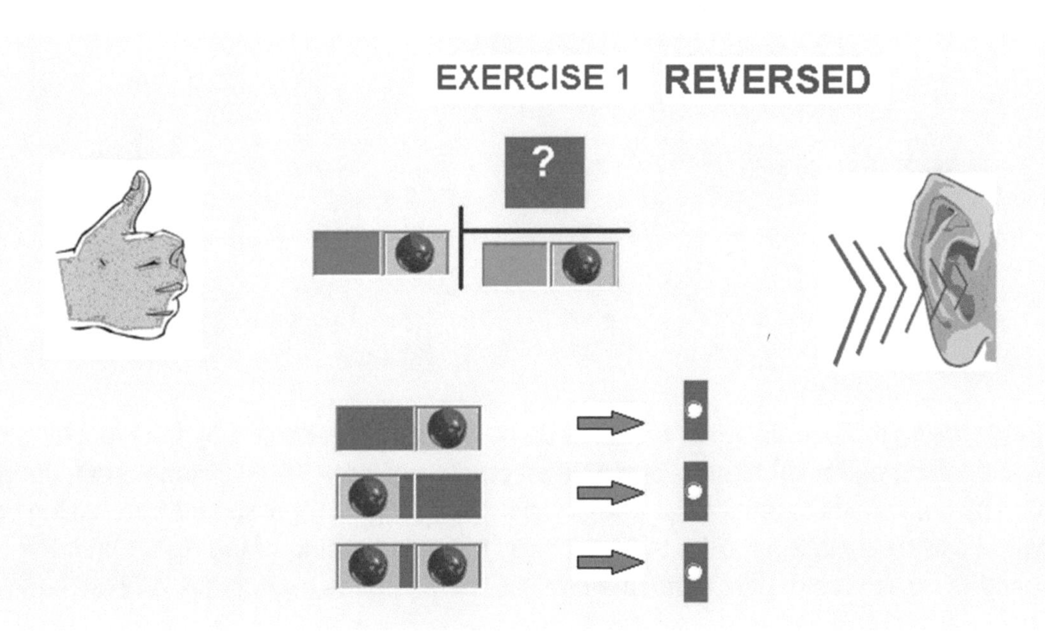

The ***kinesthetic*** representation of the filled up position is represented by the clenching of the right hand. An empty position is represented by the clenching of the left hand. The division sound, double click, is represented by both hands extended up. The equal sound, cling, is represented by crossing both hands. The kinesthetic representation for both pictures and the answer is the clenching of the left, right, both hands extended up, the clenching of the left, right, the crossing of both hands, left hand and the right hand.

MATHEMATICS REVERSE DIVISION, EXERCISE 2R

In Exercise 2R, in a ***visual*** display of division we see three pictures of boxes - the dividend, the divisor and the answer. Between the first and second pictures is the division sign. Over the dividend is the equal line, which separates the problem from the answer. On top of the dividend is the third picture, which is the answer of division. The ***audio*** representation for the division consists of **knock, double knock, double click, double knock, knock, cling, knock, and double knock.**

Exercise 2R

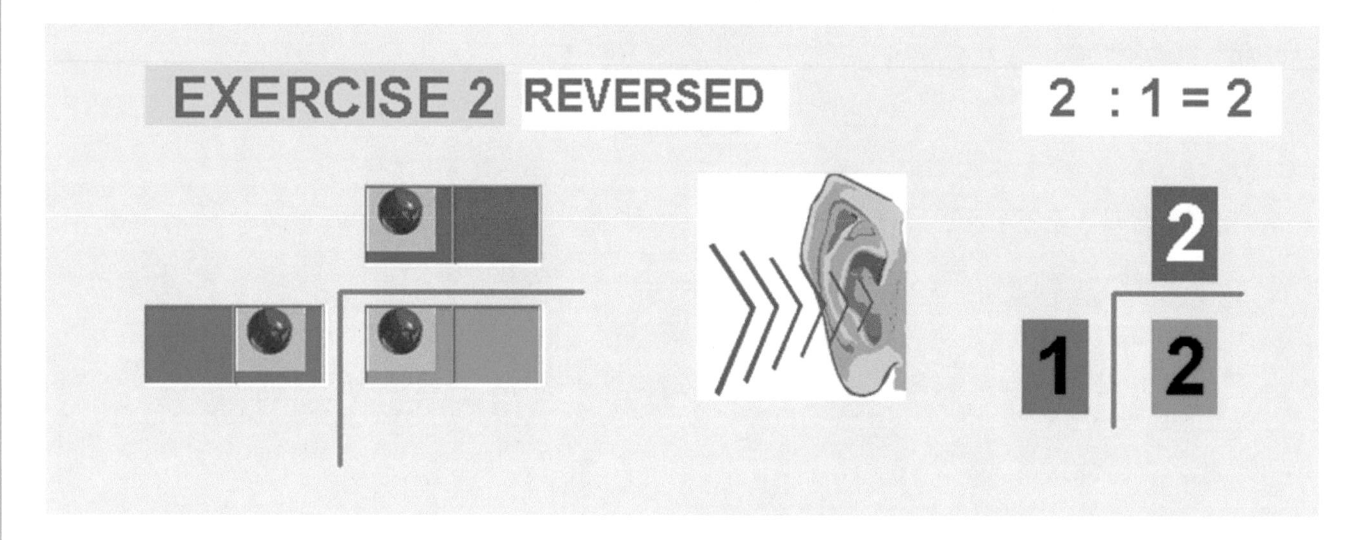

The ***kinesthetic*** representation of the filled up position is represented by the clenching of the right hand. The clenching of the left hand represents an empty position. The division sound, double click, is represented by both hands extended up. The equal sound, cling, is represented by crossing both hands. The kinesthetic representation for the three pictures is the clenching of the right hand, left hand, both hands extended up, left hand, right hand, the crossing of the hands, right hand, and left hand.

REVERSE DIVISION, EXERCISE 2.1

In Exercise 2R Step 2.1R, in a ***visual*** display of division we see three pictures of boxes – the dividend, the divisor and the answer. Between the first and second pictures is the division sign. Over the dividend is the equal line, which separates the problem from the answer. On the top of the dividend is the third picture, which is the answer of the division problem. The ***audio*** representation the division consists of a **knock, double knock, double click, knock, cling, and double knock.**

In the next part of Step 2.1R, we have a **visual** display of how we arrive to the answer. In the divisor the first empty box is automatically taken to the top.

Step 2.1

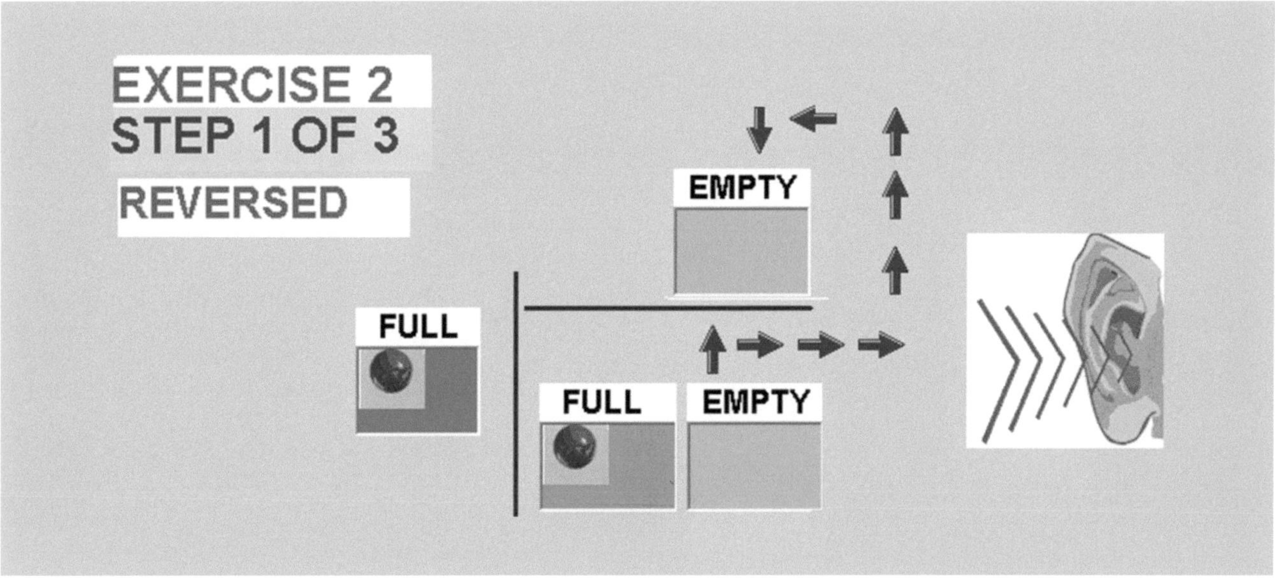

The ***kinesthetic*** representation of the filled up position is represented by the clenching of the right hand. The clenching of the left hand represents an empty position. The division sound, double click, is represented by both hands extended upwards. Both hands extended in front represent the subtraction sound, double tram. The equal sound, cling, is represented by crossing both hands. The kinesthetic representation for the pictures is the clenching of the right hand, left hand, hands extended up, right hand, crossing of the hands, and left hand.

MATHEMATICS

REVERSE DIVISION, EXERCISE 2.2R

In Exercise 2R Step 2.2R, in a ***visual*** display of division we see three pictures of boxes - the dividend, the divisor and the answer. Between the first and second pictures is the division sign. Over the dividend is the equal line, which separates the problem from the answer. On top of the dividend is the third picture, which is the answer of the division. The ***audio*** representation for the division consists of a **knock, double click, knock, cling, knock, and double knock.**

In the next part of Step 2.2R, we have a **visual** display of how we arrive to the answer. From picture two, which is the dividend, we subtract picture four. Between picture two and four is the subtraction sign. Below picture four is the equal sign which separates the problem from the answer. Picture five is the result of the subtraction. When the result of a subtraction is an empty position then a filled position is transferred to the answer of division. The ***audio*** representation consists of a **knock, double tram, knock, cling, and double knock.**

Step 2.2

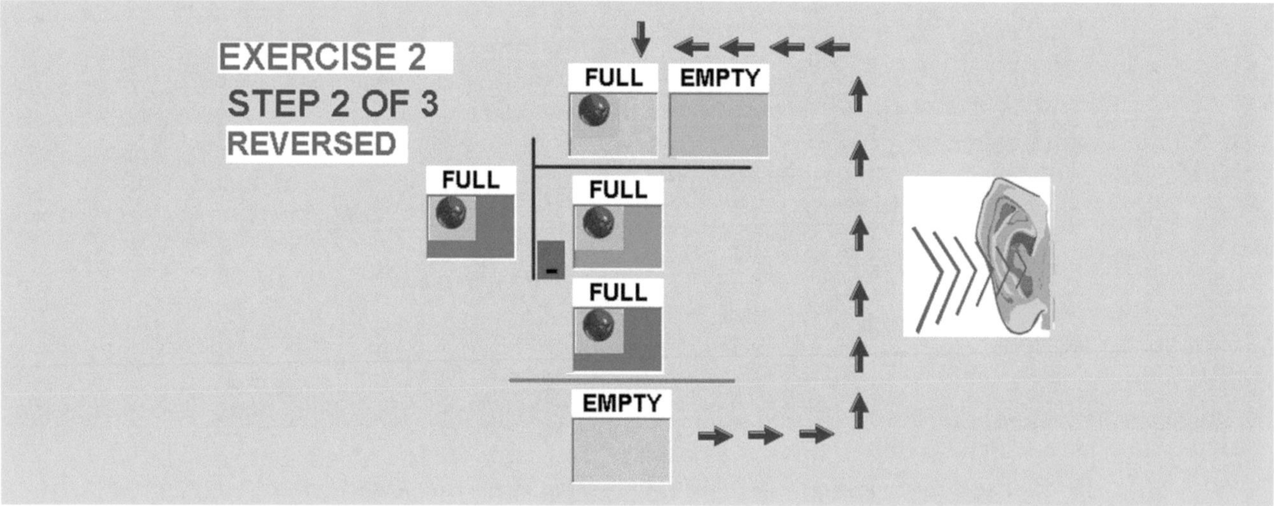

The ***kinesthetic*** representation of the filled up position is represented by the clenching of the right hand. The clenching of the left hand represents an empty position. The division sound, double click, is represented by both hands extended up. Both hands extended in front represent the subtraction sound, double tram. The equal sound, cling, is represented by crossing both hands. The kinesthetic representation for the picture is the clenching of the right hand, both hands extended up, right hand, crossing of hands, right hand, and left hand.

REVERSE DIVISION, EXERCISE 2.3

In Exercise 2R Step 2.3R, in a ***visual*** display of the last step of the division we see two pictures of boxes. In both pictures the first position is filled with a red ball while the second position is empty. The colors and the placement of the boxes and the balls are different in the pictures. The form, shape, and the color of the symbols do no change the mathematical value, which is equal to two.

The ***audio*** representation for the pictures below consists of a **knock, double knock, and a cling.**

Step 3.3

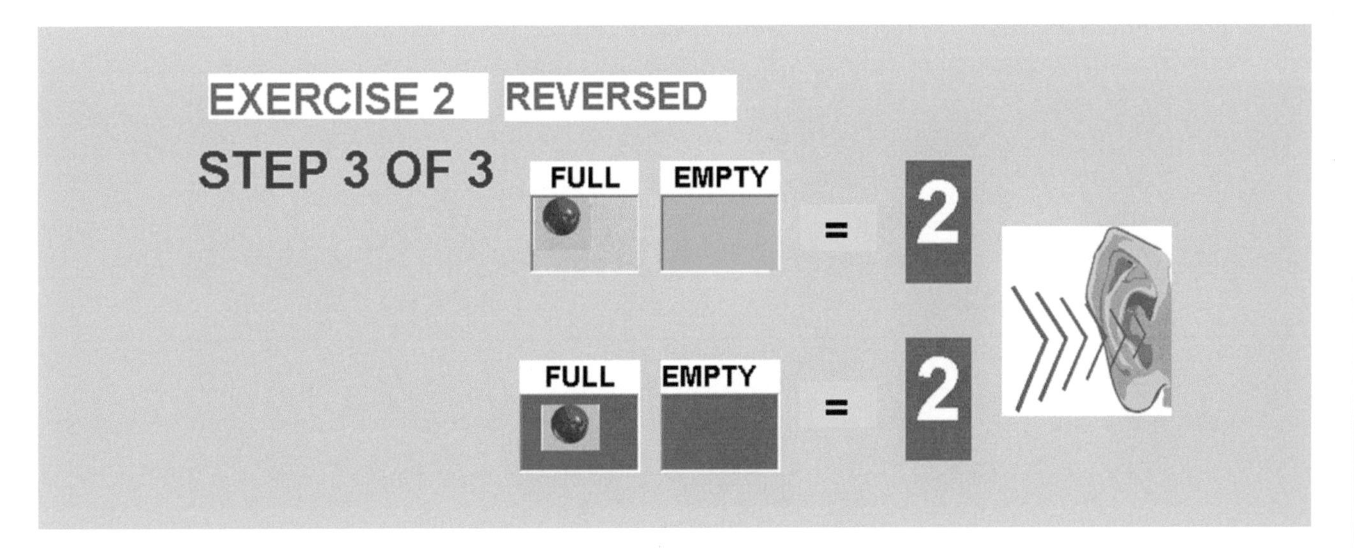

The ***kinesthetic*** representation of the filled up position is represented by the clenching of the right hand. The clenching of the left hand represents an empty position. The equal sound, cling, is represented by crossing of hands. The kinesthetic representation for the two pictures is the clenching of the right hand, left hand, the crossing of the hands, right hand, and left hand.

MATHEMATICS REVERSE DIVISION, EXERCISE 2R, ANSWER 2R

In Exercise 2R, in a ***visual*** display of division are two pictures of boxes with the division sign in-between. The first picture has the first position filled with a red ball and the second position empty. The second picture has the first position empty and the second position filled with a red ball. On the top of the second picture is the equal line, which separates the problem from the answer. The ***audio*** representation for the picture of the division consists of a **knock, double knock, double click, double knock, knock, and cling.**

At the bottom are three probable answers. From the probable answers for the division we need to find the right one. The audio signals are: 1) **knock, and double knock,** 2) **double knock and knock**, 3) **knock and knock**.

Exercise 2

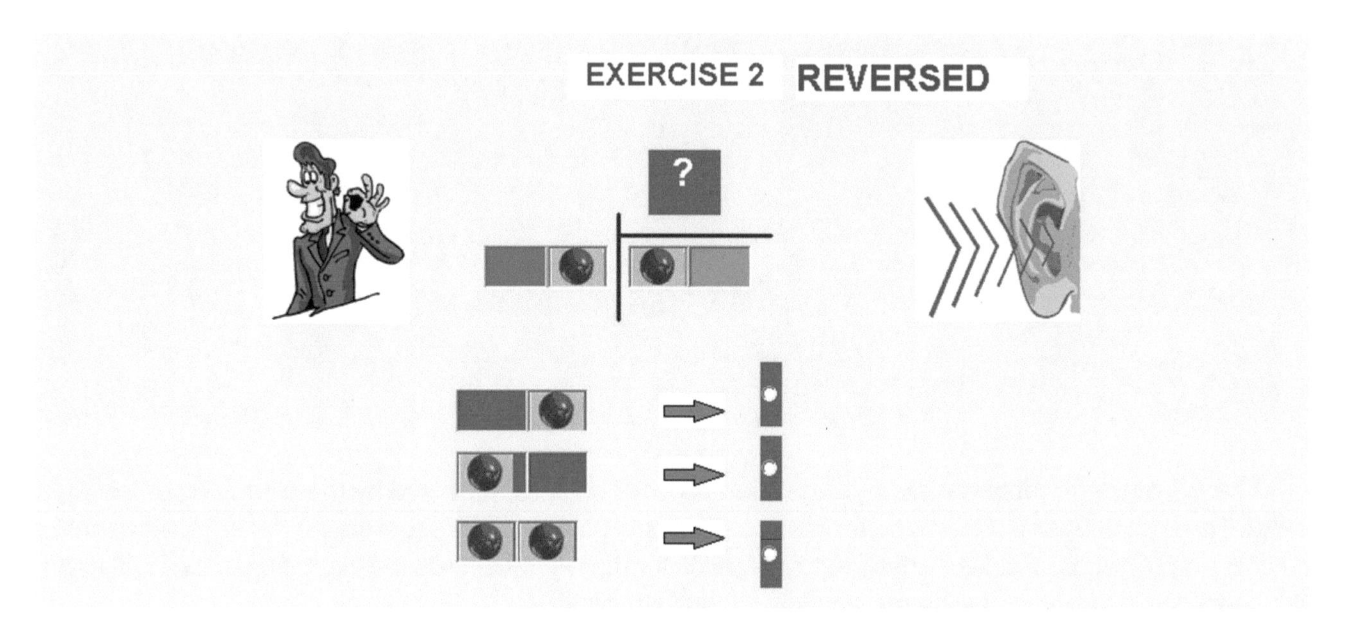

The ***kinesthetic*** representation of the filled up position is represented by the clenching of the right hand. An empty position is represented by the clenching of the left hand. The division sound, double click, is represented by both hands extended up. The equal sound, cling, is represented by crossing both hands. The kinesthetic representation for both pictures and the answer is the clenching of the right, left, both hands extended up, the clenching of the left, right, crossing both hands, right hand, and left hand

REVERSE DIVISION, EXERCISE 3R

In Exercise 3R, in a ***visual*** display of division we see three pictures of boxes - the dividend, the divisor and the answer. Between the first and second pictures is the division sign. Over the dividend is the equal line, which separates the problem from the answer. On top of the dividend is the third picture, which is the answer of the division. The ***audio*** representation the division consists of a **knock, double knock, double knock, double click, knock, double knock, knock, cling, knock, and double knock.**

Exercise 3

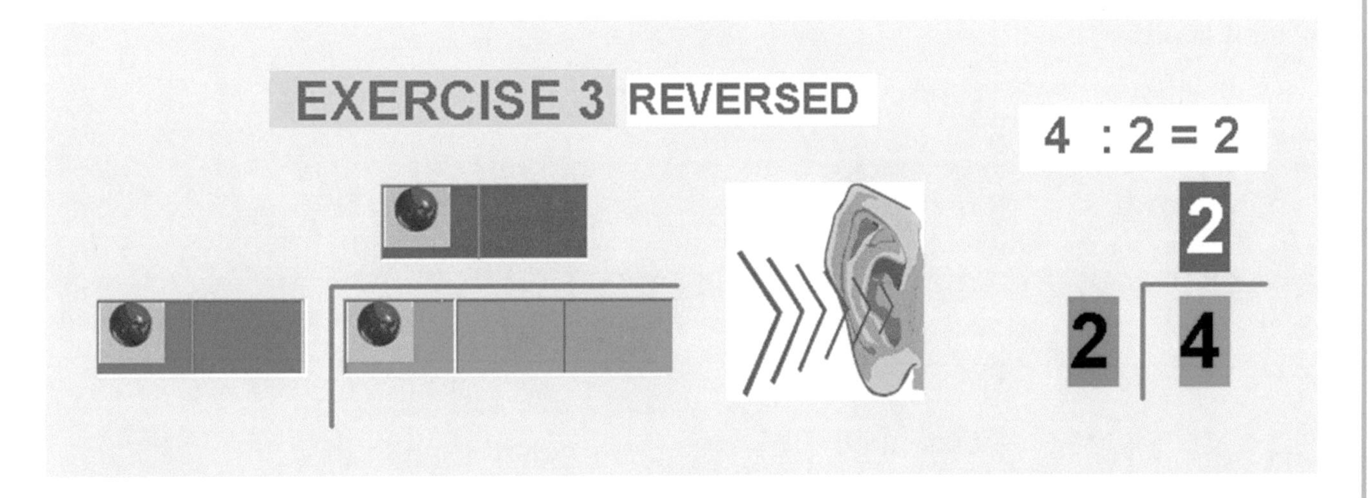

The ***kinesthetic*** representation of the filled up position is represented by the clenching of the right hand. The clenching of the left hand represents an empty position. The division sound, double click, is represented by both hands extended up. The equal sound, cling, is represented by crossing both hands. The kinesthetic representation for the pictures is the clenching of the right hand, left hand twice, both hands extended up, right hand, left hand, crossing of the hands, right hand, and left hand.

MATHEMATICS

REVERSE DIVISION, EXERCISE 3.1

In Exercise 3R Step 3.1R, in a ***visual*** display of division we see two pictures of boxes, which are the dividend and divisor. Between the first and second pictures is the division sign. Over the dividend is the equal line, which separates the problem from the answer. On top of the dividend is the third picture, which is the answer of the division problem. The ***audio*** representation for the pictures consists of a **knock, double knock, double knock, double click, knock, double knock, and cling.**

In the next part of Step 3.1R, we have a **visual** display of how we arrive at the answer. If the divisor has an empty position before the filled up position then that position cancels out the first empty position of the dividend.

Step 3.1

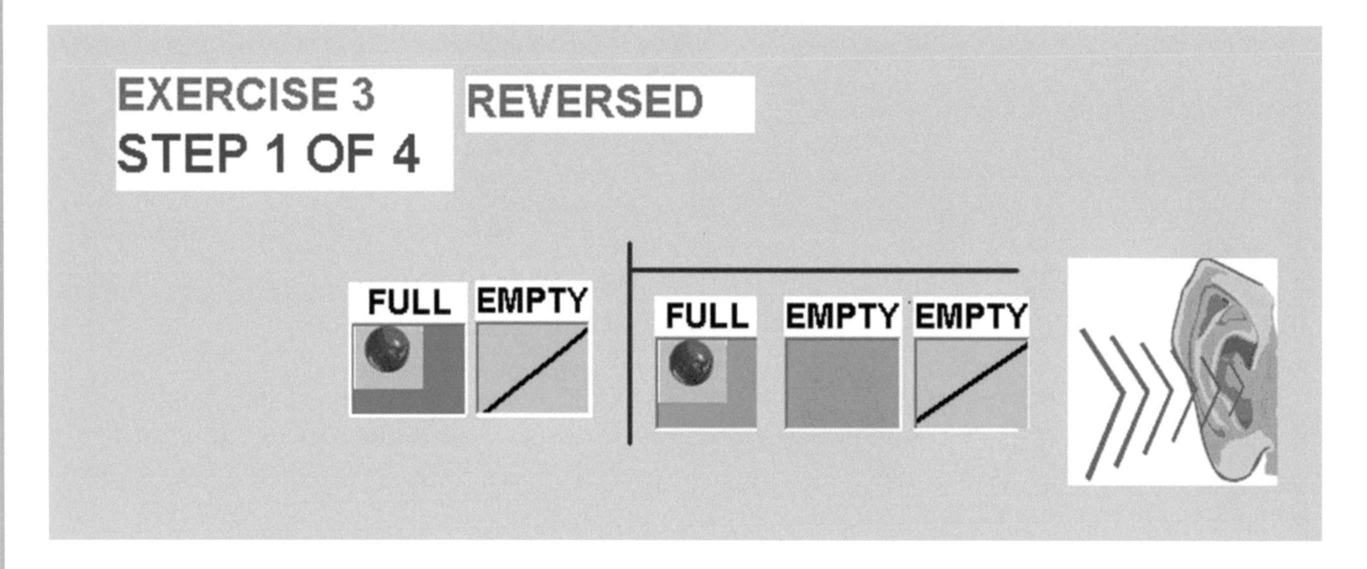

The ***kinesthetic*** representation of the filled up position is represented by the clenching of the right hand. The clenching of the left hand represents an empty position. The division sound, double click, is represented by both hands extended up. Both hands extended in front represent the subtraction sound, double tram. The equal sound, cling, is represented by crossing both hands. The kinesthetic representation for the two pictures is the clenching of the right hand, left hand twice, both hands extended up, right hand, left hand, and crossing the hands.

REVERSE DIVISION, EXERCISE 3.2

In Exercise 3R Step 3.2R, in a ***visual*** display of division we see three pictures of boxes - the dividend, the divisor and the answer. Between the first and second pictures is the division sign. Over the dividend is the equal line, which separates the problem from the answer. On top of the dividend is the third picture, which is the answer of the division problem. The ***audio*** representation for the division consists of **knock, double knock, double click, knock, cling, and double knock.**

In the next part of Step 3.2R we have a **visual** display of how we arrive at the answer. In the dividend the empty positions before the filled up position are automatically taken to the top to the answer.

Step 3.2

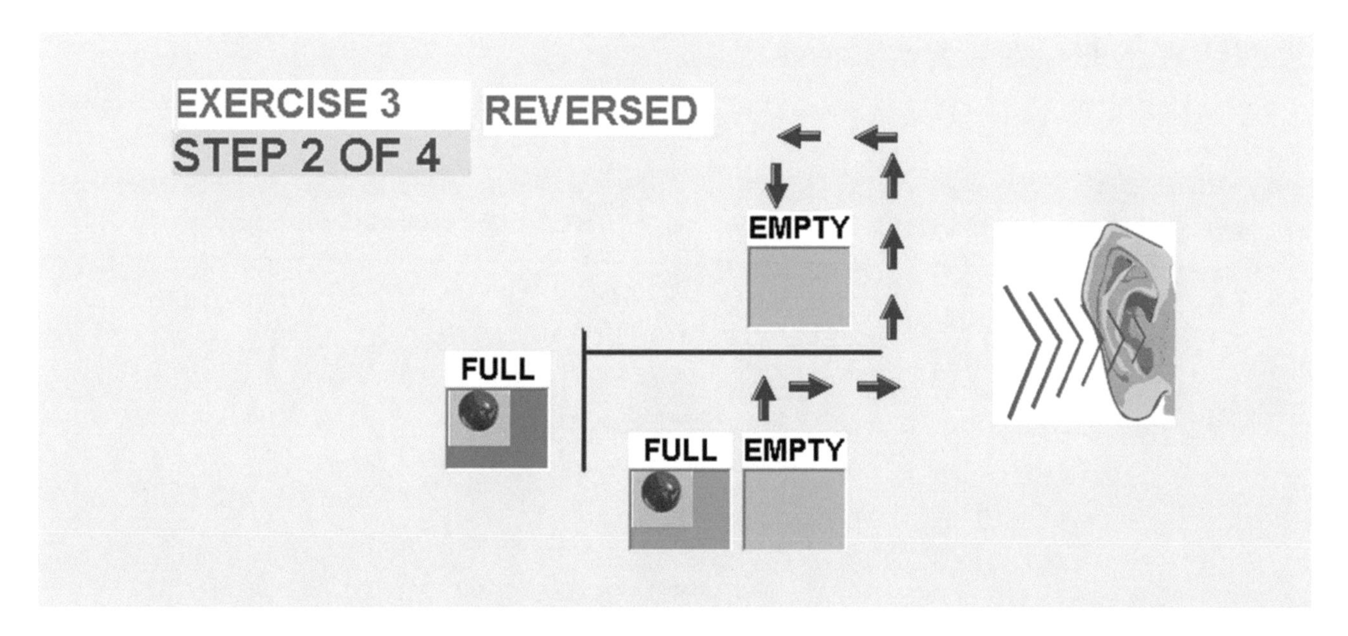

The ***kinesthetic*** representation of the filled up position is represented by the clenching of the right hand. The clenching of the left hand represents an empty position. The division sound, double click, is represented by both hands extended up. Both hands extended in front represent the subtraction sound, double tram. The equal sound, cling, is represented by the crossing of hands. The kinesthetic representation for the pictures is the clenching of the right hand, left hand, both hands extended up, right hand, crossing of the hands, and left hand.

MATHEMATICS

REVERSE DIVISION, EXERCISE 3.3

In Exercise 3R Step 3.3R, in a ***visual*** display of division we see three pictures of boxes - the dividend, the divisor and the answer. Between the first and second pictures is the division sign. Over the dividend is the equal line, which separates the problem from the answer. On top of the dividend is the third picture, which is the answer of division. The ***audio*** representation for the division consists of a **knock, double click, knock, cling, knock, and double knock.**

In the next part of Step 3.3R we also have a **visual** display of how we arrive to the answer. From picture two, which is the dividend, we subtract picture four. Between picture two and four is the subtraction sign. Below picture four we see the equal sign which separates the problem from the answer. Picture five is the result of the subtraction. When a result of a subtraction is an empty position then a filled position is transferred to the answer of the division. The ***audio*** representation for the division consists of a **knock, double tram, knock, cling, double knock.**

Step 3.3

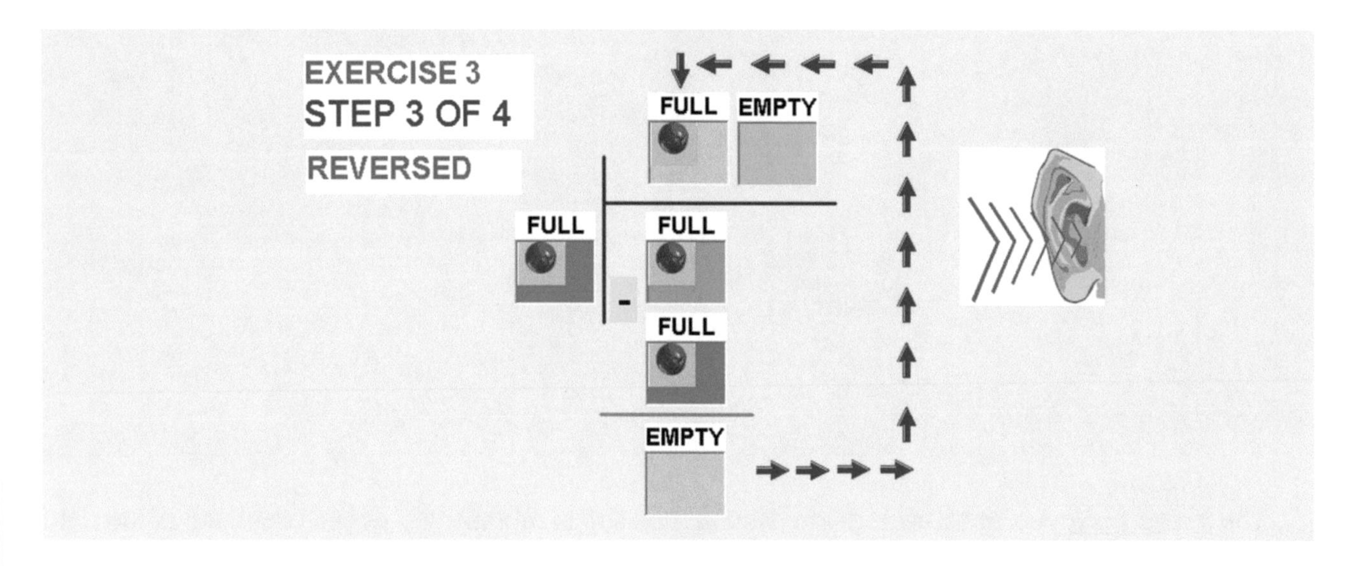

The ***kinesthetic*** representation of the filled up position is represented by the clenching of the right hand. The clenching of the left hand represents an empty position. The division sound, double click, is represented by both hands extended up. Both hands extended in front represent the subtraction sound, double tram. The equal sound, cling, is represented by crossing both hands. The kinesthetic representation for the pictures is the clenching of the right hand, both hands extended up, right hand, crossing the hands, right hand, and left hand.

REVERSE DIVISION, EXERCISE 3.4

In Exercise 3R Step 3.4R, in a ***visual*** display of the step of division we see two pictures of boxes. In both pictures we see that the first position is empty while the second position is filled with a red ball. The colors and the placement of the boxes and the balls are different in the pictures. The form, shape, and color of the symbols do no change the mathematical value, which is equal to two.

The ***audio*** representation for the two pictures consists of a **knock, double knock, and a cling.**

Exercise 3.4

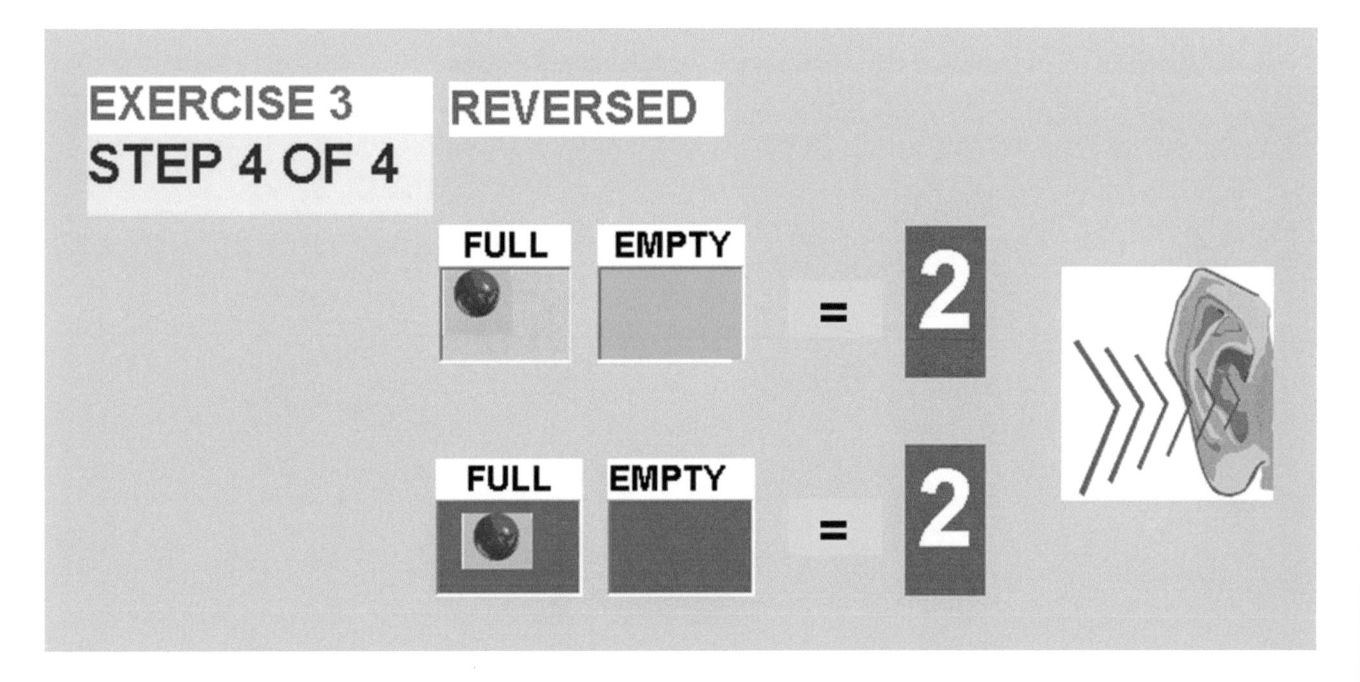

The ***kinesthetic*** representation of the filled up position is represented by the clenching of the right hand. The clenching of the left hand represents an empty position. The equal sound, cling, is represented by the crossing of hands. The kinesthetic representation for the two pictures is the clenching of the right hand, left hand, the crossing of hands, right hand, and left hand.

MATHEMATICS

REVERSE DIVISION, EXERCISE 3R, ANSWER 3R

In Exercise 3R, in a ***visual*** display of division are two pictures of boxes with the division sign is in-between. The first picture has the first position filled and the second empty. The second picture has the first filled and the next two empty. On the top of the second picture is the equal line, which separates the problem from the answer. The ***audio*** representation for the division consists of a **knock, double knock, double knock, double click, knock, double knock, and cling.**

At the bottom are the probable answers. From the probable answers for the division we need to find the right one. The audio signals are: 1) **knock and double knock,** 2) **double knock and knock**, 3) **knock and knock**.

Exercise 3

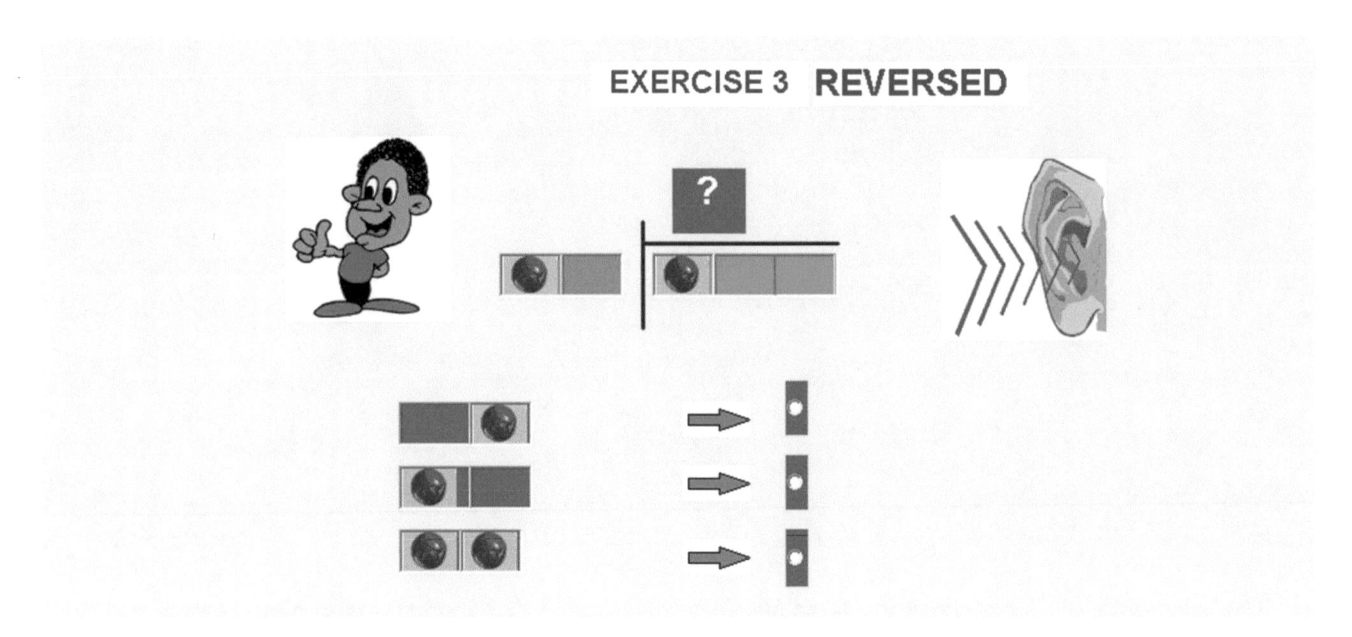

The ***kinesthetic*** representation of the filled up position is represented by the clenching of the right hand. An empty position is represented by the clenching of the left hand. The division sound, double click, is represented by both hands extended up. The equal sound, cling, is represented by crossing both hands. The kinesthetic representation for both pictures and the answer is the clenching of the right hand, left hand twice, both hands extended up, the clenching of the right hand, left hand, the crossing the hands, right hand and left hand.

REVERSE DIVISION, EXERCISE 4R

In Exercise 4R, in a ***visual*** display of division we see three pictures of boxes - the dividend, the divisor and the answer. Between the first and second pictures is the division sign. Over the dividend is the equal line, which separates the problem from the answer. On top of the dividend is the third picture, which is the answer of the division. The ***audio*** representation for the division consists of a **knock, knock, double knock, double click, knock, double knock, cling, knock, and knock.**

Exercise 4

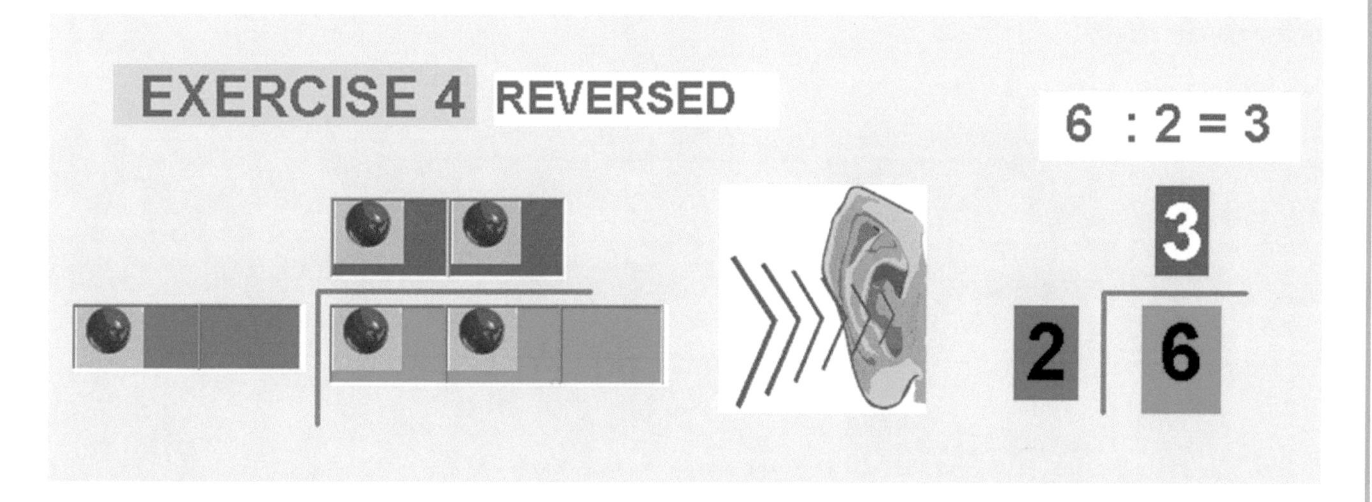

The ***kinesthetic*** representation of the filled up position is represented by the clenching of the right hand. The clenching of the left hand represents an empty position. The division sound, double click, is represented by both hands extended up. The equal sound, cling, is represented by crossing both hands. The kinesthetic representation for the three pictures is the clenching of the right hand twice, left hand, both hands extended up, right hand, left hand, crossing both hands, and right hand twice.

MATHEMATICS

REVERSE DIVISION, EXERCISE 4.1

In Exercise 4R Step 4.1R, in a ***visual*** display of division we see two pictures of boxes - the dividend and the divisor. Between the first and second pictures is the division sign. Over the dividend is the equal line, which separates the problem from the answer. On top of the dividend is the third picture, which is the answer of the division. The ***audio*** representation for the division consists of a **knock, knock, double knock, double click, knock, double knock, and cling.**

In the next part of Step 4.1R, we have a **visual** display of how we arrive at the answer. If the divisor has an empty position before the filled up position, then that position cancels out the first empty position of the dividend.

Step 4.1

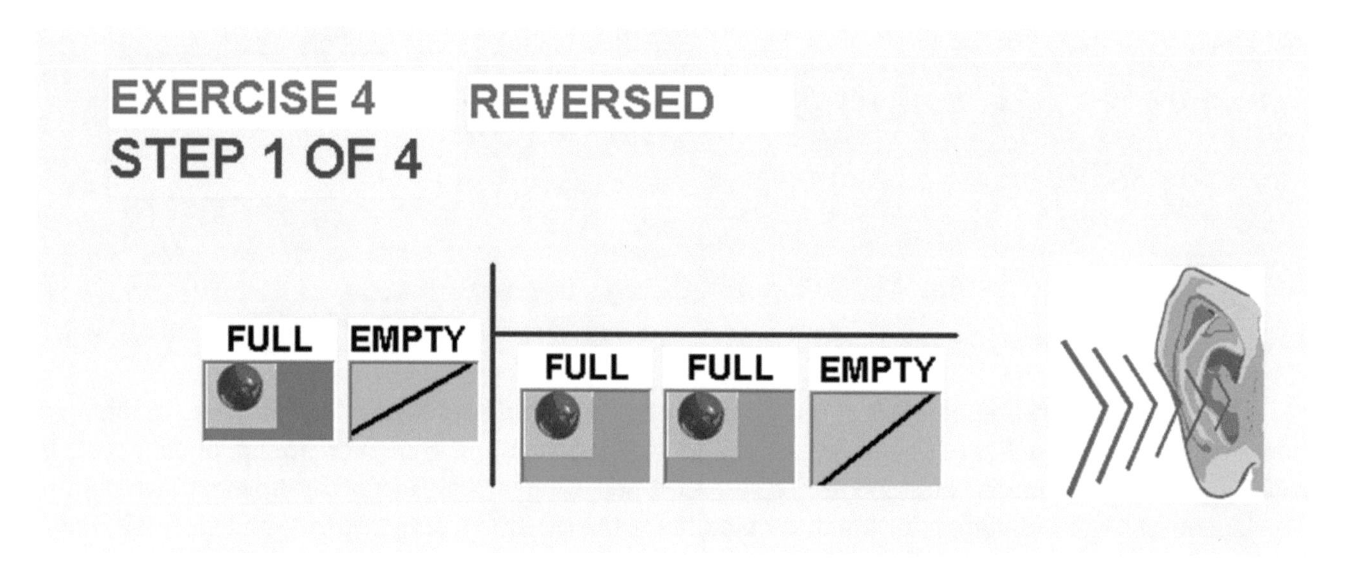

The ***kinesthetic*** representation of the filled up position is represented by the clenching of the right hand. The clenching of the left hand represents an empty position. The division sound, double click, is represented by both hands extended up. Both hands extended in front represent the subtraction sound, double tram. The equal sound, cling, is represented by crossing both hands. The kinesthetic representation for the two pictures is the clenching of the right hand twice, left hand, both hands extended up, right hand, left hand, and crossing both hands.

REVERSE DIVISION, EXERCISE 4.2

In Exercise 4R Step 4.2R, in a ***visual*** display of division we see three pictures of boxes - the dividend, the divisor and the answer. Between the first and second pictures is the division sign. Over the dividend is the equal line, which separates the problem from the answer. On top of the dividend is the third picture, which is the answer of the division problem. The ***audio*** representation for the division consists of a **knock, knock, double click, knock, cling, and knock.**

In the next part of Step 4.2R, we have a **visual** display of how we arrive at the answer. From picture two, which is the dividend, we subtract picture four. Between picture two and four is the subtraction sign. Below picture four is the equal sign which separates the problem from the answer. Picture five is the result of the subtraction. When the result of the subtraction is equal to an empty position, then a filled position is transferred to the answer of the division. The ***audio*** representation for the division consists of a **knock, knock, double tram, knock, cling, knock, and double knock.**

Step 4.2

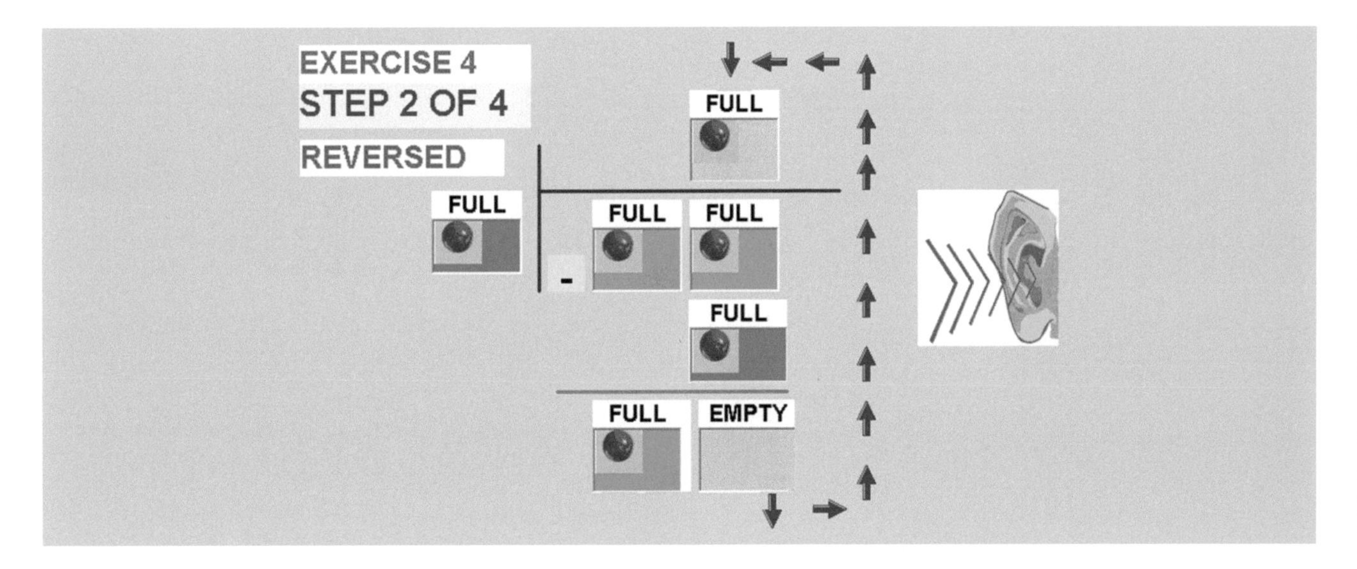

The ***kinesthetic*** representation of the filled up position is represented by the clenching of the right hand. The clenching of the left hand represents an empty position. The division sound, double click, is represented by both hands extended up. Both hands extended in front represent the subtraction sound, double tram. The equal sound, cling, is represented by crossing both hands. The kinesthetic representation for the pictures is the clenching of the right hand twice, both hands extended up, right hand, crossing both hands, and right hand.

MATHEMATICS

REVERSE DIVISION, EXERCISE 4.3

In Exercise 4R Step 4.3R, in a ***visual*** display of division we see three pictures of boxes - the dividend, the divisor and the answer. Between the first and second pictures is the division sign. Over the dividend is the equal line, which separates the problem from the answer. On top of the dividend is the third picture, which is the answer of the division. The ***audio*** representation for the division consists of a **knock, double click, knock, cling, knock, and knock.**

In the next part of Step 4.3R, we have a **visual** display of how we arrive at the answer. From picture two, which is the dividend, we subtract picture four. Between picture two and four is the subtraction sign. Below picture four is the equal sign which separates the problem from the answer. Picture five is the result of the subtraction. When the result of the subtraction is an empty position, then a filled position is transferred to the answer of division. The ***audio*** representation for the division consists of a **knock, double tram, knock, cling, double knock.**

Step 4.3

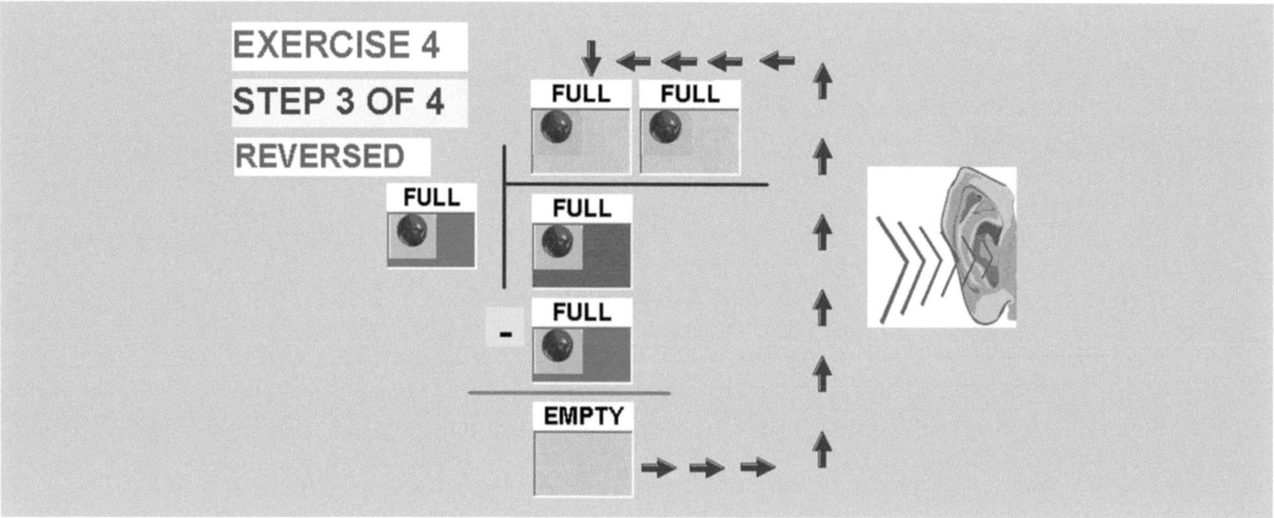

EXERCISE 4
STEP 3 OF 4
REVERSED
FULL
FULL
FULL
FULL
-
FULL
EMPTY

The ***kinesthetic*** representation of the filled up position is represented by the clenching of the right hand. The clenching of the left hand represents an empty position. The division sound, double click, is represented by both hands extended up. Both hands extended in front represent the subtraction sound, double tram. The equal sound, cling, is represented by crossing both hands. The kinesthetic representation for the pictures is the clenching of the right hand, both hands extended up, right hand, crossing both hands, right hand twice.

REVERSE DIVISION, EXERCISE 4.4

In Exercise 4R Step 4.4R, in a ***visual*** display of the last step of the division we see two pictures of boxes. In both pictures both positions are filled with a red ball. The colors and the placement of the boxes and the balls are different in the pictures. The form, shape, and color of the symbols do no change the mathematical value, which is equal to three.

The ***audio*** representation for the pictures consists of a **knock, knock, and cling.**

Step 4.4

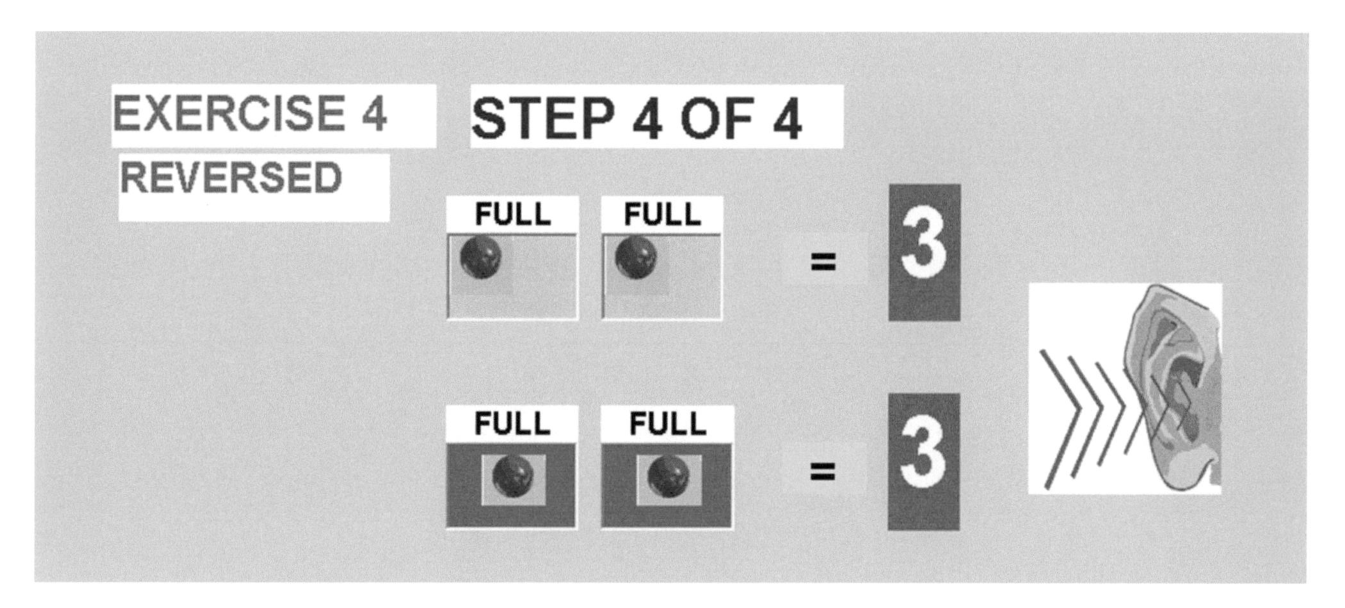

The ***kinesthetic*** representation of the filled up position is represented by the clenching of the right hand. The clenching of the left hand represents an empty position. The equal sound, cling, is represented by crossing both hands. The kinesthetic representation for the two pictures is the clenching of the right hand twice, crossing both hands, and right hand twice.

MATHEMATICS

REVERSE DIVISION, EXERCISE 4R, ANSWER 4R

In Exercise 4R, in a ***visual*** display of division we see two pictures of boxes with the division sign in-between. The first picture has the first position filled with a red ball and the second position empty. The second picture has the first two positions filled and the last one empty. On top of the second picture is the equal line, which separates the problem from the answer. The ***audio*** representation for the division consists of a **knock, knock, double knock, double click, knock, double knock, and cling.**

At the bottom are the probable answers. From the probable answers for the division we need to find the right one. The audio signals are: 1) **double knock and knock,** 2) **knock and double knock**, 3) **knock and knock**.

Exercise 4

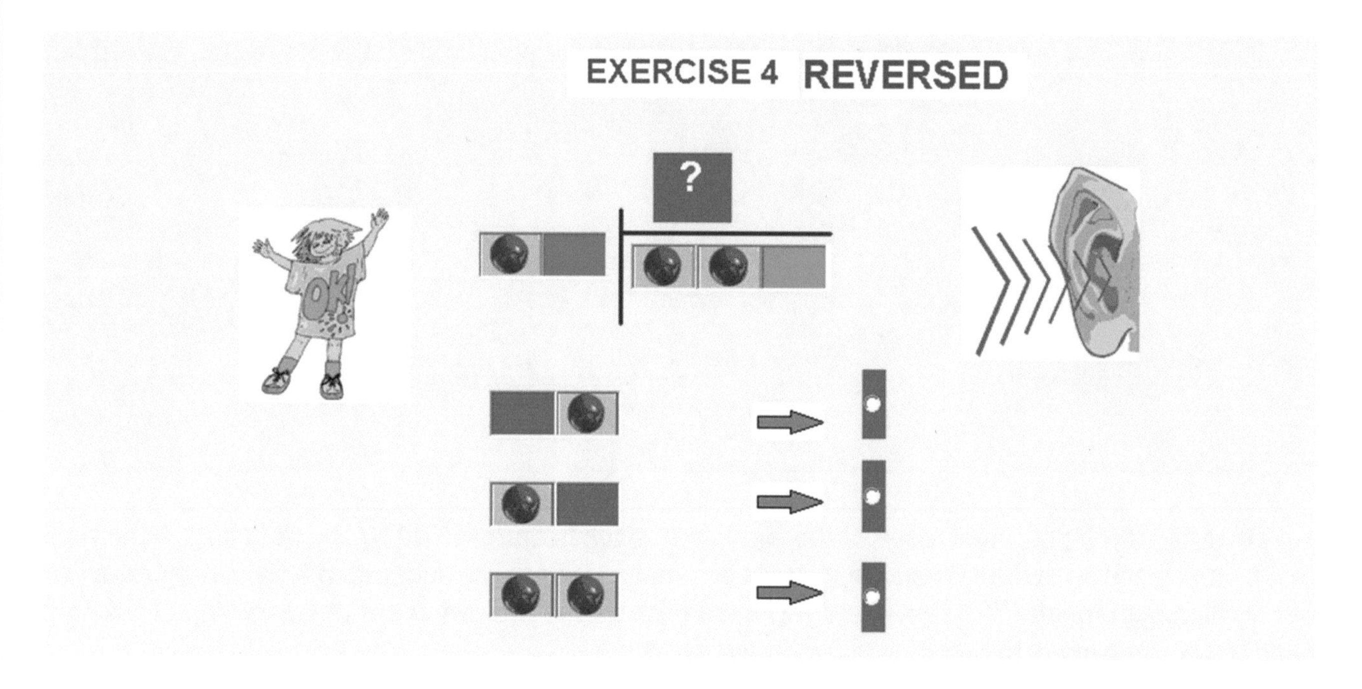

The ***kinesthetic*** representation of the filled up position is represented by the clenching of the right hand. The clenching of the left hand represents an empty position. Both hands extended up represent the division sound, double click. The equal sound, cling, is represented by crossing both hands. The kinesthetic representation for both pictures and the answer is the clenching of the right hand twice, left hand, both hands extended up, clenching the right hand, left hand, crossing of the hands, and right hand twice.

REVERSE DIVISION, EXERCISE 5R

In Exercise 5R, in a ***visual*** display of division we see three pictures of boxes. The first picture, which has a value of three, has both positions filled with red balls. The second picture, with a value of six has positions one and two filled with red balls and position three empty. The division sign is between the first and second sets of boxes. Over the second picture is the equal line, which separates the problem from the answer. The third picture, which is the answer for the division has the first position filled with a symbol, while the second position is empty.

The ***audio*** representation for the division consists of a **knock, knock, double knock, double click, knock, knock, cling, knock, and double knock.**

Exercise 5R

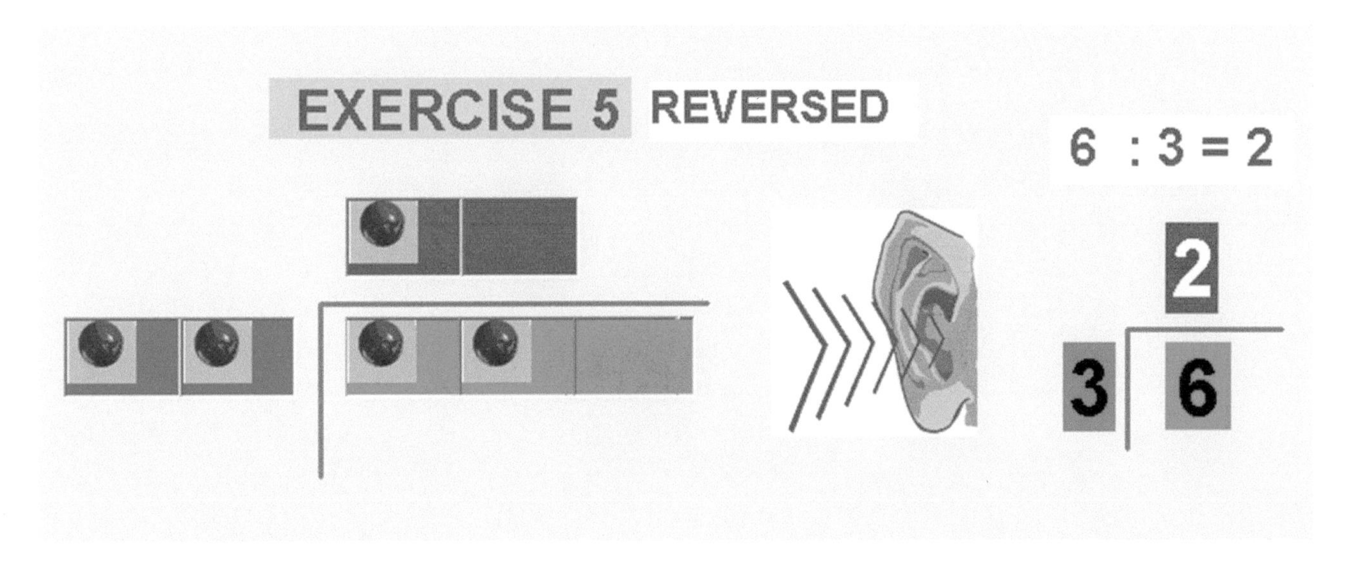

The ***kinesthetic*** representation of the filled up position is represented by the clenching of the right hand. The clenching of the left hand represents an empty position. Both hands extended up represent the division sign, double click. The equal sign, cling, is represented by crossing both hands. The kinesthetic representation for the two forms of the same division above is the clenching of the right hand twice, left hand, extending both hands up, right hand twice, crossing both hands, right hand and left hand.

MATHEMATICS REVERSE DIVISION, EXERCISE 5.1

In Exercise 5R Step 5.1R, in a ***visual*** display of division we see three pictures of boxes - the dividend, the divisor and the answer. Between the first and second pictures is the division sign. Over the dividend is the equal line, which separates the problem from the answer. On top of the dividend is the third picture, which is the answer of division. The ***audio*** representation for the division consists of a **knock, knock, double knock, double click, knock, knock, cling, and double knock.**

In the next part of Step 5.1R, we have a **visual** display of how we arrive at the answer. In the dividend, the empty positions before the filled up position are automatically taken to the top to the answer.

Step 5.1

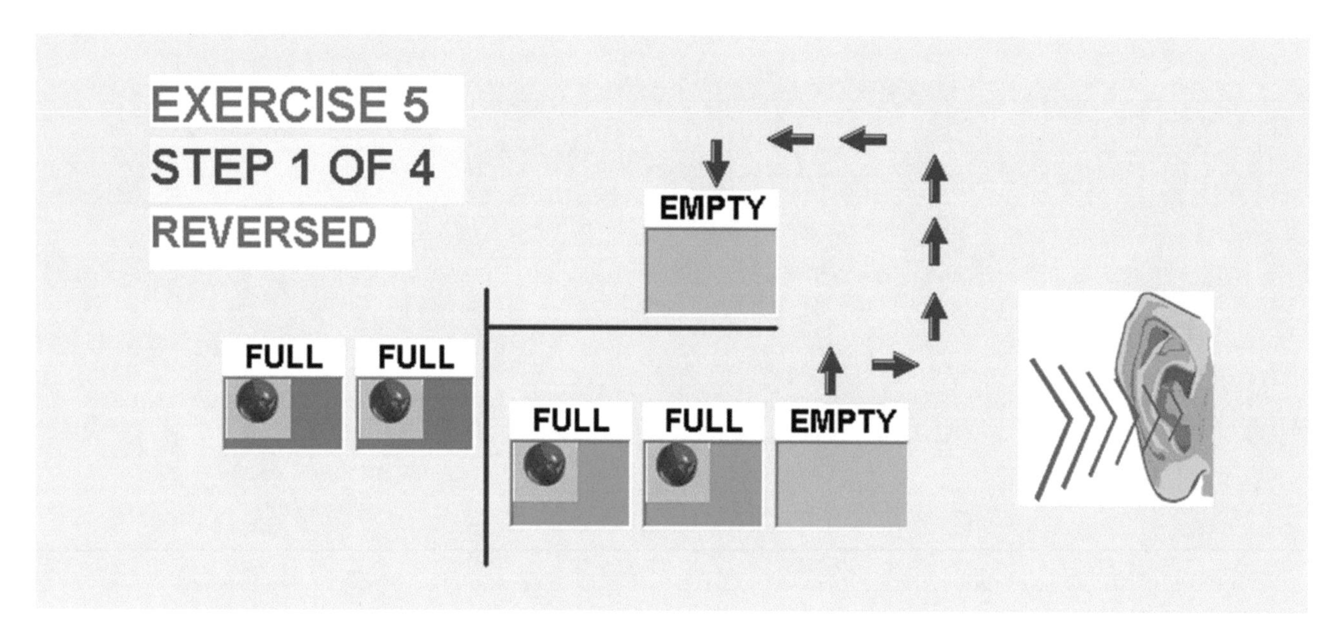

The ***kinesthetic*** representation of the filled up position is represented by the clenching of the right hand. The clenching of the left hand represents an empty position. The division sound, double click, is represented by both hands extended up. Both hands extended in front represent the subtraction sound, double tram. The equal sound, cling, is represented by crossing both hands. The kinesthetic representation for the division is the clenching of the right hand twice, left hand, both hands extended up, right hand twice, crossing of both hands, and left hand.

REVERSE DIVISION, EXERCISE 5.2

In Exercise 5R Step 5.2R, in a ***visual*** display of division we see three pictures of boxes - the dividend, the divisor, and the answer. Between the first and second pictures is the division sign. Over the dividend is the equal line, which separates the problem from the answer. On top of the dividend is the third picture, which is the answer of the division problem. The ***audio*** representation for the division consists of a **knock, knock, double click, knock, knock, cling, and double knock.**

In the next part of Step 5.2R, we have a **visual** display of how we arrive at the answer. After having eliminated the empty positions, the next step can be applied to the problem.

Step 5.2

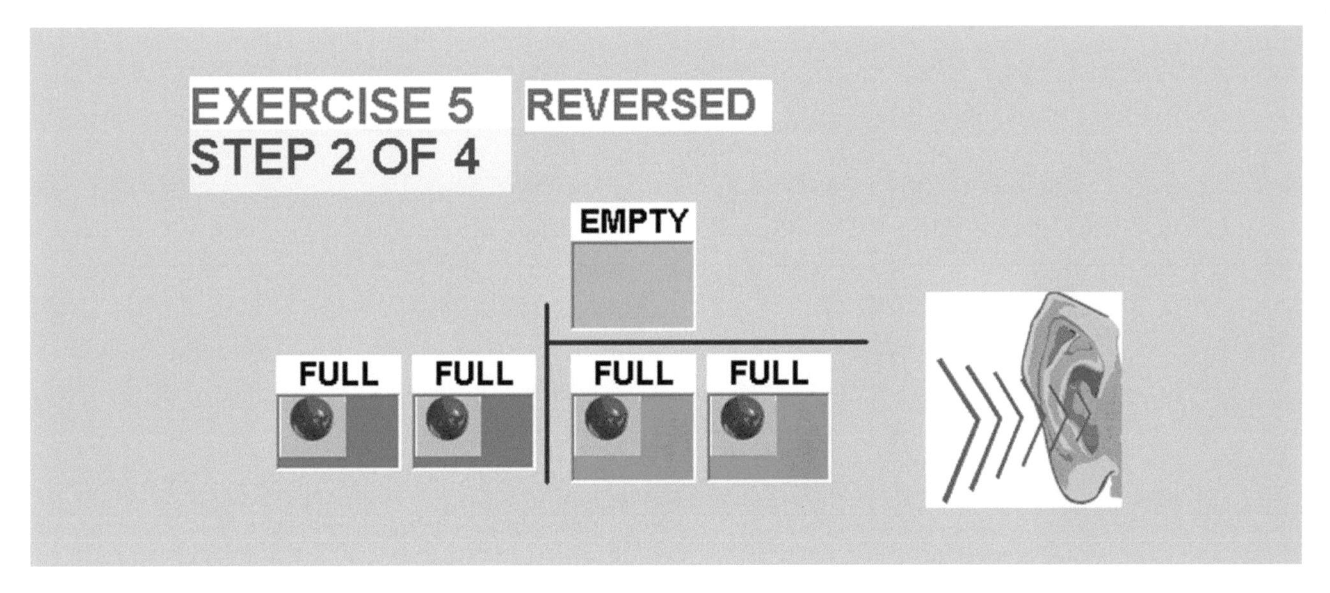

The ***kinesthetic*** representation of the filled up position is represented by the clenching of the right hand. The clenching of the left hand represents an empty position. The division sound, double click, is represented by both hands extended up. Both hands extended in front represent the subtraction sound, double tram. The equal sound, cling, is represented by crossing both hands. The kinesthetic representation for two pictures is the clenching of the right hand twice, hands extended up, right hand twice, crossing both hands, and left hand.

MATHEMATICS REVERSE DIVISION, EXERCISE 5.3

In Exercise 5R Step 5.3R, in a ***visual*** display of division we see three pictures of boxes - the dividend, the divisor and the answer. Between the first and second pictures is the division sign. Over the dividend is the equal line, which separates the problem from the answer. On top of the dividend is the third picture, which is the answer of the division problem. The ***audio*** representation for the division consists of a **knock, knock, double click, knock, knock, cling, knock, and double knock.**

In the next part of Step 5.3R, we have a **visual** display of how we arrive at the answer. From picture two, which is dividend, we subtract four. Between pictures two and four is the subtraction sign. Below picture four is the equal sign which separates the problem from the answer. Picture five is the result of the subtraction. When the result of subtraction is an empty position, then a filled position is transferred to the answer of division. The ***audio*** representations for the division consists of a **knock, knock, double tram, knock, knock, cling, double knock, and double knock.**

Step 5.3

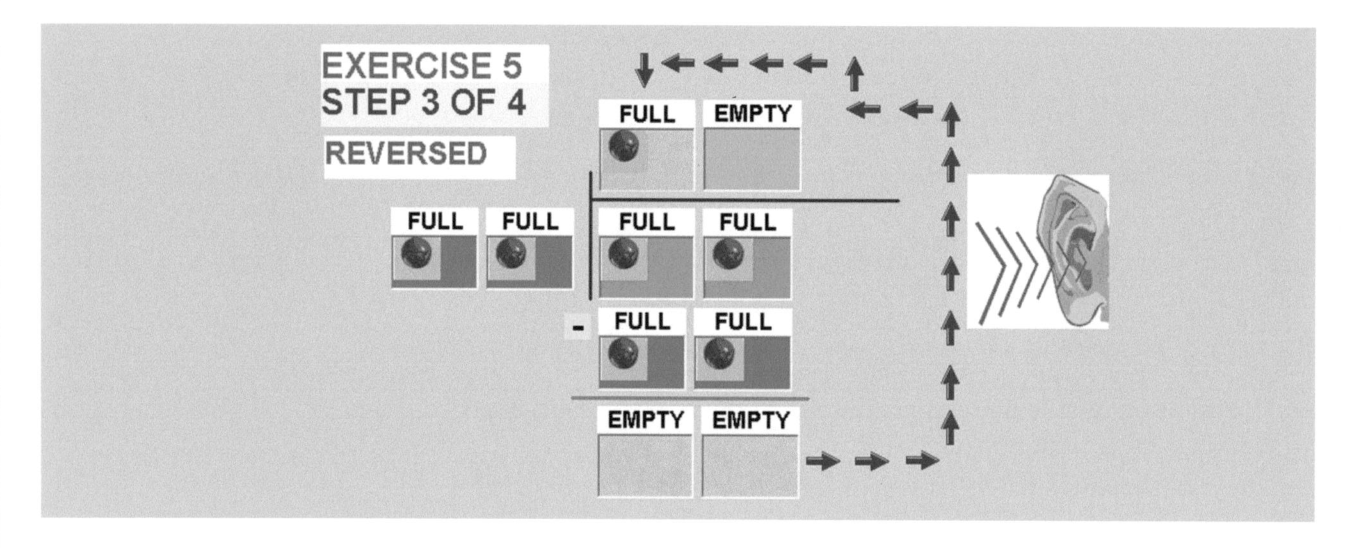

The ***kinesthetic*** representation of the filled up position is represented by the clenching of the right hand. The clenching of the left hand represents an empty position. The division sound, double click, is represented by both hands extended up. Both hands extended in front represent the subtraction sound, double tram. The equal sound, cling, is represented by crossing both hands. The kinesthetic representation for the three pictures is the clenching of the right hand twice, hands extended up, right hand twice, crossing both hands, right hand, and left hand.

REVERSE DIVISION, EXERCISE 5.4

In Exercise 5R Step 5.4R, in a ***visual*** display of the last step of the division we see two pictures of boxes. In both pictures the first position is filled with a red ball, while the second position is empty. The colors and the placement of the balls in the boxes are different in the pictures. The form, shape, and color of the symbols do no change the mathematical value, which is equal to two.

The ***audio*** representation for the picture consists of a **knock, double knock, and a cling.**

Step 5.4

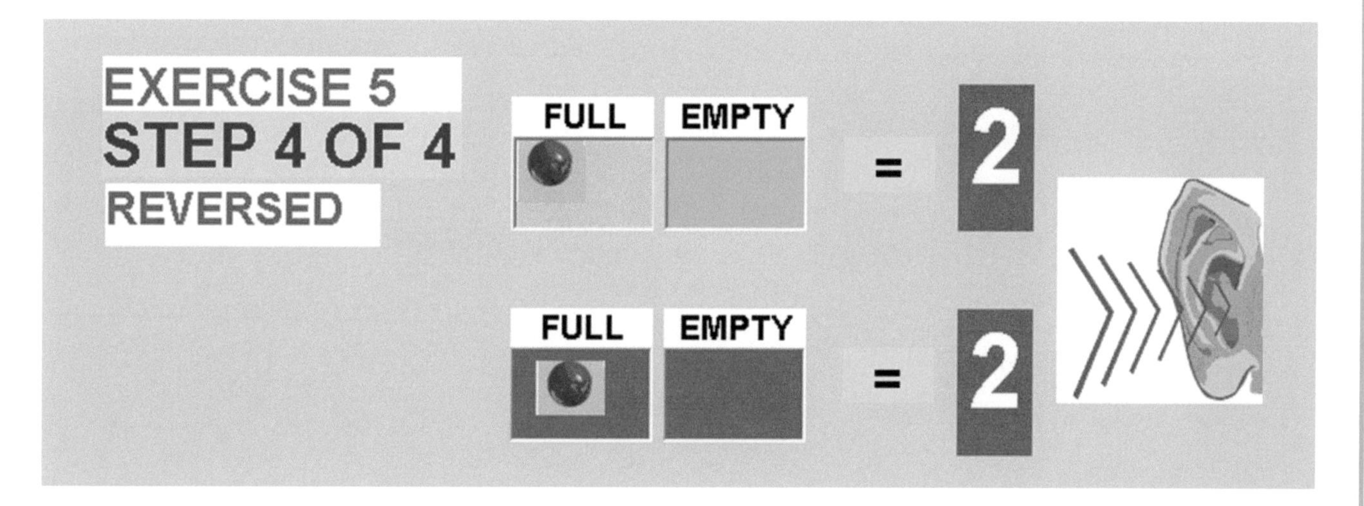

The ***kinesthetic*** representation of the filled up position is represented by the clenching of the right hand. The clenching of the left hand represents an empty position. The equal sound, cling, is represented by crossing hands. The kinesthetic representation for the two pictures is the clenching of the right hand, left hand, the crossing of hands, right hand, and left hand.

MATHEMATICS

REVERSE DIVISION, EXERCISE 5R, ANSWER 5R

In Exercise 5R, in a ***visual*** display of division we see two pictures of boxes with the division sign in-between. Both positions in the first picture are filled with red balls. In the second picture, the first two positions are filled with red balls and the third is empty. On top of the second picture is the equal line, which separates the problem from the answer. The ***audio*** representation for the division consists of a **knock, knock, double knock, double click, knock, knock, and cling.**

At the bottom are the probable answers. From the probable answers for the division we need to find the right one. The audio signals are: 1) **double knock and knock,** 2) **knock and double knock**, 3) **knock and knock**.

Exercise 5

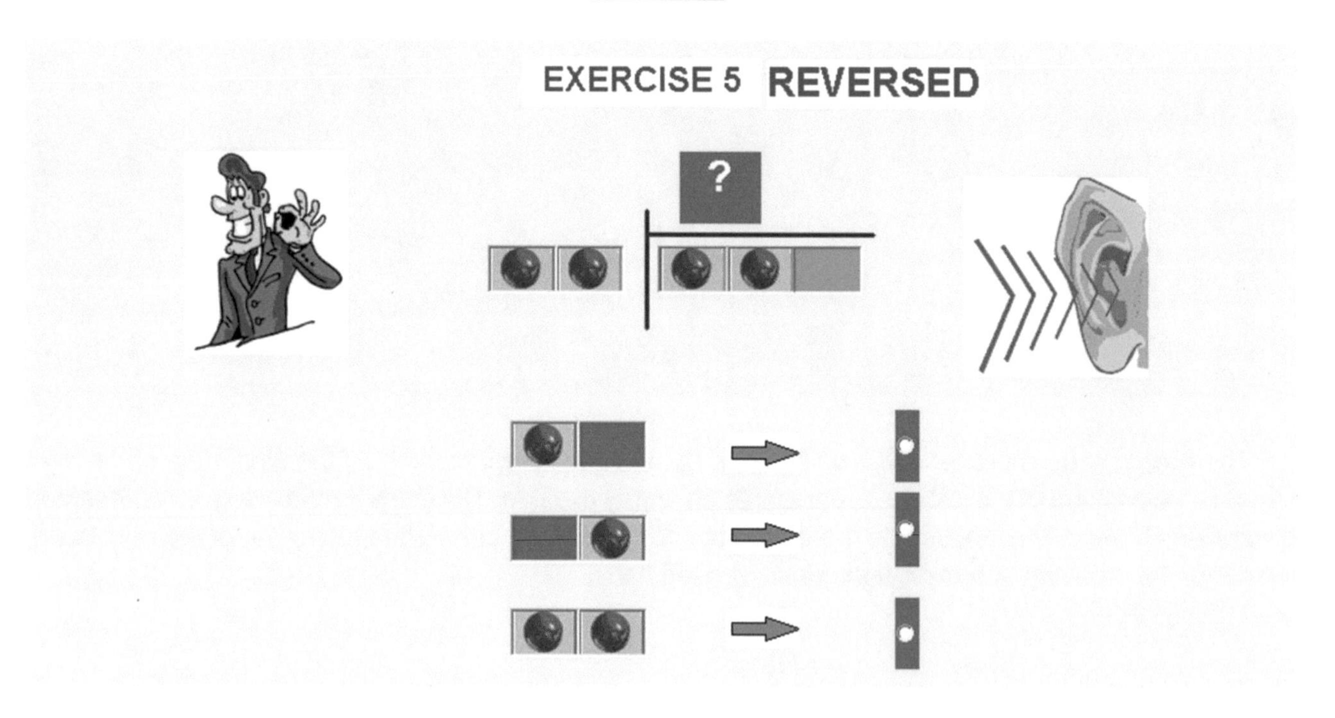

The ***kinesthetic*** representation of the filled up position is represented by the clenching of the right hand. The clenching of the left hand represents an empty position. Both hands extended up represent the division sound, double click. The equal sound, cling, is represented by crossing both hands. The kinesthetic representation for both pictures and the answer is the clenching of the right hand twice, left hand, both hands extended up, the clenching of the right hand twice, crossing the hands, right hand, and left hand.

REVERSE DIVISION, EXERCISE 6R

In Exercise 6R, in a ***visual*** display of division we see three pictures of boxes. The first picture, which has a value of five, has the first position filled with a red ball, the second position empty, and the last position has a red ball. In the second picture, which has a value of ten, position one is filled with a red ball, position two empty, position three is filled with a red ball, and position four is empty. The division sign is between the first and second sets of boxes. Over the second picture is the equal line, which separates the problem from the answer. In the third picture, which is the answer for the division and has a value of two, we see that the first position has a symbol while the second is empty.

The ***audio*** representation for the division consists of a **knock, double knock, knock, double knock, double click, knock, double knock, knock, cling, knock, and double knock.**

Exercise 6R

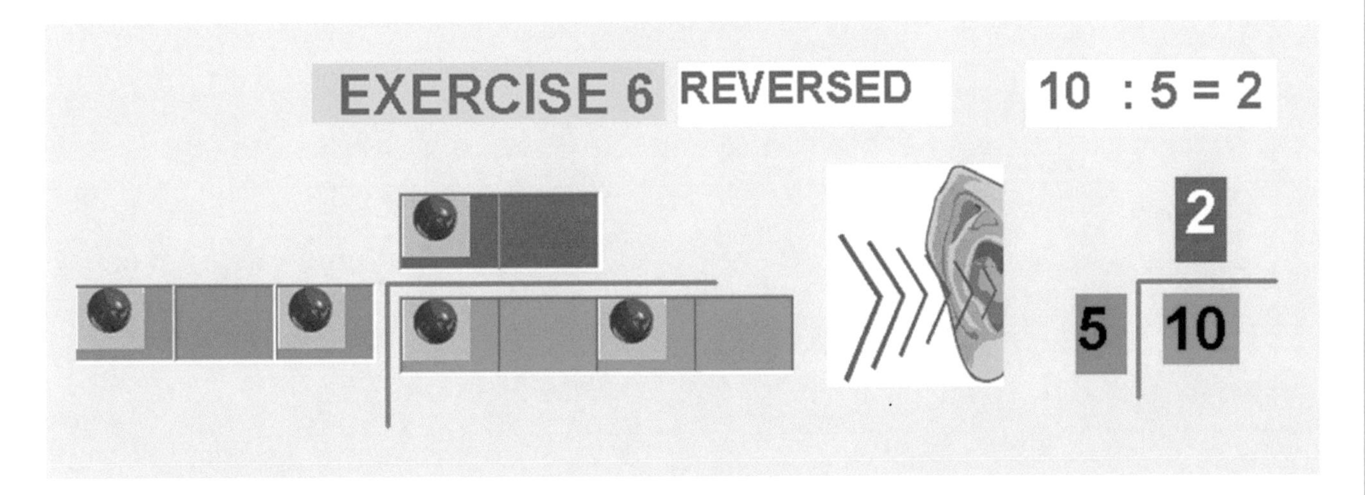

The ***kinesthetic*** representation of the filled up position is represented by the clenching of the right hand. The clenching of the left hand represents an empty position. Both hands extended up represent the division sign, double click. The equal sign, cling, is represented by crossing both hands. The kinesthetic representation for both pictures is the clenching of the right hand, left hand, right hand, left hand, extending both hands up, right hand, left hand, right hand, crossing both hands, right hand and left hand.

MATHEMATICS REVERSE DIVISION, EXERCISE 6.1

In Exercise 6R Step 6.1R, in a ***visual*** display of division we see three pictures of boxes - the dividend, the divisor and the answer. Between the first and second pictures is the division sign. Over the dividend is the equal line, which separates the problem from the answer. On top of the dividend is the third picture, which is the answer of the division problem. The ***audio*** representation for the division consists of a **knock, double knock, knock, double knock, double click, knock, double knock, knock, cling, and double knock.**

In the next part of Step 6.1R, we have a **visual** display of how we arrive at the answer. In the dividend the empty positions before the filled up position are automatically taken to the top to the answer.

Step 6.1

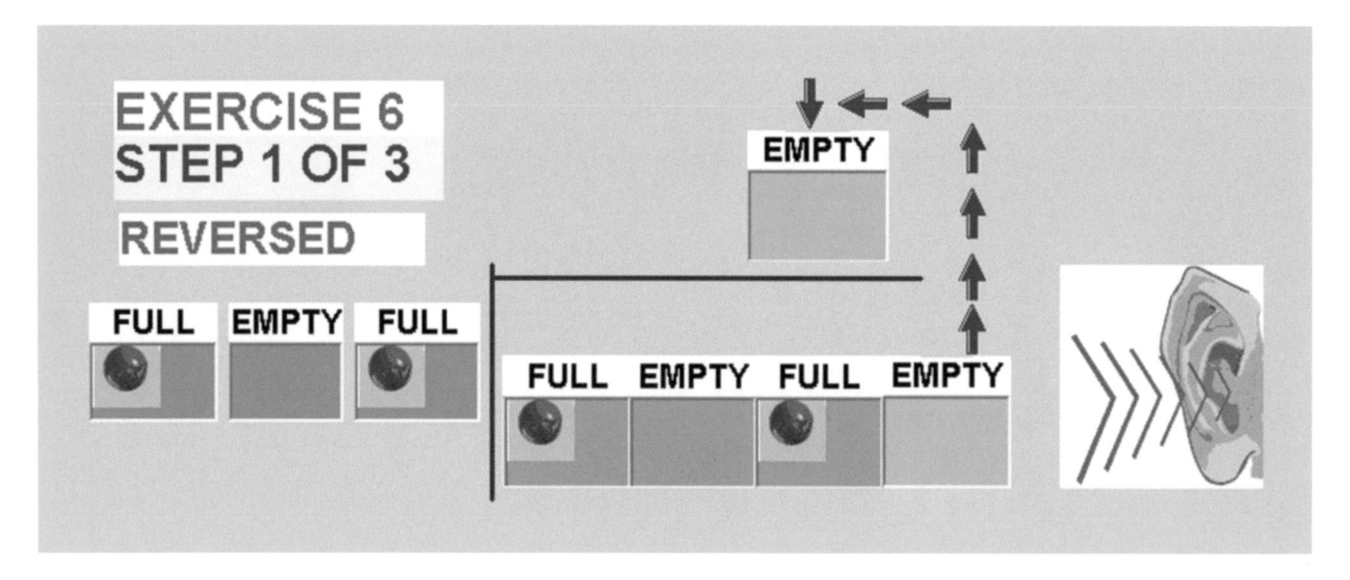

The ***kinesthetic*** representation of the filled up position is represented by the clenching of the right hand. The clenching of the left hand represents an empty position. The division sound, double click, is represented by both hands extended up. Both hands extended in front represent the subtraction sound, double tram. The equal sound, cling, is represented by crossing both hands. The kinesthetic representation for the division is the clenching of the right hand, left hand, right hand, left hand, both hands extended up, right hand, left hand, right hand, crossing the hands, and left hand.

REVERSE DIVISION, EXERCISE 6.2

In Exercise 6R Step 6.2R, in a ***visual*** display of division we see three pictures of boxes - the dividend, the divisor and the answer. Between the first and second pictures is the division sign. Over the dividend is the equal line, which separates the problem from the answer. On top of the dividend is the third picture, which is the answer of the division problem. The ***audio*** representation for the division consists of a **knock, double knock, knock, double click, knock, double knock, knock, cling, knock, and double knock.**

In the next part of Step 5.3R, we have a **visual** display of how we arrive at the answer. We subtract picture four from picture two, which is the dividend. The subtraction sign is between pictures two and four. Below picture four is the equal sign, separating the problem from the answer. The result of the subtraction is picture five. If the result of subtraction is an empty position, then a filled position is transferred to the answer of division. The ***audio*** representation for the subtraction consists of a **knock, double knock, knock, double tram, knock, double knock, knock, cling, double knock, double knock, a double knock.**

Step 6.2

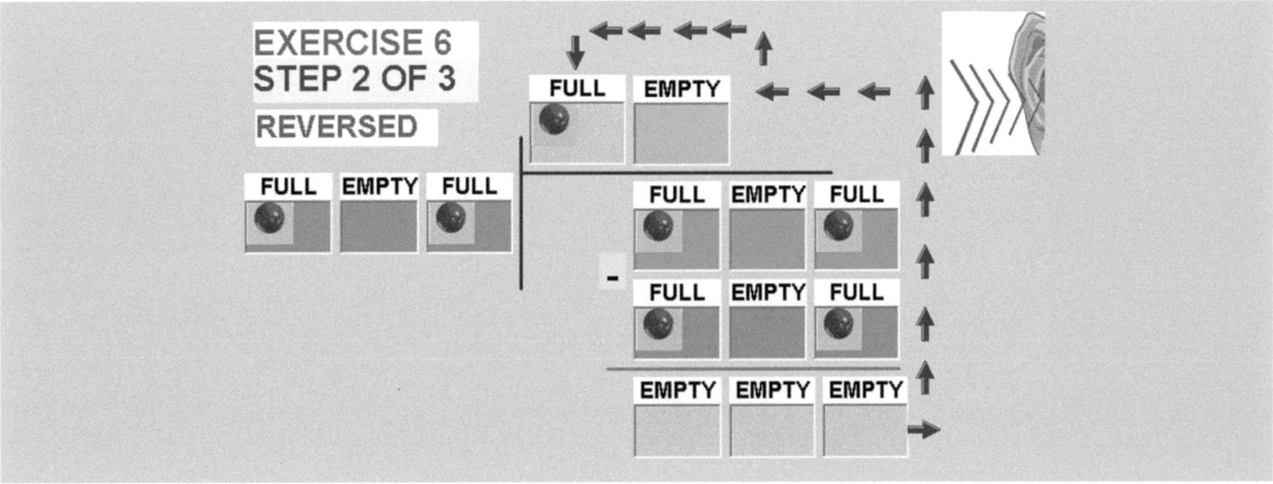

EXERCISE 6
STEP 2 OF 3
REVERSED

FULL	EMPTY

FULL	EMPTY	FULL

FULL	EMPTY	FULL

-

FULL	EMPTY	FULL

EMPTY	EMPTY	EMPTY

The ***kinesthetic*** representation of the filled up position is represented by the clenching of the right hand. The clenching of the left hand represents an empty position. The division sound, double click, is represented by both hands extended up. Both hands extended in front represent the subtraction sound, double tram. The equal sound, cling, is represented by crossing both hands. The kinesthetic representation for the picture is the clenching of the right hand, left hand, right hand, both hands extended up, right hand, left hand, right hand, crossing the hands, right hand, and left hand.

MATHEMATICS REVERSE DIVISION, EXERCISE 6R, ANSWER 6R

In Exercise 6R, in a ***visual*** display of division we see two pictures of boxes with the division sign in-between. The first position in the first picture is filled with a red ball, while the second is empty and the third is filled again. In the second picture, the first position has a red ball, the next position is empty, the third position has a red ball, and the last position is empty. On top of the second picture is the equal line, which separates the problem from the answer. The ***audio*** representation for the division is a **knock, double knock, knock, double knock, knock, double click, knock, double knock, knock, and cling.**

At the bottom are the probable answers. From the probable answers for the division we need to find the right one. The audio signals are: 1) **double knock and knock,** 2) **knock and double knock**, 3) **knock and knock.**

Exercise 6

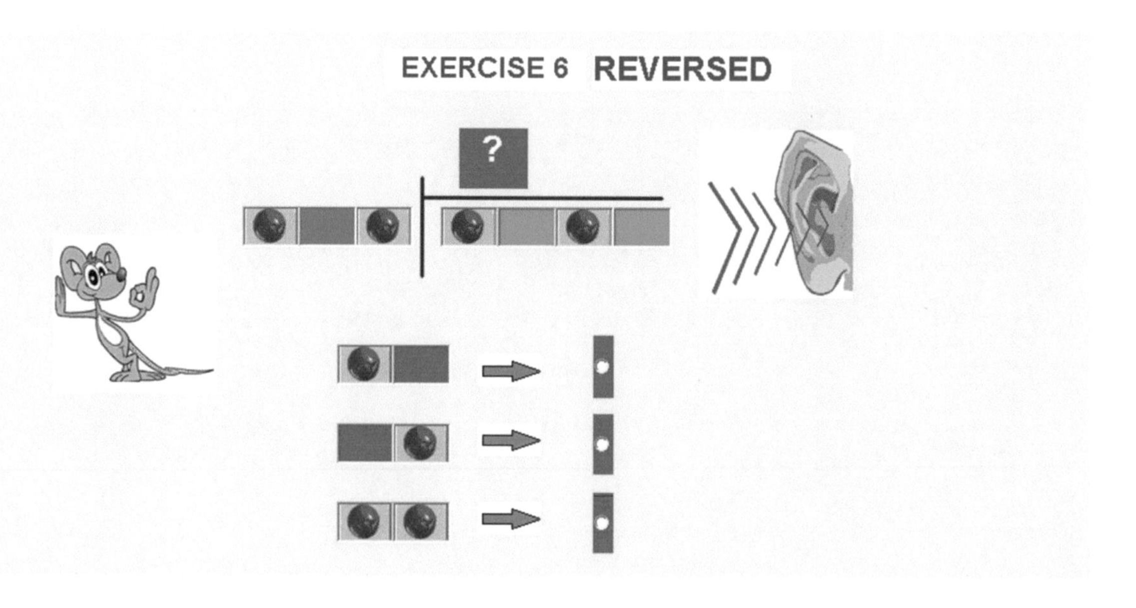

The ***kinesthetic*** representation of the filled up position is represented by the clenching of the right hand. The clenching of the left hand represents an empty position. Both hands extended up represent the division sound, double click. The equal sound, cling, is represented by crossing both hands. The kinesthetic representation for both pictures and the answer is the clenching of the right hand, left hand, right hand, left hand, both hands extended up, the clenching of the right hand, left hand, right hand, the crossing of hands, right hand, and left hand.

REVERSE DIVISION, EXERCISE 7R

In Exercise 7R, in a ***visual*** display of division we see three pictures of boxes with the division sign in-between. In the first picture, the first position is filled with a red ball and the second is empty. In the second picture, the first position is filled with a red ball and the next three are empty. In the third picture the first position is filled with a red ball and the next two are empty. On top of the second picture is the equal line, which separates the problem from the answer. The ***audio*** representation for the division consists of a **knock, double knock, double knock, double knock, double click, knock, double knock, cling, knock, double knock, and double knock.**

Exercise 7R

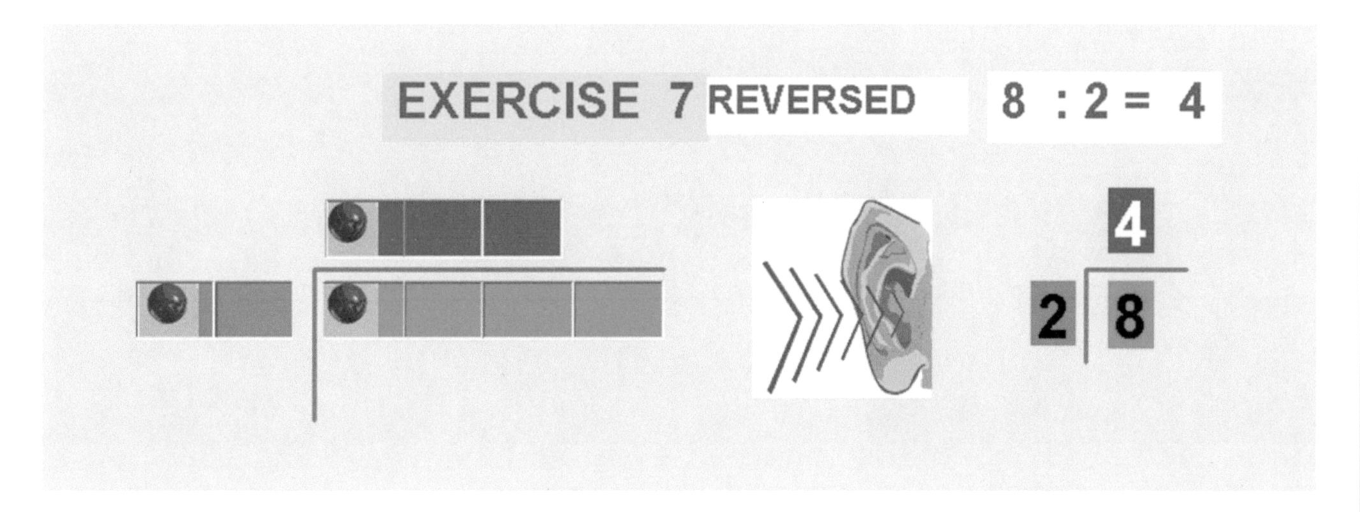

The ***kinesthetic*** representation of the filled up position is represented by the clenching of the right hand. The clenching of the left hand represents an empty position. Both hands extended up represent the division sound, double click. The equal sound, cling, is represented by crossing both hands. The kinesthetic representation for both pictures and the answer is the clenching of the right hand, left hand three times, both hands extended up, right hand, left hand, crossing of the hands, right hand, left hand twice.

MATHEMATICS

REVERSE DIVISION, EXERCISE 7.1

In Exercise 7R Step 7.1, in a ***visual*** display of division we see two pictures of boxes - the dividend and the divisor. The division sign is between the first and second pictures. Over the dividend is the equal line, which separates the problem from the answer. The ***audio*** representation for the pictures consists of a **knock, double knock, double knock, double knock, double click, knock, double knock, and cling.**

Step 7.1

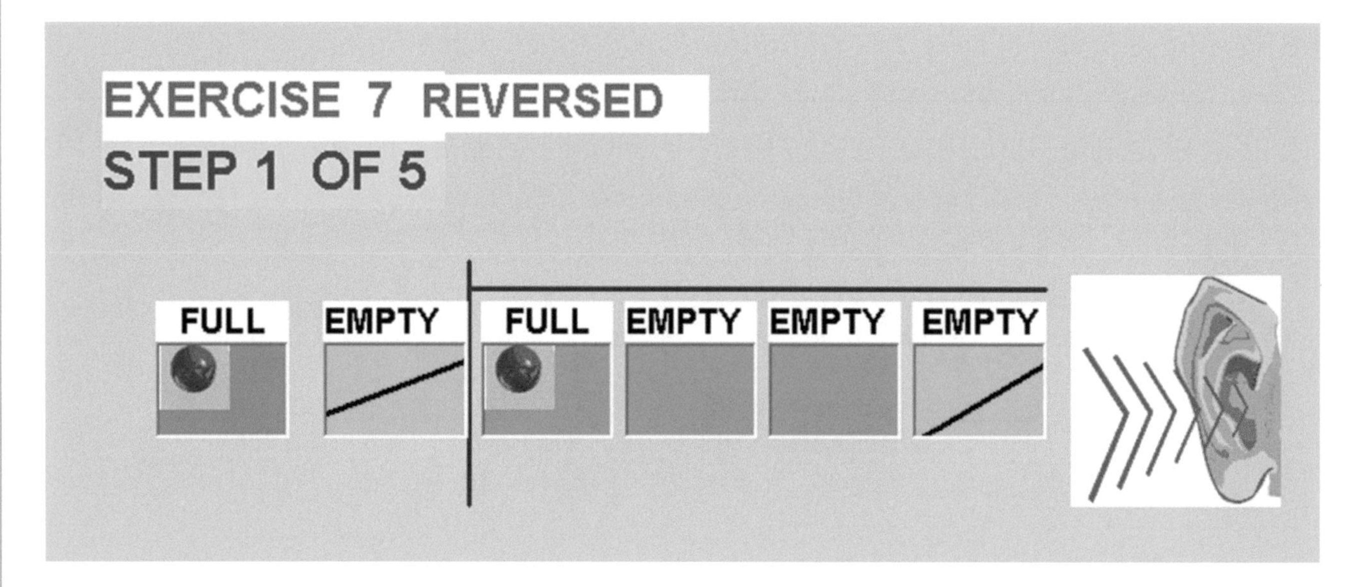

The ***kinesthetic*** representation of the filled up position is represented by the clenching of the right hand. The clenching of the left hand represents an empty position. The division sound, double click, is represented by both hands extended up. Both hands extended in front represent the subtraction sound, double tram. The equal sound, cling, is represented by crossing both hands. The kinesthetic representation for the two pictures is the clenching of the right hand, left hand three times, both hands extended up, right hand, left hand, and the crossing of both hands.

REVERSE DIVISION, EXERCISE 7.2

In Exercise 7R Step 7.2, in a ***visual*** display of division we see three pictures of boxes - the dividend, the divisor and the answer. The division sign is between the first and second pictures. Over the dividend is the equal line, which separates the problem from the answer. On top of the dividend is the third picture, which is the answer of the division. The ***audio*** representation for the pictures consists of a **knock, double knock, double knock, double click, knock, cling, and double knock.**

In the next part of Step 7.2, we have a **visual** display of how we arrive at the answer. In the dividend the empty positions before the filled up position are automatically taken to the top to the answer.

Step 7.2

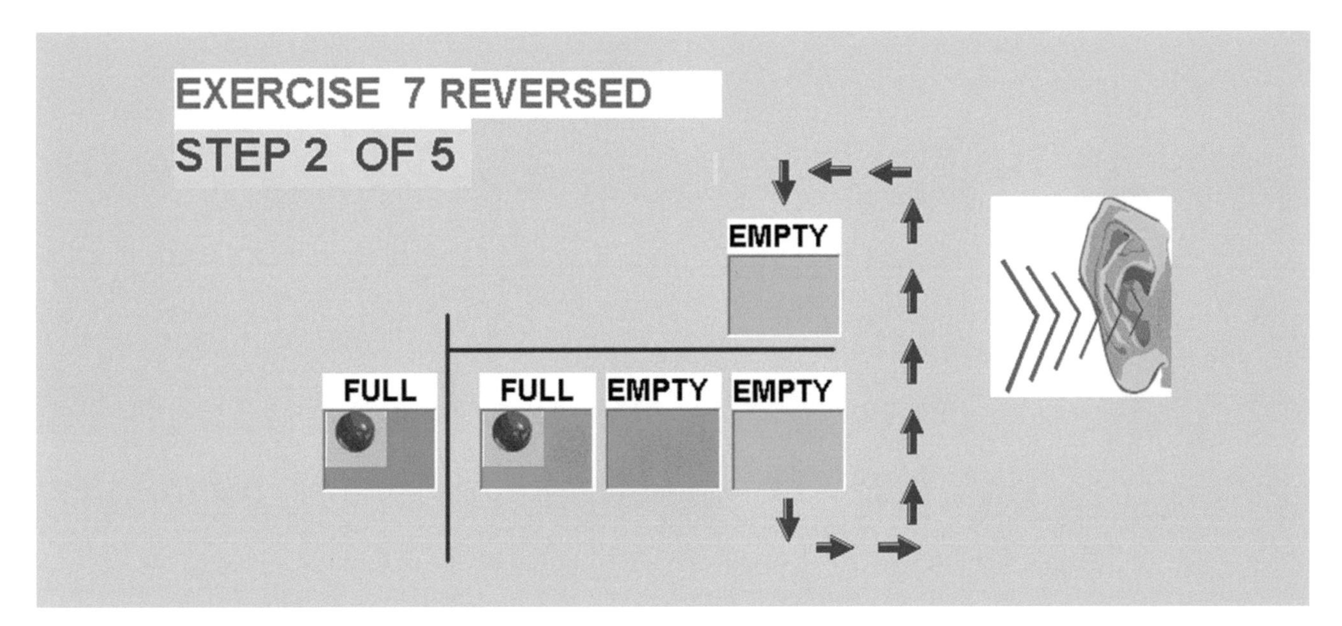

The ***kinesthetic*** representation of the filled up position is represented by the clenching of the right hand. The clenching of the left hand represents an empty position. The division sound, double click, is represented by both hands extended up. Both hands extended in front represent the subtraction sound, double tram. The equal sound, cling, is represented by crossing both hands. The kinesthetic representation for the three pictures is the clenching of the right hand, left hand twice, both hands extended up, right hand, crossing of the hands, and left hand.

MATHEMATICS

REVERSE DIVISION, EXERCISE 7.3

In Exercise 7R Step 7.3R, in a ***visual*** display of division we see three pictures of boxes - the dividend, the divisor and the answer. Between the first and second pictures is the division sign. Over the dividend is the equal line, which separates the problem from the answer. On top of the dividend is the third picture, which is the answer of the division. The ***audio*** representation for the division consists of a **knock, double knock, double click, knock cling, double knock, and double knock.**

In the next part of Step 7.3R, we have a **visual** display of how we arrive at the answer. In the dividend, the empty positions before the filled up position are automatically taken to the top to the answer.

Step 7.3

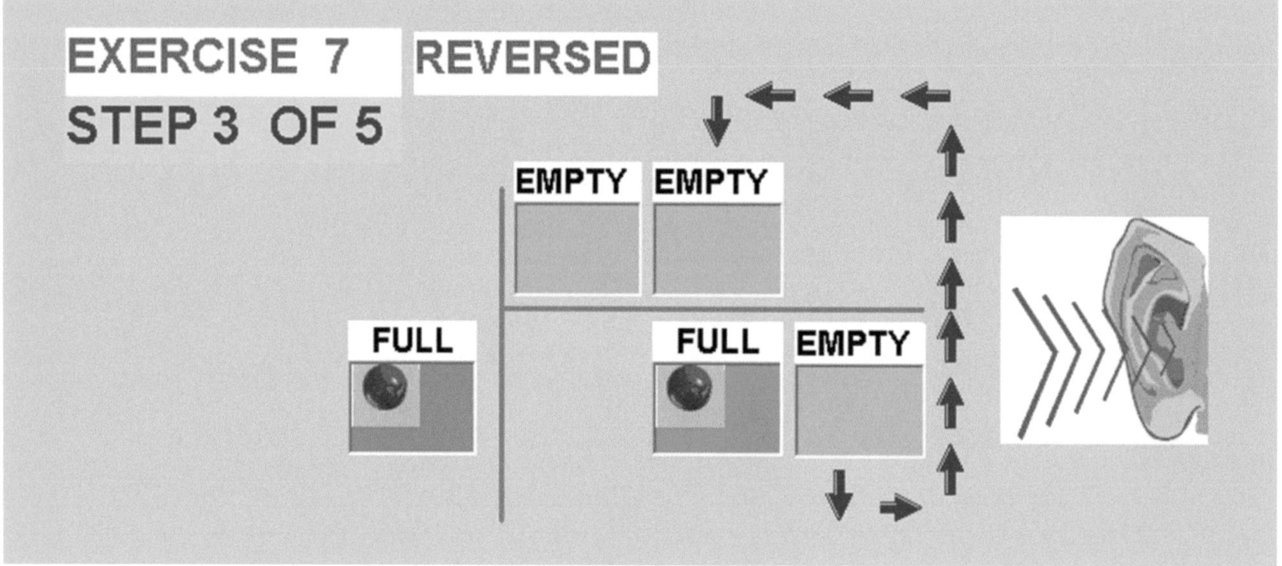

The ***kinesthetic*** representation of the filled up position is represented by the clenching of the right hand. The clenching of the left hand represents an empty position. The division sound, double click, is represented by both hands extended up. Both hands extended in front represent the subtraction sound, double tram. The equal sound, cling, is represented by crossing both hands. The kinesthetic representation for the pictures is the clenching of the right hand, left hand, both hands extended up, right hand, crossing of both hands, and left hand twice.

REVERSE DIVISION, EXERCISE 4.7

In Exercise 7R Step 7.4R, in a ***visual*** display of division we see three pictures of boxes - the dividend, the divisor and the answer. The division sign is between the first and second pictures. Over the dividend is the equal line, which separates the problem from the answer. On top of the dividend is the third picture, which is the answer of the division problem. The ***audio*** representation for the division consists of a **knock, double click, knock, cling, knock, double knock, and double knock.**

In the next part of Step 7.4R, we have a **visual** display of how we arrive at the answer. We subtract picture four from picture two, which is the dividend. The subtraction sign is between pictures two and four. Below picture four is the equal sign, which separates problem from the answer. Picture five is the result of the subtraction. When the result of the subtraction is an empty position then a filled position is transferred to the answer of the division. The ***audio*** representation for the division consists of a **knock, double tram, knock, cling, and double knock.**

Step 7.4

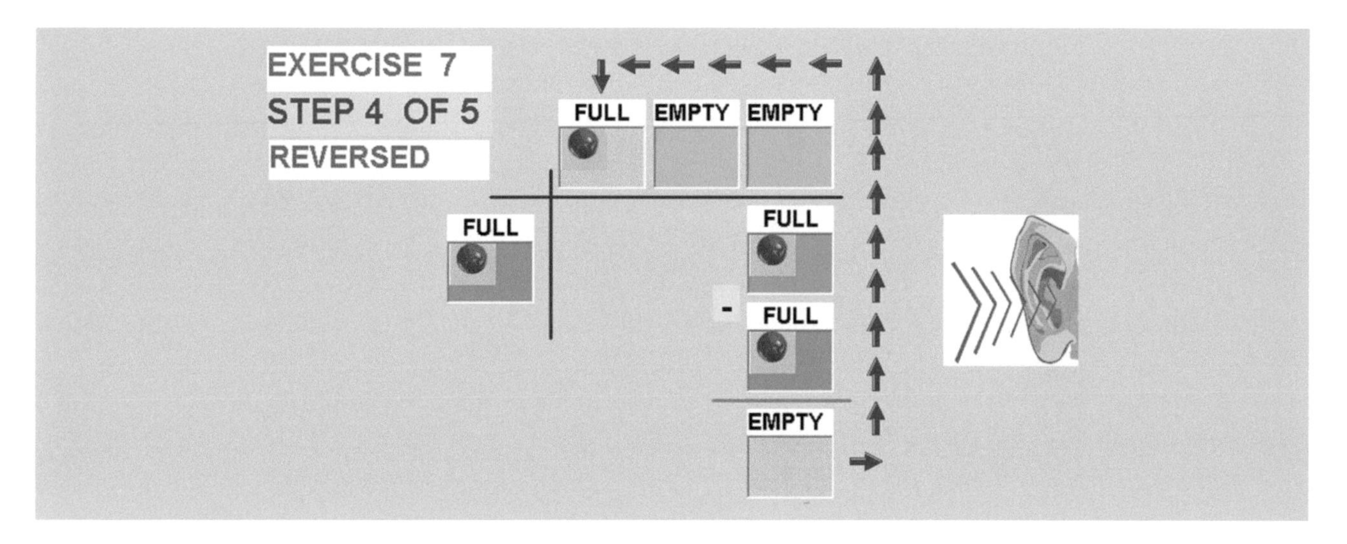

The ***kinesthetic*** representation of the filled up position is represented by the clenching of the right hand. The clenching of the left hand represents an empty position. The division sound, double click, is represented by both hands extended up. Both hands extended in front represent the subtraction sound, double tram. The equal sound, cling, is represented by crossing both hands. The kinesthetic representation for the pictures is the clenching of the right hand, both hands extended up, right hand, crossing of hands, right hand, and left hand twice.

MATHEMATICS

REVERSE DIVISION, EXERCISE 7.5

In Exercise 7R Step 7.5R, in a ***visual*** display of the last step of the division we see two pictures of boxes. In both pictures the first position is filled with a red ball and the next two positions are empty. The colors and the placement of the boxes and the balls are different in the pictures. The form, shape, and color of the symbols do no change the mathematical value, which are five.

The ***audio*** representation for the pictures consists of a **knock, double knock, double knock, and cling.**

Step 7.5

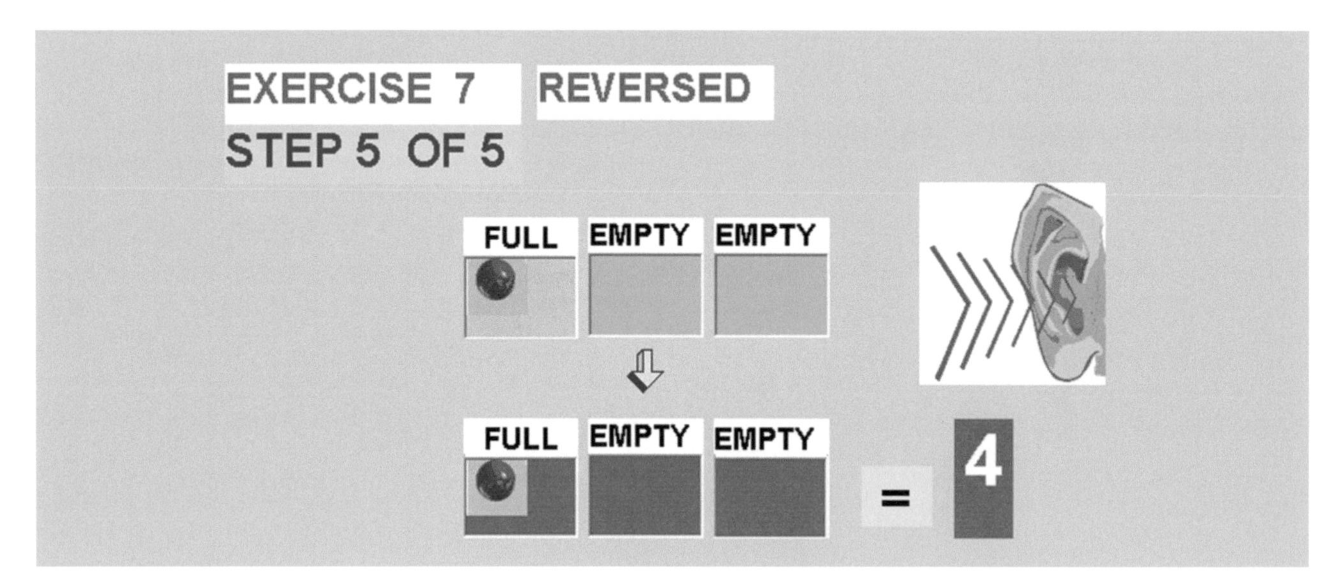

The ***kinesthetic*** representation of the filled up position is represented bvy the clenching of the right hand. The clenching of the left hand represents an empty position. The equal sound, cling, is represented by crossing the hands. The kinesthetic representation for the two pictures is the clenching of the right, left hand twice, crossing of the hands, and right hand, left hand twice.

REVERSE DIVISION, EXERCISE 7R

In Exercise 7R, in a ***visual*** display of division we see two pictures of boxes and the division sign between them. In the first picture we see one empty position and one position filled with a red ball. In the second picture the first three positions are empty while the last position is filled with a red ball. The division sign is between the first and second sets of boxes. On top of the second picture is the equal line, which separates the problem from the answer. The ***audio*** representation for the division consists of a **knock, double knock, double knock, double knock, double click, knock, double knock, and cling.**

At the bottom are the probable answers. From the probable answers for the division we need to find the right one. The audio signals are: 1) **knock, double knock, and double knock,** 2) **double knock, double knock, and knock**, 3) **double knock, knock, and double knock.**

Exercise 7R

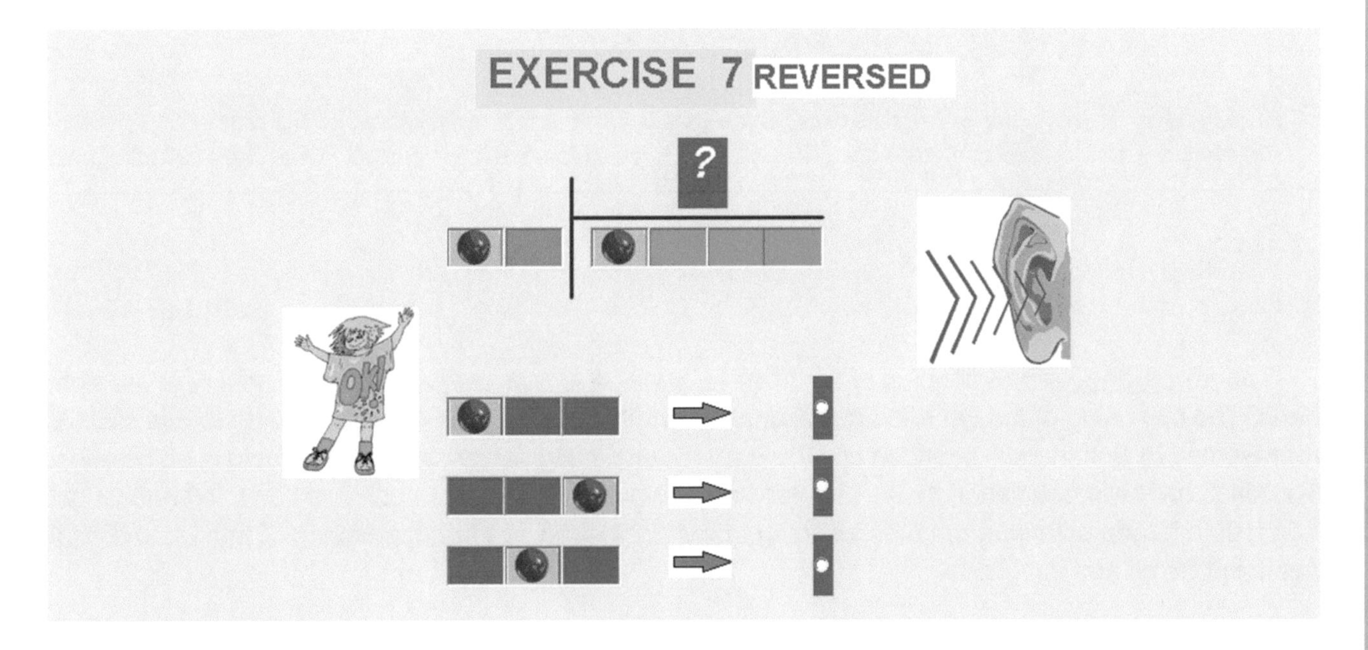

The ***kinesthetic*** representation of the filled up position is represented by the clenching of the right hand. The clenching of the left hand represents an empty position. The sound of double click is represented by both hands extended up. The equal sound, cling, is represented by crossing both hands. The kinesthetic representation for both pictures and the answer is the clenching of the right hand, left hand three times, both hands extended up, clamping of the right hand, left hand, crossing the hands, right hand, and left hand twice.

MATHEMATICS

REVERSE DIVISION, EXERCISE 8R

In Exercise 8R, in a ***visual*** display of division we see three pictures of boxes - the dividend, the divisor and the answer. The division sign is between the first and second pictures. Over the dividend is the equal line, which separates the problem from the answer. On top of the dividend is the third picture, which is the answer of the division. The ***audio*** representation for the division consists of a **knock, double knock, knock, double knock, double click, knock, double knock, cling, knock, double knock, and knock.**

Exercise 8R

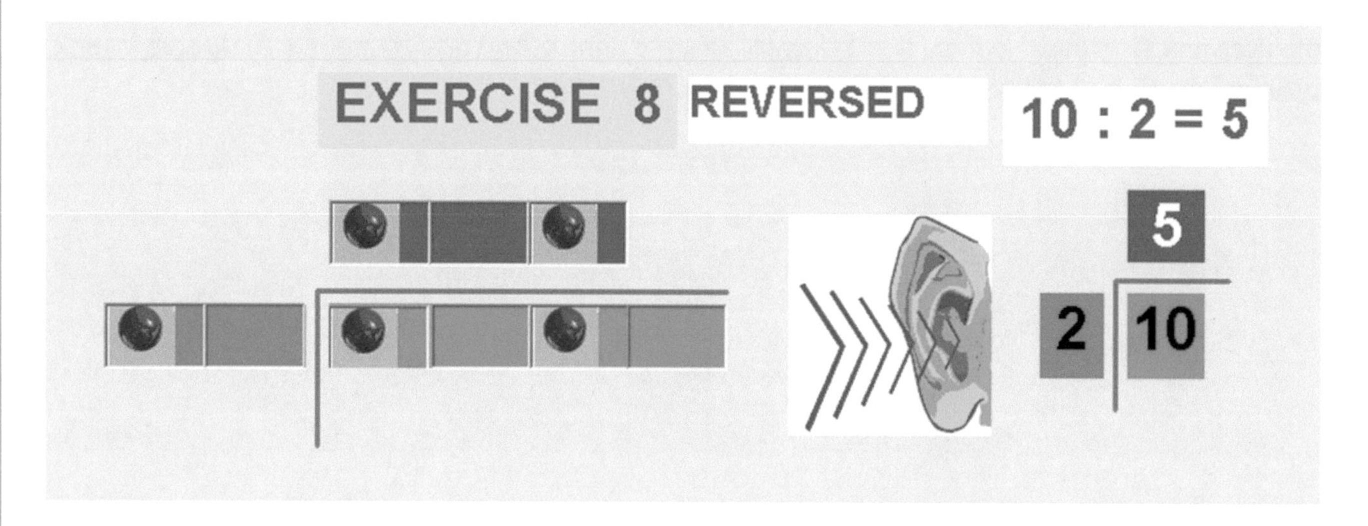

The ***kinesthetic*** representation of the filled up position is represented by the clenching of the right hand. The clenching of the left hand represents an empty position. The division sound, double click, is represented by both hands extended up. The equal sound, cling, is represented by crossing both hands. The kinesthetic representation for the pictures is the clenching of the left hand, right hand, left hand, right hand, both hands extended up, the clenching of the right hand, left hand, crossing of hands, and right hand, left hand, and right hand.

REVERSE DIVISION, EXERCISE 8.1

In Exercise 8R Step 8.1R, in a ***visual*** display of division we see two pictures of boxes - the dividend and the divisor. Between the first and second pictures is the division sign. Over the dividend is the equal line, which separates the problem from the answer. On top of the dividend is the third picture, which is the answer of division. The ***audio*** representation for the pictures consists of a **knock, double knock, knock, double knock, double click knock, double knock, and cling.**

In the next part of Step 8.1R, we have a **visual** display of how we arrive at the answer. If the divisor has an empty position before the filled up position, then that position cancels out the first empty position of the dividend.

Step 8.1

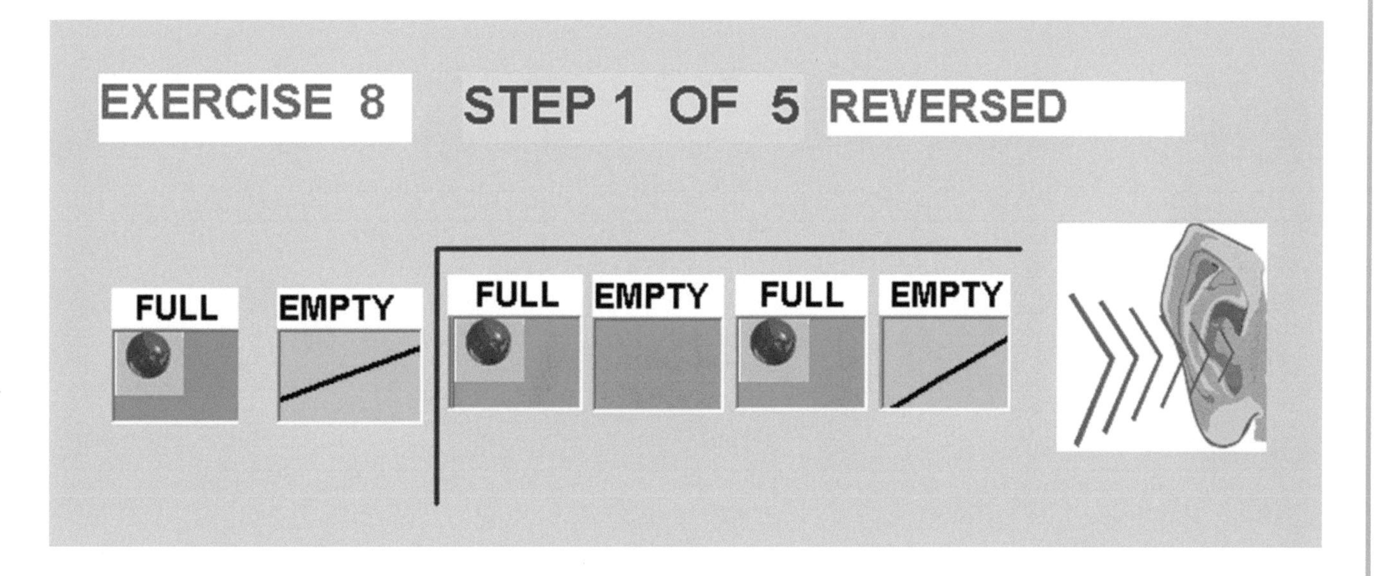

The ***kinesthetic*** representation of the filled up position is represented by the clenching of the right hand. The clenching of the left hand represents an empty position. The division sound, double click, is represented by both hands extended up. Both hands extended in front represent the subtraction sound, double tram. The equal sound, cling, is represented by crossing both hands. The kinesthetic representation for two pictures is the clenching of the right hand, left hand, right hand, left hand, both hands extended up, the clenching of the right hand, left hand, and then crossing of hands.

MATHEMATICS REVERSE DIVISION, EXERCISE 8.2

In Exercise 8R Step 8.2R, in a ***visual*** display of division we see three pictures of boxes – the dividend, the divisor and the answer. The division sign is between the first and second pictures. Over the dividend is the equal line, separating the problem from the answer. On top of the dividend is the third picture, which is the answer of division. The ***audio*** representation for the division consists of a **knock, double knock, knock, double click, knock, cling, and knock.**

In the next part of Step 8.2R, we have a **visual** display of how we arrive at the answer. We subtract picture four from picture two, which is the dividend. Between picture two and four is the subtraction sign. Below picture four is the equal sign, which separates problem from the answer. Picture five is the result of the subtraction. When the result of the subtraction is an empty position, then a filled position is transferred to the answer of the division. The ***audio*** representation for the subtraction consists of a **knock, double knock, knock, double tram, knock, cling, knock, double knock, a double knock.**

Step 8.2

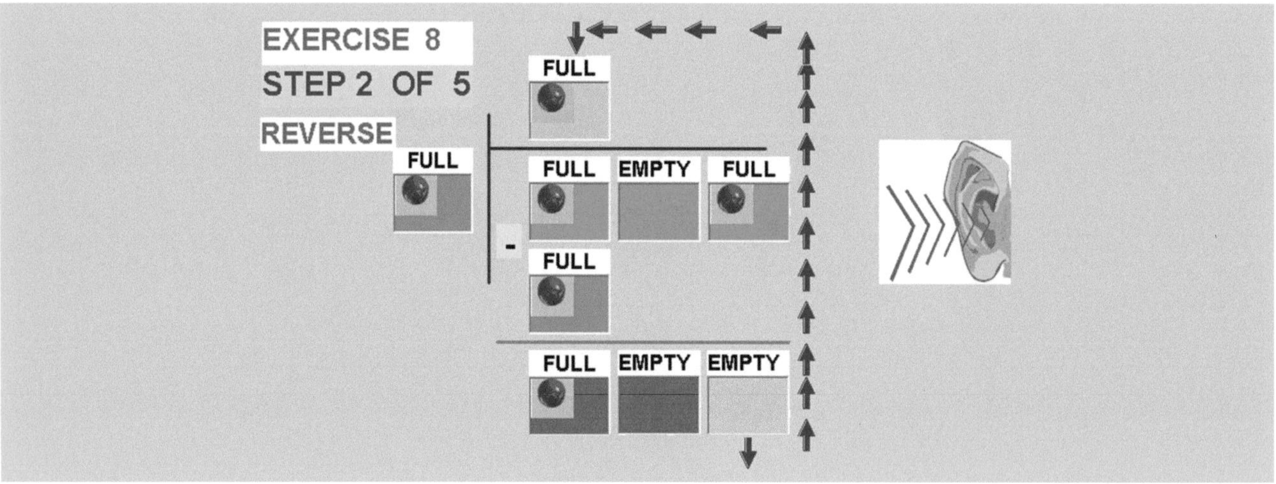

EXERCISE 8
STEP 2 OF 5
REVERSE
FULL
FULL
FULL EMPTY FULL
-
FULL
FULL EMPTY EMPTY

The ***kinesthetic*** representation of the filled up position is represented by the clenching of the right hand. The clenching of the left hand represents an empty position. The division sound, double click, is represented by both hands extended up. Both hands extended in front represent the subtraction sound, double tram. The equal sound, cling, is represented by crossing both hands. The kinesthetic representation for the pictures is the clenching of the right hand, left hand, right hand, both hands extended up, right hand, crossing of hands, and right hand.

REVERSE DIVISION, EXERCISE 8.3

In Exercise 8R Step 8.3R, in a ***visual*** display of division we see three pictures of boxes - the dividend, the divisor and the answer. The division sign is between the first and second picture. Over the dividend is the equal line, which separates the problem from the answer. On top of the dividend is the third picture, which is the answer of the division. The ***audio*** representation for the pictures consists of a **knock, double knock, double click, knock, cling, double knock, and knock.**

In the next part of Step 8.3R, we have a **visual** display of how we arrive at the answer. In the dividend, the empty positions before the filled up position are automatically taken to the top to the answer.

Step 8.3

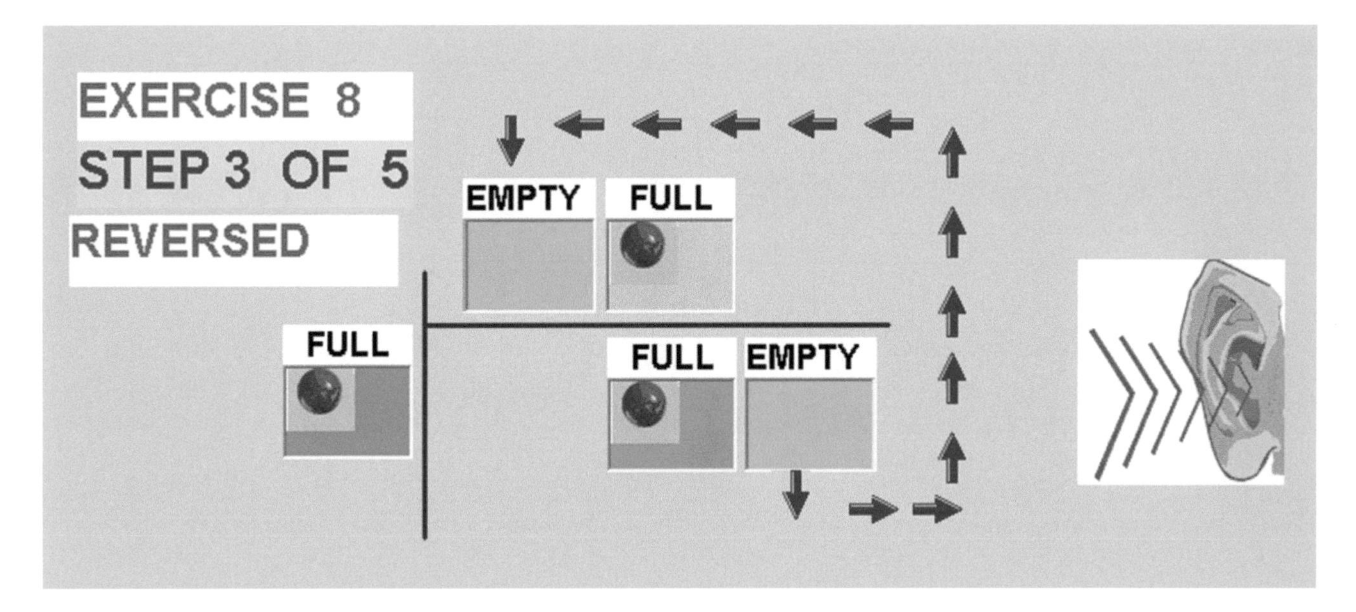

The ***kinesthetic*** representation of the filled up position is represented by the clenching of the right hand. The clenching of the left hand represents an empty position. The division sound, double click, is represented by both hands raised up. Both hands extended in front represent the subtraction sound, double tram. The equal sound, cling, is represented by crossing both hands. The kinesthetic representation for three pictures is the clenching of the right hand, left hand, both hands raised up, right hand, crossing of hands, left hand, and right hand.

REVERSE DIVISION, EXERCISE 8.4

In Exercise 8R Step 8.4R, in a ***visual*** display of division we see three pictures of boxes – the dividend, the divisor and the answer. The division sign is between the first and second pictures. Over the dividend is the equal line, separating the problem from the answer. On top of the dividend is the third picture, which is the answer of division. The ***audio*** representation for the pictures is a **knock, double click, knock, cling, knock, double knock, and knock.**

In the next part of Step 8.4R, we have a **visual** display of how we arrive at the answer. We subtract picture four from picture two, which is the dividend. Between picture two and four is the subtraction sign. Below picture four is the equal sign, which separates the problem from the answer. Picture five is the result of the subtraction. When the result of subtraction is an empty position, then a filled position is transferred to the answer of the division. The ***audio*** representation for the subtraction consists of a **knock, double tram, knock, cling, and double knock.**

Step 8.4

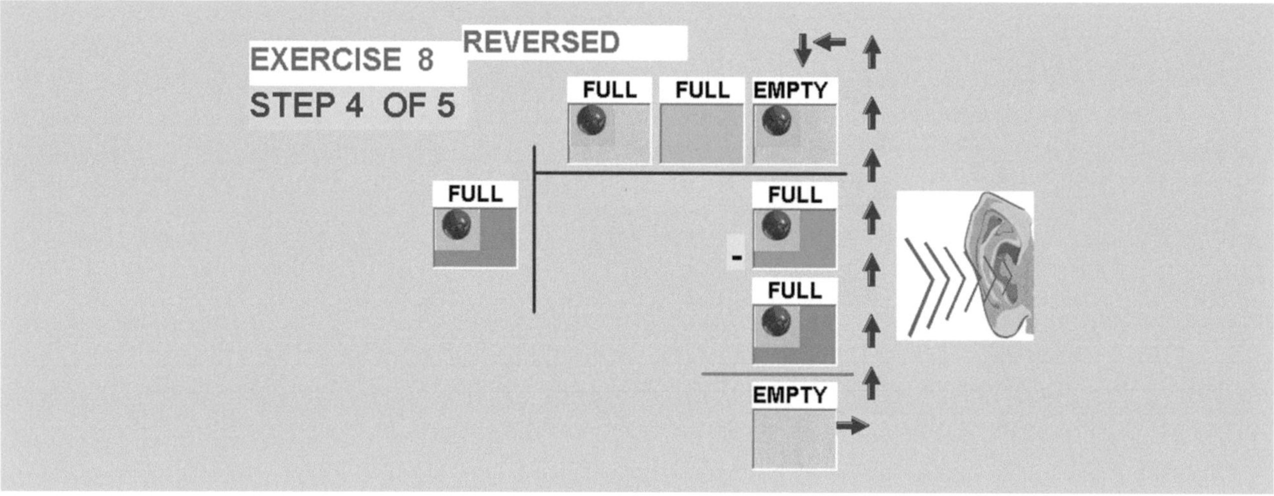

The ***kinesthetic*** representation of the filled up position is represented by the clenching of the right hand. The clenching of the left hand represents an empty position. The division sound, double click, is represented by both hands raised up. Both hands extended in front represent the subtraction sound, double tram. The equal sound, cling, is represented by crossing both hands. The kinesthetic representation for three pictures is the clenching of the right hand, both hands raised up, right hand, crossing of hands, right hand, left hand, and right hand.

REVERSE DIVISION, EXERCISE 8.5

In Exercise 8R Step 8.5R, in a ***visual*** display of the last step of the division we see two pictures of boxes. In both pictures we see that the first position is filled, the second position is empty, and the last position is filled with a red ball. The colors and the placement of the boxes and the balls are different in the pictures. The form, shape, and color of the symbols do no change the mathematical value, which is equal to five.

The ***audio*** representation for the pictures consists of a **knock, double knock, knock, and cling.**

Step 8.5

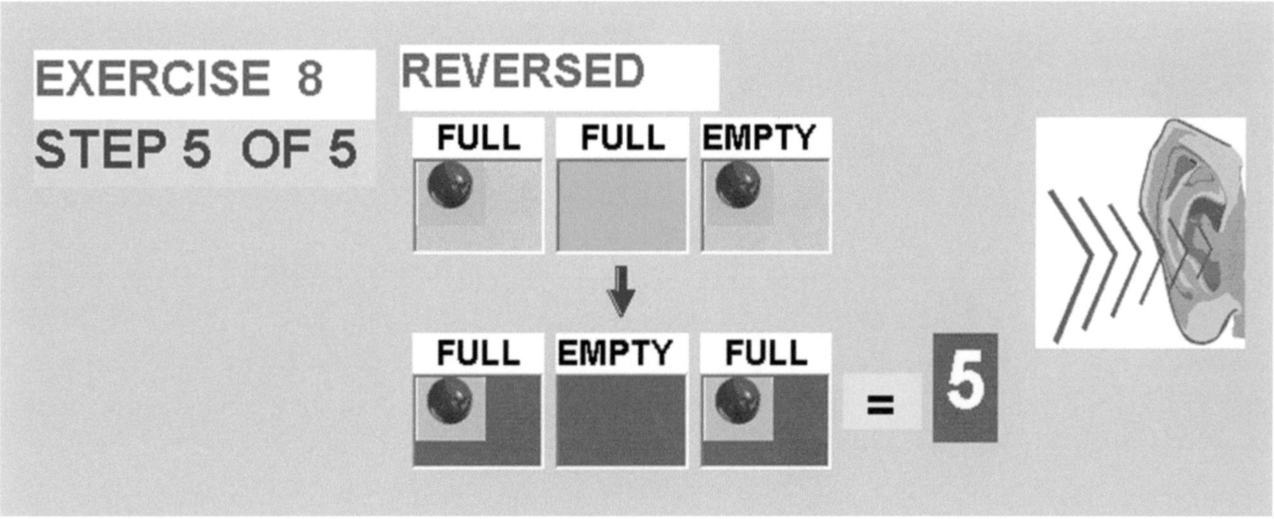

The ***kinesthetic*** representation of the filled up position is represented by the clenching of the right hand. The clenching of the left hand represents an empty position. The equal sound, cling, is represented by crossing of hands. The kinesthetic representation for the two pictures is the clenching of the right hand, left hand, right hand, crossing of hands, right hand, left hand, and right hand.

MATHEMATICS

REVERSE DIVISION, EXERCISE 8R

In Exercise 8R, in a ***visual*** display of division we see two pictures of boxes and division sign in-between. The first picture has the first position empty while the second position is filled with a red ball. In the second picture, the first position is filled with a red ball, the second position is empty, the third is filled with a red ball and the fourth is empty. On top of the second picture is the equal line, which separates the problem from the answer. The ***audio*** representation for the division consists of a **knock, double knock, knock, double knock, double click, knock, double knock, and cling.**

At the bottom are the probable answers. From the probable answers for the division we need to find the right one. The audio signals are 1) **knock, knock, and double knock**, 2) **double knock, knock, and knock**, 3) **knock, double knock, and knock**.

Exercise 8R

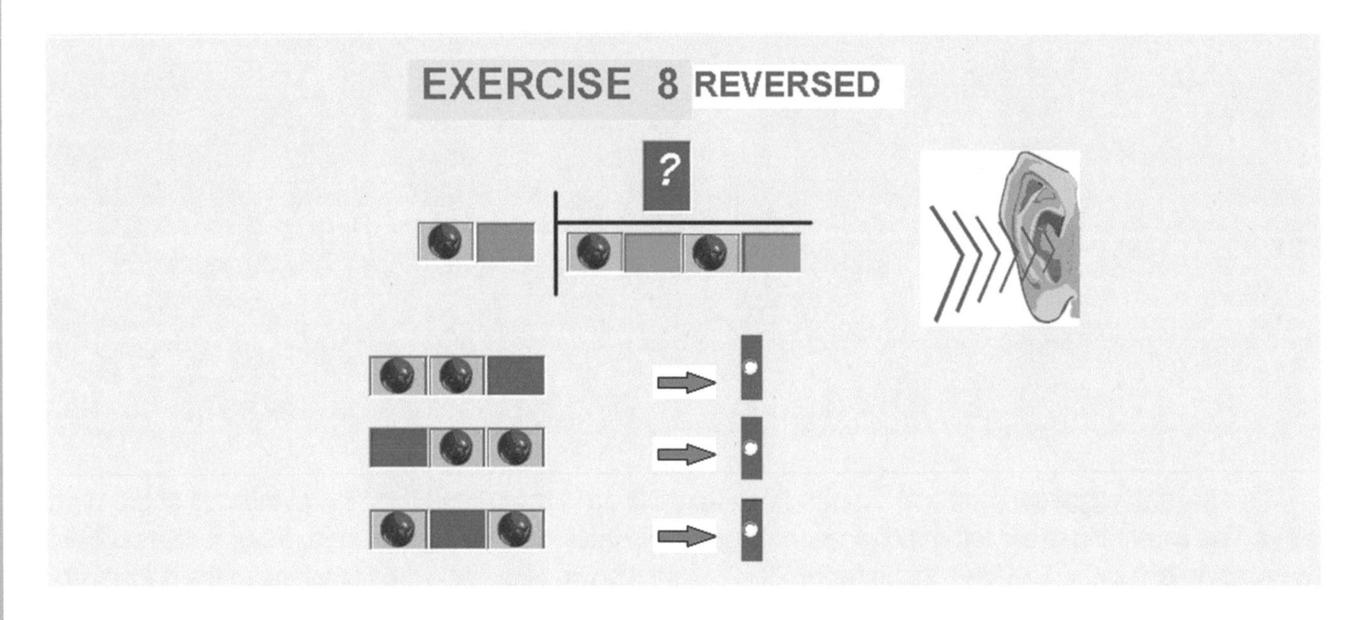

The ***kinesthetic*** representation of the filled up position is represented by the clenching of the right hand. The clenching of the left hand represents an empty position. Both hands extended up represent the division sound, double click. The equal sound, cling, is represented by crossing both hands. The kinesthetic representation for both pictures and the answer is the clenching of the right hand, left hand, right hand, left hand, extended up, the clenching of the right hand, left hand, crossing of hands, right hand, left hand, and right hand.

REVERSE DIVISION, EXERCISE 9R

In Exercise 9R, in a ***visual*** display of division we see three pictures of boxes - the dividend, the divisor and the answer. The division sign is between the first and second pictures. Over the dividend is the equal line, which separates the problem from the answer. On top of the dividend is the third picture, which is the answer of the division problem. The ***audio*** representation for the pictures consists of a **double knock, double knock, knock, knock, double click, knock, double knock, cling, double knock, knock, and knock.**

Exercise 9R

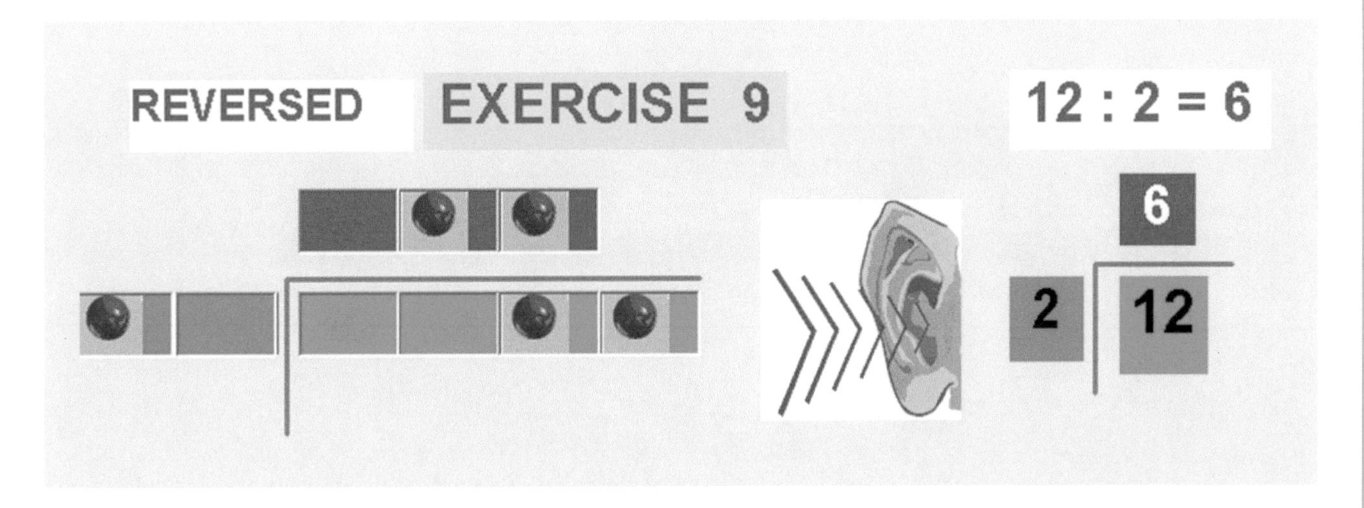

The ***kinesthetic*** representation of the filled up position is represented by the clenching of the right hand. The clenching of the left hand represents an empty position. The division sound, double click, is represented by both hands extended up. The equal sound, cling, is represented by crossing both hands. The kinesthetic representation for the division is the clenching of the left hand twice, right hand twice, both hands extended up, right hand, left hand, crossing of the hands, left hand, and right hand twice.

MATHEMATICS

REVERSE DIVISION, EXERCISE 9.1

In Exercise 9R Step 9.1R, in a ***visual*** display of division we see two pictures of boxes - the dividend and the divisor. The division sign is between the first and second pictures. Over the dividend is the equal line, which separates the problem from the answer. On top of the dividend is the third picture, which is the answer of the division. The ***audio*** representation for the pictures consists of a **knock, knock, double knock, double knock, double click, knock, double knock, and a cling.**

In the next part of Step 9.1R, we have a **visual** display of how we arrive at the answer. If the divisor has an empty position before the filled up position, then that position cancels out the first empty position of the dividend.

Step 9.1

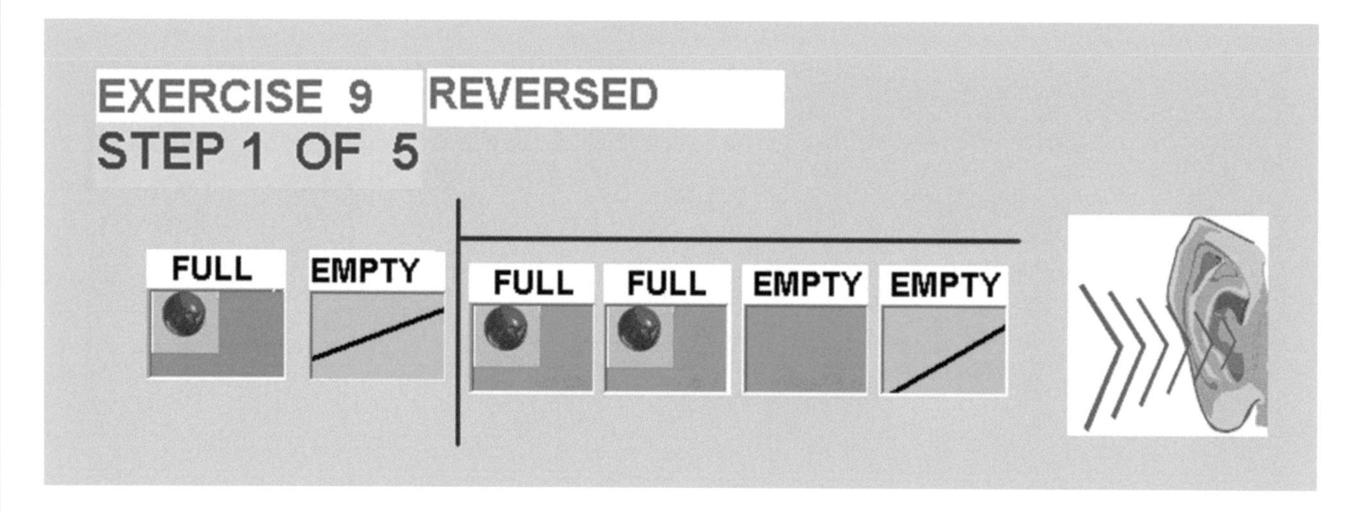

The ***kinesthetic*** representation of the filled up position is represented by the clenching of the right hand. The clenching of the left hand represents an empty position. The division sound, double click, is represented by both hands extended up. Both hands extended in front represent the subtraction sound, double tram. The equal sound, cling, is represented by crossing both hands. The kinesthetic representation for the two pictures is the clenching of the right hand twice, left hand twice, both hands extended up, right hand, left hand, and then crossing of the hands.

REVERSE DIVISION, EXERCISE 9.2

In Exercise 9R Step 9.2R, in a ***visual*** display of division we see three pictures of boxes - the dividend, the divisor and the answer. The division sign is between the first and second pictures. Over the dividend is the equal line, which separates the problem from the answer. On top of the dividend is the third picture, which is the answer of the division problem. The ***audio*** representation for the pictures consists of a **knock, knock, double knock, double click, knock, cling, and double knock.**

In the next part of Step 9.2R, we have a **visual** display of how we arrive at the answer. In the dividend, the empty positions before the filled up position are automatically taken to the top to the answer.

Step 9.2

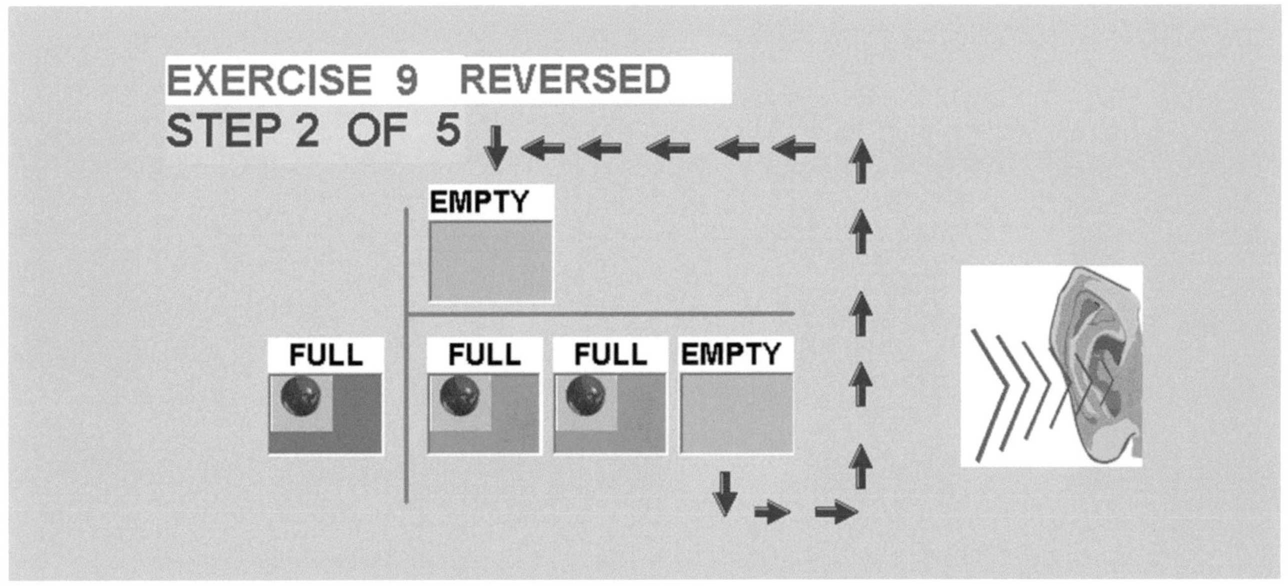

The ***kinesthetic*** representation of the filled up position is represented by the clenching of the right hand. The clenching of the left hand represents an empty position. The division sound, double click, is represented by both hands extended up. Both hands extended in front represent the subtraction sound, double tram. The equal sound, cling, is represented by crossing both hands. The kinesthetic representation for the pictures is the clenching of the right hand twice, left hand, both hands extended up, right hand, crossing of both hands, and left hand.

MATHEMATICS

REVERSE DIVISION, EXERCISE 9.3

In Exercise 9R Step 9.3R, in a ***visual*** display of division we see three pictures of boxes - the dividend, the divisor and the answer. The division sign is between the first and second pictures. Over the dividend is the equal line, which separates the problem from the answer. On top of the dividend is the third picture, which is the answer of the division problem. The ***audio*** representation for the pictures consists of a **knock, knock, double click, knock, cling, knock, and double knock.**

In the next part of Step 9.3R, we have a **visual** display of how we arrive at the answer. We subtract picture four from picture two, which is the dividend. Between picture two and four is the subtraction sign. Below picture four is the equal sign, which separates the problem from the answer. Picture five is the result of the subtraction. When the result of the subtraction is an empty position, then a filled position is transferred to the answer of the division. The ***audio*** representations for the subtraction consists of a **knock, knock, double tram, knock, cling, knock, and double knock.**

Step 9.3

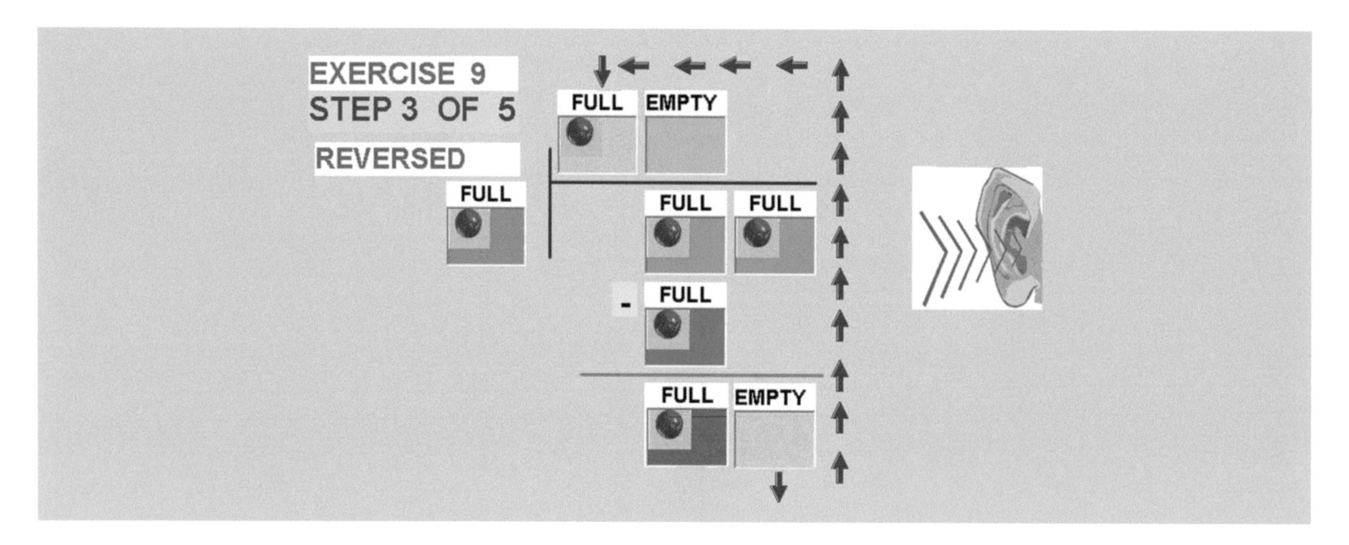

The ***kinesthetic*** representation of the filled up position is represented by the clenching of the right hand. The clenching of the left hand represents an empty position. The division sound, double click, is represented by both hands extended up. Both hands extended in front represent the subtraction sound, double tram. The equal sound, cling, is represented by crossing both hands. The kinesthetic representation for the pictures is the clenching of the right hand twice, both hands extended up, right hand, crossing of hands, right hand, and left hand.

REVERSE DIVISION, EXERCISE 9.4

In Exercise 9R Step 9.4R, in a ***visual*** display of division we see three pictures of boxes - the dividend, the divisor and the answer. The division sign is between the first and second pictures. Over the dividend is the equal line, which separates the problem from the answer. On top of the dividend is the third picture, which is the answer of the division problem. The ***audio*** representation for the pictures consists of a **knock, double click, knock, cling, knock, knock, and double knock.**

In the next part of Step 9.4R, we have a **visual** display of how we arrive at the answer. We subtract picture four from picture two, which is the dividend. Between picture two and four is the subtraction sign. Below picture four is the equal sign, which separates the problem from the answer. Picture five is the result of the subtraction. When the result of the subtraction is an empty position, then a filled position is transferred to the answer of the division. The ***audio*** representation for the subtraction consists of a **knock, double tram, knock, cling, and double knock.**

Step 9.4

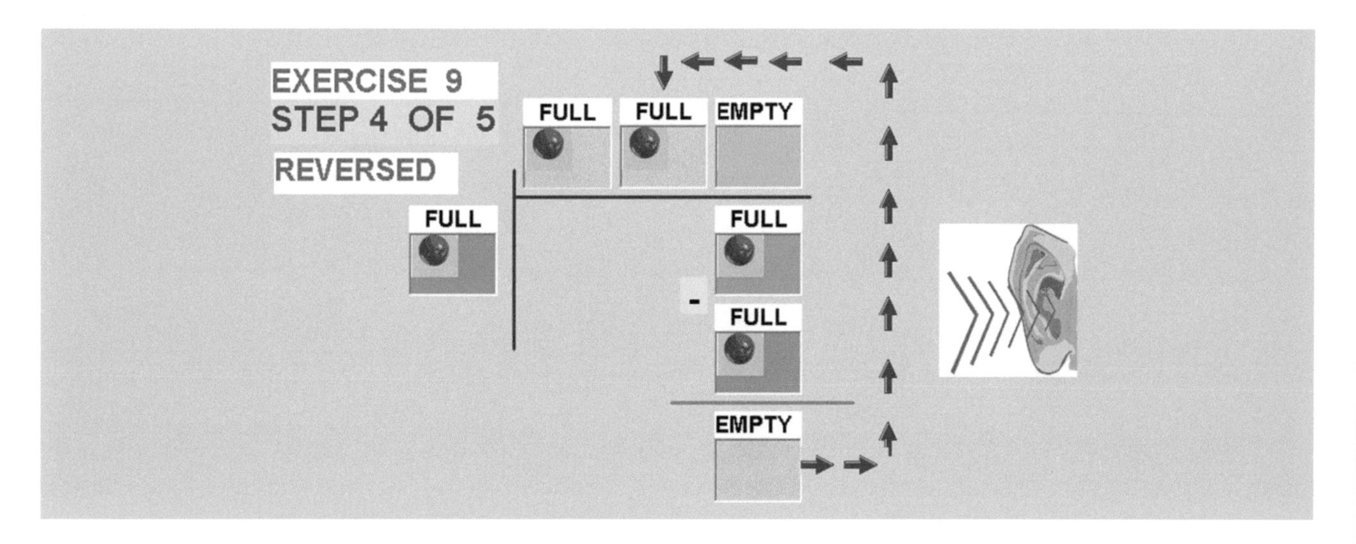

The ***kinesthetic*** representation of the filled up position is represented by the clenching of the right hand. The clenching of the left hand represents an empty position. The division sound, double click, is represented by both hands extended up. Both hands extended in front represent the subtraction sound, double tram. The equal sound, cling, is represented by crossing both hands. The kinesthetic representation for the pictures is the clenching of the right hand, both hands extended up, right hand, crossing the hands, right hand twice, and left hand.

REVERSE DIVISION, EXERCISE 9.5

In Exercise 9R Step 9.5R, in a ***visual*** display of the last step of the division we see two pictures of boxes. In both pictures, the first two positions are filled with red balls and the third position is empty. The colors and the placement of the boxes and the balls are different in the pictures. The form, shape, and color of the symbols do no change mathematical value, which is equal to six.

The ***audio*** representation for the two pictures consists of a **knock, knock, double knock, and cling.**

Step 9.5

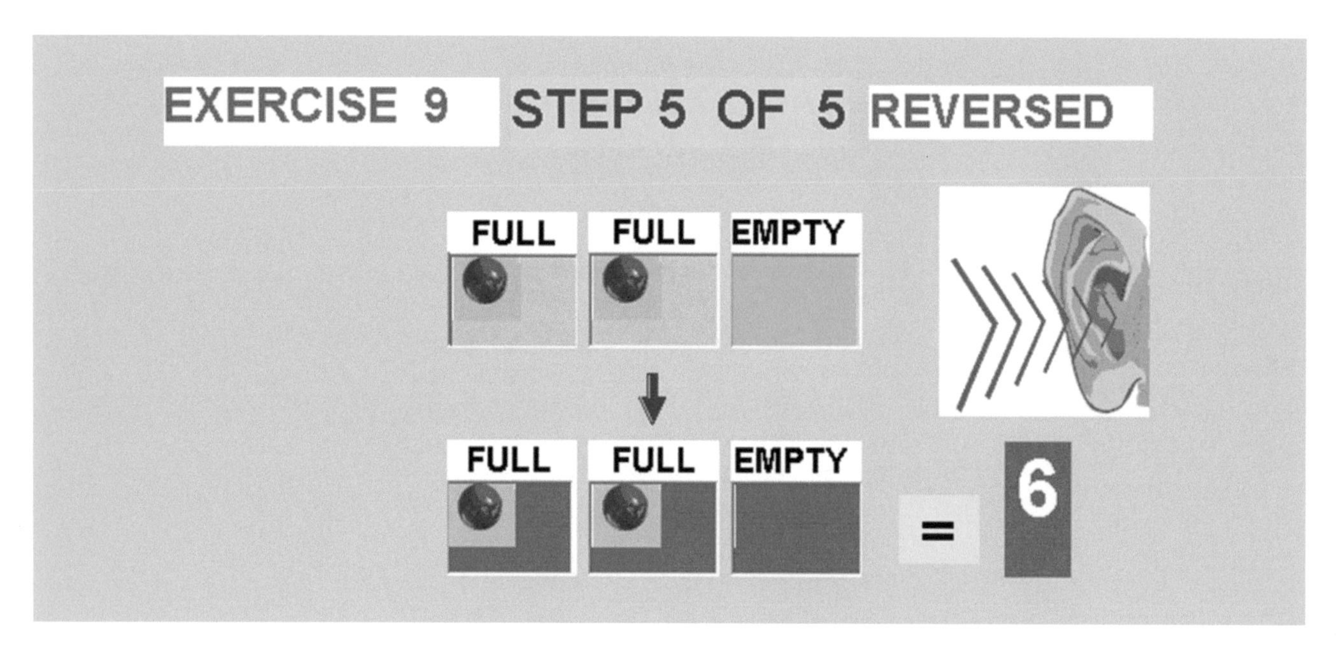

The ***kinesthetic*** representation of the filled up position is represented by the clenching of the right hand. The clenching of the left hand represents an empty position. The equal sound, cling, is represented by crossing the hands. The kinesthetic representation for the two pictures is the clenching of the right hand, twice, crossing the hands, right hand twice, and left hand.

REVERSE DIVISION, EXERCISE 9R

In Exercise 9R, in a ***visual*** display of division we see two pictures of boxes the division sign in-between. In the first picture, the first position is filled with a red ball and the second is empty. In the second picture, the first two positions are filled with a red ball and the second two are empty. On top of the second picture is the equal line, which separates the problem from the answer. The ***audio*** representation for the division consists of a **knock, knock, double knock, double knock, double click, knock, double knock, and a cling.**

At the bottom are the probable answers. From the probable answers for the division we need to find the right one. The audio signals are: 1) **double knock, knock, and knock,** 2) **knock, knock, and double knock**, 3) **knock, double knock, and knock.**

Exercise 9R

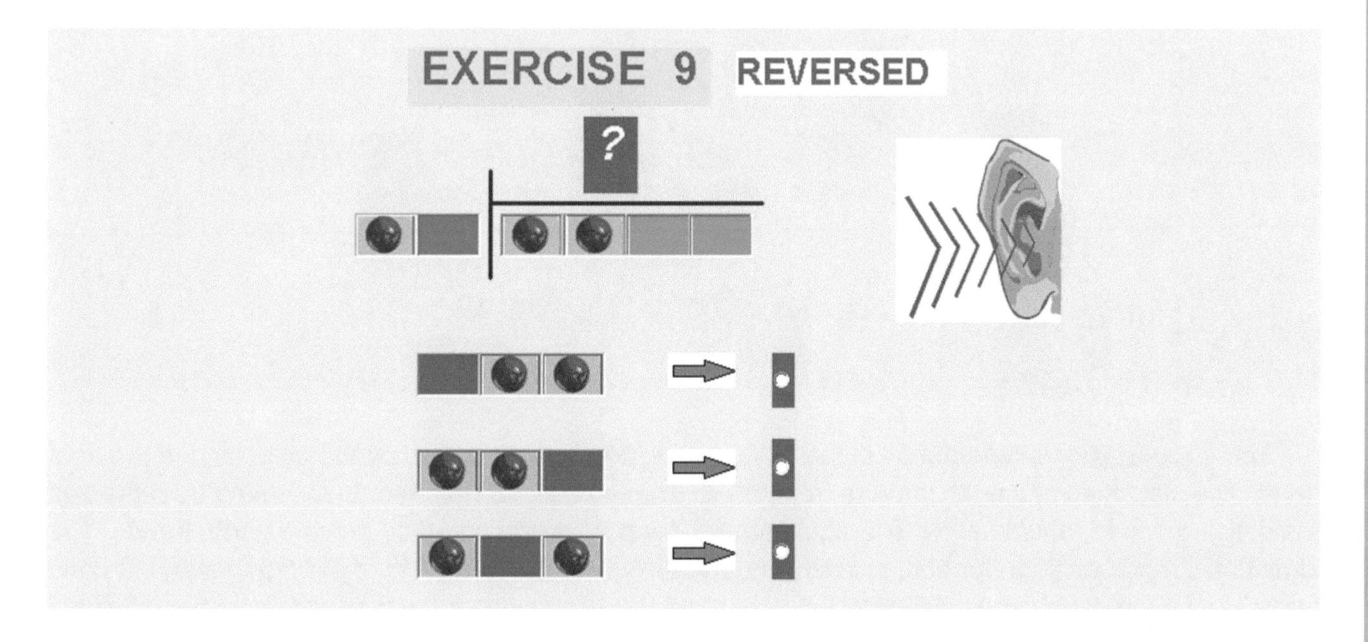

The ***kinesthetic*** representation of the filled up position is represented by the clenching of the right hand. The clenching of the left hand represents an empty position. Both hands extended up represent the division sound, double click. The equal sound, cling, is represented by crossing both hands. The kinesthetic representation for both pictures and the answer is the clenching of the right hand twice, left hand twice, both hands extended up, right hand, left hand, crossing both hands, right hand twice, and left hand.

MATHEMATICS REVERSE DIVISION, EXERCISE 10R

In Exercise 10R, in a ***visual*** display of division we see three pictures of boxes with the division sign in-between. In the first picture, both positions are filled with a red ball. In the second picture, the first position is filled with a red ball, the next two positions are empty, and the last position is filled with a red ball. The third picture has two positions filled up with red balls. On top of the second picture is the equal line, which separates the problem from the answer. The ***audio*** representation for the division consists of a **knock, double knock, double knock, knock, double click, knock, knock, cling, knock, and knock.**

Exercise 10R

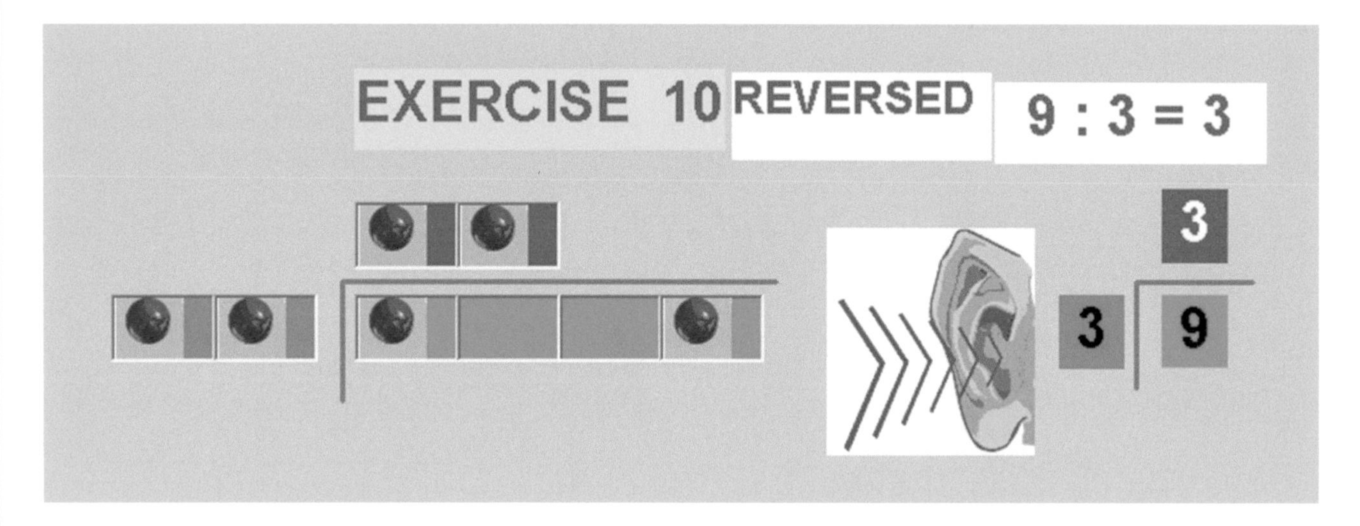

The ***kinesthetic*** representation of the filled up position is represented by the clenching of the right hand. The clenching of the left hand represents an empty position. Both hands extended up represent the division sound, double click. The equal sound, cling, is represented by crossing both hands. The kinesthetic representation for both pictures and the answer is the clenching of the right hand, left hand twice, right hand, both hands extended up, right hand twice, crossing the hands, and right hand twice.

REVERSE DIVISION, EXERCISE 10R

In Exercise 10R, in a ***visual*** display of division we see two pictures of boxes with the division sign in-between. In the first picture, both positions are filled with a red ball. In the second picture, the first position filled with a red ball, the next two positions are empty, and the last position is filled with a red ball. On top of the second picture is the equal line, which separates the problem from the answer. The ***audio*** representation for the division consists of a **knock, double knock, double knock, knock, double click, knock, knock, and cling.**

In the next part of Exercise 10R, we have a **visual** display of how we arrive at the answer. We subtract picture four from picture two, which is the dividend. Between pictures two and three is the subtraction sign. Below picture three is the equal sign which separates the problem from the answer. When the result of subtraction is an empty position, then a filled position is transferred to the answer of division. The ***audio*** representation for the subtraction consists of a **knock, double knock, double knock, knock, double tram, knock, knock, and cling.**

Exercise 10R

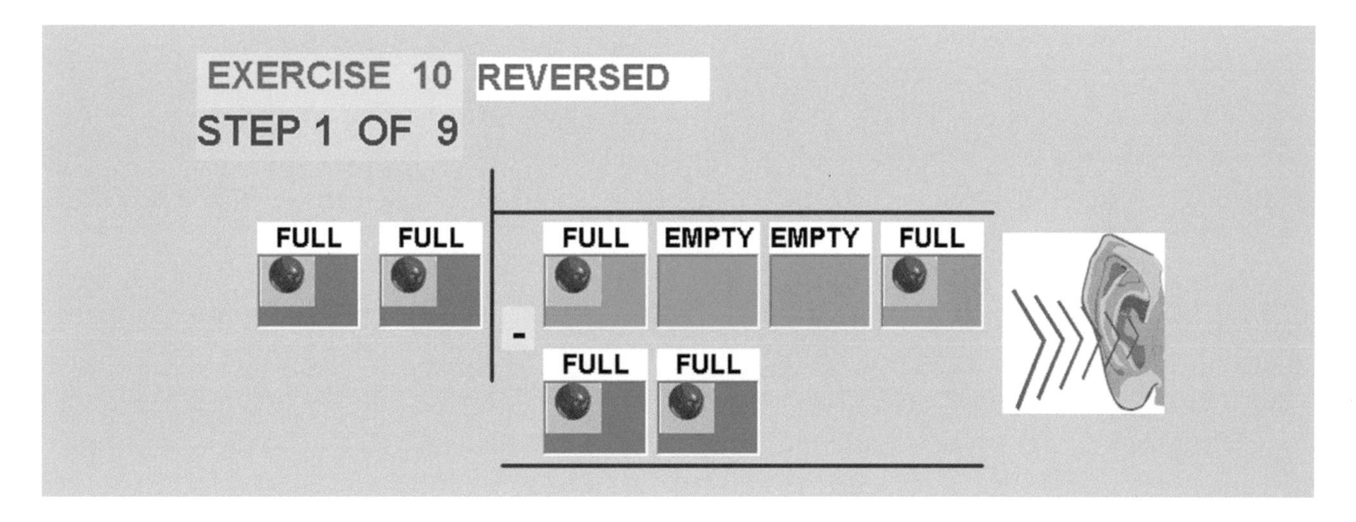

The ***kinesthetic*** representation of the filled up position is represented by the clenching of the right hand. The clenching of the left hand represents an empty position. Both hands extended up represent the division sound, double click. The equal sound, cling, is represented by crossing both hands. The kinesthetic representation for the pictures is the clenching of the right hand, left hand twice, right hand, both hands extended up, right hand twice,

MATHEMATICS

REVERSE DIVISION, EXERCISE 10.2

In Exercise 10R Step 10.2R, in a ***visual*** display of the next step of the division we see two pictures of boxes. In both pictures we see that the first position is filled with a symbol, the second and third positions are empty, and the last position is filled with a red ball. The colors and the placement of the boxes and the balls are different in the pictures. The form, shape, and color of the symbols do no change the mathematical value.

The ***audio*** representation for the pictures consists of a **knock, double knock, double knock, and knock.**

Step 10.2

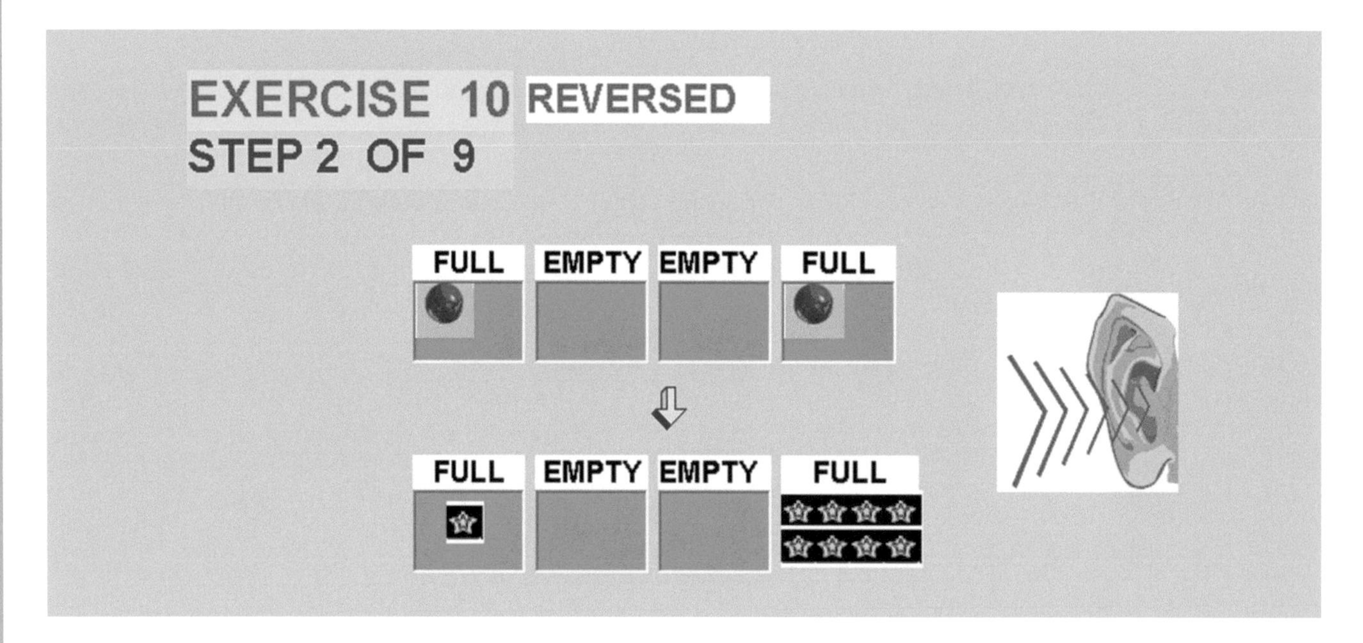

The ***kinesthetic*** representation of the filled up position is represented by the clenching of the right hand. The clenching of the left hand represents an empty position. The equal sound, cling, is represented by crossing the hands. The kinesthetic representation for the two pictures is the clenching of the right hand, left hand twice, right hand, and crossing the hands.

REVERSE DIVISION, EXERCISE 10.3

In Exercise 10R Step 10.3R, in a ***visual*** display of division we see three pictures of boxes. In the first two pictures we see the first position is filled with a symbol, the second and third positions are empty, and the last position is filled with a symbol. For the purpose of division, we introduce the concept of borrowing. We borrow from the fourth to the third position. However, the audio representation for the pictures does not change because the numerical value does not change. The ***audio*** representation for the pictures with the equal sign consists of a **knock, double knock, double knock, knock, and cling.**

Step 10.3

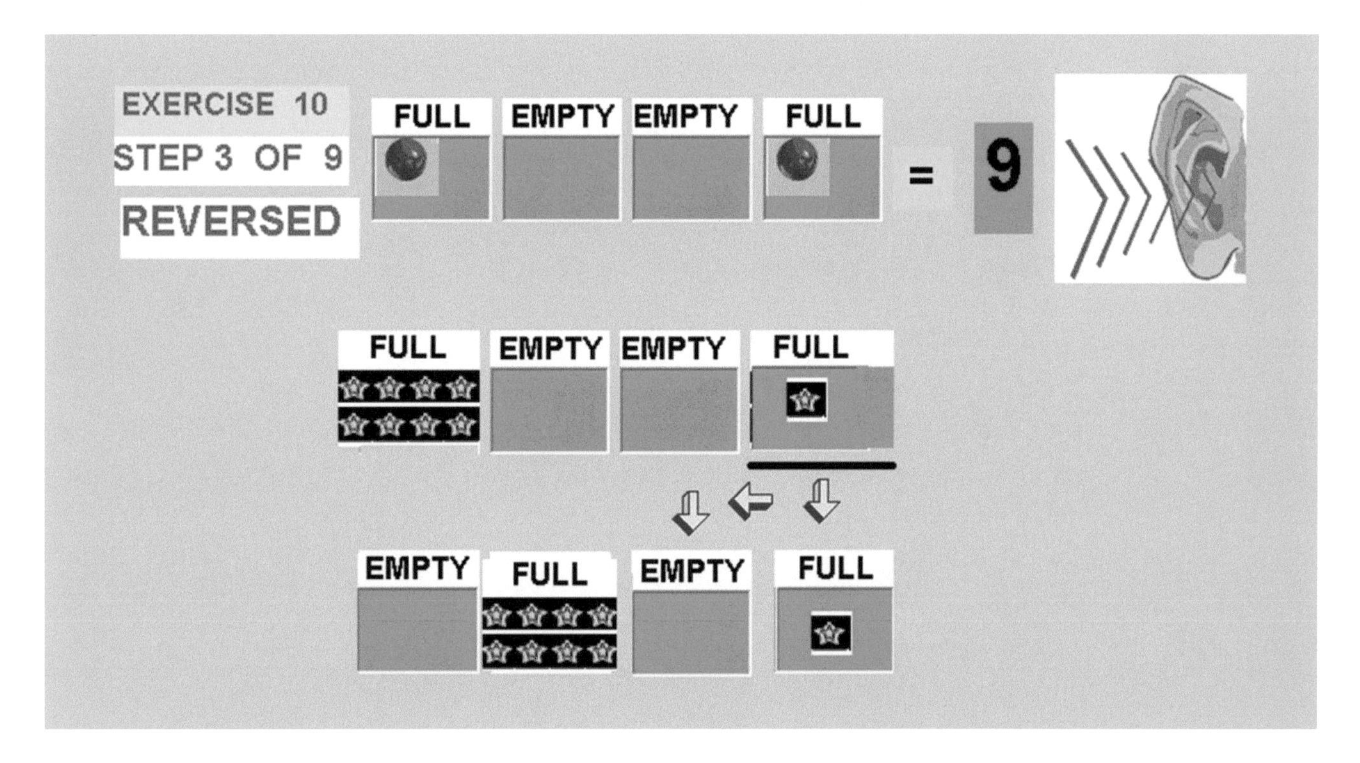

The ***kinesthetic*** representation of the filled up position is represented by the clenching of the right hand. The clenching of the left hand represents an empty position. The equal sound, cling, is represented by crossing both hands. The kinesthetic representation for both pictures and the answer is the clenching of the right hand, left hand twice, right hand, and crossing of hands.

MATHEMATICS

REVERSE DIVISION, EXERCISE 10.4

In Exercise 10R Step 10.4R, in a ***visual*** display of division we see three pictures of boxes. In first picture the first position is filled with a symbol, the second and third positions are empty, and the last position is filled with a symbol. In the second picture, the first position is empty, the second position is filled with stars, the second position is empty, and the last position is filled with a star. The third picture has the first position is empty and the next three positions filled with stars. For the purpose of division, we introduce the concept of borrowing. We borrow from the third to the second and third positions. However, the audio representation for the pictures does not change because the numerical value does not change. The ***audio*** representation for the pictures with the equal sign consists of a **knock, double knock, double knock, knock, and cling.**

Step 10.4

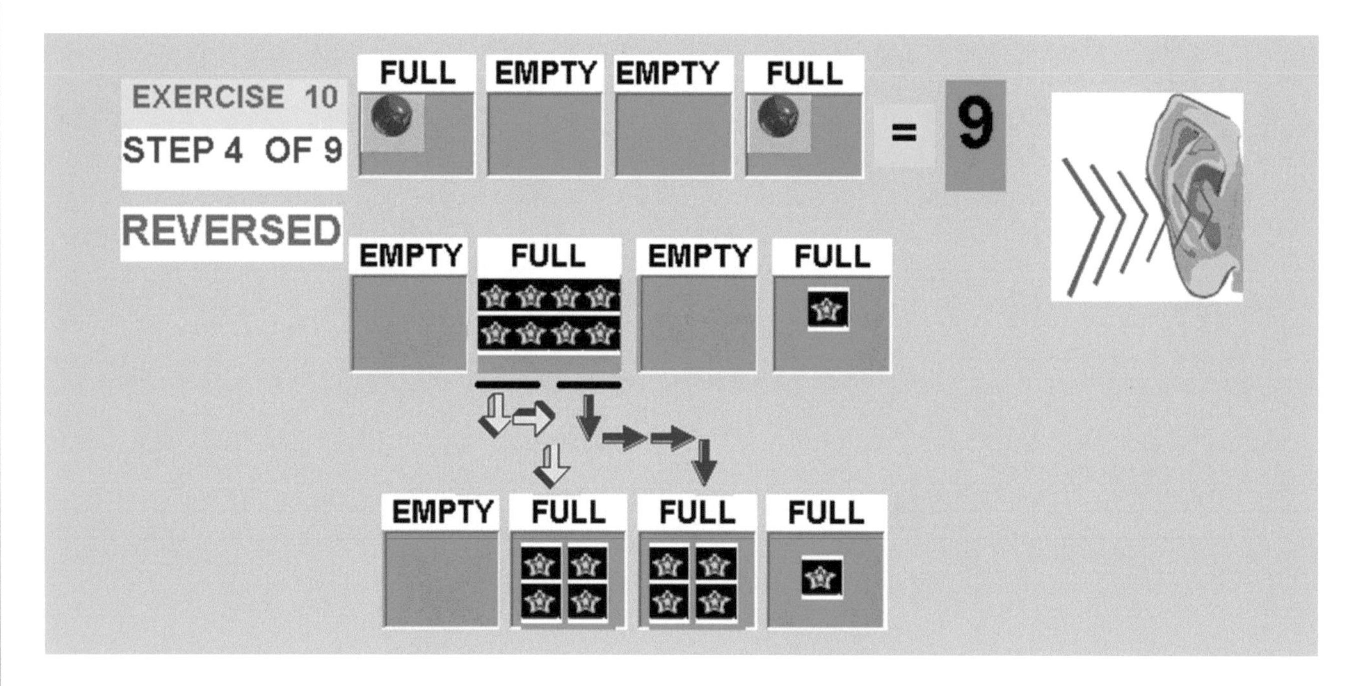

The ***kinesthetic*** representation of the filled up position is represented by the clenching of the right hand. The clenching of the left hand represents an empty position. The equal sound, cling, is represented by crossing both hands. The kinesthetic representation for both pictures and the answer is the clenching of the right hand, left hand twice, right hand, and crossing of the hands.

REVERSE DIVISION, EXERCISE 10.5

In Exercise 10R Step 10.5R, in a ***visual*** display of division we see three pictures of boxes - the dividend, the divisor and the answer. The division sign is between the first and second pictures. Over the dividend is the equal line, which separates the problem from the answer. On top of the dividend is the third picture, which is the answer of the division problem. The ***audio*** representation for the pictures consists of a **knock, knock, knock, double click, knock, knock, and a cling.**

In the next part of Step 10.5R, we have a **visual** display of how we arrive at the answer. We subtract picture four from picture two, which is the dividend. Between picture two and four is the subtraction sign. Below picture four is the equal sign, which separates problem from the answer. Picture five is the result of the subtraction. The ***audio*** representation for the subtraction consists of a **knock, knock, knock, double tram, knock, knock, cling, knock, knock, and double knock.**

Step 10.5

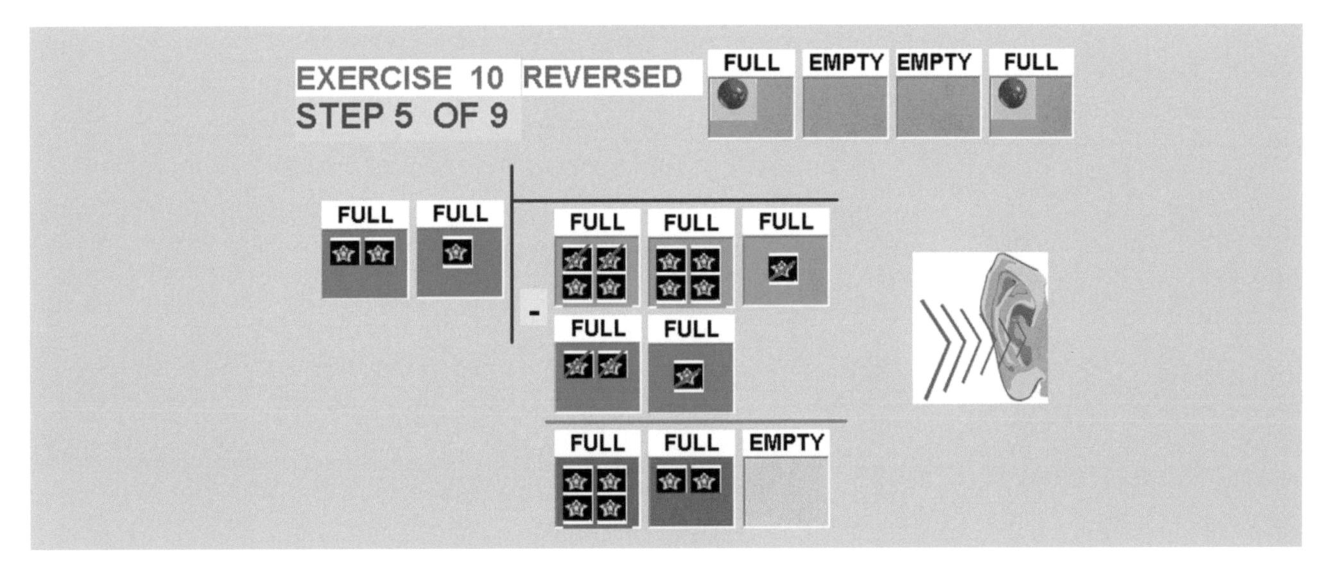

The ***kinesthetic*** representation of the filled up position is represented by the clenching of the right hand. The clenching of the left hand represents an empty position. The division sound, double click, is represented by both hands extended up in the air. Both hands extended in front represent the subtraction sound, double tram. The equal sound, cling, is represented by crossing both hands. The kinesthetic representation for the two pictures is the clenching of the right hand three times, both hands extended up, right hand twice, crossing the hands.

MATHEMATICS

REVERSE DIVISION, EXERCISE 10.5

In Exercise 10R Step 10.5R, in a ***visual*** display of division we see three pictures of boxes - the dividend, the divisor and the answer. The division sign is between the first and second pictures. Over the dividend is the equal line, which separates the problem from the answer. On top of the dividend is the third picture, which is the answer of the division problem. The ***audio*** representation for the pictures consists of a **knock, knock, knock, double click, knock, knock, and a cling.**

In the next part of Step 10.5R, we have a **visual** display of how we arrive at the answer. We subtract picture four from picture two, which is the dividend. Between picture two and four is the subtraction sign. Below picture four is the equal sign, which separates problem from the answer. Picture five is the result of the subtraction. The ***audio*** representation for the subtraction consists of a **knock, knock, knock, double tram, knock, knock, cling, knock, knock, and double knock.**

Step 10.5

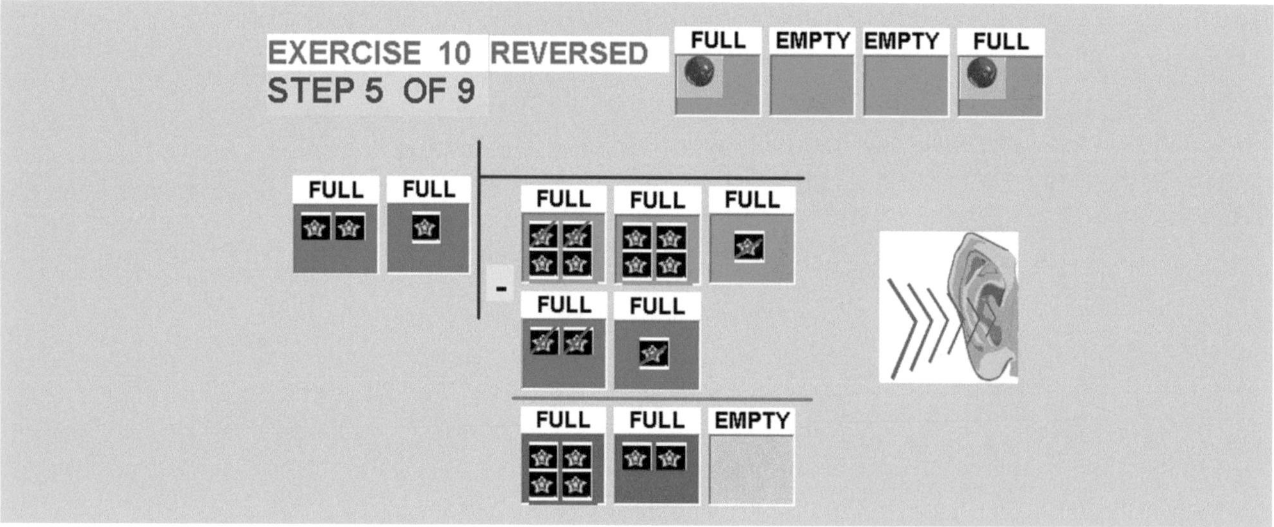

The ***kinesthetic*** representation of the filled up position is represented by the clenching of the right hand. The clenching of the left hand represents an empty position. The division sound, double click, is represented by both hands extended up in the air. Both hands extended in front represent the subtraction sound, double tram. The equal sound, cling, is represented by crossing both hands. The kinesthetic representation for the two pictures is the clenching of the right hand three times, both hands extended up, right hand twice, crossing the hands.

REVERSE DIVISION, EXERCISE 10.6

In Exercise 10R Step 10.6R, in a ***visual*** display of division we see three pictures of boxes - the dividend, the divisor and the answer. The division sign is between the first and second pictures. Over the dividend is the equal line, which separates the problem from the answer. On top of the dividend is the third picture, which is the answer of the division problem. The ***audio*** representation for the pictures consists of a **knock, knock, double knock, double click, knock, knock, cling, and knock.**

In the next part of Step 10.6R, we have a **visual** display of how we arrive at the answer. In the dividend the empty positions before the filled up position are automatically taken to the top to the answer as a filled up position.

Step 10.6

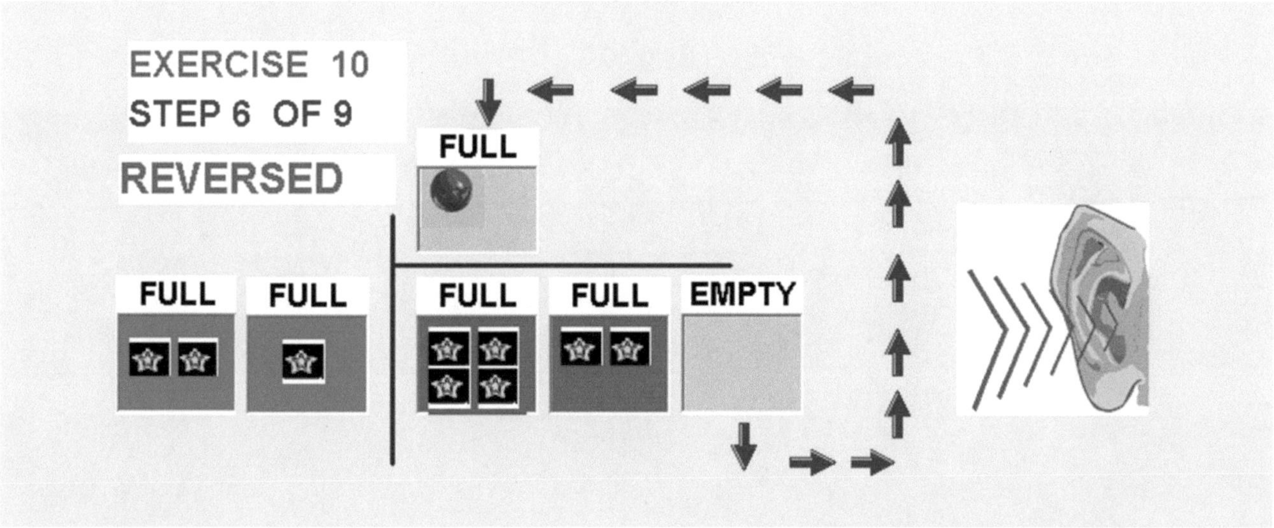

EXERCISE 10
STEP 6 OF 9
REVERSED
FULL
FULL
FULL
FULL
FULL
EMPTY

The ***kinesthetic*** representation of the filled up position is represented by the clenching of the right hand. The clenching of the left hand represents an empty position. The division sound, double click, is represented by both hands extended up in the air. Both hands extended in front represent the subtraction sound, double tram. The equal sound, cling, is represented by crossing both hands. The kinesthetic representation for the pictures is the clenching of the right hand twice, left hand, both hands extended up, right hand twice, crossing both hands, and right hand.

MATHEMATICS

REVERSE DIVISION, EXERCISE 10.8

In Exercise 10R Step 10.8R, in a ***visual*** display of division we see three pictures of boxes - the dividend, the divisor and the answer. The division sign is between the first and second pictures. Over the dividend is the equal line, which separates the problem from the answer. On top of the dividend is the third picture, which is the answer of the division problem. The ***audio*** representation for the pictures consists of a **knock, knock, double click, knock, knock, cling, knock, and knock.**

In the next part of Step 10.8R, we have a **visual** display of how we arrive at the answer. We subtract picture four from picture two, which is the dividend. Between picture two and four is the subtraction sign. Below picture four is the equal sign, which separates the problem from the answer. Picture five is the result of the subtraction. When the result of the subtraction is an empty position, then a filled position is transferred to the answer of division. The ***audio*** representation for the subtraction consists of a **knock, knock, double tram, knock, knock, cling, and double knock.**

Step 10.8

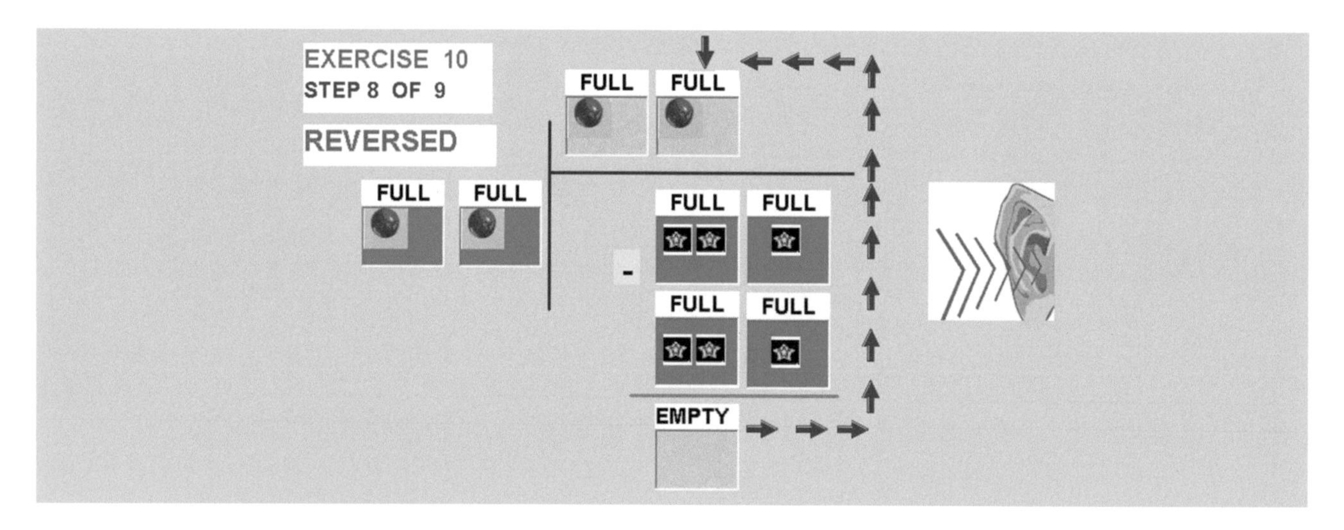

The ***kinesthetic*** representation of the filled up position is represented by the clenching of the right hand. The clenching of the left hand represents an empty position. The division sound, double click, is represented by both hands extended up. Both hands extended in front represent the subtraction sound, double tram. The equal sound, cling, is represented by crossing both hands. The kinesthetic representation for the pictures is the clenching of the right hand twice, both hands extended up, right hand twice, crossing of hands, and right hand twice.

REVERSE DIVISION, EXERCISE 10.9

In Exercise 10R Step 10.9, in a ***visual*** display of the last step of the division we see two pictures of boxes. In both pictures, both positions are filled with a red ball. The colors and the placement of the boxes and the balls are different in the pictures. The form, shape, and color of the symbols do no change the mathematical value, which is equal to three.

The ***audio*** representation for the pictures consists of a **knock, knock, and cling.**

Step 10.9

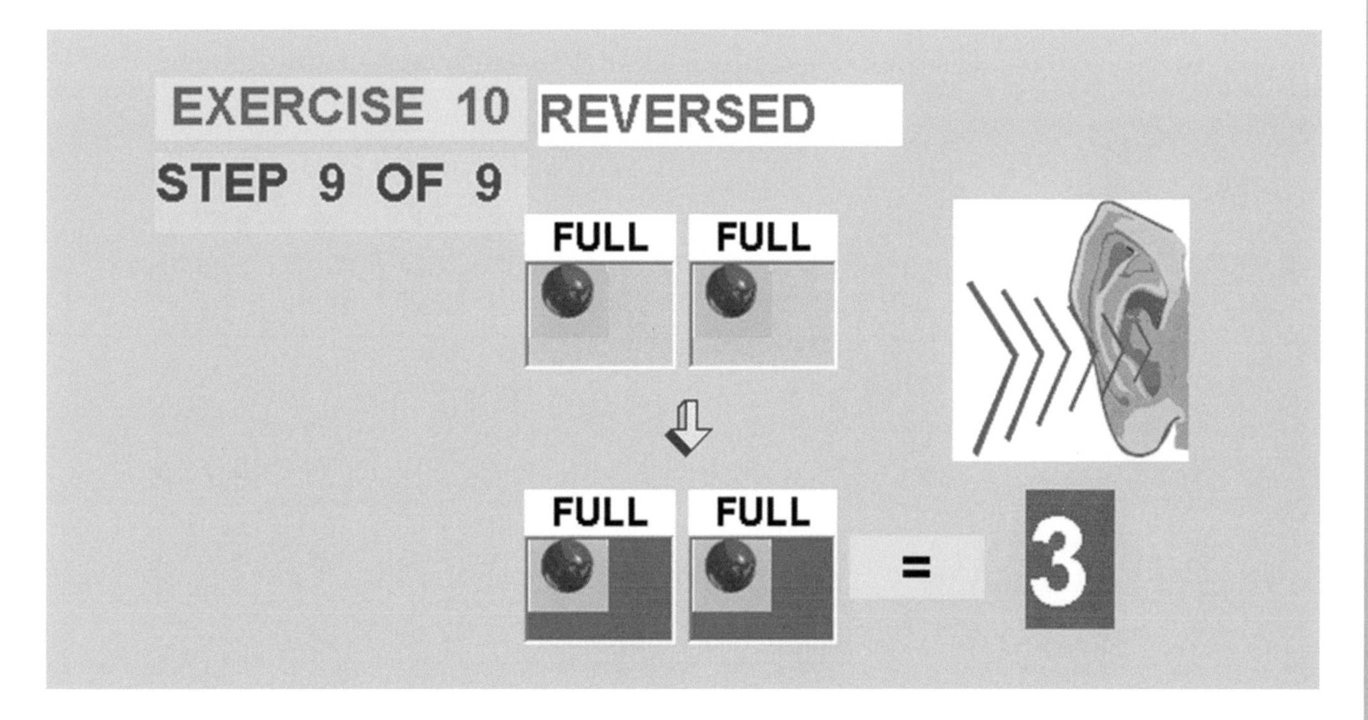

The ***kinesthetic*** representation of the filled up position is represented by the clenching of the right hand. The clenching of the left hand represents an empty position. The equal sound, cling, is represented by crossing the hands. The kinesthetic representation for the two pictures is the clenching of the right hand twice, crossing the hands, and right hand twice.

MATHEMATICS

REVERSE DIVISION, EXERCISE 10RA

In Exercise 10R Step A, in a ***visual*** display of division we see three pictures of boxes - the dividend, the divisor and the answer. The division sign is between the first and second sets of boxes. Over the dividend is the equal line, which separates the problem from the answer. On top of the dividend is the third picture, which is the answer of the division. The ***audio*** representation for the division consists of a **knock, double knock, double knock, knock, double click, knock, knock, cling, and knock.**

In the next part of Step A, we have a **visual** display of how we arrive at the answer. We subtract picture four from picture two, which is the dividend. The subtraction sign is between pictures two and four. Below picture four is the equal sign, separating the problem from the answer. Picture five is the result of the subtraction. If the result of the subtraction is empty, then a filled position is transferred to the answer of the division. The ***audio*** representation for the subtraction consists of a **knock, double knock, double knock, knock, double tram, knock, knock, cling, double knock, knock, knock, a double knock.**

Step A

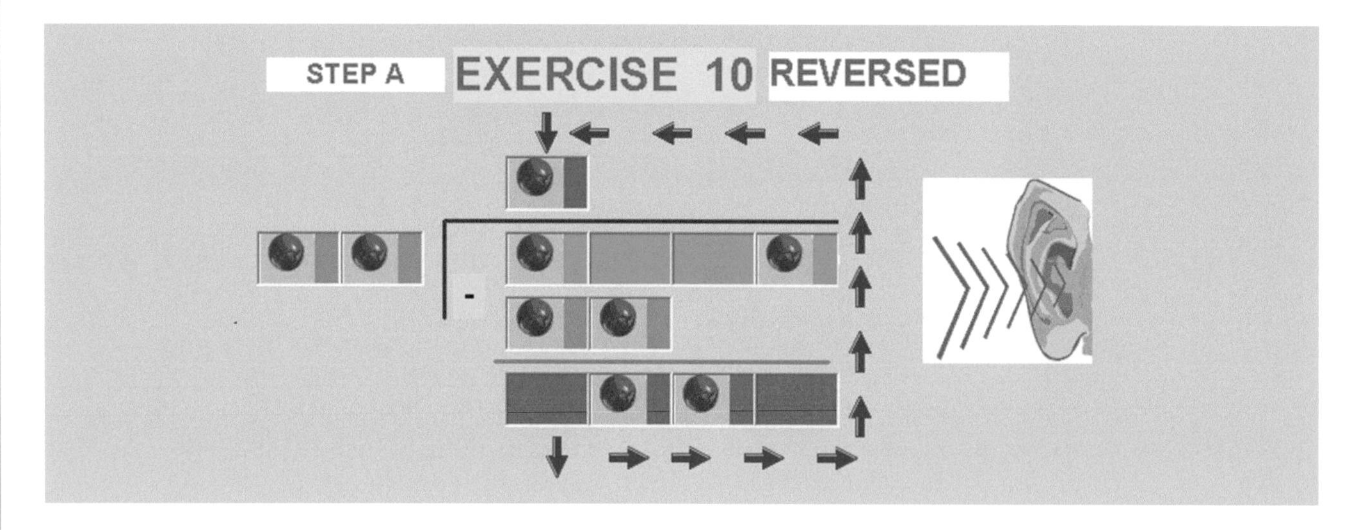

The ***kinesthetic*** representation of the filled up position is represented by the clenching of the right hand. The clenching of the left hand represents an empty position. The division sound, double click, is represented by both hands extended in up. Both hands extended in front represent the subtraction sound, double tram. The equal sound, cling, is represented by crossing both hands. The kinesthetic representation for the pictures is the clenching of the right hand, left hand twice, right hand, both hands extended up, right hand twice, crossing of the hands, and right hand.

REVERSE DIVISION, EXERCISE 10RB

In Exercise 10R Step B, in a ***visual*** display of division we see three pictures of boxes -the dividend, the divisor and the answer. The division sign is between the first and second sets of boxes. Over the dividend is the equal line, which separates the problem from the answer. On top of the dividend is the third picture, which is the answer of the division. The ***audio*** representation for the pictures consists of a **knock, knock, double click, knock, knock, cling, knock, and knock.**

In the next part of Step B, we have a **visual** display of how we arrive at the answer. We subtract picture four from picture two, which is the dividend. The subtraction sign is between the second and fourth pictures. Below picture four is the equal sign which separates problem from the answer. Picture five is the result of the subtraction. When the result of subtraction is an empty position then a filled position is transferred to the answer of the division. The ***audio*** representation for the subtraction consists of a **knock, knock, double tram, knock, knock, cling, double knock, a double knock.**

Step B

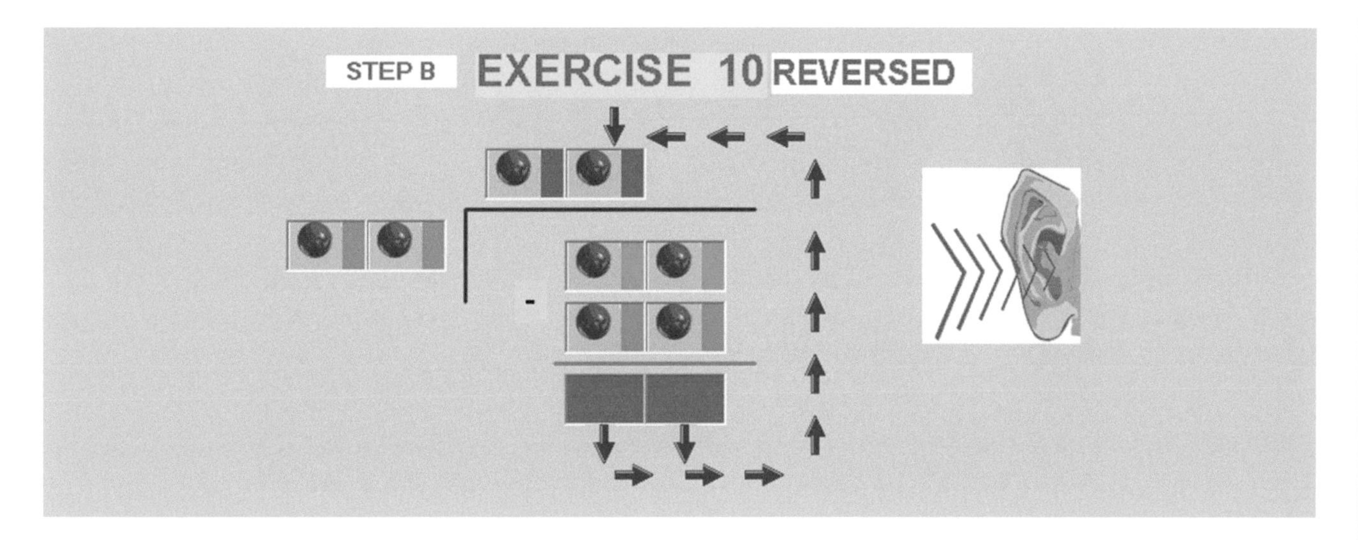

The ***kinesthetic*** representation of the filled up position is represented by the clenching of the right hand. The clenching of the left hand represents an empty position. The division sound, double click, is represented by both hands extended up. Both hands extended in front represent the subtraction sound, double tram. The equal sound, cling, is represented by crossing both hands. The kinesthetic representation for the pictures is the clenching of the right hand twice, both hands extended up, right hand twice, crossing of the hands, and right hand twice.

REVERSE DIVISION, EXERCISE 10RC

In Exercise 10R Step C, in a ***visual*** display of the last step of the division we see one picture of boxes, where both positions are filled with red balls. When the first two positions are filled with a symbol the picture represents number three.

The ***audio*** representation for the picture consists of a **knock, knock, and cling.**

Step C

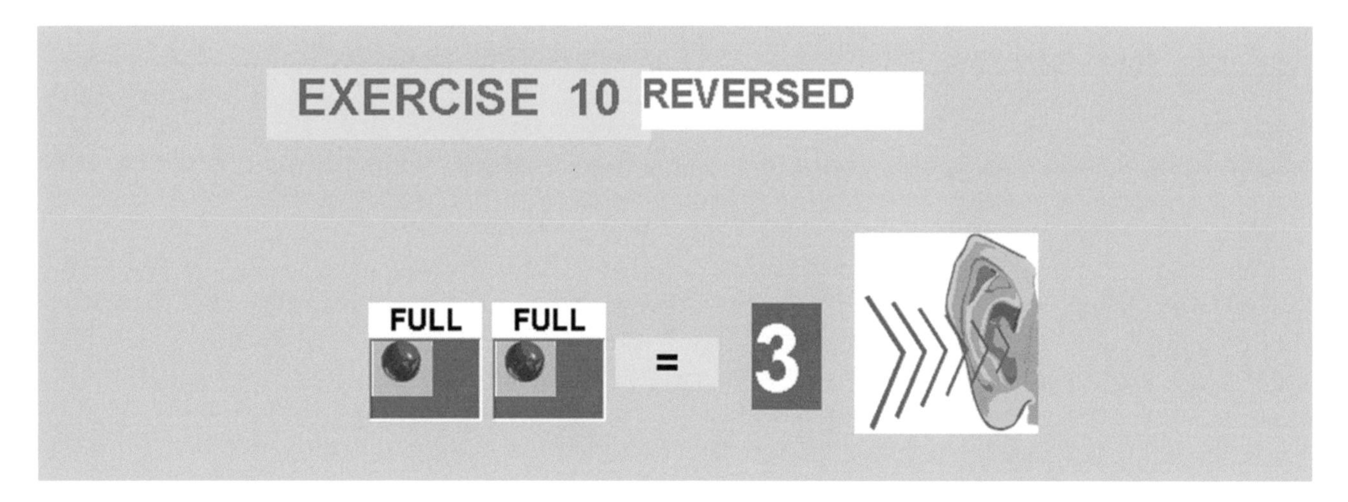

The ***kinesthetic*** representation of the filled up position is represented by the clenching of the right hand. The equal sound, cling, is represented by crossing the hands. The kinesthetic representation for the picture is the clenching of the right hand twice and cling.

REVERSE DIVISION, EXERCISE 10R

In Exercise 10R, in a ***visual*** display of division we see two pictures of boxes and the division sign in-between. In the first picture, both positions are filled with a red ball. In the second picture, the first position is filled with a red ball, the next two positions are empty, and the last position is filled with a red ball. On top of the second picture is the equal line, which separates the problem from the answer. The ***audio*** representation for the division consists of a **knock, double knock, double knock, knock, double click, knock, knock, cling, knock, and knock.**

Exercise 10R

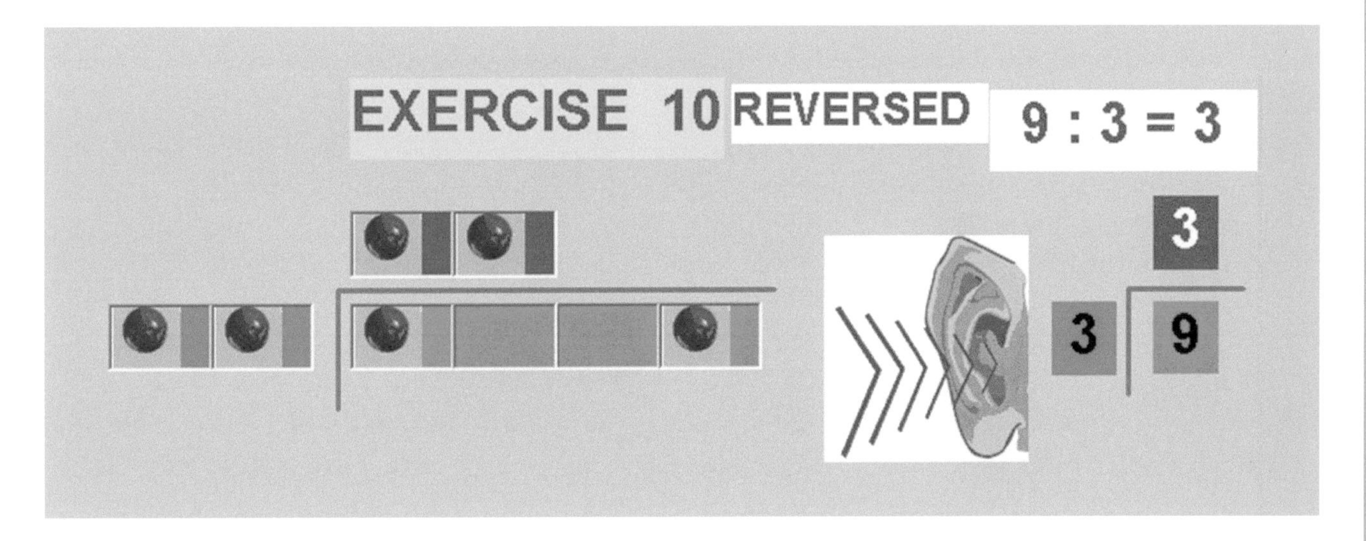

The ***kinesthetic*** representation of the filled up position is represented by the clenching of the right hand. The clenching of the left hand represents an empty position. Both hands extended up represent the division sound, double click. The equal sound, cling, is represented by crossing both hands. The kinesthetic representation for both pictures and the answer is the clenching of the right hand, left hand twice, right hand, both hands extended up, right hand twice, crossing both hands, and right hand twice.

MATHEMATICS

REVERSE DIVISION, EXERCISE 12.1

In Exercise 12R Step 12.1R, in a ***visual*** display of division we see two pictures of boxes - the dividend and the divisor. The division sign is between the first and second pictures. Over the dividend is the equal line, which separates the problem from the answer. The ***audio*** representation for the two pictures consists of a **knock, double knock, double knock, knock, double knock double click, knock, knock, double knock, cling.**

In the next part of Step 12.1R, we have a **visual** display of how we arrive at the answer. If the divisor has an empty position before the filled up position, then that position cancels out the first empty position of the dividend.

Step 12.1

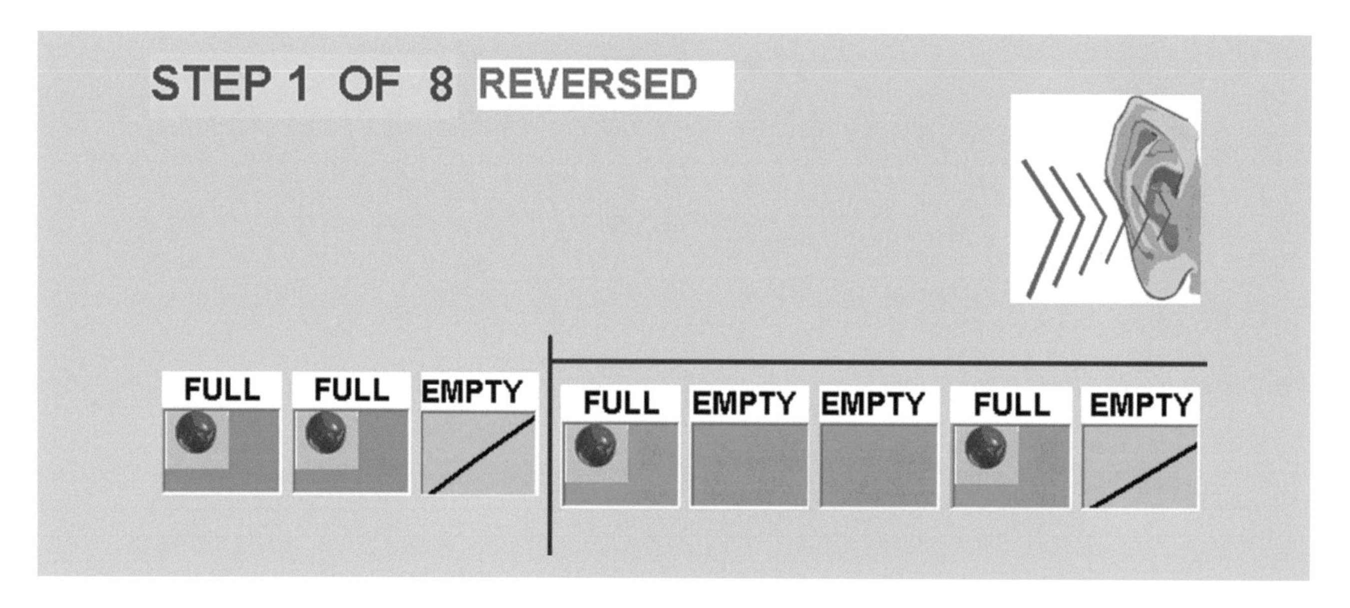

The ***kinesthetic*** representation of the filled up position is represented by the clenching of the right hand. The clenching of the left hand represents an empty position. The division sound, double click, is represented by both hands extended up. Both hands extended in front represent the subtraction sound, double tram. The equal sound, cling, is represented by crossing both hands. The kinesthetic representation for two pictures is the clenching of the right hand, left hand twice, right hand, left hand, both hands extended up, right hand twice, left hand, crossing the hands.

REVERSE DIVISION, EXERCISE 12.2

In Exercise 12R Step 12.2R, in a ***visual*** display of the next step of the division we see two pictures of boxes. In both pictures, the first position is filled with a symbol, the second and third positions are empty, and the last position is filled with a symbol. The colors and the placement of the boxes and the balls are different in the pictures. The form, shape, and color of the symbols do no change the mathematical value.

The ***audio*** representation for the two pictures consists of a **knock, double knock, double knock, and knock.**

Step 12.2

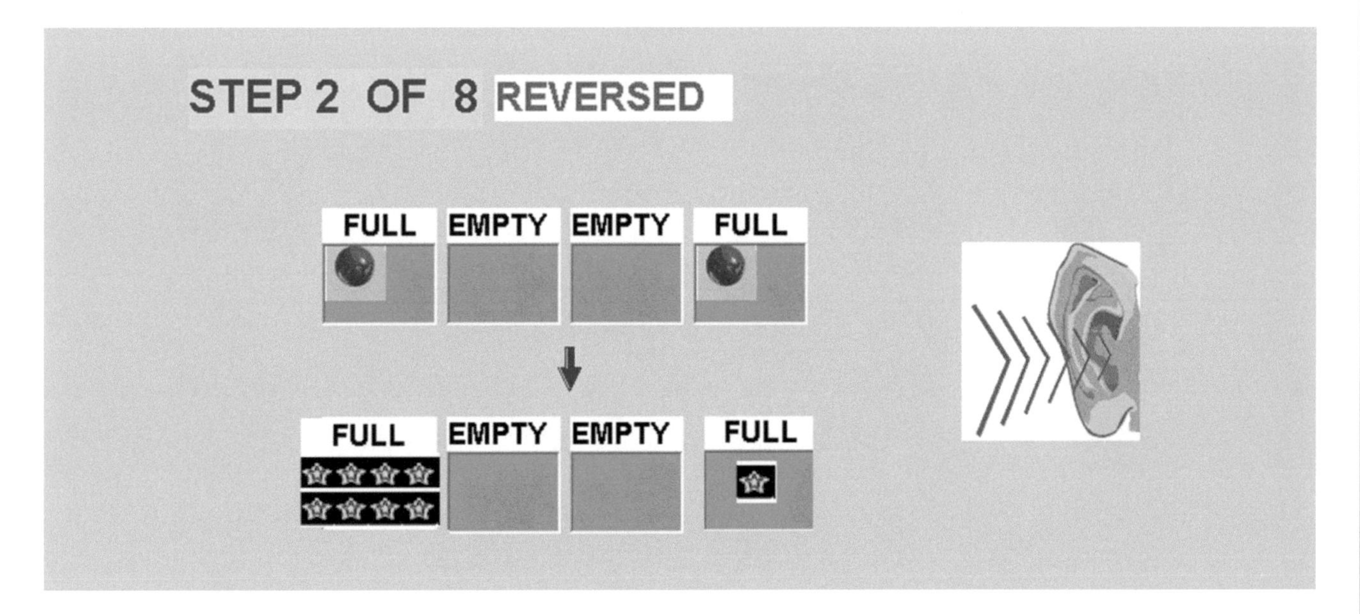

The ***kinesthetic*** representation of the filled up position is represented by the clenching of the right hand. The clenching of the left hand represents an empty position. The equal sound, cling, is represented by crossing the hands. The kinesthetic representation for the two pictures is the clenching of the right hand, left hand twice, right hand, and crossing of the hands.

REVERSE DIVISION, EXERCISE 12.3

In Exercise 12R Step 12.3R, in a ***visual*** display of division we see three pictures of boxes. In the first two pictures, the first position is filled with a symbol, the second and third positions are empty, and the last position is filled a symbol. For the purpose of division, we introduce the concept of borrowing. We borrow from the fourth to the third position. However, the audio representation for the pictures does not change because the numerical value does not change. The ***audio*** representation for the pictures with the equal sign consists of a **knock, double knock, double knock, knock, and cling.**

Step 12.3

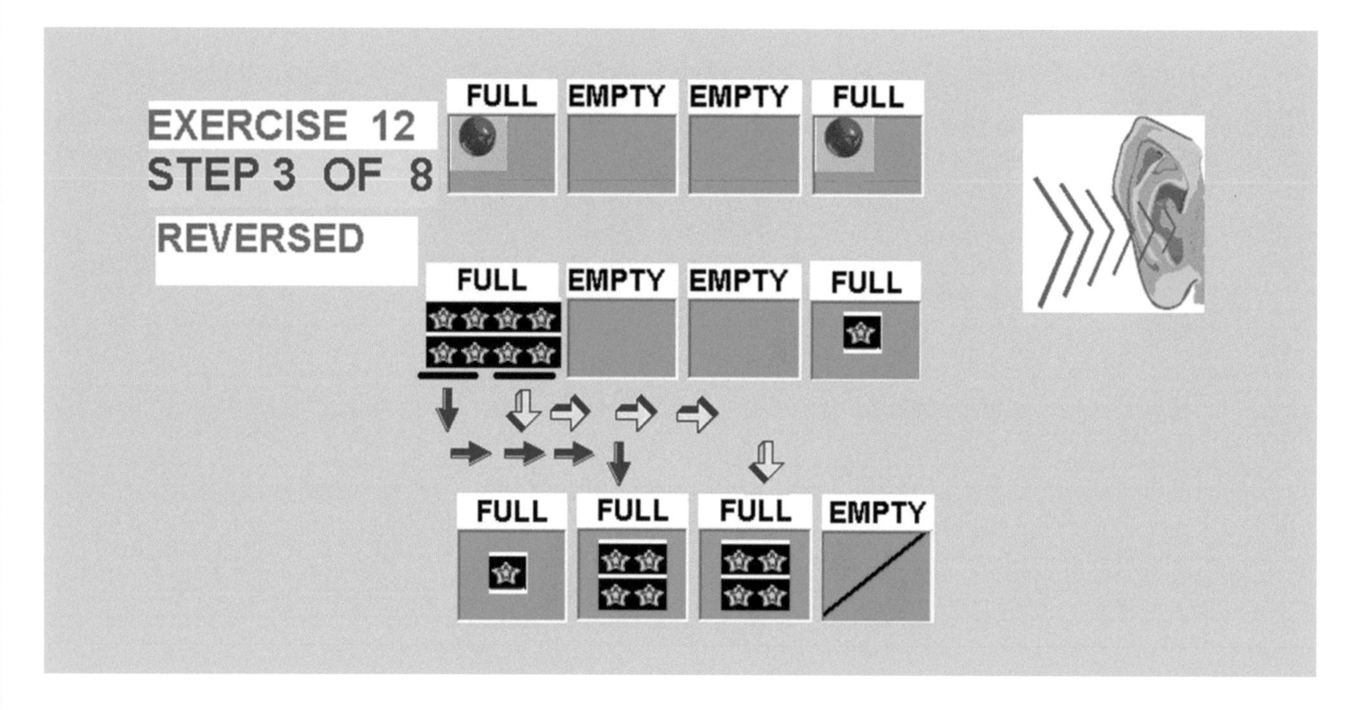

The ***kinesthetic*** representation of the filled up position is represented by the clenching of the right hand. The clenching of the left hand represents an empty position. The equal sound, cling, is represented by crossing both hands. The kinesthetic representation for both pictures and the answer is the clenching of the right hand, left hand twice, right hand, and crossing the hands.

REVERSE DIVISION, EXERCISE 12.4

In Exercise 12R Step 12.4R, in a ***visual*** display of division we see three pictures of boxes - the dividend, the divisor and the answer. The division sign is between the first and second pictures. Over the dividend is the equal line, which separates the problem from the answer. On top of the dividend is the third picture, which is the answer of the division problem. The ***audio*** representation for the pictures consists of a **knock, knock, knock double click, knock, knock, cling, and knock.**

In the next part of Step 12.4R, we have a **visual** display of how we arrive at the answer. We subtract picture four from picture two, which is the dividend. Between picture two and four is the subtraction sign. Below picture four is the equal sign which separates problem from the answer. Picture five is the result of the subtraction. The ***audio*** representation for the subtraction consists of a **knock, knock, knock, double tram, knock, knock, cling, knock, knock, and double knock.**

Step 12.4

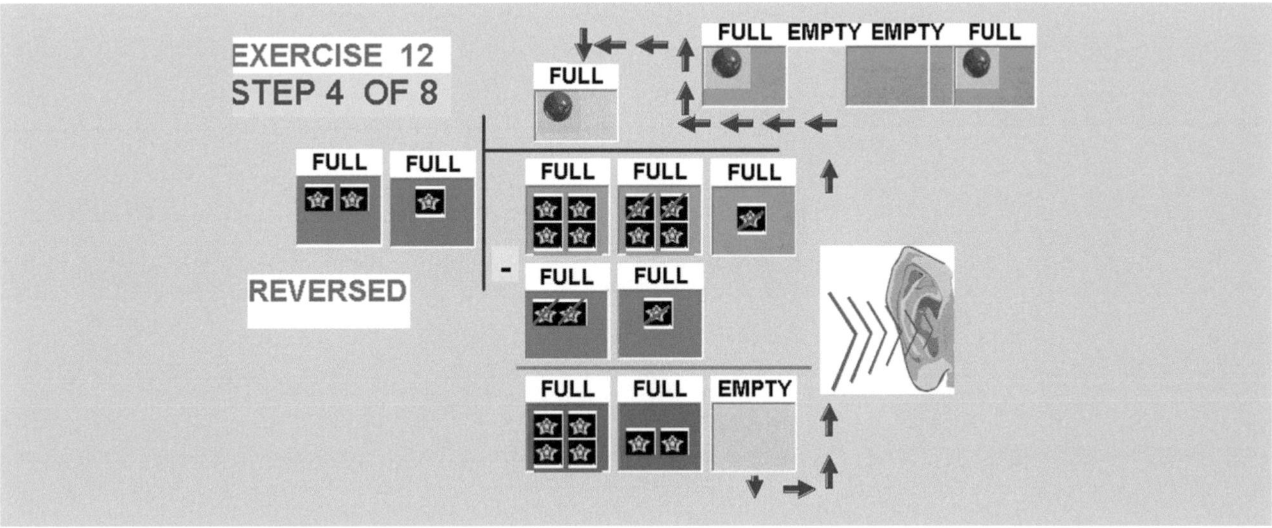

The ***kinesthetic*** representation of the filled up position is represented by the clenching of the right hand. The clenching of the left hand represents an empty position. The division sound, double click, is represented by both hands extended up. Both hands extended in front represent the subtraction sound, double tram. The equal sound, cling, is represented by crossing both hands. The kinesthetic representation for the two pictures is the clenching of the right hand three times, both hands extended up, right hand twice, crossing of the hands, and right hand.

MATHEMATICS REVERSE DIVISION, EXERCISE 12.5

In Exercise 12R Step 12.5R, in a ***visual*** display of division we see three pictures of boxes – the dividend, the divisor and the answer. The division sign is between the first and second pictures. Over the dividend is the equal line, which separates the problem from the answer. On top of the dividend is the third picture, which is the answer of the division. The ***audio*** representation for the pictures consists of a **knock, knock, double knock, double click, knock, knock, cling, and knock.**

In the next part of Step 12.1R, we have a **visual** display of how we arrive at the answer. In the dividend, the empty positions before the filled up position are automatically taken to the top to the answer as a filled up position.

Step 12.1

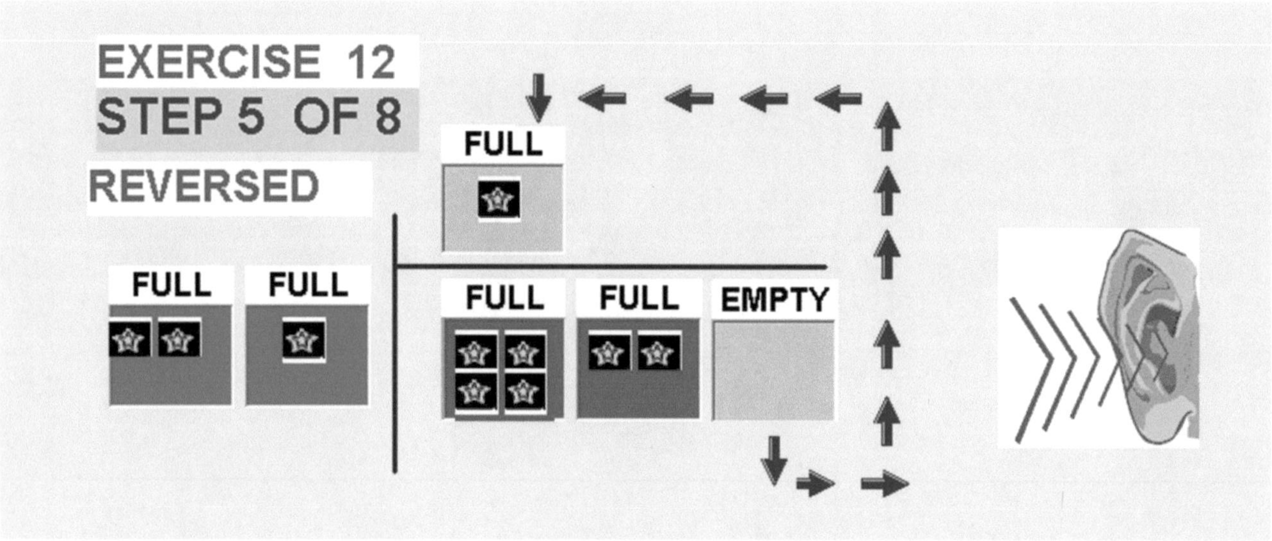

EXERCISE 12
STEP 5 OF 8
REVERSED
FULL
FULL FULL
FULL FULL EMPTY

The ***kinesthetic*** representation of the filled up position is represented by the clenching of the right hand. The clenching of the left hand represents an empty position. The division sound, double click, is represented by both hands extended up. Both hands extended in front represent the subtraction sound, double tram. The equal sound, cling, is represented by crossing both hands. The kinesthetic representation for the pictures is the clenching of the right hand twice, left hand twice, both hands extended up, right hand twice, crossing of the hands, and right hand.

REVERSE DIVISION, EXERCISE 12.6

In Exercise 12R Step 12.6R, in a ***visual*** display of the next step of the division we see two pictures of boxes. In both pictures, the two positions are filled with a red ball. The colors and the placement of the boxes and the balls are different in the pictures. The form, shape, and color of the symbols do no change the mathematical value, which is equal to five.

The ***audio*** representation for both pictures consists of a **knock, knock, cling, knock, and knock.**

Step 12.6

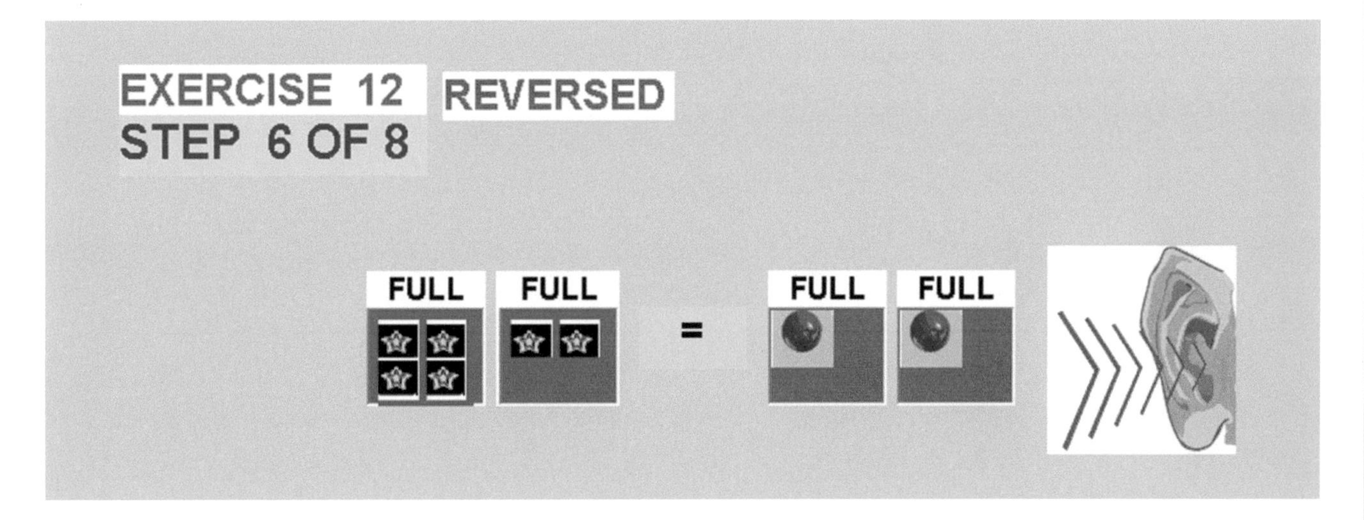

The ***kinesthetic*** representation of the filled up position is represented by the clenching of the right hand. The clenching of the left hand represents an empty position. The equal sound, cling, is represented by crossing the hands. The kinesthetic representation for the two pictures is the clenching of the right hand twice, crossing the hands, and right hand twice.

MATHEMATICS

REVERSE DIVISION, EXERCISE 12.7

In Exercise 12R Step 12.7R, in a ***visual*** display of division we see three pictures of boxes - the dividend, the divisor and the answer. The division sign is between the first and second pictures. Over the dividend is the equal line, which separates the problem from the answer. On top of the dividend is the third picture, which is the answer of the division. The ***audio*** representation for the pictures consists of a **knock, knock, double click, knock, knock, cling, knock, and knock.**

In the next part of Step 12.7R, we have a **visual** display of how we arrive at the answer. We subtract picture four from picture two, which is the dividend. Between picture two and four is the subtraction sign. Below picture four is the equal sign which separates problem from the answer. Picture five is the result of the subtraction. When the result of the subtraction is an empty position, then a filled position is transferred to the answer of division. The ***audio*** representation for the subtraction consists of a **knock, knock, double tram, knock, knock, cling, and double knock.**

Step 12.7

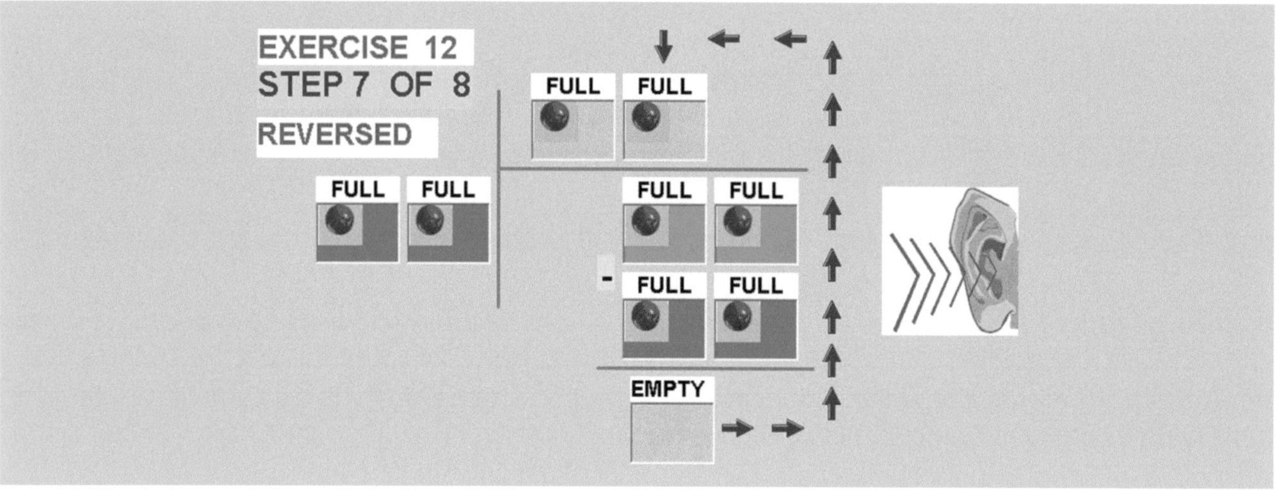

The ***kinesthetic*** representation of the filled up position is represented by the clenching of the right hand. The clenching of the left hand represents an empty position. The division sound, double click, is represented by both hands extended up. Both hands extended in front represent the subtraction sound, double tram. The equal sound, cling, is represented by crossing both hands. The kinesthetic representation for the pictures is the clenching of the right hand twice, both hands extended up, right hand twice, crossing of the hands, and right hand twice.

REVERSE DIVISION, EXERCISE 12.9

In Exercise 12R Step 12.9R, in a ***visual*** display of the next step of the division we see two pictures of boxes. In both pictures, both positions are filled with a red ball. The colors and the placement of the boxes and the balls are different in the pictures. The form, shape, and color of the symbols do no change the mathematical value, which is equal to three.

The ***audio*** representation for the two pictures consists of a **knock, knock, and cling.**

Step 12.9

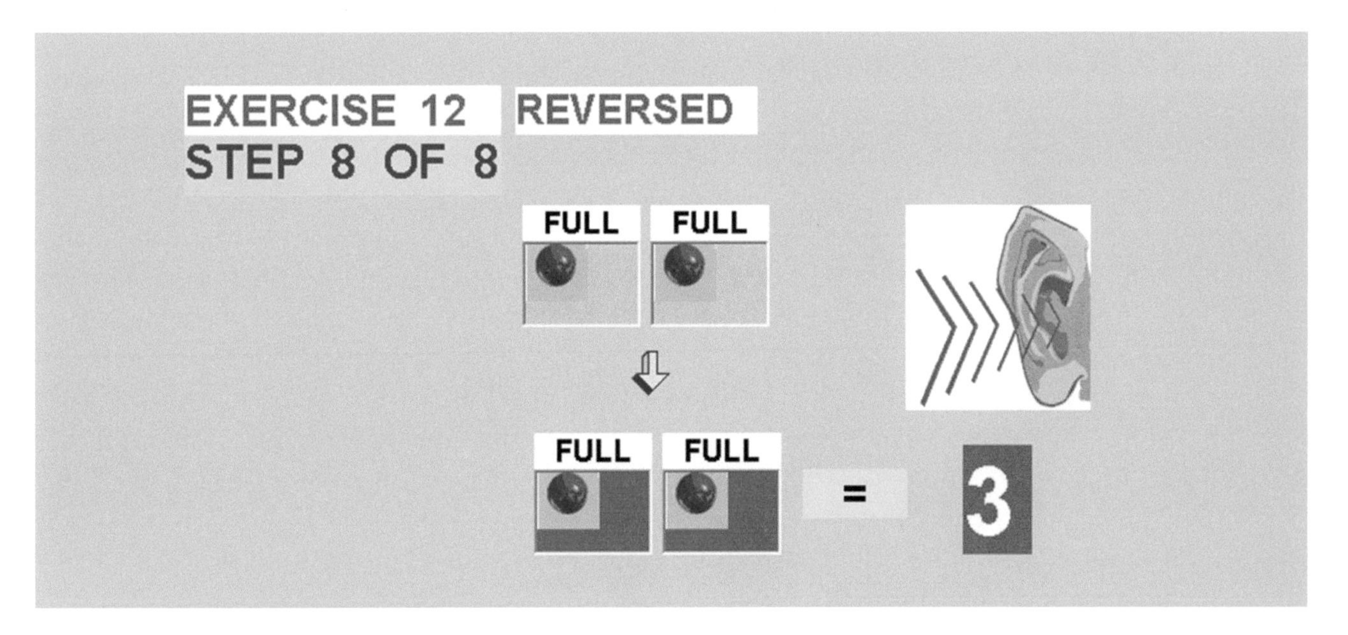

The ***kinesthetic*** representation of the filled up position is represented by the clenching of the right hand. The clenching of the left hand represents an empty position. The equal sound, cling, is represented by crossing of hands. The kinesthetic representation for the two pictures is the clenching of the right hand twice, crossing of hands and right hand twice.

MATHEMATICS

REVERSE DIVISION, EXERCISE 12R

In Exercise 12R, in a ***visual*** display of division we see two pictures of boxes with the division sign in-between. In the first picture, the first position is empty and the next two are filled with red balls. In the second picture, the first position is filled with a red ball, the next two are empty, the fourth position is filled with a red ball and the last position is empty. On top of the second picture is the equal line, separating the problem from the answer. The ***audio*** representation for the division consists of a **knock, double knock, double knock, knock, double knock, double click, double knock, knock, knock, and cling.**

At the bottom are probable answers. From the probable answers for the division we need to find the right one. The audio signals are: 1) **knock, double knock, and knock,** 2) **knock, knock, and double knock**, 3) **double knock, knock, and knock.**

Exercise 12R

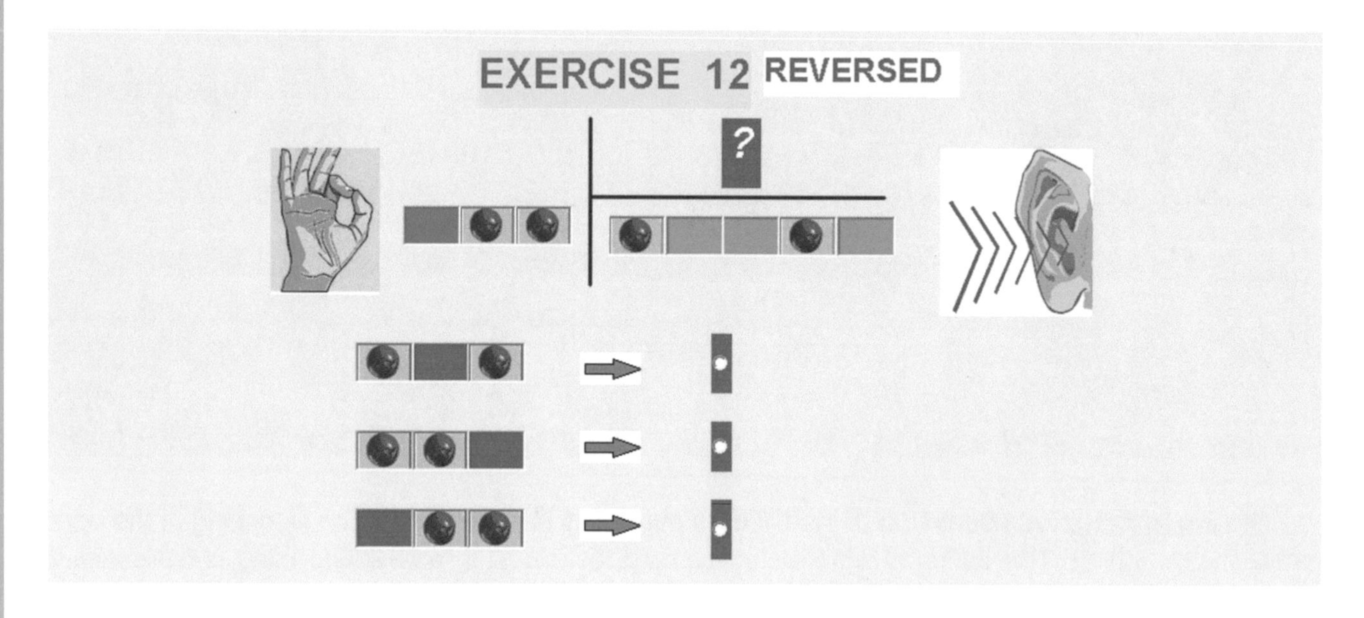

The ***kinesthetic*** representation of the filled up position is represented by the clenching of the right hand. The clenching of the left hand represents an empty position. Both hands extended up represent the division sound, double click. The equal sound, cling, is represented by crossing both hands. The kinesthetic representation for both pictures and the answer is the clenching of the right hand, left hand twice, right hand, left hand, both hands extended up, left hand, right hand twice, crossing of the hands, right hand twice, and left hand.

Method and system for psychological treatment by brain stimulation of the psychologically disordered object

Abstract:

The method and system for psychological treatment by brain stimulation of the psychologically disordered object provide a possibility to stimulate the simple cells to assist the complex cells, and/or replace the complex cells in order to correctly perform the assigned command. An improved method includes the steps providing a creation of the psychoconduction code based on the simplified symbols, which by the audio, video, kinesthetic and/or olfactory stimulation simple cells to correct or replace the complex cells.

US Patent References:

6798898	**Management of physiological and psychological state of an individual using images therapeutic imaging classification system**	2004-09-28	Fedorovskaya et al.	
6708064	**Modulation of the brain to affect psychiatric disorders**	2004-03-16	Rezai	
6556868	**Methods for improving learning or memory by vagus nerve stimulation**	2003-04-29	Naritoku et al.	
6520905	**Management of physiological and psychological state of an individual using images portable biosensor device**	2003-02-18	Surve et al.	
6443977	**Apparatus and method for changing critical brain activity using light and sound**	2002-09-03	Jaillet	

6425764	**Virtual reality immersion therapy for treating psychological, psychiatric, medical, educational and self-help problems**	2002-07-30	Lamson	434/236
6418344	**Method of treating psychiatric disorders by electrical stimulation within the orbitofrontal cerebral cortex**	2002-07-09	Rezai et al.	
6167311	**Method of treating psychological disorders by brain stimulation within the thalamus**	2000-12-26	Rezai	

Claims:

What is claimed:

1. A method for psychological treatment for brain stimulation of a psychologically disordered object comprising the steps of: creating a psychoconduction code using a simple stimuli; creating a video pattern of a video information using said psychoconduction code; providing a visual test on the basis of said video pattern of said video information; creating an audio pattern of an audio information using said psychoconduction code; providing a sonic test on the basis of said audio pattern of said audio information; creating a kinesthetic pattern of said audio information or said video information using said psychoconduction code; providing a kinesthetic test on the basis of said kinesthetic pattern of said audio information or said video information; creating an olfactory pattern of said audio information or said video information using said psychoconduction code; providing an olfactory test on the basis of said olfactory pattern of said audio information or said video information; creating a tactual pattern of said audio information or said video information using said psychoconduction code; providing a tactual test on the basis of said tactual pattern of said audio information or said video information; creating a musical pattern of said audio information or said video information using said psychoconduction code; providing a musical test on the basis of said musical pattern of said audio information or said video information; providing at least one or combination of a visual treatment, an audio treatment, a kinesthetic treatment, an olfactory treatment, a tactual treatment, a musical treatment for an appropriate complex cell correction by an appropriate simple cell of said brain, and wherein said at least one or said combination of the

treatments for said brain is provided to treat a partial damage of said complex cell of said brain; providing said at least one or said combination of said visual treatment, said audio treatment, said kinesthetic treatment, said olfactory treatment, said tactual treatment, said musical treatment for said appropriate complex cell replacement by said appropriate simple cell of said brain, and wherein said at least one or said combination of the treatments for said brain is provided to treat a complete said damage of said complex cell of said brain, and; providing said at least one or said combination of said visual treatment, said audio treatment, said kinesthetic treatment, said olfactory treatment, said tactual treatment, said musical treatment for a correction of an address of said appropriate complex cell by an assistance of said appropriate simple cell of said brain to process the submitted stimuli, and wherein said at least one or said combination of the treatments for said brain is provided to treat said partial damage of said complex cell of said brain.

2. The method of claim 1, wherein said either one of said visual treatment, said audio treatment, said kinesthetic treatment, said olfactory treatment, said tactual treatment and said musical treatment comprises the steps of: generalizing at least one or each of said pattern of said psychoconduction code providing a bank of said video information, said audio information, said kinesthetic information, said olfactory information, said tactual information and said musical information; selecting an arithmetical addition pattern from said pattern for learning of said arithmetical addition; selecting an arithmetical multiplication pattern from said pattern for learning of said arithmetical multiplication; selecting an arithmetical subtraction pattern from said pattern for learning of said arithmetical subtraction; selecting an arithmetical division pattern from said pattern for learning of said arithmetical division; selecting a reading pattern from said pattern for reading learning; selecting a writing pattern from said pattern for writing learning; transforming said arithmetical addition pattern, said arithmetical multiplication pattern, said arithmetical subtraction pattern, said arithmetical division pattern, said reading pattern and said writing pattern by said psychoconduction code in either one of said audio pattern, video pattern, kinesthetic pattern, olfactory pattern, tactual pattern, musical pattern or their combination.

Description:

FIELD OF THE INVENTION

This invention relates to the psychoconduction method and system intended for brain stimulation providing correction of the malfunctioning complex cell and/or replacement of the malfunctioning complex cell by the modified simple brain cells, and mostly relates to the calibration, balancing, alignment, and stimulation of the different parts of the brain using the different structures and pattern of the visual, audio and/or kinesthetic perception, and olfactory expressions having the same meaning.

BACKGROUND OF THE INVENTION

The various types of the methods and systems for psychological brain treatments and stimulations are well known in the medical and counseling practice. Some brain treating methods require placement of a probe and sensing, stimulating of the both areas of the brain, especially the intralaminar nuclei. Moreover, stimulation is controlled and offered when certain conditions within the area of interest are detected. Stimulation and sensing include electrical, chemical or combinations thereof. For example, the surgical principles (method) can be applied for “deep” psychological, neurological and/or psychiatric disorders. The surgical method has an extensive history. In the early 1930’s, Fulton and Jacobsen first recognized that experimentally induced neurotic behavior in chimpanzees could be abolished by frontal lobectomy. Within a few years, Freeman and Watts developed the first psychosurgical procedure for humans known as the frontal lobotomy. As the inherent physiology of the frontal lobe became more evident, the original freehand procedure of Freeman and Watts became less and less extensive and efficient. By the late 1940’s, the method of stereo-taxis, in which the patient’s brain is modeled in 3-dimensional space for exquisite targeting accuracy, merged with lesioning techniques resulting in an even more efficacious and safe psychosurgical procedure. Further developments of stereotactic equipment have combined with novel advancements in functional and anatomic imaging as well as intraoperative electrophysiological mapping to encompass the state of the art in the neurosurgical treatment of neurological and psychiatric disorders today. While technologically improved and more precise, today’s surgical lesioning techniques have the fundamental limitation of being inherently irreversible and are essentially a “one shot” procedure with little chance of alleviating or preventing potential side effects. In addition, there is a limited possibility to provide continuous benefits as the disease progresses and the patient’s symptoms evolve. Within the field of neurosurgery, the use of electrical stimulation for treating “deep” psychological, neurological and psychiatric diseases, including such disorders as, for example, movement disorders, etc. has been widely discussed in the literature. In the latest time it has been considered that electrical and/or chemical stimulation holds some advantages over lesioning, inasmuch as lesioning can only destroy nervous system tissue. In many instances, the preferred effect is to electrically stimulate the increasing, decreasing, or blocking of the psychological or neuronal activity. Electrical or chemical impact might be useful in some ways to provide the stimulation and/or modulation of the brain target neural structures.

The methods by U.S. Pat. Nos. 6,708,064; 6,418,344 and 6,167,311 provide modulation of the brain to affect psychological or psychiatric disorders. The methods provide the treating neurological conditions by proper placement of a probe and sensing, stimulating of the both areas of the brain. Generally, the methods relate to modulating the pathological electrical and chemical activity of the brain by electrical stimulation and/or direct placement of neuro-modulating chemicals within the corresponding areas of abnormal function and activity. In accordance with the inventions, the methods are the surgical treatment of psychiatric disorders (e.g. addictions/substance abuse, obsessive compulsive disorder, generalized anxiety disorder, panic attacks, social phobia, major depression, learning disorder, etc.) by implantation of stimulating electrodes and/or drug/chemical delivery micro infusion at the assigned locations.

More particularly, the methods include the following general steps of 1) implanting a device in contact with an intralaminar nuclei of the brain; sensing activity in the specific area of the brain, wherein the specific area of the brain is different than the intralaminar nuclei, and wherein the specific area of the brain is different than the intralaminar nuclei and the sensing activity occurs at a location distal from the device location; and operating the device to modulate the intralaminar nuclei in response to said activity to thereby affect the disorder associated with the specific area of the brain, or 2) placing an electrode in contact with an intralaminar nuclei of the brain and operating the device to provide stimulation to the intralaminar nuclei to thereby affect the psychiatric activity in the specific area of the brain, the specific area of the brain being different than the intralaminar nuclei.

Such methods are not completely saved considering possible malfunction of the electrical/electronic equipment (e.g., higher electrical signal), that might lead to the critical destruction of the brain nervous tissue/cell.

Another method of brain stimulation is described in the U.S. Pat. No. 6,556,868. The method provides the treatment of psychological disorder consisting the group of memory impairment, a learning disorder, impairment of cognitive processing speed, impairment of acquisition of perceptual skills, impairment of acquisition of motor skills, and impairment of perceptual processing. In general, the method comprises the steps of: selecting an appropriate human (or animal) subject and applying to the subject's vagus nerve an electrical stimulation signal having parameter values effective in modulating the electrical activity of the vagus nerve in a manner so as to modulate the activity of preselected portions of the brain. The stimulating electrical signal has to be effective to cause a physiological, structural or neuronal connective alteration in the brain. Neural function in the brain is changed as a consequence of the neuronal connective alteration; thereby changing behavior, or the capacity for behavior, in the human or animal subject.

This method has the same deficiency as the previous patented method, i.e. the electrical stimulation is not completely saved considering possible malfunction of the electrical/electronic equipment (e.g., higher electrical signal), that might lead to the critical destruction of the brain nervous tissue/cell.

The other known devices (U.S. Pat. Nos. 6,520,905 and 6,798,898) also use the portable biosensor. The described in these patents methods classify an individual's personal preference for an image. In common the methods comprise the steps of: viewing an image for a period of time; ranking the image on a scale extending between a "detached" feeling and an "attached" feeling, where "detached" is a feeling of not being able to personally connect to the object or situation depicted in the image, and "attached" is a feeling of a personal connection to the object or situation depicted in their image; providing a portable device having at least one sensor for monitoring a physiological state of an individual carrying the device; recording at least one sensed physiological state over a period of time; analyzing the recorded physiological data to predict the individual's psychological and physiological state; and alerting the individual if the predicted state is determined to require management of said state. A portable biometric device is worn or carried by a user and which senses and records physiological parameters on a continuous basis. A biometric analyzer extracts the physiological activation state of user from one or more measured physiological parameters. A cognitive analyzer extracts cognitive state from cognitive responses to images. A personal profiler combines the physiological and cognitive measures obtained from the biometric analyzer and cognitive analyzer to generate an individual's personal image profile for a given state response.

These methods using the principles of the wearing/carrying of the portable device (sensors) can bring the additional stress and discomfort to the psychologically disordered individual, and again include in some way the mentioned above deficiency inherent in electrical contact of the electrical equipment (sensors) with the human body, that might be not completely saved considering the possible malfunction of the electrical/electronic equipment, that might lead to the possible injury. Also, these methods can provide recognition of the individual's physiological and psychological state, but does not provide brain (cell) stimulation (treatment), e.g. such as improving memory and learning capabilities.

The U.S. Pat. No. 6,433,977 discloses the apparatus and method for changing critical brain activity using light and sound, exposing the patient to one or more lights placed in close proximity to a patient's eyes wherein the one or more of that lights selectively stimulate the non-dominant eye connected to the non-dominant cerebral hemisphere. The apparatus is presented either by a device that covers the patient's eyes, such as a pair of sunglasses, or devices including sports helmets that are used to protect players' craniums and may be integrated into the protective head gear (e.g., football, bicycle helmets, etc.). Also, the device can include the computer monitors and televisions, which are able to encompass one or more oscillating lights set-up in a proscribed manner on a person's monitor. The light may be displayed in a subliminal alternating checkerboard pattern that would be set to the individual user. The light stimulates the non-dominant cerebral hemisphere greater than the dominant cerebral hemisphere. In this version, the non-dominant cerebral hemisphere is stimulated to a greater degree than the dominant cerebral hemisphere. It is the coordinated stimulation of the non-dominant hemisphere that helps create a balance of integration of excitatory post synaptic potentials (EPSP). The apparatus for selectively stimulating the non-dominant cerebral includes a surface placed in close proximity to a patient's eyes and one or more lights disposed on the surface. The one or more lights stimulate the eye connected to the non-dominant cerebral hemisphere to a greater extent than the eye connected to the dominant cerebral hemisphere at a rate of approximately, e.g., 60/40. By over stimulating the non-dominant hemisphere there is an increase in the patient's ability to maintain heighten mental status, and in turn sets up for a globosity of the increased muscular activity. Alternatively, the surface may be sleeping goggles or the glasses reflecting the light from a source next to the eye (light is reflected from the glass surface) into the patient's eyes.

These method and device are not able to provide an improvement of the individual's learning processes, such as mathematics, alphabet, etc.

Thus, there is a great need in the art for the improved method and system for psychological treatment by brain stimulation of the psychologically disordered object, employing at the same time the correction of the malfunctioning complex cell and/or replacement of the malfunctioning complex cell by the modified simple brain cells, and providing possibility to balance, align and stimulate of the different parts of the brain using different structures and pattern of visual, audio and/or kinesthetic perception, and olfactory expressions having the same meaning.

OBJECT AND ADVANTAGES OF THE INVENTION

Accordingly, several objects and advantages of the present invention are to provide an improved brain psychological treatments, such as calibration, balancing, alignment and stimulation of the different parts of the brain of the psychologically disordered object.

It is another object of the invention to modify the simple brain cells in order to provide correction or replacement of the complex cell.

It is still another object of the invention to provide video patterns (images) expressed in the visual symbols stimulating brain cells.

It is further object of the invention to provide audio patterns expressed in the adequate sound symbols stimulating brain cells.

It is still further object of the invention to provide kinesthetic patterns expressed in the individual's body and/or object movements stimulating brain cells.

It is still further another object of the invention to provide olfactory patterns expressed in the smells stimulating the adequate brain cells.

It is another further object of the invention to provide the enhancement of the existing structure, restructuring of the tune, calibration and recalibration of the different brain areas without any surgical and/or pharmacological/chemical interventions.

It is still another further object of the invention to improve learning possibility and skills of the psychologically disordered object.

Still, further objects and advantages will become apparent from a consideration of the ensuing description accompanying drawings.

SUMMARY OF THE INVENTION

The method and system for psychological treatment by brain stimulation of the psychologically disordered object provide a possibility to stimulate the simple cells to assist the complex cells, and/or replace the complex cells in order to correctly perform the assigned command.

This approach is based on the congruent processing of the verbal and emotional information. When a psychologically disordered object is experiencing anger and frustration, the body sensations or kinesthetic functions are taking over of all other functions. It manifests itself in the heart palpitation increasing, muscles tension and rapid breathing. In the condition of anger and frustration, the object's movement, verbal, and audio communications in this state of mind get down to a very simple level. An object (e.g. an individual or a person) in this state uses the simple movements, has difficulty using the verbal expressions, and has difficulty understanding audio information. In such condition the object reacts only on simple audio or video signals. In this state the complex cells do not work and the communication is possible on a very simple level only.

The psychologically disordering (and sometimes psychotic) symptoms are mainly related to different hallucinations. The audio or visual hallucinations take over the object's life. The stressed or depressed object may not have hallucinations, but have the flash backs and nightmares. The hyperactive objects have the exaggerated kinesthetic functions. They are not in control of their desired fast movements, restless appearance, fidgeting, and the rapid speech activities. For instance, if kinesthetic response does not support the exaggerated reaction, visual and/or audio information can be adjusted by kinesthetic response.

An improved method includes the steps providing a creation of the psychoconduction code based on the simplified symbols, which by the audio, video, kinesthetic and/or olfactory stimulation simple cells to correct or replace the complex cells. An improved system comprises a controlling system comprising a processing system, including a controller and a memory, and a terminal means comprising at least one of a compact disk means, a floppy disk means, a printing means and a control panel. Also, the improved system comprises an auxiliary equipment and a display connected to the controller, which is connected to the speaker(s).

An improved method provides a psychological brain stimulation treatment of the psychologically disordered object by psychoconduction. The psychoconduction method is a sequence of the operations (steps) intended for correction and enhancement of the brain capacity. An improved method (process) provides the enhancement of the existing structure and restructuring of the tune, calibration and recalibration of the different brain areas. Psychoconduction allows simple cells act as the complex cells within the different areas of the brain. Also, the improved method (psychoconduction) reinforces the addresses to provide access to the complex cells for the appropriate stimuli processing. The brain structure uses the simple cells of the brain in order to process the simple stimuli (symbol), and the more complex cells are used to process the more complex stimuli. Psychoconduction provides the simple stimuli utilizing the simple brain cells. Many disorders are caused by deficiency of the brain to process the visual, audio, kinesthetic and olfactory information correctly. The results of such disorder might be the confusion, misunderstanding, anxiety, despair anger overreaction, and/or learning disabilities. The described difficulties lead to the inability of the impaired or underdeveloped complex brain cells to process complex information correctly. Normally, the visual, audio, kinesthetic and/or olfactory stimulus causes a nerve impulse to travel down the axon to the synapses where the pulse triggers the release of the chemical messenger molecules, for example, such as dopamine, SSRI, and non-epinephrine for a correct reaction. If the stimulus is complicated (e.g. because of the ambivalent information) the impulse travels down the axon to the incorrect location. This is the reason why the impaired complex cells are not provided with the authentic information.

The areas of the brain that are involved for the vision are the occipital, parietal and temporal cortexes. The brain area processing the sonic (audio) information includes the primary auditory cortex (PAC) in the superior temporal gyrus of the temporal lobe. The kinesthetic stimulus is processed in the cerebellum that is responsible for the appropriate action according to the motor commands.

The olfactory stimulus is processed in the olfactory mucosa (OM). The olfactory mucosa processes the different odors. The improved method (process) uses only the simple cells in the areas that are responsible for the simple stimulus processing. Psychoconduction stimulates the different brain areas by the simple symbol only. That simple symbol can be the visual, audio, kinesthetic and/or aroma. Psychoconduction provides processing of the simple symbol instead of the complex symbol. Psychoconduction enhances the brain capacity by stimulation of the brain and gradually increases different patterns of the simple symbol using the simple cells of the brain, that usually processes the simple information. Psychoconduction provides stimulation of the different parts/areas of the brain by the same patterns. For example, the psychoconduction method provides translation (interpretation) of the kinesthetic symbols to the audio and/or visual symbols. During translation the areas of the brain, which process the kinesthetic symbol, is attuned with the ones in the brain area that reproduces the visual and/or audio symbol. In the improved method (psychoconduction), the same pattern of the simple symbol processing provides the congruence of the chemicals released by the different areas of the brain. This part of the improved method is similar to the brain biofeedback while the brain corrects itself and creates the equilibrium between different brain areas in response to the same stimulus. Psychoconduction is intended for correction of the inappropriate responses to the stimulus. For instance, when one area of the brain used for information (e.g. sonic information) processing is affected by the distortion, there is a high probability that the information might be distorted and improperly processed. As a consequence, an inappropriate response may lead to the overreaction and/or underestimation of the original information. In order to correct the problem, the psychoconduction uses the pattern that is transformed in the codogram (Litvin's code). The simple symbol presented by such codogram stimulates the different parts of the brain and enhances the brain learning capability and capacity. The codogram with the same patterns of the simple symbol creates the connection between different parts of the brain and utilizes the similar molecular structure for stimulation of the different parts of the brain with the similar patterns of the molecular messenger.

The psychoconduction code (codogram) creates the different value (weight) of the codogram positions, for example, it is conditionally assumed that "a full" position is equal to "X" and "an empty" position is equal to "0 (zero)". Additionally it is assumed, that for visual perception of the psychologically disordered object the "full" position can be represented, for instance as ("x"), ("1"), dot or as a dark geometrical figure, such as dark color circle, square or rectangle, etc. and the "empty" position can be represented, for instance as a circle ("o"), "0 (zero)", or as a light geometrical figure (e.g. such as light color circle, square or rectangle, etc.).

TABLE 1

1	A					x	Or				0	x
2	B				x		Or				x	0
3	C				x	x	Or				x	x
4	D			x			Or			x	0	0
5	E			x		x	Or			x	0	x
6	F			x	x		Or			x	x	0
7	G			x	x	x	Or			x	x	x
8	H		x				Or		x	0	0	0
9	I		x			x	Or		x	0	0	x
10	J		x		x		Or		x	0	x	0

11	K		x		x	x	Or		x	0	x	x
12	L		x	x			Or		x	x	0	0
13	M		x	x		x	Or		x	x	0	x
14	N		x	x	x		Or		x	x	x	0
15	O		x	x	x	x	Or		x	x	x	x
16	P	x					Or	x	0	0	0	0
17	Q	x				x	Or	x	0	0	0	x
18	R	x			x		Or	x	0	0	x	0
19	S	x			x	x	Or	x	0	0	x	x
20	T	x		x			Or	x	0	x	0	0
21	U	x		x		x	Or	x	0	x	0	x
22	V	x		x	x		Or	x	0	x	x	0
23	W	x		x	x	x	Or	x	0	x	x	x
24	X	x	x				Or	x	x	0	0	0
25	Y	x	x			x	Or	x	x	0	0	x
26	Z	x	x		x		Or	x	x	0	x	0

Psychoconduction consists the four different parts. The first part is the calibration of the brain, with the opportunity to increase the brain's intellectual capacity. This approach deals with the training of the brain to achieve congruency in responses on visual, audio, kinesthetic, and olfactory stimuli. The brain needs to be exposed to different pattern of stimuli without explanation or the logical connection in the patterns design. The second part of the method provides the changes in the structure of the brain cells and allows the non-functional cells to be replaced by others, which are able to function in desired capacity. The third part includes the therapy, that reduces the discrepancies in the perceived and real emotional experiences and increases the equilibrium of the emotional experiences. The fourth part of psychoconduction is as mentioned above called the psychoconduction code (codogram or Litvin's code) and provides the unlimited amount of the patterns to balance and restructure the brain cells. The codogram can be for example based, as it is mentioned above, on binary arithmetic and provides the easy absorption by brain and is easily translated into audio, visual, kinesthetic and olfactory stimuli.

Psychoconduction creates the congruence between different modes of communication, which are the sonic, visual, kinesthetic and olfactory, and provides the training for the different parts of the brain in order to release the congruent amount of chemicals in reaction on the sonic, visual, kinesthetic and olfactory stimuli. The psychoconduction training provides the use of the coded patterns and translation of them to the different stimuli. In this stage of the treatment the psychologically disordered object only intuitively understands the patterns, which are clear in the next stage of treatment. The complication of the patterns is increased from the stage to stage. The psychoconduction method translates the patterns from the one of the modes of expression to the other. The recognition and processing of the same information transmitted in the audio, video, kinesthetic and olfactory form provide the calibration of the different parts of the brains. The purpose of the improved training/treatment is to achieve the congruence of the different parts of the brain response on the same patterns, but in the different forms of representation.

Commonly, the simple brain cells respond to the stationary stimuli, such as lines, bars, and edges, whereas the complex cell responds to more complex stimuli and more sophisticated forms. The difficulty of the brain to process complex information is mainly attributed to difficulty of complex cells to process complex information due to physiological issues receiving incomplete or distorted request or the limitation of the capacities of the complex cell. The main purpose of the codogram in processing of the information by brain cells is to clear request with the not distorted data to process an acquired information and find the right address for requested information. Psychoconduction is simplifying the request to process information, and, by modifying the simple cells for processing the stimuli of the same information as the complex cells do.

The codogram creates the new functional cells, that process and congruence the complex information in the response to the sonic (audio), visual (video), kinesthetic and/or olfactory stimuli of the psychotherapy by the improved psychoconduction method (codogram).

The psychoconduction code provides the better (more effective) activity of simple cells by using simple stimuli in complex patterns to substitute the use of injured or underdeveloped complex brain cells. The improved method uses the simple cells, which are responsible for processing the simple stimulus. Psychoconduction enhances the process of information, stimulating the brain with the simple symbols and transferring them between the different areas of the brain. These transfers are between the areas responsible for different functions: audio to the visual, audio to kinesthetic, visual to audio, visual to kinesthetic, kinesthetic to visual, kinesthetic to audio. For instance, the olfactory functions can be used for the brain information processing. Benefits of processing the simple stimuli in the different brain areas are in creating additional axons for the impulse traveling, and by enhancing the brain capacity for complex patterns of simple symbols processing. The same pattern of the simple symbol translated by the brain to the different expression creates the internal communication and enhancing the brain's equilibrium. By using a different part of the brain to process the same pattern, codogram compensates the inadequate information and assists the object to avoid psychological difficulties. The brain also provides the clarity of the logical patterns of the simple symbol and opportunity for all different parts of the brain to enhance performance using the congruent responses presented to the object in the psychoconduction code utilizing the complex pattern in the form of the simple visual, sonic, kinesthetic, olfactory, tactual (not shown), etc. symbols. It is important, for the object with the difficulties, for example, in processing of the visual information to process visual symbols as the last and other symbols before to have valid references. .

CONCLUSION, RAMIFICATION AND SCOPE

Accordingly the reader will see that, according to the invention, I have provided the improved method and system for psychological treatment by brain stimulation of the psychologically disordered object. The improved method and system have various possibilities, considering variety of the psychologically disordered object possible improvements.

While the above description contains many specificities, these should be not construed as limitations on the scope of the invention, but as exemplification of the presently-preferred embodiments thereof. Many other ramifications are possible within the teaching to the invention. For example, in the present time society is more aware that illiteracy or inadequate academic performance is not only caused by social disadvantages. As well known, it is also caused by the neuro-psychological barriers. Those barriers are creating the difficulties to recognize or learn complex patterns needed to be read or to do mathematical calculations, etc. For example, the same way as Braille's alphabet helps people with the visual impairments, the psychoconduction assists people with brain deficiencies. If one of testing results (by either one or combination of a visual treatment, an audio treatment, a kinesthetic treatment, an olfactory treatment, tactual treatment, and a musical treatment) reveals a partial damage of the complex cell of the brain, the improved treatment (method) provides a correction (calibration, balancing, alignment, stimulation) of an appropriate complex cell of the brain by an appropriate simple cell, or provides the correction of an address of the complex cell of the brain by an assistance of the simple cell to process the submitted stimuli. If one of testing results (by either one or combination of a visual treatment, an audio treatment, a kinesthetic treatment, an olfactory treatment, tactual treatment, and a musical treatment) reveals a complete damage of the complex cell of the brain, the improved treatment (method) provides the replacement of an appropriate complex cell of the brain by an appropriate simple cell. The improved method increases the disordered object brain capability and efficiency to perform at least the standard intellectual operations.

Thus, the scope of the invention should be determined by the appended claims and their legal equivalents, and not by examples given.